Semiclassical Physics

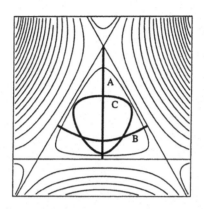

Semiclassical Physics

Matthias Brack
University of Regensburg, Germany

Rajat K. Bhaduri
McMaster University, Canada

Advanced Book Program

CRC Press
Taylor & Francis Group
Boca Raton London New York

CRC Press is an imprint of the
Taylor & Francis Group, an **informa** business

Frontiers in Physics Series
Advanced Book Program

First published 2003 by Westview Press

Published 2018 by CRC Press
Taylor & Francis Group
6000 Broken Sound Parkway NW, Suite 300
Boca Raton, FL 33487-2742

CRC Press is an imprint of the Taylor & Francis Group, an informa business

Visit the Taylor & Francis Web site at
http://www.taylorandfrancis.com

and the CRC Press Web site at
http://www.crcpress.com

A Cataloging-in-Publication data record for this book is available from the
Library of Congress.

This book was set in LaTeX by the authors.

ISBN 13: 978-0-8133-4084-5 (pbk)

To Lis, Roma and Raja

Frontiers in Physics
David Pines, Editor

Volumes of the Series published from 1961 to 1973 are not officially numbered. The parenthetical numbers shown are designed to aid librarians and bibliographers to check the completeness of their holdings.
Titles published in this series prior to 1987 appear under either the W. A. Benjamin or the Benjamin/Cummings imprint; titles published since 1986 appear under the Westview Press imprint.

1. N. Bloembergen   Nuclear Magnetic Relaxation: A Reprint Volume, 1961
2. G. F. Chew   S-Matrix Theory of Strong Interactions: A Lecture Note and Reprint Volume, 1961
3. R. P. Feynman   Quantum Electrodynamics: A Lecture Note and Reprint Volume
4. R. P. Feynman   The Theory of Fundamental Processes: A Lecture Note Volume, 1961
5. L. Van Hove,   Problem in Quantum Theory of Many-Particle Systems:
    N. M. Hugenholtz   A Lecture Note and Reprint Volume, 1961
    L. P. Howland
6. D. Pines   The Many-Body Problem: A Lecture Note and Reprint Volume, 1961
7. H. Frauenfelder   The Mössbauer Effect: A Review—With a Collection of Reprints, 1962
8. L. P. Kadanoff   Quantum Statistical Mechanics: Green's Function Methods
    G. Baym   in Equilibrium and Nonequilibrium Problems, 1962
9. G. E. Pake   Paramagnetic Resonance: An Introductory Monograph, 1962
                         [cr. (42)—2nd edition]G. E. Pake
10. P. W. Anderson   Concepts in Solids: Lectures on the Theory of Solids, 1963
11. S. C. Frautschi   Regge Poles and S-Matrix Theory, 1963
12. R. Hofstadter   Electron Scattering and Nuclear and Nucleon Structure: A Collection of Reprints with an Introduction, 1963
13. A. M. Lane   Nuclear Theory: Pairing Force Correlations to Collective Motion, 1964
14. R. Omnès   Mandelstam Theory and Regge Poles: An Introduction
                         M. Froissart for Experimentalists, 1963
15. E. J. Squires   Complex Angular Momenta and Particle Physics: A Lecture Note and Reprint Volume, 1963
16. H. L. Frisch   The Equilibrium Theory of Classical Fluids: A Lecture
    J. L. Lebowitz   Note and Reprint Volume, 1964
17. M. Gell-Mann   The Eightfold Way (A Review—With a Collection of
    Y. Ne'eman   Reprints), 1964
18. M. Jacob   Strong-Interaction Physics: A Lecture Note Volume, 1964
    G. F. Chew
19. P. Nozières   Theory of Interacting Fermi Systems, 1964
20. J. R. Schrieffer   Theory of Superconductivity, 1964 (revised 3rd printing, 1983)
21. N. Bloembergen   Nonlinear Optics: A Lecture Note and Reprint Volume, 1965

Frontiers in Physics

22.	R. Brout	Phase Transitions, 1965
23.	I. M. Khalatnikov	An Introduction to the Theory of Superfluidity, 1965
24.	P. G. deGennes	Superconductivity of Metals and Alloys, 1966
25.	W. A. Harrison	Pseudopotentials in the Theory of Metals, 1966
26.	V. Barger	Phenomenological Theories of High Energy Scattering:
	D. Cline	An Experimental Evaluation, 1967
27.	P. Choquàrd	The Anharmonic Crystal, 1967
28.	T. Loucks	Augmented Plane Wave Method: A Guide to Performing.Electronic Structure Calculations—A Lecture Note and Reprint Volume, 1967
29.	Y. Ne'eman	Algebraic Theory of Particle Physics: Hadron Dynamics In Terms of Unitary Spin Current, 1967
30.	S. L. Adler	Current Algebras and Applications to Particle Physics,
	R. F. Dashen	1968
31.	A. B. Migdal	Nuclear Theory: The Quasiparticle Method, 1968
32.	J. J. J. Kokkede	The Quark Model, 1969
33.	A. B. Migdal	Approximation Methods in Quantum Mechanics, 1969
34.	R. Z. Sagdeev	Nonlinear Plasma Theory, 1969
35.	J. Schwinger	Quantum Kinematics and Dynamics, 1970
36.	R. P. Feynman	Statistical Mechanics: A Set of Lectures, 1972
37.	R. P. Feynman	Photon-Hadron Interactions, 1972
38.	E. R. Caianiello	Combinatorics and Renormalization in Quantum Field Theory, 1973
39.	G. B. Field	The Redshift Controversy, 1973
	H. Arp	
	J. N. Bahcall	
40.	D. Horn	Hadron Physics at Very High Energies, 1973
	F. Zachariasen	
41.	S. Ichimaru	Basic Principles of Plasma Physics: A Statistical Approach, 1973 (2nd printing, with revisions, 1980)
42.	G. E. Pake	The Physical Principles of Electron Paramagnetic
	T. L. Estle	Resonance, 2nd Edition, completely revised, enlarged, and reset, 1973 [cf. (9)—1st edition

Volumes published from 1974 onward are being numbered as an integral part of the bibliography.

43.	C. Davidson	Theory of Nonneutral Plasmas, 1974
44.	S. Doniach	Green's Functions for Solid State Physicists, 1974
	E. H. Sondheimer	
45.	P. H. Frampton	Dual Resonance Models, 1974
46.	S. K. Ma	Modern Theory of Critical Phenomena, 1976
47.	D. Forster	Hydrodynamic Fluctuation, Broken Symmetry, and Correlation Functions, 1975
48.	A. B. Migdal	Qualitative Methods in Quantum Theory, 1977
49.	S. W. Lovesey	Condensed Matter Physics: Dynamic Correlations, 1980
50.	L. D. Faddeev	Gauge Fields: Introduction to Quantum Theory, 1980
	A. A. Slavnov	
51.	P. Ramond	Field Theory: A Modern Primer, 1981 [cf. 74—2nd ed.]
52.	R. A. Broglia	Heavy Ion Reactions: Lecture Notes Vol. I, Elastic and
	A. Winther	Inelastic Reactions, 1981
53.	R. A. Broglia	Heavy Ion Reactions: Lecture Notes Vol. II, 1990
	A. Winther	

Frontiers in Physics

54. H. Georgi — Lie Algebras in Particle Physics: From Isospin to Unified Theories, 1982

55. P. W. Anderson — Basic Notions of Condensed Matter Physics, 1983

56. C. Quigg — Gauge Theories of the Strong, Weak, and Electromagnetic Interactions, 1983

57. S. I. Pekar — Crystal Optics and Additional Light Waves, 1983

58. S. J. Gates
M. T. Grisaru
M. Rocek
W. Siegel — Superspace or One Thousand and One Lessons in Supersymmetry, 1983

59. R. N. Cahn — Semi-Simple Lie Algebras and Their Representations, 1984

60. G. G. Ross — Grand Unified Theories, 1984

61. S. W. Lovesey — Condensed Matter Physics: Dynamic Correlations, 2nd Edition, 1986.

62. P. H. Frampton — Gauge Field Theories, 1986

63. J. I. Katz — High Energy Astrophysics, 1987

64. T. J. Ferbel — Experimental Techniques in High Energy Physics, 1987

65. T. Appelquist
A. Chodos
P. G. O. Freund — Modern Kaluza-Klein Theories, 1987

66. G. Parisi — Statistical Field Theory, 1988

67. R. C. Richardson
E. N. Smith — Techniques in Low-Temperature Condensed Matter Physics, 1988

68. J. W. Negele
H. Orland — Quantum Many-Particle Systems, 1987

69. E. W. Kolb
M. S. Turner — The Early Universe, 1990

70. E. W. Kolb
M. S. Turner — The Early Universe: Reprints, 1988

71. V. Barger
R. J. N. Phillips — Collider Physics, 1987

72. T. Tajima — Computational Plasma Physics, 1989

73. W. Kruer — The Physics of Laser Plasma Interactions, 1988

74. P. Ramond — Field Theory: A Modern Primer, 2nd edition, 1989 [cf. 51—1st edition]

75. B. F. Hatfield — Quantum Field Theory of Point Particles and Strings, 1989

76. P. Sokolsky — Introduction to Ultrahigh Energy Cosmic Ray Physics, 1989

77. R. Field — Applications of Perturbative QCD, 1989

80. J. F. Gunion
H. E. Haber
G. Kane
S. Dawson — The Higgs Hunter's Guide, 1990

81. R. C. Davidson — Physics of Nonneutral Plasmas, 1990

82. E. Fradkin — Field Theories of Condensed Matter Systems, 1991

83. L. D. Faddeev
A. A. Slavnov — Gauge Fields, 1990

84. R. Broglia
A. Winther — Heavy Ion Reactions, Parts I and II, 1990

85. N. Goldenfeld — Lectures on Phase Transitions and the Renormalization Group, 1992

86. R. D. Hazeltine
J. D. Meiss — Plasma Confinement, 1992

87. S. Ichimaru — Statistical Plasma Physics, Volume I: Basic Principles,

Frontiers in Physics

1992

88. S. Ichimaru   Statistical Plasma Physics, Volume II: Condensed Plasmas, 1994

89. G. Grüner   Density Waves in Solids, 1994

90. S. Safran   Statistical Thermodynamics of Surfaces, Interfaces, and Membranes, 1994

91. B. d'Espagnat   Veiled Reality: An Analysis of Present Day Quantum Mechanical Concepts, 1994

92. J. Bahcall   Solar Neutrinos: The First Thirty Years
R. Davis, Jr.
P. Parker
A. Smirnov
R. Ulrich

93. R. Feynman   Feynman Lectures on Gravitation
F. Morinigo
W. Wagner

94. M. Peskin   An Introduction to Quantum Field Theory
D. Schroeder

95. R. Feynman   Feynman Lectures on Computation

96. M. Brack   Semiclassical Physics
R. Bhaduri

97. D. Cline   Weak Neutral Currents

98. T. Tajima   Plasma Astrophysics
K. Shibata

99. J. Rammer   Quantum Transport Theory

100. R. Hazeltine   The Frameworkof Plasma Physics
F. Waelbroeck

101. P. Ramond   Journeys Beyond the Standard Moel

102. Nutku   Conformal Field Theory: New Non-Perturbative
Saclioglu   Methods in String and Feild Theory
Turgut

103. P. Philips   Advanced Solid State Physics

Editor's Foreword

The problem of communicating in a coherent fashion recent developments in the most exciting and active fields of physics continues to be with us. The enormous growth in the number of physicists has tended to make the familiar channels of communication considerably less effective. It has become increasingly difficult for experts in a given field to keep up with the current literature; the novice can only be confused. What is needed is both a consistent account of a field and the presentation of a definite "point of view" concerning it. Formal monographs cannot meet such a need in a rapidly developing field, while the review article seems to have fallen into disfavor. Indeed, it would seem that the people who are most actively engaged in developing a given field are the people least likely to write at length about it.

Frontiers in Physics was conceived in 1961 in an effort to improve the situation in several ways. Leading physicists frequently give a series of lectures, a graduate seminar, or a graduate course in their special fields of interest. Such lectures serve to summarize the present status of a rapidly developing field and may well constitute the only coherent account available at the time. One of the principal purposes of the *Frontiers in Physics* series is to make notes on such lectures available to the wider physics community.

As *Frontiers in Physics* has evolved, a second category of book, the informal text/monograph, an intermediate step between lecture notes and formal text or monographs, has played an increasingly important role in the series. In an informal text or monograph an author has reworked his or her lecture notes to the point at which the manuscript represents a coherent summation of a newly developed field, complete with references and problems, suitable for either classroom teaching or individual study.

Semiclassical Physics is just such a volume. The authors provide for the non-specialist an excellent introduction to recent developments including quantum chaos, in which Feynman's path integral formalism often plays a central role. They succeed admirably in their goal of forging links between classical periodic orbits and quantum mechanics, and of demonstrating that semiclassical physics is not only useful, but fun.

Their emphasis throughout on a pedagogical presentation makes their book accessible to graduate students and experienced researchers alike. It gives me pleasure to welcome the authors to *Frontiers in Physics*.

David Pines
Urbana, Illinois
January, 1997

Preface to the Paperback Edition

This is an essentially unchanged paperback version of our book. We have taken the opportunity to correct a large number of misprints, some errors and unclear formulation, and some inconsistencies of notation. We are very grateful to Ch. Amann, S. Fedotkin, H. Friedrich, E. Heller, G. Ingold, S. Kümmel, J. Law, A. Magner, P. Meier, M. V. N. Murthy, W. Plaum, J. Sakhr, M. Sieber, D. Sprung, F. Steiner, K. Tanaka, R. Winkler, and B. van Zyl for drawing our attention to many of them.

While it would have been impossible to keep up with the rapidly expanding literature on all of the topics covered in this book, we have included some recent references that are of direct relevance to the examples which we discuss in detail. A new section 4.5 on Bose-Einstein condensation in a harmonic trap has been added in the context of the Thomas-Fermi model. Some passages in Sections 5.3 and 6.2 concerning bifurcations have been improved, and a new Section 6.3 on uniform approximations has been added. The Appendix C.4 has been rewritten, and Appendix D has received a short general introduction. We have also included six new problems (3.6, 3.7, 4.7, 4.8, 6.5, and C.3) in order to illustrate the newly added material.

We acknowledge the warm hospitality of Stephanie Reimann-Wacker and the Division of Mathematical Physics at Lund Institute of Technology, where parts of the preparations to this edition were done during a research visit.

In the event that we should discover further misprints or errors in the paperback edition, these will be made available on M. Brack's homepage, like the errata list of the first edition: http://homepages.uni-regensburg.de/~brm04014/book/index.html

We hope that our book will continue to convey to the reader our fascination with semiclassical physics, and that the present more economical edition will reach a larger class of students and scholars.

Matthias Brack
Rajat K. Bhaduri
Matting and Dundas
July 2002

xiii

Preface

This book attempts to convey to the reader that semiclassical physics can be fun, as well as useful for understanding quantum fluctuations in interacting many-body systems. The deep connection between classical motion and quantal fluctuations is fascinating and worth studying. The overall thrust of this book is to show the close link of the shorter classical periodic orbits with the partly resolved shell fluctuations. Applications to finite fermion systems in diverse areas of physics are discussed, culminating in Chapter 8 where manifestations of periodic orbits in atomic nuclei, metal clusters and semiconductor quantum dots are presented. We also pay considerable attention to the Thomas-Fermi model and its extensions, applied to a many-fermion system at zero and also at finite temperature. This is partly due to our personal bias: we met 22 years ago, when one of us (R. K. B.) had embarked on the extensions of the Thomas-Fermi model and the other had become engaged in the Strutinsky method. We think that it is important to understand the average self-consistent behavior of the system before attempting to unravel its quantal fluctuations.

We have taken care to derive Gutzwiller's trace formula starting from an elementary level. A challenging problem of a fundamental nature is the semiclassical quantization of a non-integrable (one-body) system with some accuracy. In this category, we find the Riemann zeta function to be beautiful and at the same time mysterious. The extensions of the trace formula to continuous families of orbits and their applications are rewarding. We present this without going into the powerful formalism of Lagrangian manifolds. Some of the simpler scattering problems in two dimensions are developed in detail, with a view to include diffractive orbits in the extended trace formula.

This book is addressed to graduate students with a basic knowledge of classical and quantum mechanics, and to the scientists with an interest in semiclassical methods. We have left the formal and mathematical subtleties of the theory to the experts. Our approach is informal, largely guided by simple solvable models and by practical applications to real physical phenomena. The emphasis is on a pedagogical presentation with many examples, often solving the same model by different techniques at different places in the book. Some of the problems at the end of each chapter may be difficult, but all have been worked out by one or the other author. The serious reader may find them useful. Although we have tried to present the material in a unified scheme, the level of difficulty varies considerably. Chapters 1, 2, and 8 are by and large for the general reader without much interest in learning the technical aspects. Chapter 3 presents analytically solvable quantum-mechanical models to set the stage for links with the semiclassical methods developed later. Chapters 4 and 5 form the core of the book, presenting the basics of the extended Thomas-Fermi model and some of its applications, and Gutzwiller's periodic orbit theory. Chapters 6 and 7 describe extensions of the trace formula encompassing continuous symmetries, and illustrations thereof, as well as zeta function techniques and the inclusion of diffractive orbits.

Special thanks are due to Stephen Creagh, David Goodings, Avinash Khare, Jimmy Law, Alexander (Sasha) Magner, M. V. N. Murthy, Stephanie Reimann, Diptiman Sen, Nina Snaith, Franz Stadler, and last but not least, Kaori Tanaka. Stephen, David, and Sasha patiently coached one or both of us and went through large parts of the manuscript. So did Kaori, in addition to her immense editorial help. Franz devoted uncounted hours to his professional drawings. Avinash, Diptiman, Jimmy, Murthy, Nina and Steffi are co-researchers who helped in our understanding of the subject. Thanks are due also to our postgraduate students, who corrected many mistakes in the text, and whose enthusiasm for the subject has been a great stimulus for us. M. B. takes this opportunity to express his gratitude to the late V. M. Strutinsky from whom he learned a lot, and whose early contributions to the periodic orbit theory he took far too long to appreciate. Indeed, the Kiev school taught us to focus on the coarse-grained shell structure, which has become a recurring theme in this book.

We are grateful to the reviewers of the draft manuscript: David Brink, Stephen Creagh, and Byron Jennings, for their patience and constructive criticism.

But all of this would not have worked, were it not for the support by our closest collaborators: our beloved families who helped in creating an ideal atmosphere for both work and living. Matthias ('uncle') is indebted to Manju and Ronnie, Ranju, Mallika, and Sharmila, and Rajat to Lis and Helene, Marion, Christian, and Sonja. Many pleasant hours were spent with them on our mutual visits playing music and games and cooking exotic dishes. When our collaboration began, most of the children were not yet born—while we are finishing this book, the oldest have already left our homes and are about to create their own.

M. B. is grateful to McMaster University for the invitation as a Hooker Visiting Professor. He thanks Don Sprung, the chair of the Department of Physics and Astronomy, for his friendly support and hospitality over many years. We acknowledge travel support from the DAAD, Germany, and from the Commission of the European Communities, Bruxelles, under contract nos. SC1-CT92–0770 and CHRX-CT94–0612. The work reported here has also been supported by the Natural Sciences and Engineering Research Council of Canada, by the Deutsche Forschungsgemeinschaft, and by INTAS, grant no. 93–151.

Matthias Brack
Rajat K. Bhaduri
Matting and Dundas
December 1996

Contents

1 Introduction   **1**
  1.1 The quantum propagator   4
  1.2 Old quantum theory   9
     1.2.1 A ball bouncing off a moving wall   10
     1.2.2 A pendulum with variable string length   11
     1.2.3 The phase space of a simple harmonic oscillator   13
     1.2.4 Three-dimensional anisotropic harmonic oscillator   16
  1.3 Wave packets in Rydberg atoms   19
     1.3.1 The large-n limit in the Bohr atom   19
     1.3.2 Where are the periodic orbits in quantum mechanics?   20
  1.4 Chaotic motion: atoms in a magnetic field   27
     1.4.1 Scaling of classical Hamiltonian and chaos   27
     1.4.2 Quasi-Landau resonances in atomic photoabsorption   32
  1.5 Chaos and periodic orbits in mesoscopic systems   36
     1.5.1 Ballistic magnetoresistance in a cavity   37
     1.5.2 Scars in the wave function   39
     1.5.3 Tunneling in a quantum diode with a tilted magnetic field . . .   42
     1.5.4 Electron transport in a superlattice of antidots   44
  1.6 Problems   49

2 Quantization of integrable systems   **57**
  2.1 Introduction   57
  2.2 Hamiltonian formalism and the classical limit   59
  2.3 Hamilton-Jacobi theory and wave mechanics   63
  2.4 The WKB method   67
     2.4.1 WKB in one dimension   68
     2.4.2 WKB for radial motion   75
  2.5 Torus quantization: from WKB to EBK   78
  2.6 Examples   83
     2.6.1 The two-dimensional hydrogen atom   83
     2.6.2 The three-dimensional hydrogen atom   86
     2.6.3 The two-dimensional disk billiard   88
  2.7 Connection to classical periodic orbits   89
     2.7.1 Example: The two-dimensional rectangular billiard   94
  2.8 Transition from integrability to chaos   98
     2.8.1 Destruction of resonant tori   98
     2.8.2 The model of Walker and Ford   100
  2.9 Problems   106

3 The single-particle level density   **111**
  3.1 Introduction   111
     3.1.1 Level density and other basic tools   112
     3.1.2 Separation of $g(E)$ into smooth and oscillating parts   117
  3.2 Some exact trace formulae   118

		3.2.1	The linear harmonic oscillator	118
		3.2.2	General spectrum depending on one quantum number	120
		3.2.3	One-dimensional box	122
		3.2.4	More-dimensional spherical harmonic oscillators	122
		3.2.5	Harmonic oscillators at finite temperature	124
		3.2.6	Three-dimensional rectangular box	126
		3.2.7	Equilateral triangular billiard	128
		3.2.8	Cranked or anisotropic harmonic oscillator	132
	3.3		Problems	136

4 The extended Thomas-Fermi model — **143**

	4.1		Introduction	143
	4.2		The Wigner distribution function	148
	4.3		The Wigner-Kirkwood expansion	151
	4.4		The extended Thomas-Fermi model	155
		4.4.1	The ETF model at zero temperature	155
		4.4.2	The ETF density variational method	162
		4.4.3	The finite-temperature ETF model	169
	4.5		Bose-Einstein condensation in a trap	176
		4.5.1	BEC in an ideal trapped bose gas	177
		4.5.2	Inclusion of interactions in a dilute gas	179
	4.6		$\hbar$ expansion for cavities and billiards	180
		4.6.1	The Euler-MacLaurin expansion	180
		4.6.2	The Weyl expansion	183
		4.6.3	Black-body radiation in a small cavity	186
	4.7		The Strutinsky method	188
		4.7.1	The energy averaging method	189
		4.7.2	The shell-correction method	195
		4.7.3	Relation between ETF and Strutinsky averaging	197
	4.8		Problems	200

5 Gutzwiller's trace formula for isolated orbits — **213**

	5.1		The semiclassical Green's function	215
	5.2		Taking the trace of $G_{scl}(\mathbf{r}, \mathbf{r}'; E)$	220
	5.3		The trace formula for isolated orbits	224
	5.4		Stability of periodic orbits	227
	5.5		Convergence of the periodic orbit sum	229
	5.6		Examples	235
		5.6.1	Applications to chaotic systems	235
		5.6.2	The irrational anisotropic harmonic oscillator	235
		5.6.3	The inverted harmonic oscillator	236
		5.6.4	The Hénon-Heiles potential	238
	5.7		Problems	243

6 Extensions of the Gutzwiller theory   **249**

6.1 Trace formulae for degenerate orbits 251

     6.1.1 Two-dimensional systems, singly degenerate orbits 252

     6.1.2 Example 1: The equilateral triangular billiard 253

     6.1.3 Example 2: The two-dimensional disk billiard 262

     6.1.4 More general treatment of continuous symmetries 266

     6.1.5 Example 3: The 2-dimensional rectangular billiard 273

     6.1.6 Example 4: The three-dimensional spherical cavity 275

6.2 The problem of symmetry breaking 278

     6.2.1 A trace formula for broken symmetry 279

     6.2.2 Example 1: The two-dimensional elliptic billiard 280

     6.2.3 Example 2: Inclusion of weak magnetic fields 287

     6.2.4 Example 3: The quartic Hénon-Heiles potential 293

6.3 Uniform approximations 300

     6.3.1 U(1) symmetry breaking 300

     6.3.2 An example of SU(2) symmetry breaking 303

     6.3.3 Uniform approximations for bifurcations 306

6.4 Problems 306

7 Quantization of nonintegrable systems   **311**

7.1 The Riemann zeta function 312

     7.1.1 The zeros of the Riemann zeta function 312

     7.1.2 A trace formula for the zeros 315

     7.1.3 Nearest-neighbor spacings and chaos 318

     7.1.4 The Riemann-Siegel relation 319

7.2 The quantization condition 322

     7.2.1 The Selberg zeta function 322

     7.2.2 Pseudo-orbits and the Selberg zeta function 326

7.3 The scattering matrix method 329

7.4 The transfer-matrix method of Bogomolny 335

7.5 Diffractive Corrections to the Trace Formula 343

     7.5.1 Introduction 343

     7.5.2 Quantum theory of scattering 345

     7.5.3 Scattering by a hard disk 347

     7.5.4 The scattering amplitude and the Green's function 354

     7.5.5 Modification to the trace formula 357

     7.5.6 The circular annulus billiard 362

7.6 Problems 370

8 Shells and periodic orbits in finite fermion systems   **377**

8.1 Shells and shapes in atomic nuclei 377

     8.1.1 Nuclear ground-state deformations 379

     8.1.2 The double-humped fission barrier 384

     8.1.3 The mass asymmetry in nuclear fission 388

8.2 Shells and supershells in metal clusters 394

8.3 Conductance oscillations in a circular quantum dot 403

9 Concluding remarks   **415**

A The self-consistent mean field approach   **419**
A.1 Hartree-Fock theory 420
A.2 Density functional theory 423
A.3 The Strutinsky energy theorem 425

B Inverse Laplace transforms   **429**

C More about the monodromy matrix   **431**
C.1 Linear differential equations with periodic coefficients 431
C.2 Hamiltonian equations 433
    C.2.1 Example: Two-dimensional harmonic oscillator 433
C.3 Non-linear systems and the Poincaré variational equations 434
C.4 Calculation of the monodromy matrix M 435
C.5 Calculation of $\widetilde{M}$ for two-dimensional billiards 436
    C.5.1 Example: Elliptic billiard 439
C.6 Problems 440

D Calculation of Maslov indices for isolated orbits   **443**
D.1 Isolated orbits in smooth potentials 444
    D.1.1 Unstable orbits 444
    D.1.2 Stable orbits 445
    D.1.3 Example: Two-dimensional harmonic oscillator 447
D.2 Isolated orbits in billiards 449

Index   **451**

Chapter 1

Introduction

Quantum mechanics has played a pivotal role in our understanding of the microscopic physical world. It differs fundamentally from classical physics, incorporating the concept of a complex probability amplitude in its dynamics. This amplitude is determined by a linear equation, leading to the principle of superposition, and the consequent interference phenomenon in measurement. The wave-particle duality is inherent in quantum mechanics, and strange effects like the self-interference of an electron in a two-slit experiment have been verified experimentally [1]. In spite of these subtleties, as we well know, quantum mechanics can be applied in a relatively straight-forward manner to physical problems. The recipe is to use either the wave equation of Schrödinger [2], or the matrix-mechanics of Heisenberg, Born and Jordan [3], after defining the Hamiltonian of the system. Although only a handful of such problems can be solved exactly, there are powerful approximation methods and computational algorithms for obtaining relevant solutions: these include bound state and scattering problems in atomic, subatomic, and condensed matter physics. Physical insight in a diverse variety of natural phenomena involving a large number of particles is gained through approximate quantum mechanical solutions, as in the spontaneous fission of a nucleus, or the superfluidity of liquid Helium. The success stories are too numerous to mention. So one may well ask, *what is the need for doing semiclassical physics*, specially since numerical calculations in the computer are more feasible now than ever before? This is an important question, and it must be answered in a meaningful way. The answer rests, in our opinion, on the following points:

(a) The most important point to note is that we are *not* abandoning the essentials of quantum mechanics while doing semiclassics. Semiclassical physics, in the sense used in this book, is just a way of doing quantum mechanics using a simplified path-integral formalism, with a focus on the classical motion of the particle. In the path-integral formalism of Feynman [4], *a probability amplitude is associated with an entire motion of a particle as a function of time, rather than simply with a position of the particle at a particular time*. This is called the quantum mechanical propagator, and has a phase which is Hamilton's principal function, defined as the time integral of the Lagrangian along the path of the motion. In classical mechanics, only the path along which this is an extremum is allowed. In quantum mechanics, all paths, classical and nonclassical, that connect the initial and the final points are considered. The total amplitude is an

1

appropriate superposition of the amplitudes of all these paths. In the simplest semi-classical approximation, on the other hand, the probability amplitude is approximated only by its contribution from the classical path. This approximation to the quantum propagator is due to Van Vleck [5], who not only proposed the phase, but also its amplitude. The superposition principle is still maintained in a limited sense, meaning that if there are two or more alternative classical paths between two points (say one direct, and the other reflected from a wall), then the separate contributions of these paths are still superposed in semiclassics. Gutzwiller [6], while deriving the semiclas-sical propagator from the path-integral method, made an important modification to the result of Van Vleck by adding a term to the phase that made it consistent at the conjugate points of classical motion, to be discussed at length later. The Van Vleck-Gutzwiller propagator has also been generalized to include diffractive paths. Thus semiclassical calculations embrace the concepts of quantum mechanics, but have direct links to classical trajectories.

(b) A quantum system like an atom, a nucleus, a metal cluster, or a quantum dot consists of a few (or many) fermions moving in a potential well that arises from the mutual interactions between the particles, or the imposition of a device potential. To a good approximation, an electron (or a nucleon) in the well may be regarded to move independently of the others in such systems. The quantum density of states of a particle in the well has a fluctuating part as a function of energy, and plays a crucial role in the response of the system to an external stimulus (like a magnetic or electric field). For a self-bound system, this shell-effect also determines the stability, as in the atom or the nucleus. Using the semiclassical propagator or the Green's function, we know from the work of Balian and Bloch [7], and Gutzwiller [8], that this quantum density of states is expressible largely in terms of the classical periodic orbits of the finite system. The classical periodic orbits, therefore, forge a link to quantum mechanics, and form a practical basis for semiclassical calculations. These calculations, particularly in many-body systems, give an intuitive understanding of quantum phenomena.

(c) Even before the advent of modern quantum mechanics, in the old quantum theory formulated by Niels Bohr [9] (see also Ref. [10]), classical periodic orbits played a central role in the quantization of the energy spectrum of the hydrogen atom. It was the classical *action*, $\oint p\, dq$, along the circular trajectory over a complete cycle that was quantized in units of the Planck's constant h to yield the discrete energy spectrum. Most systems, including the classical Hamiltonian for the Helium atom, however, are non-integrable. In non-integrable classical motion, the number of constants of motion in involution is less than the spatial degrees of motion, and the Bohr-Sommerfeld quantization procedure is not applicable directly [11]. In such a system, moreover, nonlinear coupling may give rise to very unstable classical trajectories. An unstable trajectory diverges exponentially from its initial path with a slight alteration in the initial conditions of motion. The simplest examples of this extreme sensitivity to initial conditions may be seen in the trajectories of a particle moving in a force-free region, but being reflected by the walls of an enclosure of a certain shape. This is shown in Fig. 1.1 in a two-dimensional enclosure of a stadium shape, often referred to as a stadium billiard [12]. Two groups of trajectories with slightly different initial conditions end up with different paths, and will diverge greatly after a very few reflections. This would not be the case, however, for a circular billiard which conserves angular momentum.

As will be explained in the next chapter, the trajectories in chaotic motion are not constrained within the narrow torus-structure of the phase space. Even in such chaotic motion, there are isolated unstable periodic orbits [13, 14]. These unstable periodic orbits are embedded in a sea of non-periodic orbits, and are of zero measure in the integrated area of the phase space. Nevertheless, Gutzwiller's work has shown that these periodic orbits play a crucial role in the quantum description of the system. Note in Fig. 1.1 that trajectories that diverge from a given point (due to different initial conditions) may again converge at a different point. Such points (that are also possible in regular motion) are called *conjugate* points. The old quantum theory failed because the role of the conjugate points along the classical trajectory was not appreciated, nor was the importance of the periodic orbits in non-integrable classical motion known. The modern developments in semiclassical physics have given a new impetus to the formulation of quantization rules from the periodic orbits of a classically chaotic system and, indeed, of obtaining the entire quantum spectrum from such a procedure. This is a challenging endeavor but not, in our view, the center-piece of the periodic orbit theory. The real strength of the periodic orbit theory is in our view that often only a few short classical orbits of high degeneracy are important to determine global characteristics of a many-body system, such as stability or deformation. These properties depend on the quantal nature of the single-particle degrees of freedom in the system, but only the gross features of this shell structure are often needed to determine them.

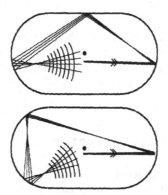

Figure 1.1:   Classical orbits in the stadium billiard. (From [12].) The classical trajectories are very sensitive to the initial conditions. Following Heller, a quantum wave-front associated with the group of trajectories is shown schematically, to underline their close connection in a semiclassical description.

(d) Are there signatures of chaos in quantum mechanical systems? This is often referred to as quantum chaos [12, 15, 16, 17, 18, 19]. Technological advances in laser and semiconductor fabrication have made possible experimental studies on quantum systems that would be chaotic if they were governed by classical laws. The periodic orbit theory of Gutzwiller is ideally suited for the analysis of such systems. We mention a few examples in two different categories: the spectra of the highly excited states of atoms, and the transport properties in man-made solid-state devices. With the advent of fast pulse lasers, highly excited Rydberg states of atomic systems may be studied with precision. These states, in the spirit of the correspondence principle, are specially amenable to the semiclassical methods under discussion. Particularly interesting are the experiments on Rydberg atoms in the presence of a magnetic field (see a later section in this chapter). The classical motion of an electron in such a system is chaotic, but of course the electron obeys quantum laws. One may then study what remnants of classical chaos survive in the quantum behavior. Rydberg atoms also show rapid ionisation above a certain threshold of microwave radiation. Analysis of classical chaotic motion appears to shed light on this behavior [20, 21]. Nano-structures in the mesoscopic regime are fabricated that have spatial lengths much less than the mean-free path of collisions (from impurities). Such ballistic transport in simple quantum systems may be studied experimentally. Of particular interest is the study of the transport properties of electrons in two-dimensional cavities (often called billiards) of different shapes to find "signatures" of classical chaos [22]. The tunneling characteristics of electrons in a quantum well under a bias voltage, and a magnetic field, appear to relate to certain unstable periodic orbits [23]. This will be discussed in the last part of this chapter.

The above four arguments constitute our view on semiclassical physics, and why it is worth studying. Of course, semiclassical physics is a vast subject, touching on the borderline of quantum and classical physics in all areas of physics. In this book, we shall only focus on the topics relating to the points presented above, but not necessarily in the same order. In the rest of this chapter, we shall first elaborate more on point (a), and in particular why Hamilton's principal function plays such an important role in the quantum propagator. Following this, we shall discuss in more detail the role that periodic orbits play in experiments relating to point (d). The theme of this chapter is to motivate the reader by touching on the fundamental aspects, as well as by describing some of the exciting experiments in the field.

1.1 The quantum propagator

As already discussed above, the quantum-mechanical propagator plays a central role in the path-integral formulation of quantum mechanics. It was Dirac [24] who first attempted to formulate quantum mechanics in terms of the Lagrangian. Writing the quantum propagator as the overlap of two position state vectors at different times, he discovered that this transformation function, as he called it, "corresponds" to the classically computed quantity $\exp(i \int \mathcal{L} dt)$, where $\mathcal{L}$ is the Lagrangian. Even before this, inspired by Dirac's transformation brackets in his representation theory [25, 26], Van Vleck [5] had obtained the factor of proportionality by appealing to the classical limit. But it was Feynman [4] who made it into a precise formulation, stressing the interference aspect of the contributions from the different non-classical paths, and showing its equivalence to Schrödinger's wave equation. The reader will enjoy reading the classic paper by Feynman on this subject. The path-integral method has been generalized to relativistic field theories with great success. In this book, we are interested only in nonrelativistic problems. Several excellent texts [27, 28, 29] are available that describe clearly the path-integral formulation of quantum mechanics. In spite of this, we have decided to give a derivation of the connection between the quantum propagator and the phase-factor containing Hamilton's principal function. We do this because it is fundamental to the semiclassical methods described in this book. The Gutzwiller-Van Vleck approximation will be discussed at length in a later chapter. Our intention here is to start from conventional quantum mechanics, and show how Hamilton's principal function arises naturally in the propagator. This treatment is by no means new, and a more complete discussion, for example, will be found in the texts mentioned above.

Consider the quantum-mechanical description of a particle in one spatial dimension. This is primarily to simplify the notation; the formalism is completely general. In one spatial dimension, the coordinate of a particle is denoted by q. (In higher dimensions, this will be replaced by a vector $\mathbf{q}$.) Let us first define the (time-independent) position operator $\hat{q}$, generating the complete set of eigenkets

$$\hat{q}|q\rangle = q|q\rangle. \tag{1.1}$$

In the Heisenberg picture, the time-dependent position operator is defined as

$$\hat{q}(t) = e^{it\hat{H}/\hbar}\hat{q}e^{-it\hat{H}/\hbar}, \tag{1.2}$$

assuming that the Hamiltonian $\hat{H}$ has no explicit time dependence. We may now define the instantaneous eigenstates $|q,t\rangle$ of the Heisenberg position operator $\hat{q}(t)$,

$$\hat{q}(t)|q,t\rangle = q|q,t\rangle. \tag{1.3}$$

It is easily verified that the time evolution of the state $|q,t\rangle$, (unlike the Schrödinger picture, which has a negative sign before i), is given by

$$|q,t\rangle = e^{it\hat{H}/\hbar}|q\rangle. \tag{1.4}$$

We know from quantum mechanics that if $\hat{A}$ and $\hat{B}$ are two hermitian operators corresponding to dynamical variables A and B, obeying $\hat{A}|\psi_a\rangle = a|\psi_a\rangle$, and $\hat{B}|\xi_b\rangle = b|\xi_b\rangle$, then the probability that a measurement of B in the state $|\psi_a\rangle$ will yield b is given by $|\langle\xi_b|\psi_a\rangle|^2$. In particular, we may take for $\hat{A}$, $\hat{B}$ the position operator $\hat{q}(t)$ at two different times. Then the quantum propagator for the particle from a point q_1 at time t_1 to a point q_2 at time t_2 is given by the overlap function

$$\langle q_2, t_2|q_1, t_1\rangle = \langle q_2|e^{-i(t_2-t_1)\hat{H}/\hbar}|q_1\rangle. \tag{1.5}$$

Recalling that

$$|q_2\rangle = e^{-i(q_2-q_1)\hat{p}/\hbar}|q_1\rangle, \tag{1.6}$$

where $\hat{p}$ is the linear momentum operator, we get

$$\langle q_2, t_2|q_1, t_1\rangle = \langle q_1|e^{i/\hbar[(q_2-q_1)\hat{p}]}e^{-i/\hbar[(t_2-t_1)\hat{H}]}|q_1\rangle. \tag{1.7}$$

We now insert the unit operator $\int_{-\infty}^{\infty} dp_1|p_1\rangle\langle p_1| = 1$ to obtain

$$\langle q_2, t_2|q_1, t_1\rangle = \int_{-\infty}^{\infty} dp_1\, e^{i(q_2-q_1)p_1/\hbar}\langle q_1|p_1\rangle\,\langle p_1|e^{-i/\hbar(t_2-t_1)\hat{H}(\hat{q},\hat{p})}|q_1\rangle. \tag{1.8}$$

Taking infinitesimal intervals $t_2 = t_1 + \delta t_1$ and $q_2 = q_1 + \delta q_1$, we may expand the exponential into $\exp(-i\delta t_1 \hat{H}/\hbar) \simeq 1 - i\delta t_1 \hat{H}(\hat{q}, \hat{p})/\hbar$. We next assume that the Hamiltonian structure is such that we may order the $\hat{q}$'s to the right, and the $\hat{p}$'s to the left in $\hat{H}(\hat{q}, \hat{p})$. The latter, occurring in the exponent in Eq. (1.8), may then be taken out of the matrix element and replaced by its classical counterpart $H(q, p)$. Noting further that $\langle q_1|p_1 \rangle = (2\pi\hbar)^{-1/2} \exp(ip_1 q_1/\hbar)$, and $\langle q_1|p_1\rangle\langle p_1|q_1 \rangle = (2\pi\hbar)^{-1}$, we obtain

$$\langle q_2, t_2|q_1, t_1 \rangle = \frac{1}{2\pi\hbar} \int_{-\infty}^{\infty} dp_1 \, e^{i[-\delta t_1 H(q_1, p_1) + p_1 \delta q_1]/\hbar} \,. \tag{1.9}$$

In the infinitesimal time interval δt_1, we may assume a constant velocity $\dot{q}_1$, such that $\delta q_1 = \dot{q}_1 \delta t_1$, obtaining

$$\langle q_2, t_2|q_1, t_1 \rangle = \frac{1}{2\pi\hbar} \int_{-\infty}^{\infty} dp_1 \, e^{i\delta t_1[-H(q_1, p_1) + p_1 \dot{q}_1]/\hbar} \,. \tag{1.10}$$

This is the quantum propagator for an infinitesimal time interval, already expressed in terms of the classical Lagrangian function times the time interval in the exponent. To be more specific, assume that $H(q_1, p_1) = p_1^2/2m + V(q_1)$. The p_1 integral in Eq. (1.10) may then be immediately performed by completing the square to give

$$\langle q_2, t_2|q_1, t_1 \rangle = \sqrt{\frac{m}{2\pi i\hbar\delta t_1}} \, e^{i\delta t_1 \mathcal{L}(q_1, \dot{q}_1)/\hbar} \,, \tag{1.11}$$

where the Lagrangian is $\mathcal{L}(q_1, \dot{q}_1) = m\dot{q}_1^2/2 - V(q_1)$. This is a remarkable result. A *finite* time interval from t' to t, may be divided into n equal intervals δt, with $t' = t_0$, and $t = t_n$:

$$t - t' = n\,\delta t \,. \tag{1.12}$$

Note that the instantaneous eigenstates $|q, t\rangle$ also form a complete set:

$$\int_{-\infty}^{\infty} dq|q, t\rangle\langle q, t| = e^{it\hat{H}/\hbar} \left(\int_{-\infty}^{\infty} dq|q\rangle\langle q| \right) e^{-it\hat{H}/\hbar} = 1 \,. \tag{1.13}$$

We may therefore write

$$\langle q, t|q', t' \rangle = \int dq_1 \int dq_2 \ldots \int dq_{n-1} \langle q, t|q_{n-1}, t_{n-1} \rangle \ldots \langle q_2, t_2|q_1, t_1 \rangle\langle q_1, t_1|q', t' \rangle \,. \tag{1.14}$$

This implies a sum over all intermediate paths. For infinitesimal time intervals, we may use Eq. (1.11), so the exponents all add up to yield the time integral over the Lagrangian. The final result for the propagator, which is also denoted by $K(q, q'; (t - t'))$, may be written as

$$K(q, q'; (t - t')) = \langle q, t|q', t' \rangle = \int \mathcal{D}q \, e^{\frac{i}{\hbar} \int_{t'}^{t} dt'' \mathcal{L}(q, \dot{q})} \,, \tag{1.15}$$

where the measure is formally defined as

$$\mathcal{D}q = \lim_{n \to \infty} \left(\frac{m}{2\pi i\hbar\delta t} \right)^{n/2} dq_1 dq_2 \ldots dq_{n-1} \,. \tag{1.16}$$

The exponent occurring in Eq. (1.15) for the propagator is called Hamilton's principal function, and is denoted by the symbol R:

$$R(q, q', (t - t')) = \int_{t'}^{t} \mathcal{L}(q, \dot{q}) \, dt'' \,. \tag{1.17}$$

The principal function depends on the time interval $(t - t')$, which by causality is positive, and may therefore be expressed as $R(q, q'; (t - t'))$. This basically completes our derivation of the path integral formula (1.15), which is a superposition of $\exp(iR_j/\hbar)$ from the various paths j. The exponent is highly oscillatory if $R_j \gg \hbar$, and neighboring paths having different phases cancel out in general. Only along the classical paths (in general, there may be more than one classical path connecting two points; we denote these by $\{cl\}$) is R stationary, and the phase remains unchanged by small variations in the path, with the end-points fixed. Thus the classical limit is naturally obtained in this case.

The semiclassical propagator that Van Vleck used is a superposition of terms involving the classical action in the exponent, and allowing for more than one possible classical paths between two specified points in a given time interval $(t - t')$:

$$K_V(q, q'; (t - t')) = \sum_{j\{cl\}} \mathcal{A}_j \exp\left[iR_j(q, q'; (t - t'))/\hbar\right]. \tag{1.18}$$

The subscript V is for Van Vleck. The principal function R_j above is along the classical path j, and all such classical paths have been summed over. By appealing to the correspondence principle, Van Vleck also obtained the amplitude $\mathcal{A}_j$ in the limit $\hbar \to 0$. This is discussed in Sect. 2.2, and in more detail in Chapter 5, see Eq. (5.5). Gutzwiller [6] added important phases of quantum origin to K_V to obtain the semiclassical propagator K_{scl}; see Eq. (5.8) that plays a central role in this book. At this point, we only want to emphasize that it is often more useful in time-independent problems to work in the energy domain, for a fixed energy E, rather than a fixed time t (we set $t' = 0$). To go to the energy domain, we take the Fourier transform of the propagator by multiplying it with $\exp(iEt/\hbar)$, and integrating over the time interval from zero to infinity. This is the Green's function $G(q, q'; E)$:

$$G(q, q'; E) = \int_0^{\infty} dt e^{iEt/\hbar} \langle q, t | q', 0 \rangle \,. \tag{1.19}$$

The quantity in the exponent of the off-diagonal Green's function therefore comes in the combination $R(q, q'; t) + Et$. For classical trajectories at a fixed energy, t may be eliminated in favour of E, and the following relation is obeyed:

$$R(q, q'; t) + Et = \int_{q'}^{q} p(q'', E) \, dq'' = S_{cl}(q, q'; E) \,. \tag{1.20}$$

We shall call $S_{cl}(q, q'; E)$ the classical action. For example, denote the principal function for a free particle moving from q' to q in a time interval t by $R_0(q, q'; t)$. Since the free particle moves with a constant velocity along a straight line, R_0 is easily calculated using Eq. (1.17):

$$R_0(q, q'; t) = \frac{m}{2} \frac{(q - q')^2}{t} \,. \tag{1.21}$$

We may then verify Eq. (1.20) for the free case:

$$R_0(q, q'; t) + Et = m\dot{q}^2 t = m\dot{q}^2 \frac{(q - q')}{\dot{q}} = \int_{q'}^{q} p \, dq'' \,. \tag{1.22}$$

We should remember, however, that Eq. (1.20) is valid for any classical trajectory in a time-independent potential. Quite generally, we may define an action $S(q, q'; E)$ along *any* path from $q' \to q$ at energy E:

$$S(q, q'; E) = \int_{q'}^{q} p(q'', E) \, dq'' \,. \tag{1.23}$$

It also follows at once from Eq. (1.23) that

$$\frac{\partial S}{\partial q} = p, \qquad \frac{\partial S}{\partial q'} = -p' \,. \tag{1.24}$$

If $S(q, q'; E)$ is along a classical trajectory, then we also have, from Eq. (1.20)

$$\frac{\partial S_{cl}}{\partial E} = t \,. \tag{1.25}$$

In this book, we shall be using throughout (except for diffractive orbits in Chapter 7) the S_{cl} along classical trajectories. The subscript will generally be dropped for simplicity of notation. Similarly, we have

$$\frac{\partial R}{\partial q} = p, \qquad \frac{\partial R}{\partial q'} = -p', \qquad \frac{\partial R}{\partial t} = -E \,. \tag{1.26}$$

These equations will be used in Chapter 5.

It is useful at this stage to write an alternative expression for the quantum propagator to underline the importance of its diagonal elements, when $q = q'$. From Eq. (1.5), by inserting a complete set of states $|\psi_n\rangle$ that are eigenstates of the Hamiltonian $\hat{H}$, we have

$$
\begin{aligned}
\langle q(t)|q'(0)\rangle &= \langle q|e^{-\frac{it}{\hbar}\hat{H}}|q'\rangle \\
&= \sum_n \langle q|\psi_n\rangle\langle\psi_n|q'\rangle \, e^{-iE_n t/\hbar} \\
&= \sum_n \psi_n(q)\psi_n^*(q')e^{-iE_n t/\hbar} \,.
\end{aligned}
\tag{1.27}
$$

It is clear from the above expression that all the information about the energy spectrum E_n is contained in the trace of the diagonal elements of K. In the semiclassical approximation in the energy domain, this amounts to taking the q-integral of the Green's function $G(q, q'=q; E)$. This involves the rapidly oscillating function $\exp[iS(q, q; E)/\hbar]$, where the action $S(q, q'; E)$ is defined by Eq. (1.23), and can be done by stationary phase integration. In writing the stationary phase condition, it is important to vary both limits of the integral (1.23) which gives

$$\left(\frac{\partial S}{\partial q} + \frac{\partial S}{\partial q'}\right) = 0 \,. \tag{1.28}$$

Using Eq. (1.24), we see that this implies the condition that $p = p'$. Thus, on taking the trace in the semiclassical approximation, only those orbits contribute dominantly that are periodic, i.e., when the particle comes back to the same initial point with the same initial momentum. Of course, in a bound one-dimensional system, all orbits are periodic. A calculation for the more realistic higher dimensional system, done in Chapter 5, gives the same essential result. A quantitative semiclassical formula for the Green's function will be developed in Chapter 5; here we have only argued to underscore the importance of the classical periodic orbits.

1.2 Old quantum theory

Before the advent of quantum mechanics, the classical action over a periodic orbit played a dominant role in the development of the old quantum theory. As pointed out earlier, the method faltered when nonintegrable motion was encountered [11]. We have stressed that semiclassical physics is not old quantum theory. But its modern version, embracing chaotic motion, puts the old quantum theory in the proper perspective. The connection of the old quantum theory to quantum mechanics is limited to the role of the periodic orbits in the quantization condition. Quantum mechanics, in addition, includes the revolutionary concept of the superposition principle of probability amplitudes. Nevertheless, it is very instructive to go back to the early years of the old quantum theory, and try to appreciate the role that the periodic orbits played.

Niels Bohr developed the quantum theory on the edifice of classical mechanics. In his theory of the hydrogen atom [9, 10] an electron in a periodic orbit obeyed Newton's laws of motion, but would not emit radiation unless making a transition to another orbit of a lower energy. Such orbits, with certain integral units of classical action are said to be *quantized*, and labeled by some quantum numbers that are constants of motion. We may ask if classical mechanics gives some clue as to why the action over a periodic orbit is a robust quantity well-suited for quantization. For this, we discuss the behavior of the classical *action* in some simple examples where a parameter of the dynamical system is slowly changing with time. The idea is to test if the action over a periodic orbit remains invariant, or almost so, in spite of the changing Hamiltonian. The relevance of adiabatic invariance to the quantization condition of Planck and Bohr is discussed at length in a paper by Ehrenfest [30]. The first example given below (of a wall moving with a constant velocity) is spectacular, but it comes about because the variation in the parameter of the Hamiltonian may be transformed away with new space and time scales [31] (see Problem 1.1). The second example of the pendulum with a slowly changing string length exemplifies what is called adiabatic invariance. In our view, these examples lend some intuitive justification for asserting that the *action* of an allowed periodic orbit in quantum theory cannot be changed arbitrarily.

1.2.1   A ball bouncing off a moving wall

The idealized situation is depicted in Fig. 1.2. The left wall is fixed, while the right wall moves inwards with a speed V. At $t = 0$, the spacing between the two walls is x_0, and a perfectly elastic ball starts horizontally with a velocity v_0 from the left end. At the time t_1 of the first collision with the moving wall, the position of the right wall is x_1. After the first collision with the moving wall, the ball rebounds with a speed $v_1 = v_0 + 2V$. This sequence continues, and the nth collision with the moving wall takes place at time t_n, when the right wall is at x_n. Clearly,

$$t_1 = \frac{x_1}{v_0} = \frac{(x_0 - Vt_1)}{v_0} . \tag{1.29}$$

Therefore $t_1 = x_0/(v_0 + V)$. Note further that $x_1 = (x_0 - Vt_1)$, $x_2 = (x_0 - Vt_2)$, etc., so that

$$(t_2 - t_1) = \frac{(x_1 + x_2)}{v_1} = \frac{(2x_0 - V(t_1 + t_2))}{(v_0 + 2V)} . \tag{1.30}$$

A little algebra then yields

$$t_2 = \frac{3x_0}{(v_0 + 3V)} . \tag{1.31}$$

In a similar manner, using the relations $v_2 = v_1 + 2V = v_0 + 4V$, etc., it is straightforward to show that

$$t_3 = \frac{5x_0}{(v_0 + 5V)} , \qquad t_n = \frac{(2n - 1)x_0}{v_0 + (2n - 1)V} . \tag{1.32}$$

Recall that

$$x_1 = x_0 - Vt_1 = \frac{x_0 v_0}{(v_0 + V)} . \tag{1.33}$$

Similarly,

$$x_2 = x_0 - Vt_2 = \frac{x_0 v_0}{(v_0 + 3V)} , \tag{1.34}$$

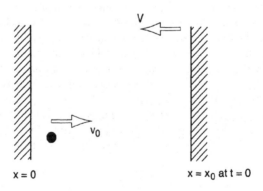

Figure 1.2: Ball rebounding elastically between a fixed and a moving wall. The wall at the right is moving in with a constant speed V.

$$x_n = x_0 - Vt_n = \frac{x_0 v_0}{(v_0 + (2n-1)V)}. \tag{1.35}$$

Starting with $t = 0$, the speed times the distance for the first cycle is then given by

$$S_1 = (x_1 v_0 + x_1 v_1) = 2x_1(v_0 + V) = 2x_0 v_0. \tag{1.36}$$

For the subsequent cycles, the action remains exactly the same:

$$S_2 = (x_2 v_1 + x_2 v_2) = 2x_2(v_0 + 3V) = 2x_0 v_0, \tag{1.37}$$

$$S_n = (x_n v_{n-1} + x_n v_n) = 2x_n(v_0 + (2n-1)V) = 2x_0 v_0. \tag{1.38}$$

Note, however, that the motion is not quite periodic over each cycle as defined, since the velocity of the ball is higher in the second half of each cycle. An alternate definition of a cycle would be over a period of time during which the speed of the ball remains the same, i.e., between successive collisions with the moving wall [32]. Instead of the action remaining constant (irrespective of the wall velocity V) as in our example above, it then changes very slowly, provided $V \ll v_0$.

As mentioned earlier, this is an example where the action is invariant because of a scaling property of transformed variables. This is so only if the velocity of the moving wall is constant. This is further discussed in Problem 1.1. A more appropriate example of the action being an "adiabatic invariant" is discussed below.

1.2.2 A pendulum with variable string length

This is an example showing that the classical action is an adiabatic invariant [33]. Consider an ideal pendulum consisting of a massless string of length l attached to a mass m, executing simple harmonic motion of small amplitude A, see Fig. 1.3. The linearized equation of motion is

$$\frac{d^2\theta}{dt^2} = -\frac{g}{l}\theta, \tag{1.39}$$

and the total energy is

$$E = \frac{1}{2}ml^2\left(\frac{d\theta}{dt}\right)^2 + \frac{1}{2}mgl\,\theta^2, \tag{1.40}$$

where the second term above is the potential energy with reference to the lowest point $\theta = 0$. The tension F on the string balances the component of the gravitational pull along the string and the centrifugal force, so

$$F = mg\left(1 - \frac{\theta^2}{2}\right) + ml\left(\frac{d\theta}{dt}\right)^2. \tag{1.41}$$

Figure 1.3: An ideal simple pendulum with a length that is
shortened slowly.

The solution of Eq. (1.39) is simple harmonic motion,

$$\theta = A \cos(2\pi\nu t + \delta), \tag{1.42}$$

where ν is the frequency,

$$\nu = \frac{1}{2\pi}\sqrt{\frac{g}{l}}. \tag{1.43}$$

Note that the tension F is time dependent, being maximum at times when $\theta = 0$. The
average $\langle F \rangle$ over one complete time period of the motion is easily seen to be

$$\langle F \rangle = mg + \frac{1}{4}mgA^2. \tag{1.44}$$

We consider an external agency to shorten the length of the string slowly by pulling
it in smoothly. The string is shortened by a length Δl over a time spanning many
periods, such that the change in the magnitude of the length over one period, δl, obeys
the condition $\delta l/l \ll 1$. Note, however, that the total change Δl need not be small in
the same sense. Such a slow change in a parameter of the system is termed adiabatic,
and the system itself is executing harmonic motion throughout the time. Under this
assumption, the work done by the external agency over one period is

$$\delta W = \delta l \langle F \rangle = mg\,\delta l + \frac{1}{4}mgA^2\,\delta l. \tag{1.45}$$

Here the first term on the right-hand side is the work expended against gravitation,
whereas the last term goes into increasing the energy of the oscillating pendulum by

$$\delta E = \frac{1}{4}mgA^2\,\delta l. \tag{1.46}$$

From Eqs. (1.40) and (1.42), one deduces that the energy of the system for a fixed
length is $E = mgA^2 l/2$, so the fractional change over one period is

$$\frac{\delta E}{E} = \frac{1}{2}\frac{\delta l}{l}. \tag{1.47}$$

From Eq. (1.43), it is also clear that shortening the length l of the pendulum by δl would result in an increase $\delta\nu$ in the frequency of oscillations, given by

$$\delta\nu = \frac{1}{2}\frac{\delta l}{l}\nu. \tag{1.48}$$

Comparing this with Eq. (1.47), we note that although the adiabatic change in the parameter l of the pendulum alters both the energy of oscillations as well as the frequency, it respects the relation

$$\frac{\delta E}{E} = \frac{\delta\nu}{\nu}, \tag{1.49}$$

which on integration yield the adiabatic invariant

$$\frac{E}{\nu} \simeq \text{const.} \tag{1.50}$$

This relation is an approximate one, since an average over a period was made to derive it. Nevertheless, this remarkable result (and the similar one in the earlier example) suggests that such adiabatic invariants have a robust stability that may even persist in a quantum theory. Indeed, we shall now see that the adiabatic invariants in the two examples given above may be identified as the classical "action", that plays a pivotal role in the quantization rule of the semiclassical theory, as well as in the description of quantum mechanics [26, 27]. In the above example, note that

$$\oint p_\theta d\theta = \frac{E}{\nu}, \tag{1.51}$$

where $p_\theta = ml^2\dot{\theta}$ is the angular momentum, and the integration is over one complete period of motion. This is the action for the angular motion. Similarly, $\oint p_x dx$ is the action over one period in the example (a) for linear motion. In quantum theory, the motion of a particle is associated with a de Broglie wave, of wave length $\lambda = h/p$. A periodic orbit in this picture may be regarded as a self-interfering standing wave, with the number of complete wave lengths equal to the integral quantum number [34]. The right-hand side of Eq. (1.51) is then equated to nh, where n is an integer, and h the Planck's constant. In this sense, quantization links the wave- and particle-pictures together. The quantization condition in these examples is simply the quantization of the action over a periodic orbit in one dimension. We shall see later that the classical periodic orbits will play a crucial role in our understanding of dynamical systems.

1.2.3   The phase space of a simple harmonic oscillator

Quite generally, for a given Lagrangian $\mathcal{L}(q_1, q_2, ..q_N; \dot{q}_1, \dot{q}_2, ..\dot{q}_N; t)$, the canonical momentum to the generalized coordinate q_i is defined by $p_i = \partial\mathcal{L}/\partial\dot{q}_i$, and the Hamiltonian is $H = \sum_i(p_i\dot{q}_i - \mathcal{L})$. One may envisage a $2N$-dimensional "phase space" whose coordinates are (q_i, p_i), and a point in this phase space describes the state of the classical system at an instant of time. The time evolution of the state is depicted by a unique non-intersecting trajectory in the phase space. For the one-dimensional systems described in (a) and (b), the phase space is only a two-dimensional space. For such cases, the quantity $\oint p_x dx$ or $\oint p_\theta d\theta$ may be interpreted as the area of a closed path in the two-dimensional phase space, whose two axes are given by the position coordinate and the corresponding canonical momentum.

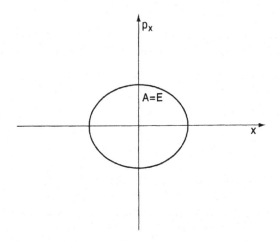

Figure 1.4: Ellipse in phase space. A simple harmonic oscillator, for a given energy E, traces the perimeter of the ellipse.

Since we have noted that the action over a periodic orbit plays an important role, it is instructive to plot $\oint p_x dx$ for the one-dimensional harmonic oscillator in the phase space. Note that any conservative system with one degree of freedom is necessarily integrable, since it has one constant of motion, the energy E. For the simple harmonic oscillator,

$$H = \frac{p_x^2}{2m} + \frac{1}{2}m\omega^2 x^2 = E. \tag{1.52}$$

This is just the equation of an ellipse in the phase space, as shown in Fig. 1.4, with the semi-axes given by $(2E/m\omega^2)^{1/2}$ and $(2mE)^{1/2}$. The particle is constrained to move on the circumference of the ellipse by energy conservation. The area of the ellipse, delineated by the trajectory of the particle over one complete period, is given by

$$S = \oint p_x dx = \pi\sqrt{2mE}\sqrt{\frac{2E}{m\omega^2}} = \frac{E}{\nu}. \tag{1.53}$$

This area is the action, denoted by the symbol S, and having the dimension of angular momentum. Quantizing S has the implication that only those orbits in the phase space are allowed whose areas have specified discrete values in units of the Planck's constant h. In Bohr's theory, the action may only take values

$$S_n = nh, \qquad n = 1, 2, 3, \ldots \tag{1.54}$$

For numerical work, it is useful to remember that

$$\hbar c = 1973 \, eV \, \mathring{A}, \tag{1.55}$$

where $1\mathring{A} = 10^{-10}\,\mathrm{m}$, and $1\,eV = 1.602 \times 10^{-19}\,J$. For the harmonic motion under consideration, the above quantization condition is actually incorrect, and it must be modified to $S_n = (n + 1/2)h$. The origin of the constant $1/2$ in the above equation is nontrivial, and is related to the single-valuedness of the wave function in the semiclassical approximation [35, 36]. This will be discussed in Sect. 2.4 on the WKB method [37, 38, 39] (named after Wentzel, Brillouin, and Kramers). Irrespective of this, it is clear from Eq. (1.54) that if one changes n by one unit, the extra area in the phase space is

$$\Delta S = \Delta \oint p\,dx = h, \qquad (1.56)$$

which is an annular area between two ellipses with energies $E_2 = (n + 1)\hbar\omega$ and $E_1 = n\hbar\omega$. On the average, we may now state that the number of quantum states between energies 0 and E is given by

$$\widetilde{N}(E) = \frac{1}{h} \oint p\,dx = \frac{E}{h\nu} = \frac{E}{\hbar\omega}. \qquad (1.57)$$

In writing the above equation, we have made use of the fact that there is, on the average, one quantum state in in an area h of the phase space [see Eq. (1.56)]. The tilde over N indicates that it is a semiclassical expression where the quantum discontinuities have been smoothed out. We may denote by $N(E)$ the number of quantum states between energy 0 and E. Both are plotted in Fig. 1.5.

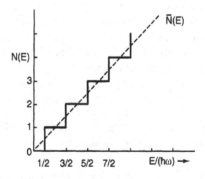

Figure 1.5: The quantum stair-case function $N(E)$. Its smooth part, $\widetilde{N}(E)$, is shown by the dashed curve.

The quantity $N(E)$ is a stair-case like function that jumps in discrete steps, and is 0 for $E/(\hbar\omega) < 1/2$. From this figure, it is clear that $\widetilde{N}$ alternately under- and overestimates the stair-case function N at the position of the eigenvalues. The oscillatory pattern in the difference $\delta N = N - \widetilde{N}$ is a general characteristic in any quantum spectrum, and not just for harmonic oscillators. The smooth part $\widetilde{N}(E)$, obtained from the global phase space average, is called the Thomas-Fermi term, and the corrections $\delta N(E)$ represent the quantum fluctuations. In Chapter 5 we shall see that the latter may be expressed as a Fourier-like sum of oscillatory terms, each term representing the contribution of a classical periodic orbit to the level density. In higher-dimensional systems, there is often some "bunching" of energy levels due to additional symmetries, leading to quantum shells. These "shell effects" have important implications in the stability of physical systems like atoms, nuclei, or metal clusters. The origin of the shell effects is rooted to quantum fluctuations of the density of states and the total energy around their smooth behaviour [40]. Some of their appearances in finite many-fermion systems will be studied in Chapter 8.

1.2.4 Three-dimensional anisotropic harmonic oscillator

The close link between classical periodic orbits and quantum energy gaps comes out spectacularly in the following three-dimensional simple example. The Hamiltonian of the particle is given by

$$ H = \left(\frac{p_x^2}{2m} + \frac{1}{2}\, m\, \omega_x^2\, x^2 \right) + \left(\frac{p_y^2}{2m} + \frac{1}{2}\, m\, \omega_y^2\, y^2 \right) + \left(\frac{p_z^2}{2m} + \frac{1}{2}\, m\, \omega_z^2\, z^2 \right) . \qquad (1.58) $$

Classically, there is periodic rectilinear motion along the x, y and z directions with periods $\frac{2\pi}{\omega_x}$, $\frac{2\pi}{\omega_y}$, and $\frac{2\pi}{\omega_z}$, respectively. But a closed periodic orbit in the three-dimensional space is only possible if the three frequencies ω_x, ω_y and ω_z have integer ratios, i.e.,

$$ \omega_x : \omega_y : \omega_z = p : q : s , \qquad (1.59) $$

where p, q and s are positive integers. For example, if $p = 1$, $q = 2$, and $s = 3$, there will be one closed "Lissajous" figure of a three-dimensional orbit with a period $2\pi/\omega_x$, whose precise shape will be governed by the phase differences between the three modes. The quantized spectrum of the above Hamiltonian (putting in by hand the zero-point energy $\hbar\omega_i/2$ for each degree of freedom) is given by

$$ E_{n_x,n_y,n_z} = \left(n_x + \frac{1}{2} \right) \hbar\omega_x + \left(n_y + \frac{1}{2} \right) \hbar\omega_y + \left(n_z + \frac{1}{2} \right) \hbar\omega_z , \qquad (1.60) $$

with n_x, n_y and n_z taking the values $0, 1, 2, \dots$. When the oscillator parameters are such that Eq. (1.59) is satisfied, Eq. (1.60) reduces to

$$ E_{n_x,n_y,n_z} = \frac{\hbar\omega_x}{p} \left[\frac{1}{2}\, (p + q + s) + (p\, n_x + q\, n_y + s\, n_z) \right] . \qquad (1.61) $$

For a given set of (p, q, s), it follows that the energy levels are degenerate, since there may be many combinations of (n_x, n_y, n_z) satisfying the equation $\mathcal{N} = (p n_x + q n_y + s n_z)$, with $\mathcal{N} \neq 0$, resulting in energy gaps in the spectrum of $\hbar\omega_x/p$ as $\mathcal{N}$ changes by unity.

This may be illustrated by an elegant diagram [41] for the axially symmetric case of $\omega_x = \omega_y = \omega_\perp \neq \omega_z$. In this situation, Eq. (1.60) reduces to

$$ E_{n_\perp,n_z} = (n_\perp + 1) \hbar\omega_\perp + \left(n_z + \frac{1}{2} \right) \hbar\omega_z . \qquad (1.62) $$

It is convenient to express the energy levels in terms of a deformation parameter ϵ and an average frequency ω_0 which is kept constant, defined by

$$ \epsilon = \left(\frac{\omega_\perp - \omega_z}{\omega_0} \right) , \qquad \omega_0 = \frac{(2\omega_\perp + \omega_z)}{3} . \qquad (1.63) $$

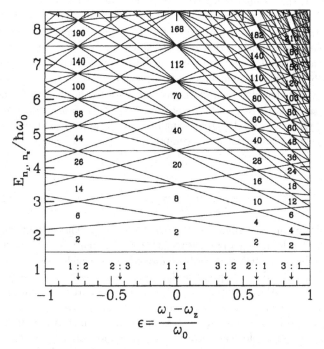

Figure 1.6: The single-particle states (1.62) of an axially symmetric deformed harmonic oscillator potential as a functions of the deformation parameter ϵ defined in Eq. (1.63). The indicated "magic numbers" correspond to the numbers of particles obtained by filling the states up to a given Fermi energy, respecting the Pauli principle (including a spin degeneracy factor 2). (From [41].)

A positive (negative) value of ϵ gives a prolate (oblate) shape, whereas $\epsilon = 0$ corresponds to spherical symmetry. The energy eigenvalues (1.61) may now be written as

$$E_{n_\perp, n_z} = \hbar\omega_0 \left[N + \frac{3}{2} - \frac{1}{3}\epsilon \left(2n_z - n_\perp \right) \right], \qquad (1.64)$$

where $N = (2n_\perp + n_z)$ is the total number of excitation quanta, i.e., the principal quantum number.

The single-particle energy level diagram is shown in Fig. 1.6 as a function of the deformation parameter ϵ. Note the pronounced energy gaps, and the corresponding "magic numbers" of particles obtained by filling the states from below according to the Pauli principle, at the frequency ratios

$$\frac{\omega_\perp}{\omega_z} = \frac{1}{2}, \frac{2}{3}, 1, \frac{3}{2}, 2, \dots, \qquad (1.65)$$

that span the potential shape from oblate to prolate. At each of these rational values of $\omega_\perp/\omega_z$, there are families of three-dimensional classical periodic orbits, which in phase space are confined to a three-torus (see Sect. 2.5). For irrational values of this ratio, however, the three-dimensional trajectories are not closed, even though they are confined to the torus. (Lower-dimensional periodic orbits still exist for irrational ratios, such as two-dimensional Lissajous orbits in the plane perpendicular to the symmetry axis, and one-dimensional orbits that oscillate along the symmetry axis. These have, however, much less overall weight in their contributions to the level density, as we will see in Chapter 6 in a general semiclassical treatment.)

The axially symmetric deformed harmonic oscillator potential with cylindrical co-ordinates (ρ, z, ϕ), where $\rho = \sqrt{x^2 + y^2}$, and $\tan \phi = y/x$, is often used in nuclear and metal-cluster physics. The single-particle angular momentum l^2 is no longer a constant of motion for such a system, but its z component, $l_z = \hbar l$, is. The other two quantum numbers n_z, n_ρ count the number of nodes in the z and the ρ directions. The eigenenergies are then written as

$$E_{n_\rho, n_z, l} = (2n_\rho + |l| + 1) \hbar \omega_\perp + (n_z + 1/2) \hbar \omega_z \,, \qquad (1.66)$$

and the principal quantum number is $N = 2n_\rho + |l| + n_z$. These quantum numbers play an important role in realistic nuclear shell-model potentials such as the Nilsson model [42], introduced in 1955 to describe deformed atomic nuclei. There, a spin-orbit interaction must be included to reproduce the experimentally observed magic numbers. Then, instead of l_z, it is the z component of the total single-particle angular momentum (including the intrinsic spin), $\mathbf{j} = \mathbf{l} + \mathbf{s}$, that is a constant of motion. Its spectrum is $j_z = \hbar \Omega = \pm 1/2, \pm 3/2, \pm 5/2$, etc., where Ω is the conserved quantum number. Nilsson also included a phenomenological correction term proportional to l^2 that simulates a steeper surface than that of the simple harmonic oscillator. For such a realistic potential, the above quantum numbers are no longer exactly conserved (e.g., states with different values of N or n_z are coupled). Nevertheless, the so-called "Nilsson quantum numbers" $[Nn_z\Lambda\Omega]$ (whereby only the absolute values of $\Lambda = l$ and Ω are used in the notation) have turned out [41, 42] to be very helpful in characterizing the asymptotic behavior of the single-particle wavefunctions, in particular for large deformations, where they indicate the leading components (see Sect. 8.1.3 for an example).

Due to the different radial forms of more realistic potentials used in the mean-field description of an atomic nucleus, and to the presence of the spin-orbit interaction, the level scheme is less regular than in Fig. 1.6, and the magic numbers are different. But the physical principles remain the same. An energy gap in the single-particle spectrum above the last occupied level lends extra stability against removal of particles or against shape changes. Such gaps (and corresponding magic numbers) also exist in deformed systems. Both for nuclei and for metal clusters, these gaps lead to the existence of deformed ground states, as we shall discuss in Chapter 8. The general occurrence of shells in deformed nuclei was also emphasized by Strutinsky [40], who designed a powerful method to calculate the total binding energy of a bound interacting fermion system from a phenomenological single-particle spectrum. We shall discuss this method in Chapter 4 and some of its applications in Chapter 8. Typical ground-state deformations of large fermion systems correspond to frequency ratios much less than $\omega_\perp : \omega_z = 2$. However, strong shell effects at frequency ratios ~ 2 and ~ 3 have also been observed. For example, "superdeformed" ($\omega_\perp/\omega_z \simeq 2$), and "hyperdeformed" ($\omega_\perp/\omega_z \simeq 3$) nuclear shapes have been discovered by studying the γ-rays emitted from highly excited nuclei in heavy-ion reactions [43]. The so-called fission isomers, which will be discussed in Sect. 8.1.2, also have shapes that come close to an axis ratio $\omega_\perp : \omega_z = 2$.

1.3 Wave packets in Rydberg atoms

In the previous sections, it was noted that the classical action for the periodic orbit played a very important part in quantization. A periodic orbit is characterized by a time period (or the frequency). One may naturally ask what is the connection of this classical frequency to the frequency of an emitted photon in a quantum transition. Moreover, although in the Bohr model of the atom the periodic orbits appear naturally, it is not obvious what their role is in quantum mechanics, where one may solve the Schrödinger equation, or diagonalize the Hamiltonian, to obtain the eigenvalues and the eigenstates. We attempt to answer these questions in the present section.

1.3.1 The large-n limit in the Bohr atom

In the Bohr-Sommerfeld atom, the allowed energy levels are obtained by quantizing the classical action. The emitted (or absorbed) photon frequency is then related naturally to the transition energy between the energy levels. We now show that the photon frequency has indeed a close connection to the classical frequency of the orbiting electron in the limit of large quantum number n. To see this connection simply, consider first a one-dimensional potential $V(x)$ with many bound states. As before, the action over a periodic orbit is given by

$$S(E) = 2 \int_{x_1}^{x_2} p_x(E, x)\, dx,\qquad(1.67)$$

where $p_x = \sqrt{2m(E - V(x))}$, and x_1, x_2 are the classical turning points where $V(x) = E$. The classical period in this case is $T = \partial S/\partial E$. Now consider the Bohr model, according to which the transition frequency of an emitted photon from a state with quantum number n to a state $(n - k)$ is

$$\nu_{n \to (n-k)} = \frac{E_n - E_{(n-k)}}{h}.\qquad(1.68)$$

For small n, this frequency bears little resemblance to the frequency of the classical periodic orbit. But let us consider large n, such that $n \gg k$. Then the right-hand side of Eq. (1.68) may be approximated by

$$\frac{E_n - E_{(n-k)}}{h} \simeq \frac{1}{h}k\left(\frac{\partial E}{\partial n}\right)_{E=E_n} = \frac{1}{h}k\left(\frac{\partial E}{\partial S}\right)_{E=E_n}\left(\frac{\partial S}{\partial n}\right)_{S=nh}.\qquad(1.69)$$

In the last step, we recognize the quantization rule that E_n itself is determined by the condition that $S(E) = nh$. Therefore, in this continuum limit, we obtain the relation

$$\nu_{n \to (n-k)} = \frac{E_n - E_{(n-k)}}{h} \simeq k\left(\frac{\partial E}{\partial S}\right)_{E=E_n} = k\nu_n^{cl},\qquad(1.70)$$

where ν_n^{cl} is precisely the frequency of the classical periodic orbit (in one dimension) with energy $E = E_n$. In this simple case, the transition from $n \to (n-1)$ in quantum theory corresponds to the fundamental frequency ν_n^{cl} of the classical periodic orbit, while the second, third, and higher harmonics match the transitions $n \to (n-2)$, $n \to (n-3)$ etc. Note that Eq. (1.70) holds irrespective of the details of the spectrum, provided $n \gg k$ (see Problem 1.2). This result, Eq. (1.70), may also be generalized to more degrees of freedom for the transition from a state labeled by quantum numbers $(n_1, n_2, ..)$ to $(n_1 - k_1, n_2 - k_2, ..)$, where each quantum number is for a separable degree of freedom. Provided $n_1 \gg k_1, n_2 \gg k_2, ..$, one gets

$$\nu_{(n_1,n_2,..)\to(n_1-k_1,n_2-k_2,..)} = \sum_i k_i \nu_i^{cl} , \qquad (1.71)$$

where $\nu_i^{cl} = (\partial E/\partial S_i)$, and $S_i = n_i h$. We have adapted the treatment of Tomonaga [33] in this section, and the reader should consult his book for an in-depth treatment of the old quantum theory. This link between quantum and classical physics for very large quantum numbers is part of the "correspondence principle" that was formulated by Niels Bohr [44]. It holds good in the modern formulation of quantum mechanics also, and will be discussed from a different angle in Sect. 2.2. The above discussion shows that in the large-n limit, one may associate the frequencies of quantum *transitions* with the Fourier components of classical motion according to Eq. (1.71). In general, it is required to follow the time evolution of an ensemble of quantum states, superposed coherently, to trace the classical behavior of a particle, and *vice versa*. We go on to discuss this next. The reader will profit by reading two general articles [45], [46] on this subject. The latter includes an extensive bibliography.

1.3.2   Where are the periodic orbits in quantum mechanics?

In post-Bohr quantum mechanics, formulated by the matrix mechanics of Heisenberg and the wave equation of Schrödinger [47], the connection to classical periodic orbits is not at all apparent. For example, when one solves the Schrödinger equation for the eigenstates of the H-atom, it is not clear what an eigenstate has to do with the Kepler orbit of the classical particle. This question was posed by Lorentz [48] to Schrödinger, and led the latter to the discovery of the coherent state in the harmonic oscillator [49] (see Problem 1.3). Consider first the classical motion of a point particle moving freely with a well-defined momentum. The corresponding description in quantum mechanics has inherent limitations, since canonical variables like position and momentum cannot have simultaneous eigenvalues. The best one can do is to construct a localized wave packet by an appropriate superposition of plane waves, spread over a band of momentum eigenvalues. Here localization in space is achieved by the destructive interference of the plane waves in all but a small region of space, through a suitable combination of initial phases. It is not surprising that when such a wave packet evolves in time, the different momentum components travel with different velocities, and there is a spreading in the spatial width. This spreading will take place in a classical wave packet also. In the quantum mechanics of a one-dimensional harmonic oscillator, however, Schrödinger [49] was able to construct a *coherent* state, which did not spread in time. This was a wave packet made from an appropriate superposition of its eigenstates of the Hamiltonian. Since the time evolution of a stationary state with eigenvalue E_n is governed by the factor $\exp(-iE_n t/\hbar)$, and since in the harmonic oscillator potential all the E_n's are equidistant, it is possible to retain the initial phase coherence in the wave packet, so there is no spreading. The localized wave packet may bounce back and forth in the potential, just like a classical particle.

Coming back to the H-atom, one may attempt to construct a localized wave packet (representing the classical particle), and see if it goes round the Kepler orbit. Schrödinger was unable to construct such a wave packet in the form of a non-spreading coherent wave. This is not surprising, since the energy eigenvalues E_n's are not multiples of each other. Note, however, from Eqs. (1.70) and (1.71), that in the Rydberg limit of the atom [50], the energy spacings bear definite relations to the harmonics of the classical frequencies. It may be possible, therefore, to preserve the initial phase differences with increasing time, at least approximately, in a wave packet made up of such Rydberg states. In the last few years, such localized wave packets have been studied theoretically, and generated experimentally [51] using short-pulsed lasers on hydrogen and other simple atoms. These wave packets exhibit classical features, but also some essential quantum characteristics like "recurrence" [51] due to the nonlinear dependence of the eigenvalues on the principal quantum number n.

Consider the hydrogen atom wave function $\psi_{nlm}(r, \theta, \phi)$ in spherical polar coordinates (an eigenstate of the Coulomb potential), where the spin of the electron is ignored. The notation is standard, with the principal quantum number

$$n = (n_r + l + 1).$$

The localization in the radial coordinate r is achieved by taking a wave function that peaks sharply at a single distance r, (i.e., it has only one maximum, with no nodes except at $r = 0$ and $r = \infty$, and large l to ensure sharpness). Further, to localize the electron in the equatorial plane ($\theta = \pi/2$) as best as possible, choose the l_z component to to be aligned to l, i.e., $m = l$. This is a "circular orbit wave function" $\psi^{(c)}$, with (see Problem 2.2) $n_r = 0$, i.e., $l = (n - 1)$. It has the form

$$\psi_{nlm}^{(c)}(r, \theta, \phi) = \mathcal{N} \left(\frac{r}{a}\right)^l \exp(-r/na) \sin^l \theta \exp(il\phi). \tag{1.72}$$

Here $\mathcal{N}$ is the normalization constant, $a = \hbar^2/(e^2 m_e)$ is the atomic length, and the superscript (c) is for the circular orbit characterization of the eigenfunction. By denoting the electron mass by m_e, we distinguish it from the azimuthal quantum number in this section. From Eq. (1.72), it is easy to check that $r^2 |\psi^{(c)}|^2$ peaks at

$$\left(\frac{r}{a}\right) = (l + 1)^2 = n^2, \tag{1.73}$$

so a Rydberg state with $n = 10^4$, for example, has a radius of almost 1 cm! Furthermore, for such large n values, the uncertainties in the radial and θ coordinates are almost minimal [52]:

$$\Delta r \Delta p_r = \frac{\hbar}{2}\left(1 + \frac{1}{2n}\right) + O\left(\frac{1}{n^2}\right), \quad \Delta\theta \Delta p_\theta = \frac{\hbar}{2}\left(1 + \frac{1}{4n}\right) + O\left(\frac{1}{n^2}\right). \tag{1.74}$$

Thus an electron in a high Rydberg state has a macroscopic radius and some almost classical characteristics. Note, however, that $|\psi^{(c)}|^2$ is independent of ϕ for a given l, and it has no localization in this coordinate. To do this, it is necessary to superpose circular wave functions of different l values. This was first suggested by L. Brown [53] and its time evolution was computer-simulated by Gaeta and Stroud [52]. For circular wave functions, superposition of different l values amount to superposing differing n values. But the window of n-mixing should have a narrow width in comparison to the average $\bar{n}$, so as not to destroy the radial peaking. Gaeta and Stroud [52] consider a wave packet of the form

$$\Psi(t) = \frac{1}{\sqrt{2\pi\sigma^2}} \sum_{n=1}^{\infty} \exp\left[-\frac{(n-\bar{n})^2}{4\sigma^2}\right] \psi_{nlm}^{(c)} \exp(-iE_n t/\hbar), \qquad (1.75)$$

where $\bar{n}$ is the mean, and σ is the standard deviation of the distribution, with $\sigma \ll \bar{n}$. The assumption of a Gaussian distribution is not crucial, but simplifies the analysis. The parameters $\bar{n}, \sigma$ are to be chosen to match the spread of the states in the wave packet produced by a laser in an experiment.

The eigenvalues of the Coulomb potential are $E_n = -m_e e^4/(2\hbar^2 n^2)$. The time evolution of such a wave packet with $\bar{n} = 320$, and $\sigma = 2.5$ is shown in Fig. 1.7, with the time given in units of the Kepler period,

$$T_K = \frac{2\pi r_0}{\langle v \rangle} = \frac{2\pi n^2 a}{\frac{\alpha}{n}c} = \frac{2\pi n^3 a}{\alpha c}. \qquad (1.76)$$

In the above, $\alpha = e^2/\hbar c$ is the fine structure constant and c the velocity of light. It is evident from Fig. 1.7 that the circulation of the classical particle along the Kepler orbit is being mimicked by the quantum mechanical ensemble of wave functions, constructed according to Eq. (1.75). Although there is some amount of spreading of the wave packet, its effect is small over the time of one revolution around the Kepler orbit.

The complete quantum behavior is, however, more complicated than the classical case. In fact, after spreading of the wave packet, there is self-interference and *revival*, as seen from the following consideration. Let us write

$$\omega_n = -\frac{E_n}{\hbar} = \frac{e^2}{2\hbar a n^2}, \qquad (1.77)$$

corresponding to the "frequency" of the eigenvalue E_n. Unlike the harmonic oscillator, these are not multiples of each other. But we may expand ω_n about its mean value $\omega_{\bar{n}}$, to obtain

$$\omega_n = \omega_{\bar{n}} + (n-\bar{n})\frac{\partial \omega_n}{\partial n}\Big|_{\bar{n}} + \frac{1}{2}(n-\bar{n})^2 \frac{\partial^2 \omega_n}{\partial n^2}\Big|_{\bar{n}} + \dots. \qquad (1.78)$$

Using Eqs. (1.76) and (1.78), we obtain

$$\omega_{\bar{n}} = \frac{\bar{n}}{2}\frac{2\pi}{T_K^{(\bar{n})}}, \quad \frac{\partial \omega_n}{\partial n}\Big|_{\bar{n}} = -\frac{2\pi}{T_K^{(\bar{n})}}, \qquad (1.79)$$

and

$$\frac{\partial^2 \omega_n}{\partial n^2}\Big|_{\bar{n}} = \frac{3}{\bar{n}}\frac{2\pi}{T_K^{(\bar{n})}} = \frac{2\pi}{T_R}, \qquad (1.80)$$

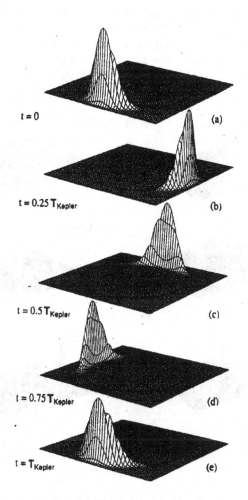

Figure 1.7: Computer simulated initial time evolution of a quantum wave packet, Eq. (1.75), in the hydrogen atom (from [52]). Here $\bar{n} = 320$ and $\sigma = 2.5$. This is the quantum analogue of an electron in a circular Kepler orbit.

where the *recurrence* period T_R is defined as

$$T_R = \frac{\bar{n}}{3} T_K^{(\bar{n})} . \tag{1.81}$$

Writing Eq. (1.75) as

$$\Psi(t) = \sum_n a_n \psi_n \exp(-i\omega_n t) , \tag{1.82}$$

it may be expressed, using Eqs. (1.79) - (1.81) and setting $n = \bar{n} + k$, as

$$\Psi(t) = \exp\left(-i\omega_{\bar{n}} t\right) \sum_k a_k \psi_{(\bar{n}+k)} \exp\left(2\pi i \left[kt/T_K^{(\bar{n})} - k^2 t/(2T_R)\right]\right) . \tag{1.83}$$

In the above expression, the first exponential in the sum produces the periodic oscillations over the Kepler period, $T_K^{(\bar{n})}$, as shown in Fig. 1.7. The second exponential is governed by the recurrence period T_R; it causes the dispersion in the wave packet and its subsequent revival after time T_R, but with a phase shift of π. Some typical numerical results of computer simulations are shown in Fig. 1.8.

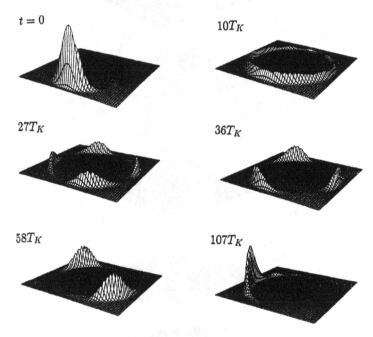

Figure 1.8: The long-time evolution of the same wave packet as in the previous figure, showing spreading, and the subsequent regrouping after the revival period T_R. (From [51].)

The spreading of a wave packet in a dispersive medium appears already on the classical level. The revival, however, is a pure quantum phenomenon; it arises from the nonlinear n-dependence of the energy eigenvalues of the coherently superposed states. Quite generally, if the energy E_n of the superposed waves is expanded about the mean $\bar{n}$ (a prime denoting differentiation with respect to n),

$$E_n = E_{\bar{n}} + E'_{\bar{n}}(n - \bar{n}) + \frac{1}{2} E''_{\bar{n}}(n - \bar{n})^2 + \ldots \tag{1.84}$$

then the Kepler and recurrence (or revival) periods are defined by

$$T_K = \frac{2\pi\hbar}{|E'_{\bar{n}}|} \, , \; T_R = \frac{2\pi\hbar}{\frac{1}{2}|E''_{\bar{n}}|} \, . \tag{1.85}$$

There is no recurrence for a wave packet made up from the eigenstates of a simple harmonic oscillator, since $E''_n = 0$. Also, if the quantum spectrum were random, there would be no recurrence and the wave packet would continue to spread indefinitely. The correspondence principle as such does not imply recurrence of the wave packet. In alkali metals, the energies E_{n^*} given by the Rydberg series depend on quantum defects, and n^* is shifted from the integer value. This is expected to modify the long-term recurrence pattern of the wave-packet [54].

Experimentally, the recurrence of Rydberg wave packets has been observed [55]. But the wave packets formed in these experiments are not composed of the circular wave functions discussed above. Rather, they are only localized radially, and not in the angular coordinate ϕ. The atoms are excited by the coherent optical radiation of a pulsed laser (of pico- or femto-second pulse duration), and the resulting quantum mechanical state of the electron is a superposition of the Rydberg states about an average $\bar{n}$ of order 100 in a narrow band of n values. However, because of the dipole selection rule, the angular momentum of all these states increases only by $\Delta l = 1$, so there is no localization in the azimuthal direction ϕ. If the electron is excited from the s-state, then the final p-state probability (with $m = 1$) has a $\sin^2\theta$ distribution, peaked in the equatorial plane. The quantum wave packet is like a spherical shell that oscillates in a radial breathing mode about a mean radius determined by the relation

$$\langle r_0 \rangle = \bar{n}^2 a \simeq 10^{-4} \text{cm} \,. \tag{1.86}$$

Following Stroud, the quantum motion of the oscillating wave packet may be pictured as an ensemble of classical configurations, and not just by one periodic orbit. This is depicted schematically in Fig. 1.9. The electronic motion in each of the classical orbits is in phase, coming to the nearest point of the nucleus at one end, and at the farthest turning point at the other end, in coherence. Experimentally, this picture has been verified by different groups. Note that the electron velocity, and therefore the electric current **J**, has a maximum when it is nearest to the nucleus, and a minimum when it is at the outer edge. Therefore, when the electron has been excited by the laser "pump" to this breathing mode state, if a further "probe" in the form of an electric pulse with field **E** is applied to the atom, the electron is subjected to a power supply **J · E** that varies with the magnitude of the current **J**. It is maximum

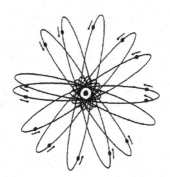

Figure 1.9: A radially excited wave packet describes an ensemble of classical Kepler orbits with the electron in each orbit in the same phase as shown above. (From [51].)

near the nucleus, and by adjusting **E** one can cause ionization at this end, while failing to do so at the outer turning point. Therefore a measurement of the count of the ions, synchronized with delayed probe pulses, can help map the breathing mode. The experimental result [56], performed on atomic potassium, is shown in Fig. 1.10. The experimental details will not be given here. Very interestingly, the oscillations in the ionization pattern has a beating structure that arises with the spreading and the subsequent reassembling (or recurrence) of the wave packet, justifying the discussion given earlier in this section. A wave packet localized radially as well as in angular coordinates has not been obtained experimentally yet. Recently, the Rochester group [57] has proposed a method of achieving this in a two-step process. In the first step, the atom is excited to a circular state with a large n value. Next, neighboring n values are populated by subjecting the atoms to a strong electric pulse of short duration. It is suggested that the resulting wave packet will have localization in all three coordinates, and in this case will be in a circular orbit. The role of the transitory electric field may be understood from a classical analogue. Imagine an ensemble of electrons distributed evenly around a circle and going round it with a uniform angular velocity. The short electric pulse accelerates the electrons on one side of the circle and decelerates them at the opposite end. After a certain time, there is therefore a bunching of the electrons, when the faster of them catch up with the slower ones. This classical picture can be recast in the quantum language in terms of spatial localization of a wave packet.

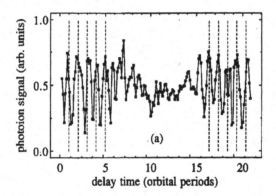

Figure 1.10: The experimental photoionization signal in the long-time evolution of a radially localized wave packet in atomic potassium. The wave packet is formed by single-photon excitation by the pump laser pulse tuned to $n = 65$ p-state. The probe pulse delay time is synchronized to the orbital period. The damping of the signal is due to the spreading of the wave packet. The signal gets strong again after about 15 orbital periods due to its subsequent revival. (From [56].)

1.4 Chaotic motion: atoms in a magnetic field

Till now we have been examining the role of periodic orbits in classically integrable systems. There is, however, more interest in studying the quantum behavior of systems that are classically chaotic, and in particular the part that periodic orbits play. In a classically chaotic system that is governed by a time-independent non-dissipative Hamiltonian, the motion of the particle spans all the phase space available energetically. Moreover, its trajectory diverges exponentially from its original path if the initial conditions are changed even slightly. The quantum behavior of the same particle appears to be more muted. Specially suitable for this is the study of highly excited Rydberg states of the hydrogen and other simple alkali atoms in a uniform magnetic field. This is because the classical motion of the electron near the ionisation threshold is highly chaotic, and the quantum behavior may be studied by precise experiments. Moreover, for the hydrogen atom in a uniform magnetic field, the Hamiltonian is well-known and simple. Even for a more complex atom, an electron in an excited Rydberg orbital (of large radius) sees a net positive charge of one unit and is almost hydrogen-like. In this section, we first describe the classical motion of the electron in the magnetic field, displaying its ergodic character on what is called a 'Poincaré surface of section' of the phase-space. The quantum aspects are revealed by a variety of experiments. We describe some photo-absorption experiments, specially for states near the ionization limit, where 'quasi-Landau resonances' are found. These are specially interesting, because very near the threshold, an approximate separation of the Hamiltonian may be made for motion along, and motion transverse to the magnetic field, allowing for a simple interpretation of the resonances in terms of the classical closed orbits [58]. We shall also see that many of the irregularly spaced peaks in the energy spectrum may be Fourier-analyzed in the time domain in terms of a few classical orbits.

1.4.1 Scaling of classical Hamiltonian and chaos

The motion of a charged particle in a uniform magnetic field is discussed in Problem (1.4). It is found that its single-particle spectrum collapses into Landau levels with an energy spacing of

$$\hbar\Omega_c = \frac{eB\hbar}{mc}, \tag{1.87}$$

where Ω_c is called the cyclotron frequency. This idealized result was obtained for strictly two-dimensional motion, with the magnetic field B perpendicular to the plane of the motion. This is also shown in Fig. 3.4 in Sect. 3.2.8. In three dimensions, the motion along the direction of the magnetic field remains unaffected by the field. The problem is classically nonintegrable, however, if a uniform magnetic field is applied to an electron bound to an atom. This problem has been studied in great detail both experimentally and theoretically [59, 60, 61]. Here we only touch on those topics that have a direct bearing on the nature of the periodic orbits, and the quantization of the action discussed earlier. First consider the motion of an electron in an attractive Coulomb potential and a uniform magnetic field along the z-axis, neglecting spin. In the symmetric gauge, the Hamiltonian in spherical polar coordinates is given by

(compare with the cranked oscillator, Sect. 3.2.8)

$$H = \frac{p^2}{2m} - \frac{e^2}{r} + \frac{1}{2} m \Omega_L^2 r^2 \sin^2\theta + \Omega_L L_z , \tag{1.88}$$

where $\Omega_L = \Omega_c/2$ is called the Larmor frequency. In this three-dimensional problem, $p^2 = (p_x^2 + p_y^2 + p_z^2)$, and $r = \sqrt{x^2 + y^2 + z^2}$. The system has only two constants of motion in involution, the energy E, and the z-component of the angular momentum, L_z. L^2 is not a constant of motion, since the term proportional to Ω^2 breaks the spherical symmetry. For a magnetic field of a few Tesla, this term is negligible compared to the Zeeman term, provided the atom is in its ground state, or a low-lying excited state. But for a highly excited Rydberg state, (since the atomic radius grows as n^2) the nonspherical term becomes dominant. This results in the classical motion being chaotic for large radii near the ionisation threshold. The quantal spectrum also exhibits some signatures of this. Such behavior is expected when the noncentral term in the Hamiltonian (1.88) is comparable in magnitude to the Coulomb potential:

$$\left(\frac{B^2}{8mc^2}\right) r^2 \sin^2\theta \simeq \left(\frac{1}{r}\right) . \tag{1.89}$$

This may be rewritten conveniently (with the sine term $= 1$) in units of the Bohr radius a and the magnetic length l_B as

$$\left(\frac{r}{a}\right)^3 \simeq 8 \left(\frac{l_B}{a}\right)^4 . \tag{1.90}$$

In the above, the two important length scales, a and l_B are defined as

$$a = \frac{\hbar^2}{me^2} , \qquad l_B = \sqrt{\frac{\hbar c}{eB}} . \tag{1.91}$$

The magnetic length l_B is just the radius of the quantized circular orbit of the electron in a transverse magnetic field B, with no Coulomb potential. For numerical work, it is convenient to remember that for an electron

$$l_B = \frac{256.5}{\sqrt{B}} \, \text{\AA} , \qquad a = 0.5292 \, \text{\AA} , \tag{1.92}$$

where B is in Tesla. Let us write the dimensionless parameter

$$\gamma = \frac{a^2}{l_B^2} = \frac{B}{(2.349 \times 10^5 \, T)} . \tag{1.93}$$

Using Eq. (1.90), and $(r/a) = n^2$ from Eq. (1.73), we get the rough criterion for chaotic behavior to be

$$n \simeq \gamma^{-1/3} . \tag{1.94}$$

For a laboratory field strength of a few Tesla, this corresponds to a principal quantum number $n \simeq 50 - 100$, and a radius of a few thousand Å.

We now show that the Hamiltonian (1.88) may be written in a *scale invariant* form, such that the equations of motion depend on a single parameter proportional to $(E\,B^{-2/3})$, rather than on the energy E and the magnetic field B separately. Denoting the fine-structure constant by $\alpha = e^2/\hbar c$, the Hamiltonian may be written as

$$H = \frac{p^2}{2m} - \frac{e^2}{r} + \frac{1}{8}\alpha^2\gamma^2(mc^2)\frac{r^2}{a^2}\sin^2\theta + \frac{\alpha c}{2a}\gamma L_z\,. \tag{1.95}$$

By making the transformation

$$\tilde{x} = \gamma^{\frac{2}{3}}x\,, \quad \tilde{t} = \gamma t\,, \tag{1.96}$$

we obtain

$$\tilde{r} = \gamma^{\frac{2}{3}}r\,, \quad \tilde{p} = \gamma^{-\frac{1}{3}}p\,, \quad \tilde{L}_z = \gamma^{\frac{1}{3}}L_z\,. \tag{1.97}$$

A little algebra then shows that we may write the *scaled* Hamiltonian as

$$\widetilde{H} = \gamma^{-\frac{2}{3}}H = \frac{\tilde{p}^2}{2m} - \frac{e^2}{\tilde{r}} + \frac{1}{8}\alpha^2 mc^2\frac{\tilde{r}^2}{a^2}\sin^2\theta + \frac{\alpha c}{2a}\tilde{L}_z\,. \tag{1.98}$$

This scaled Hamiltonian obviously does not depend on B and E separately. For a given energy $H = E$, and field strength B, it is the scaled energy

$$\tilde{\epsilon} = \gamma^{-\frac{2}{3}}E \tag{1.99}$$

that determines the classical motion. Due to scaling, the motion of a high energy electron in a weak magnetic field is similar to that of a low energy one in a strong field. It is interesting to note that

$$[\tilde{x}, \tilde{p}] = i\,\gamma^{1/3}\hbar\,. \tag{1.100}$$

In scaled coordinates, it is as if the Planck's constant is variable, getting weaker for smaller γ, i.e., for weaker magnetic fields. From Eq. (1.99), we note that a smaller γ also samples a higher scaled energy. This is manifested in the semiclassical behavior of Rydberg states for relatively weak magnetic fields.

The classical motion is generally described in the literature by introducing the so-called semi-parabolic (scaled) coordinates that eliminates the singular Coulomb term in the Hamiltonian. We explain this here in order that the reader may appreciate the phase plots that are shown in the literature. These coordinates [62] are denoted by u, v, and τ, and are defined in terms of the scaled cylindrical coordinates $\tilde{\rho} = \sqrt{\tilde{x}^2 + \tilde{y}^2}$, and $\tilde{z}$. The transformation equations are (the tildes are omitted henceforth for simplicity):

$$\rho = uv\,, \quad z = \frac{1}{2}\left(v^2 - u^2\right)\,, \quad dt = (u^2 + v^2)\,d\tau\,. \tag{1.101}$$

It is easily checked that

$$p_\rho = \frac{(vp_u + up_v)}{(u^2 + v^2)}\,, \quad p_z = \frac{(vp_v - up_u)}{(u^2 + v^2)}\,, \tag{1.102}$$

where $p_u = (du/d\tau)$, $p_v = (dv/d\tau)$. The scaled Hamiltonian in the cylindrical coordinates is give by Eq. (1.106) in the next subsection. For a fixed $L_z = m_l$, the Hamiltonian transforms to

$$\left(\widetilde{H} - \frac{m_l}{2}\right) = \frac{1}{2(u^2+v^2)}\left[p_u^2 + \frac{m_l^2}{u^2} + p_v^2 + \frac{m_l^2}{v^2} + \frac{1}{4}u^2v^2(u^2+v^2) - 4\right]. \qquad (1.103)$$

For a scaled energy of $\widetilde{H} = \tilde{\epsilon}$, we may rearrange the equation as

$$\left[\frac{1}{2}p_u^2 + \frac{m_l^2}{2u^2} + \frac{1}{2}p_v^2 + \frac{m_l^2}{2v^2} - (\tilde{\epsilon} - \frac{m_l}{2})(u^2+v^2) + \frac{1}{8}u^2v^2(u^2+v^2)\right] = 2, \qquad (1.104)$$

and denote the left-hand side by $\tilde{h}$. The coordinate singularity due to the Coulomb term has been eliminated by this transformation. The last term in the above, $u^2v^2(u^2+v^2)/8$, represents the nonlinear coupling $\Omega^2 r^2 \sin^2\theta$ due to the magnetic field, also called the diamagnetic term [see Eq. (1.88)]. This term in the above equation makes the classical motion nonintegrable even when the magnetic field B is very weak. In the absence of the magnetic field, the effective Hamiltonian $\tilde{h}$ above is separable in the coordinates (u, v). Note that the harmonic potential gets inverted if $(\tilde{\epsilon} - m_l/2) > 0$. Although there is a one-to-one correspondence between the classical trajectories of $\widetilde{H}$ given by Eq. (1.106) and $\tilde{h}$ of Eq. (1.104), the shapes of the periodic orbits are not the same. For more information on the orbits, the reader should consult Ref. [60].

Consider the classical motion generated by the Hamiltonian $\tilde{h}$ for a fixed value of $\tilde{\epsilon}$ and $m_l = 0$. Note that only three of the four coordinates (u, v, p_u, p_v) of the phase space are independent. To specify the motion completely, it is therefore sufficient to choose any three of the four coordinates, say $u, v,$ and p_u. The solution of Newton's equations of motion as functions of time for some initial conditions is then given by a set of numbers $u(\tau), v(\tau)$ and $p_u(\tau)$ at a time τ. The three-dimensional evolution in time is too complicated to plot. It is more convenient to take a two-dimensional slice of the phase space, e.g., the plane with $v = 0$, and mark all the points of intersection of the three-dimensional trajectory on this plane, going downwards. Such a visual representation of the dynamical motion in the (u, p_u) hyperplane of the phase space is called the *Poincaré surface of section*. This is shown in Fig. 1.11 on the following side. The boundary of this surface is easily obtained from Eq. (1.104) by putting $m_l = v = p_v = 0$, and is given by

$$p_u^2 - 2\tilde{\epsilon}u^2 = 4, \qquad (1.105)$$

which is a circle of radius 2 in the coordinates $(\sqrt{-2\tilde{\epsilon}}u, p_u)$. In Fig. 1.11, following [59], we depict the surfaces of section for a few selected (negative) values of $\tilde{\epsilon}$. A detailed discussion of the plots will be found in this review article. It is clear from Eq. (1.104) that bounded motion results for $m_l = 0$ and a negative ϵ. As pointed out earlier, a weaker magnetic field yields an $\tilde{\epsilon}$ of larger magnitude. To appreciate the significance of the various patterns shown in Fig. 1.11, note that a periodic orbit on the Poincaré surface is represented by a fixed point or by a sequence of a finite number of points, whereas a quasi-periodic motion traces an infinite sequence of non-repetitive points on a curve. Fully *ergodic* motion, on the other hand, fills the entire circular boundary with isolated points, as for example seen in the figure for $\tilde{\epsilon} = -0.1$, all the points arising from the time development of a single trajectory.

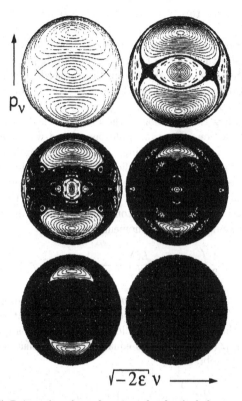

Figure 1.11: Poincaré surface of section of a classical electron trajectory in a hydrogen atom in the presence of a strong magnetic field. The plots (from left to right, and top to bottom) are for $\tilde{\epsilon}$ = -0.8, -0.5, -0.4, -0.3, -0.2, and -0.1. (After [59].)

Ergodic motion, accompanied by extreme sensitivity to initial conditions, is (classical) *chaos*. The example in hand, for $\tilde{\epsilon} < -0.35$, gives a majority of chaotic trajectories. A few qualitative features of the trajectories shown in the figure are worth mentioning. Note that the single parameter $\tilde{\epsilon} = \gamma^{-2/3} E$ determines the degree of chaos. We also see that islands of regular motion coexist with a "stochastic sea" in the phase space. It is not possible from the figure to identify the periodic orbits which are present even when the motion turns chaotic. These periodic orbits, however, determine the quantum density of states, which in turn yields the the photo-absorption spectrum. We now turn our attention to this phenomenon.

1.4.2   Quasi-Landau resonances in atomic photoabsorption

It is first useful to recapitulate some relevant properties of the Rydberg states again, using the Bohr model of the atom as a guide. We have noted that the radius of the Bohr orbit is proportional to n^2, so the area traced by the orbit goes up as n^4. For $n = 100$, this amounts to a 10^8-fold increase in the area compared to the ground state. The induced current in the atom due to the external magnetic field is proportional to the magnetic flux, and therefore to the area of the orbit perpendicular to the field. The diamagnetic interaction that arises from this induced current therefore increases as n^4, whereas the Coulomb energy goes down as n^{-2}. Thus, for large n, the behavior of the electron in the x-y plane is mainly governed by the diamagnetic term, while in the z direction (along the direction of the magnetic field), the motion is dictated by the Coulomb interaction. The time scales associated with the two motions, for a magnetic field of a few Tesla, are very different near the ionization threshold. Whereas the rectilinear motion along the magnetic field is rather slow, the circular motion in the x-y plane is fast. This may be understood by noting that although the magnetic field confines the motion in the x-y plane, near the ionization limit the electron may travel far from the nucleus. Consider the Hamiltonian (1.88) in cylindrical scaled coordinates $\tilde{\rho}$, $\tilde{z}$, and the corresponding canonical momenta:

$$\widetilde{H} = \gamma^{-\frac{2}{3}} H = \frac{1}{2} \left(\tilde{p}_\rho^2 + \tilde{p}_z^2 + \frac{\widetilde{L}_z^2}{\tilde{\rho}^2} \right) - \frac{1}{\sqrt{\tilde{\rho}^2 + \tilde{z}^2}} + \frac{1}{8} \tilde{\rho}^2 + \frac{1}{2} \widetilde{L}_z . \tag{1.106}$$

The frequently used convention $e = \hbar = m = c = 1$ is being adopted here to comply with the literature. The discussion above about the different time scales, and the observation that on the average $z \gg \rho$, allows us to make an approximate separation of variables [63] in Eq. (1.106). In this equation, if we replace $\sqrt{\tilde{\rho}^2 + \tilde{z}^2}$ by $|z|$, and specialize to $\tilde{L}_z = 0$, we obtain

$$\widetilde{H} \simeq \frac{1}{2} \left(\tilde{p}_\rho^2 + \frac{1}{8} \tilde{\rho}^2 \right) + \frac{1}{2} \left(\tilde{p}_z^2 - \frac{1}{|z|} \right) .$$

Thus the semiclassical quantization condition

$$\frac{1}{2\pi} \oint_i (\tilde{p}_\rho d\tilde{\rho} + \tilde{p}_z d\tilde{z}) = n \gamma^{1/3} \tag{1.107}$$

over the closed (but not necessarily periodic) orbits i yields the resonant states that show up in the photo-absorption of the Rydberg atom [64].

For free electrons in a magnetic field, it was noted earlier that there are Landau levels at intervals of the cyclotron frequency $\Omega_c = eB/mc$. Garton and Tomkins [65], on the other hand, discovered resonances spaced at intervals of $1.5\,\Omega_c$ in the photo-absorption cross section near the continuum in barium (see Ref. [66] for more accurate measurements). These are termed quasi-Landau resonances. More recently, photo-absorption experiments have also been performed on hydrogen [64] and lithium atoms [63]. Garton and Tomkins obtained the absorption spectrum in atomic Ba vapor in a magnetic field of 0.25 Tesla for states up to $n = 75$. For much lower values of n, dipole transition shows normal Zeeman triplets, with $\Delta m_l = \pm 1$ symmetrically placed about the $m_l = 0$ line. For larger n, the quadratic diamagnetic term becomes more important, and is responsible for the quasi-Landau resonances. Modern experiments use tunable lasers of highly monochromatic electromagnetic radiation to excite atoms to Rydberg states of the desired polarization, often using two- or three-step ladders of excitation. The Bielefeld group [64, 67] apply the magnetic field along a well-collimated hydrogen beam. They use a pulsed laser (of pulse length 16 ns) to first excite the atoms from the ground state to the *Lyman-α* line ($\lambda \approx 1216$ Å) with $\Delta m_l = 0, \pm 1, \pm 2$.

Fig. 1.12 shows the resulting spectrum with a field of $B = 6$ Tesla, and a low resolution ($\Delta \nu \approx 30$ GHz). The Garton-Tomkins resonances, spaced about $1.5 \, \hbar\Omega_c$ apart, are clearly visible near the (free-field) ionization limit and into the continuum. These resonances may be reproduced in a very simple model by considering the vibrational motion of the electron in the two-dimensional x-y plane only, and quantizing the corresponding action for $E = 0$ (see Problem 1.4). A comprehensive analysis of this phenomenon in terms of closed classical orbits was developed by Du and Delos [58].

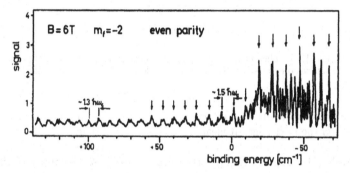

Figure 1.12: The Rydberg spectrum of a hydrogen atom in a magnetic field of $6\,T$ at low resolution. The arrows indicate the Garton-Tomkin quasi-Landau resonances. (From [67].)

Reinhardt [68] has interpreted the quasi-Landau resonances in terms of the wave packet dynamics in an insightful way. He uses the time-dependent wave packet approach developed originally by Heller [69] in the study of molecular physics. In this theory, the formula for the total photo-absorption cross section at a given photon frequency ω is given by

$$\sigma(\omega) = 4\pi^2 a^2 \frac{\omega}{2\pi} \int_{-\infty}^{\infty} d\tau \exp(i\omega\tau) \langle 0|\mu(\tau)\mu(0)|0\rangle. \qquad (1.108)$$

Here a is the Bohr radius, and $\mu(\tau)$ is the time-dependent dipole operator in the Heisenberg representation. The state $\mu(0)|0\rangle$ may be regarded as the initial wave packet formed by the absorption of the photon. The expectation value on the right-hand side of the above equation is then just the overlap of this initial wave packet with its time-evolved form at time τ. With the evolution in time, the initial wave packet moves forward, spreading both in the ρ and the z directions, so the overlap decreases quickly. In the presence of the strong magnetic field, however, there is a harmonic wall in the x-y plane, which the traveling wave packet encounters at some later time. Consequently, there is a sharp increase in the overlap after a *recurrence time* τ_R when the reflected wave comes back. According to Reinhardt, it is the first recurrence of the propagating wave packet that determines the observed oscillating pattern of the observed spectrum. This was tested by noting that the auto-correlation function of the dipole operator, $\langle 0|\mu(\tau)\mu(0)|0\rangle$, may be recovered by taking the inverse Fourier transform of $\sigma(\omega)$. The experimental $\sigma(\omega)$ data of Ba near the ionization threshold at $B = 4.7\,T$ are shown in Fig. 1.13 from [68].

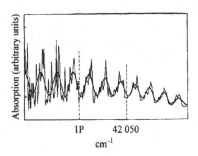

Figure 1.13: The experimental photoionization cross section of Ba near the ionization threshold in a magnetic field of 4.7 Tesla. (After [68].)

The real part of the inverse Fourier transform of this in the time domain is shown in Fig. 1.14 on the next page. It clearly shows the initial sharp drop in the overlap of the wave packet, followed by its rise due to the first recurrence at a later time. In the inset of the Fig. 1.14, the overlap function is simplified by setting it arbitrarily to zero at intermediate times. The resulting Fourier transform of it is very interesting, and is shown as the smooth oscillating (heavy) curve superimposed on the experimental data for $\sigma(\omega)$ in Fig. 1.13. This is just what one would obtain for the absorption cross section at low resolution.

With higher resolution, other families of resonances are also found experimentally [64, 67]. These may be obtained semiclassically by considering the motion of an electron not confined to the x-y plane, and applying the semiclassical quantization condition given by Eq. (1.107). This is a generalization of the method given in Problem 1.9, where only the vibrational motion in the x-y plane is considered. The Bielefeld group used it to understand the origin of these resonances. To apply this method, one has to isolate the relevant closed orbits from the experimental quantum spectrum directly. A neat connection can be made between the quantum excitations to the Rydberg states (in energy scale) and the classical closed orbits (in time scale), by employing the technique of Fourier transforms. This was done by Wintgen [70] for the diamagnetic problem under discussion. The connection is formally based on the periodic orbit theory developed by Gutzwiller [8] and will be discussed in detail in Chapter 5. We anticipate here one of its main results, in order to allow the reader to appreciate its value in analyzing the above experiments. The quantum density of states $g(E)$ can be generally written in the form of a "trace formula" [8]

$$g(E) = \tilde{g}(E) + \sum_{po} \mathcal{A}_{po}(E) \cos\left[S_{po}(E)/\hbar - \sigma_{po}\pi/2\right], \qquad (1.109)$$

where $\tilde{g}(E)$ is its smooth part and the sum on the right-hand side represents a Fourier decomposition of its oscillating part. Hereby the sum runs over all classical periodic orbits 'po' of the system (that are found by solving Newton's equations of motion), and $S_{po}(E)$ are the classical actions along the periodic orbits. (For more details, in particular the calculation of the amplitudes $\mathcal{A}_{po}$ and the phases σ_{po}, see Chapter 5.)

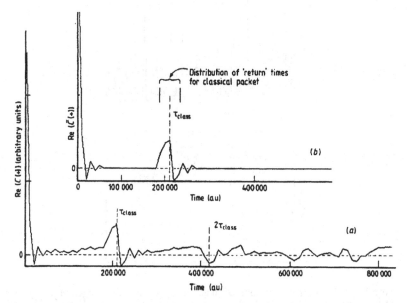

Figure 1.14: The inverse Fourier transform of the photoionization cross section given in the previous figure. (From [68].)

Now, in the photo-absorption spectrum of atomic hydrogen, we have a direct access to the density of states $g(E)$. It is shown (schematically) as a function of energy in Fig. 1.15 on the next page. Its Fourier transform in the time domain, shown in the same figure, shows clear peaks which can be directly related to the actions of very specific classical orbits (indicated near the corresponding peaks of the Fourier spectrum). To be a little more quantitative: using the scaled variables given by Eqs. (1.93) and (1.99), the argument of the cosine function in Eq. (1.109) is $\gamma^{-1/3}\tilde{S}_r(\tilde{\epsilon})/\hbar$. If the energy E and the magnetic field B are varied in such a way that $\tilde{\epsilon}$ remains a constant, then the variable in $\cos[\gamma^{-1/3}\tilde{S}_r(\tilde{\epsilon})]$ is $\gamma^{-1/3}$ (instead of E), with respect to which the Fourier transform is taken to go over to the time domain.

In Fig. 1.15a, then, the horizontal scale is for the variable $\gamma^{-1/3}$, with $\tilde{\epsilon}$ kept constant. Similarly, for the Fourier transform given in Fig. 1.15b, the horizontal axis represents the generalized action $n\gamma^{1/3}$ defined in Eq. (1.107). Recall that for the Rydberg states, there is a one-to-one correspondence with these peaks to the classical closed orbits, whose (ρ, z) projected shapes are also shown in Fig. 1.15b. In Ref. [64], the Bielefeld group use this novel technique of constant $\tilde{\epsilon}$ spectroscopy in experimental scanning, varying E (via the laser beam wavelength) and the magnetic field B

simultaneously, correlating the observed families of quasi-Landau resonances with the semiclassical condition given by Eq. (1.107). We thus recognize the important role that classical orbits play in the understanding of the quantum spectra even in a situation where the classical motion is chaotic.

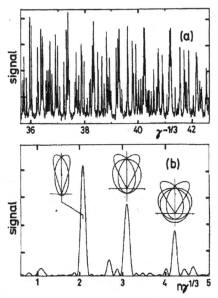

Figure 1.15: The scaled photoionization energy spectrum of hydrogen in a magnetic field (in the range $3.03 \leq B \leq 5.19\,T$) and its Fourier transform. (From [64].)

1.5 Chaos and periodic orbits in mesoscopic systems

We have emphasized the role of unstable periodic orbits in the analysis of the quantum spectrum of a system that is classically chaotic. In this section we describe three recent experiments involving the transport properties of electrons in high-mobility semiconductor microstructures that show the imprint of chaotic orbits in conductance measurements. In all these experiments, an applied magnetic field plays a crucial role. In the first experiment [22] we describe electron transport in cavities of different shapes, where a weak magnetic field of a few Gauss is applied to introduce a Aharonov-Bohm type of phase [71] in the wave functions and destroy the so-called weak localization. The second experiment [23] involves electron transport through a tunneling diode in a voltage gradient, but in the presence of a very strong tilted magnetic field of several Tesla that induces chaotic motion. Before describing this experiment, we explain what is meant by scars in wave functions. The experimental set-up and the results are then given. The current-voltage characteristic of the device shows peaks that may be associated with a subset of states in the quantum well that are 'scarred' by a few classical orbits. We conclude by describing electron transport in a two-dimensional antidot array of scatterers [72, 73], again in a magnetic field applied perpendicular to the plane of the electron gas. The antidots are holes that are etched on to the semiconductor structure, and give rise to repulsive posts that form a 'superlattice', with a regular spacing of a few hundred nanometer. Both the longitudinal and the transverse Hall resistances show novel behavior at low magnetic fields that may be understood by studying the classical electron orbits.

The field of mesoscopic transport is growing rapidly with the advance of technology, and we expect more experimental results in this area. Theoretical study of these experimental results in terms of classical orbits (regular and chaotic) is rejuvenating the area of semiclassical physics. Some other experimental developments related to quantum dots will be described later in Chapter 8, after the theoretical background in periodic orbit theory has been given.

1.5.1 Ballistic magnetoresistance in a cavity

The first experiment that we describe is the *two-probe* measurement of electrical resistance in the transport of electrons that suffer multiple reflections within the walls of a tiny cavity of size $\simeq 1.25\,\mu m$ at low temperatures $(T = 50\,mK)$ [22]. The resistance in a wire is normally due to the scattering of electrons from impurities and lattice vibrations. In the high-mobility $(GaAs/Al_xGa_{1-x}As)$ heterostructure crystal with an electron density of $3.3 \times 10^{11}/cm^2$, however, the mean free path between collisions at low temperatures is an order of magnitude larger than the cavity dimensions. The size of the cavity is also small compared to the "phase-coherence" length, which is determined by the distance between inelastic scatterings. The resistance, therefore, is solely due to the collisions that the trapped electron makes with walls of the cavity before escaping through an opening. It is the size of the aperture that determines the mean escape time τ_0. If the aperture is very small, then τ_0 may be large enough that $\hbar/\tau_0 \ll \Delta E$, where ΔE is the mean level spacing in the cavity. In this regime, the system is like a quantum dot, with many resonances that may be resolved, and which give rise to peaks in the conductance. In the experiment that we describe here, however, it is the other limit where τ_0 is small enough for $\hbar/\tau_0 \gg \Delta E$. In this case there are still weak resonances, but they overlap so much that the spectrum is blurred, with only the single energy scale $\hbar/\tau_0$ governing it. This regime has been studied extensively in nuclear physics under the name Ericson fluctuations [74]. We are not interested here, however, in the Ericson fluctuations, but in another phenomenon called weak localization. The latter is a quantum effect that causes the electrical resistance of the device to be above the classical value. The theoretical description of the ballistic conductance of microstructures in this case, with emphasis to the concept of *weak localization*, is given by Baranger *et al.* [75]. The conduction process from the injecting end to the outgoing one may be viewed as a coherent quantum transmission of independent electrons [76]. Semiclassically, the quantum transmission amplitude from one (cavity) mode to another may be expressed in terms of a superposition of terms, each term with an amplitude and a phase factor coming from a classical orbit from the en-

trance to the exit channel. The transmission *coefficient* for the current is proportional
to the squared absolute value of the transmission amplitude, and may be expressed
by sums over pairs of trajectories that include their time reversed partners. The reflected back-scattered amplitude from the wall of the cavity interferes constructively
with its time-reversed partner, thus increasing the resistance from the classical value.
Direct trajectories are not included in this argument. This constructive interference
may be destroyed by applying a (weak) magnetic field [77]. This is because a magnetic field introduces an additional Aharonov-Bohm phase that is proportional to the
vector potential integrated over the line of the trajectory, which is of opposite sign for
its time-reversed partner. We therefore expect a decrease in the electrical resistance
on applying a magnetic field due to destruction of the weak localization. This also
gives us a handle to differentiate between regular and chaotic scattering, depending
on the shape of the cavity. In a circular cavity, for example, the orbits of an electron
are non-chaotic, and the conservation of angular momentum ensures circulation of the
orbit in one direction before it escapes. This gives rise to a steadily increasing area,
and a correspondingly large Aharonov-Bohm phase in the presence of a magnetic field.
As already mentioned, the corresponding time-reversed partners gather a phase of opposite sign, leading thereby to a rapid destruction of weak localization with increasing
magnetic field.

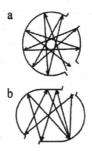

Figure 1.16:   Typical orbits that bounce eight times before escaping
through the aperture in (a) circle billiard with regular trajectories, and
(b) in a stadium with chaotic scatterings. (From [78].)

The situation is different in a stadium billiard with chaotic zig-zag orbits with no
angular momentum conservation, as shown in Fig. 1.16. From these considerations,
we expect the fall in the resistance with increasing magnetic field to be steeper for the
circular billiard. The experimental results [22] for the resistance for these two shapes are
shown in Fig. 1.17. In this experiment, the Ericson fluctuations (also called universal
conductance fluctuations) are averaged out by simultaneously doing the measurements
on a large number (48 in this case) of identical devices fabricated on a single GaAs
heterostructure crystal. The experimental line shapes shown in Fig. 1.17 confirm the
theoretical calculations. We thus have a beautiful example of the imprint of classical
trajectories on a quantum phenomenon, with measurable effects. Moreover, it is an
example that shows how a semiclassical analysis may be useful in understanding the
underlying physics.

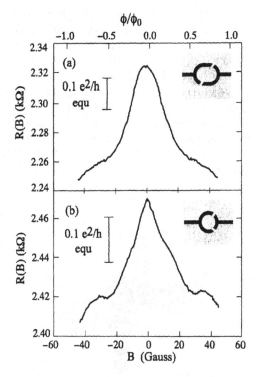

Figure 1.17: The line-shape of the decrease in the resistance of (a) a stadium cavity, and (b) a circle cavity at $T = 50$ mK. The universal conductance fluctuations have been filtered out (see text). (From [22].)

1.5.2 Scars in the wave function

The other experiment that we describe [23] supports the "scarring" of wave functions in a quantum well with chaotic dynamics. Before describing the experiment and its interpretation, it is appropriate to outline what is meant by scarring. In quantum mechanics, it is the wave function of the system that is the probability amplitude, and contains all the available information. One may well ask if classical dynamics, and chaos in particular, have some effect on the wave function. A striking example of the quantum probability density being enhanced along the track of a classical periodic orbit is found in the calculation of the helium atom by Wintgen and collaborators [79]. Earlier in this chapter, we noted the failure of the old quantum theory when it was applied to more complicated atoms than hydrogen. Even the three-body problem in a Coulomb potential is non-integrable, and the classical motion in the helium atom is mostly chaotic. Nevertheless, in recent times, considerable progress has been made in the semiclassical theory of the helium atom. For the angular momentum $L = 0$ states, Wintgen *et al.* applied the modern version of the periodic orbit theory of Gutzwiller to obtain beautiful agreement with a class of excited states. They found that the classical

dynamics of the atom in the collinear configuration, with the two electrons on either side
of the nucleus, is fully chaotic. In the quantum mechanical calculation of states with
orbital angular momentum zero, a subset of states with the same (principal) quantum
numbers N, n for the two electrons has a particularly simple structure. In Fig. 1.18,
we show the probability distribution for $|\Psi_{Nn}|^2$ with $N = n = 6$ in a two-dimensional
(r_1, r_2) plot in the collinear plane. Here r_1 is the magnitude of the distance of one
electron from the nucleus at the origin, and likewise for r_2, with the restriction that
$r_1 + r_2 = r_{12}$, the distance between the two electrons. We clearly see in Fig. 1.18 the
concentration of the quantum probability of this excited state along the fundamental
periodic orbit marked AS. This is the orbit in which the two electrons stretch about
the nucleus asymmetrically.

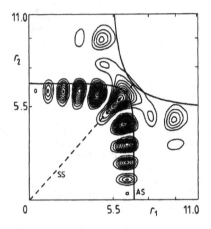

Figure 1.18: The quantum probability distribution of the state $|\Psi_{N,n}(x, y, z)|^2$
with $N = n = 6$ in the collinear plane (see text). The classical periodic orbit
with the asymmetric stretch of the two electrons about the nucleus is marked
AS, and the orbit with the symmetric oscillations along the Wannier bridge
$(r_1 = r_2)$ is marked SS. The nucleus is fixed at the origin, and r_1, r_2 are the
distances of the two electrons (in atomic units) on either side of the nucleus
on a straight line. The distance scale is quadratic since the nodal distance
increases quadratically in a Coulomb system. (From [79].)

The above example in a chaotic system is surprising, since from the point of view
of ergodicity, i.e., equal *a priori* probability in phase space, one may think that there
would be no structure left in the quantum probability density due to chaotic orbits.
Long before the above calculation was done, however, McDonald reported in his Ph.D.
thesis [80] the structure in some wave functions being connected to the tracks of classical
periodic trajectories in a stadium billiard. This was discussed in detail by Heller [81]
who compared the probability distribution of some highly excited states of the stadium
billiard with the tracks of unstable periodic orbits. In Fig. 1.19, we show comparisons
of the probability density $|\psi(q)|^2$ of an excited quantum state with a classical orbit in
a lemon-shaped billiard [12]. In this enclosure, both chaotic (uppermost diagram) and

regular motion (lower three) are possible. Such systems are called *mixed*. We see a remarkable overlap in the patterns formed in the quantum probability density, and the classical paths.

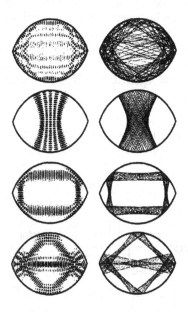

Figure 1.19: Probability density of some quantum states (left) and classical orbits (right) in a lemon-shaped billiard. (From [12].)

We shall come back to an example involving the phenomenon of asymmetric fission of an atomic nucleus in Chapter 8 where again a close correlation will be found in the probability density of a class of quantum states, and a few periodic orbits (see Sect. 8.1.3). In general, however, there is no clear-cut criterion for the choice of a group of classical orbits scarring a quantum wave function. A very readable, and pioneering paper on this topic is by Bogomolny [82]. In this paper, Bogomolny examines the *energy averaged* eigenfunction squared, $\langle|\psi(q)|^2\rangle$, over a small interval ΔE of energy. This is defined as

$$\langle|\psi(q)|^2\rangle = \frac{1}{N}\sum_{\{n\}} |\psi_n(q)|^2, \tag{1.110}$$

where the summation is performed over all N eigenfunctions of energy in the range $E_0 - \Delta E/2 \le E_n \le E_0 + \Delta E/2$. This is evaluated by using the semiclassical expansion of the quantum Green's function given later in Eq. (3.29). The Green's function is just the Fourier transform of the propagator, defined by Eq. (1.27). The semiclassical expansion includes leading contributions of terms as $\hbar \to 0$, one varying smoothly with energy, and the others oscillating. The latter are given in terms of the classical action $S(q, q; E)$ (1.23) of orbits that start and end at the same point q. The number of these orbits is of the order $h/\Delta E$. This is the same Gutzwiller approximation [8] developed in detail in Chapter 5, but adapted for the smoothed probability density. The difference is that there is no integration (trace) over q, so it is actually the nonperiodic trajectories in the vicinity of the periodic ones that contribute. The relative weights of the different periodic orbits depend sensitively on the energy. For some optimum choice of energy E, this smoothed wave function (squared) may show considerable enhancement in the vicinity of a few classical periodic trajectories. One may then say that the wave function is scarred by these classical trajectories.

1.5.3   Tunneling in a quantum diode with a tilted magnetic field

A recent experiment in the current-voltage characteristics of a mesoscopic tunneling diode [23] lends credibility to the scarring of certain quantum states.

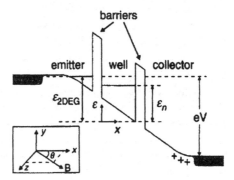

Figure 1.20:   Electron tunneling through a resonant-tunneling diode under an applied bias voltage V and a strong magnetic field, tilted at an angle θ from the x axis. (From [23].)

The experimental set-up is shown schematically in Fig. 1.20. On the left, there is an accumulation layer of electrons in the x-y plane (in the plane of the paper). Under a bias voltage, the electrons tunnel into a quantum well formed between two (AlGa)As layers. The width of the quantum well in the x-direction is 22 nm, and is narrow enough for the individual single-particle quantum energy levels to be resolved at low temperatures. A magnetic field B, which could be as strong as 37 Tesla, is applied at an angle $\theta = 40$ degrees. The classical motion of an electron in the well under such a strong tilted field is chaotic. In the experiment, the electrons are collected at the other end, and the current-voltage characteristic, (dI/dV), is measured for a constant B. In Fig. 1.21, the energy levels of the quantum well are shown in (a), and the partially smoothed density of states show dips corresponding to the gaps in the energy spectrum. These are shown by the arrows pointing downwards. The probability of an electron tunneling into a state n of the well with energy ϵ_n is proportional to the absolute value squared of the matrix element M_n, which is determined by the overlap of the entrance wave function with the eigenfunction in the quantum well. A model calculation shows that $|M_n|^2$ is strongly peaked for certain states, shown by *upward* arrows in Fig. 1.21. A semiclassical calculation also shows that these are precisely the states that get strongly scarred by classical orbits making two or three successive collisions with the right edge of the barrier for each collision on the left wall. This seems reasonable, since the current coming out on the right depends not only on the tunneling probability through the barrier, but also on the frequency of the hits.

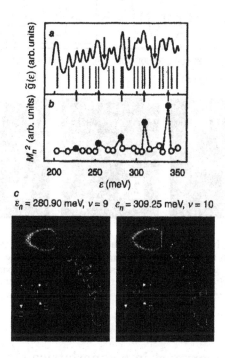

Figure 1.21: (a) The energy levels in the quantum well. The oscillations in the smoothed density of states, which mainly come from classical periodic orbits, are also shown. A subset of the quantum states, whose energy levels are marked by the upward arrows, are strongly scarred by classical orbits spiraling about the magnetic field. (b) The matrix-element squared for resonant tunneling is strongly peaked for the scarred states. (From [23].)

In the uppermost part (a) of Fig. 1.22 on the next page, the experimental peaks in the dI/dV characteristic are shown. The average spacing between the peaks is about 87 mV, which is much larger than the average spacing between the energy levels in the quantum well, suggesting that only a few of the states contribute dominantly to the current. The result of the theoretical calculation is shown in the middle (b) of Fig. 1.22, together with the probability density plots in the x-z plane (below, part c) of the scarred states that contribute dominantly to the current. The classical orbits follow the probability density closely, spiraling about the magnetic field in the same inclination. For each calculated current peak, the contribution of one state is dominant, though the other scarred states also contribute. Note also that the spacing between the conductance peaks is much larger than the spacing between the individual scarred states.

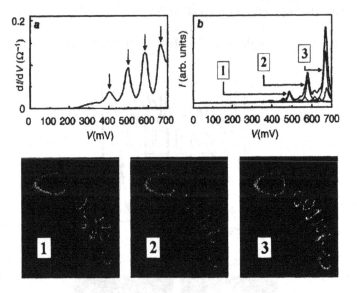

Figure 1.22: (a) The experimental conductance peaks are plotted against the bias voltage V. (b) The theoretical calculation of the same, showing also the dominant contributions coming from the different scarred states. The probability distributions of these scarred states in the x-z plane are plotted in the lower part of the figure. (After [23].)

1.5.4   Electron transport in a superlattice of antidots

We shall briefly describe the experimental results of Weiss *et al.* [72, 73] and their interpretation in terms of classical orbits. A non-technical account is given in the article by Weiss and Richter [83]. The earlier experiment [72] was performed at higher temperatures (a few K), while the later one [73] recorded the magnetoresistance at a much lower temperature (0.4 K) that showed novel quantum oscillations as a function of the applied magnetic field B. The experimental set-up before patterning the array of antidots is basically the same as in quantum Hall effect measurements. The high-mobility two-dimensional electron gas (say in the x-y plane) is at the interface of the GaAs-(AlGaAs) heterostructure. An electric voltage is applied along the (longitudinal) x-direction, and a uniform magnetic field in the z-direction. The resulting electron current may be measured both along the longitudinal x- and the transverse y-directions.

Before describing the antidot array results, we first explain what is found in conventional quantum Hall measurements [84]. This sets the stage for appreciating the role of the periodic and chaotic orbits in the antidots experiments. The important results in the latter will actually show up at low magnetic fields ($B < 1$ T), where the classical Drude theory is applicable in the absence of the superlattice. This is because for such low values of B, thermal broadening wipes out the quantum effects. In the absence of the magnetic field, the current density is

$$\mathbf{j} = \sigma_0 \mathbf{E}\,, \tag{1.111}$$

where σ_0 is the conductivity, and $\mathbf{E}$ the applied electric field. The simple Drude expression for the conductivity is given by

$$\sigma_0 = \frac{ne^2\tau_0}{m^*}\,, \tag{1.112}$$

where n is the number density of electrons, τ_0 the mean-free time between the collisions, and $m^* \simeq 0.07m$ is the effective electron mass. Note that the incremental velocity that the electron picks up between successive collisions is $-e\mathbf{E}\tau_0/m^*$, and it is this *drift* velocity that contributes to the net electric current. Adding up the contribution from all the electrons, and multiplying by $-e$, we obtain Eq. (1.112). When a magnetic field is also applied, the Lorentz force causes the electrons to bend from their otherwise straight line paths, and gives rise to a transverse (y-component) of the current. The current vector is now given by

$$\mathbf{j} = \sigma_0 \mathbf{E} - \sigma_0 \frac{\mathbf{j} \times \mathbf{B}}{nec}\,. \tag{1.113}$$

Using the two-dimensional geometry that we have specified, we may write the above equation in the form $j_x = \sigma_0 E_x - \sigma_0 j_y B/(nec)$, and $j_y = \sigma_0 j_x B/(nec)$. Defining

$$E_x = \rho_{xx} j_x + \rho_{xy} j_y\,, \tag{1.114}$$

we immediately see that

$$\rho_{xx} = \frac{1}{\sigma_0}\,, \qquad \rho_{xy} = \frac{B}{nec}\,. \tag{1.115}$$

The main point of this exercise has been to define the resistance tensor, and to show that in this simple model, ρ_{xx} is *independent* of the magnetic field B. We also note from Eq. (1.115) that ρ_{xy} is linearly dependent on B. We are now ready to describe the very different results obtained in this regime of B when the electrons move in the presence of the two-dimensional array of the antidots. The antidots are prepared by etching microscopic holes in the two-dimensional conductor. The lattice spacing in the square array of the antidots is $a = 200 - 300$ nm, and the effective diameter of a hole is typically $d \simeq 100$ nm. As in the other experiments described earlier in this section, these lengths are much smaller than the transport mean free path ($\simeq 10\,\mu$m). We see from Fig. 1.23 that for small values of the magnetic field, there are two spectacular peaks in the measured ρ_{xx}, accompanied by corresponding plateaux in ρ_{xy}. These are *not* due to the orbit quantization in the vector potential that gives rise to the Landau level structure and the quantum Hall effect peaks and plateaux shown in the same diagram for larger values of the applied magnetic field. The dashed curves in Fig. 1.23 show that without the patterning, the low-B behavior of the ρ's are more in line with the Drude results outlined above. The peaks in ρ_{xx} arise for those values of B for which the classical cyclotron radius R_c is $1/2$ and $3/2$ of the lattice spacing a. Recall that classically

$$\frac{m^* v_F^2}{R_c} = \frac{eB v_F}{c}, \qquad \text{so} \quad R_c = \frac{v_F}{\Omega_c^*}, \tag{1.116}$$

where $\Omega_c^* = eB/m^*c$ is the cyclotron frequency. Here, to disallow blocking due to the Pauli principle, we have taken the electron velocity corresponding to the Fermi energy.

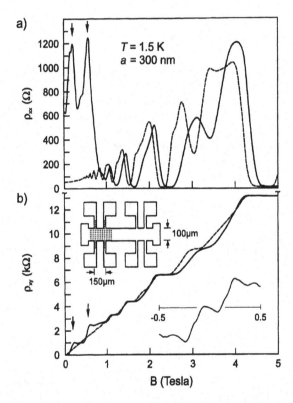

Figure 1.23:   (a) Magnetoresistance and (b) Hall resistance in patterned (solid line) and unpatterned (dashed line) sample segments at a temperature of 1.5 K. The arrows mark magnetic field positions where $R_c/a \sim 1/2$ and $3/2$, respectively. Top insert in (b): Sketch of the sample geometry with patterned (left) and unpatterned (right) segments of the 2DEG. Lower insert in (b): magnification of quench in ρ_{xy} about $B = 0$. (From [72].)

The Fermi energy of a two-dimensional (unpolarized) gas may be expressed in terms of the number density by noting that

$$n = 2\frac{\pi p_F^2}{h^2} = \frac{k_F^2}{2\pi}, \qquad \text{so} \quad v_F = \frac{\hbar}{m^*}\sqrt{2\pi n}. \tag{1.117}$$

Substituting this in the expression for R_c in Eq. (1.116), we obtain

$$R_c = \frac{\hbar c}{eB}\sqrt{2\pi n}. \tag{1.118}$$

The carrier density n in these samples is $\sim (1.4 - 2.8) \times 10^{11} \, \text{cm}^{-2}$. The peaking of ρ_{xx} at the commensurate ratios led Weiss *et al.* to suggest that it is the pinning of the circular orbits, with their centers localized about the potential posts, that causes a net depletion in the flow of electrons and an increase in the resistance.

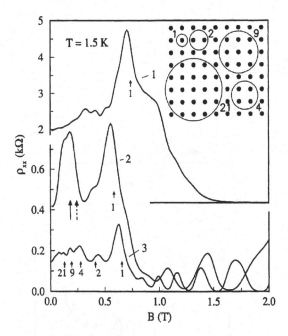

Figure 1.24: Low-B anomalies from three different samples. For smaller d/a, more structure in ρ_{xx} evolves. All peaks in trace 3 can be associated to commensurate orbits around 1, 2, 4, 9, and 21 antidots, as is sketched in the insert. Corresponding values of R_c/a, marked by arrows, are 0.5, 0.8, 1.14, 1.7, and 2.53, respectively. The dashed arrow for trace 2 marks the position of an unperturbed cyclotron orbit around four antidots ($R_c/a = 1.14$). The shift of the corresponding resistance peak towards lower B indicates the deformation of the cyclotron orbit in a "soft" potential. (From [83].)

The radius of the circular cyclotron orbit, for some particular values of B, is commensurate with the lattice spacing a, and encloses a certain number of antidots. This is shown in the right insert of Fig. 1.24. It is at these B values that the peaks in ρ_{xx} are found. The exact number of antidots enclosed depend sensitively on the ratio d/a of the sample, and three different cases are shown in Fig. 1.24. This simple 'Pin-ball model', although very illuminating, does not tell the full story. For example, in curve 2 of Fig. 1.24, the peak coming at $B = 0.18 \, \text{T}$ for $R_c/a = 3/2$ is not quite consistent with the model. The Pin-ball model works better for hard-wall potentials, whereas there are softer slopes in the potentials around the antidots. Actual model calculations in such potentials show that there are current-carrying *whirling chaotic trajectories* orbiting around four antidot peaks give rise to the shift of the peak towards lower B as seen in curve 2 of Fig. 1.24. More details about the classical calculations will be found in the clearly written article of Weiss and Richter [83]. The results of the experiment [73] at a lower temperature $(T = 0.4\,\text{K})$ are displayed in Fig. 1.25.

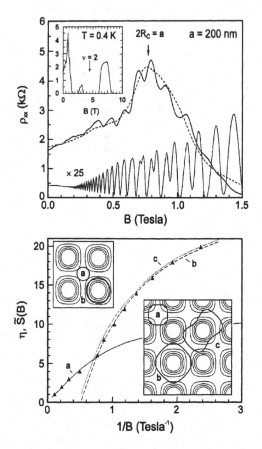

Figure 1.25: *Upper part:* ρ_{xx} measured in the patterned (top traces) and unpatterned (bottom trace) segment of the same sample for temperatures $T = 0.4$ K (solid lines) and 4.7 K (dashed line). The insert displays the ρ_{xx} trace from the patterned segment up to 10 Tesla; the filling factor $\nu = 2$ is marked. At high field B, the emergence of Shubnikov-de Haas oscillations reflects the quantization of essentially unperturbed orbits. *Lower part:* The triangles mark all $1/B$ positions (up to 10 Tesla) of the ρ_{xx} minima. At high B the resistance minima lie equidistant on the $1/B$ scale; at low B the spacing becomes periodic in B. Solid, dashed and dotted lines are calculated reduced actions $\tilde{S}(1/B)$ of orbits (a), (b), and (c), respectively. These orbits are shown for $1/B = 0.6$ T^{-1} (top insert) and 2.7 T^{-1} (bottom insert). (From [83]; after [73].)

The ρ_{xx} peak shows a fine structure with clear oscillations that are periodic in B, in contrast to the usual Shubnikov-de Haas oscillations, shown on a magnified scale ($\times 25$) in the same figure, that are periodic in $1/B$. To appreciate what is happening, see the Lagrangian given by Eq. (1.128) in Problem 1.4. The conjugate momentum calculated from this Lagrangian is $\mathbf{p} = m^*\mathbf{v} + (e/c)\mathbf{A}(\mathbf{r})$. To obtain the quantized energy levels, one calculates the action as in Eq. (1.53) and equates it to an integral multiple of $\hbar$, as in Eq. (1.54). Actually, more care with the phases is necessary in such semiclassical quantization (see Chapter 2), but this is not crucial for our point here. We see that the action $S(B)$ is given by

$$S(B) = \oint \mathbf{p} \cdot d\mathbf{r} = m^* \oint \mathbf{v} \cdot d\mathbf{r} - \frac{|e|B}{c} \times (Area). \tag{1.119}$$

A circular cyclotron orbit, as we see from Eq. (1.118), traces an area proportional to $1/B^2$, so the B-dependence of $S(B)$ goes like $1/B$. The trace formula given in Eq. (1.109) shows that the oscillatory part of the level density $\delta g(E)$ is governed by periodic orbits, with the periodicity of $S(B)$. Therefore when the electronic motion is governed by the magnetic field alone, the oscillatory part of the density of states has a periodic variation with $1/B$. This is the case in a free electron gas, or in an antidot array for strong magnetic fields. Otherwise, for a very weak magnetic field, the isolated chaotic orbits responsible for the conductivity are governed more by the dynamics of the periodic potential posts of the antidots. To a first approximation, the area of such an orbit is independent of B, and therefore $S(B) \propto B$. The extent of chaoticity in the orbits depends, of course, on the detailed lithography of the posts.

To reproduce the finer details in the oscillatory behavior of ρ_{xx}, which depends on the square of the level density at the Fermi energy, more detailed calculation is necessary. Such calculations have been done in [73]. In Fig. 1.25, three periodic orbits are shown that reproduce the minima in the oscillatory behavior of ρ_{xx} adequately. For more details on this aspect, the reader should consult Ref. [73]. Further work on the role of the trace formula on the transport properties of antidot lattices will be found in Refs. [85] and [86]. We point out that quantum mechanical band-structure calculations [87] also reproduce the main profile of the resistivity of the antidot array at low B. More applications of the periodic orbit theory to mesoscopic systems (quantum dots) will be found in Sect. 8.3 of Chapter 8.

1.6 Problems

(1.1) Scaling in the bouncing ball problem

This is a problem on a time-dependent scale-parameter in the classical equations of motion, and is adapted from a paper by M. V. Berry and G. Klein, J. Phys. **17**, 1805 (1984). Consider classical motion of a particle in one dimension. The Hamiltonian is given by

$$H(x, p, l(t)) = \frac{p^2}{2m} + \alpha(l(t)) V(x/l(t)), \tag{1.120}$$

Here the strength α of the potential is time-dependent through an overall scale factor l. Because of this time-dependence, the Hamiltonian is not conservative. Show that Newton's equation of motion may still be written in the form

$$m\frac{d^2\rho}{d\tau^2} = -\frac{\partial}{d\rho}\left[V(\rho) + \frac{1}{2}k\rho^2\right], \tag{1.121}$$

where

$$\rho = x/l(t), \quad \tau(t) = \int_0^t \frac{dt'}{l^2(t')}, \quad \alpha = 1/l^2. \quad (1.122)$$

The parameter

$$k = ml^3 d^2 l/dt^2 \quad (1.123)$$

is taken to be time-independent. Thus, in the transformed coordinates ρ and τ, the motion appears to be conservative, with the addition of a harmonic inertial force, provided $l(t)$ is a solution of Eq. (1.123). Solve the differential equation (1.123) to show that $l(t) = (at^2 + 2bt + c)^{1/2}$, where a, b, and c are constants. Show that $ac - b^2 = k/m$.

(1.2) Large-n transitions in a square-well

Consider a one-dimensional infinite square-well potential of width L. Find the action for the fundamental periodic orbit, and the eigenvalues by the quantization condition given by Eq. (1.54). Verify, by an explicit calculation, that Eq. (1.70), relating the quantum transitions to the frequencies of the classical orbits for the large-n states, is satisfied.

(1.3) Squeezed and coherent states

The Hamiltonian of a particle in a one-dimensional harmonic oscillator is given by Eq. (1.52). In quantum mechanics, the operators x and p may be expressed in terms of the annihilation and creation operators a and $a^\dagger$:

$$x = \sqrt{\frac{\hbar}{2m\omega}}(a + a^\dagger), \quad p_x = i\sqrt{\frac{\hbar m\omega}{2}}(a^\dagger - a).$$

A coherent state $|\alpha\rangle$ is an eigenstate of the annihilation operator a, $a|\alpha\rangle = \alpha|\alpha\rangle$, where the eigenvalue α is in general complex. First consider the properties of a 'squeezed' state, which is a generalization of the coherent state in a harmonic oscillator. Define the operators

$$Q = \mu\, a + \nu\, a^\dagger, \quad Q^\dagger = \mu^* a^\dagger + \nu^* a, \quad (1.124)$$

where μ and ν are complex, with the constraint $|\mu|^2 - |\nu|^2 = 1$.

(a) Show that the commutator $[Q, Q^\dagger] = 1$.

(b) A squeezed state $|\alpha\rangle$ is an eigenstate of Q, i.e., $Q|\alpha\rangle = \alpha|\alpha\rangle$. Let us denote the expectation value of any operator O with respect to $|\alpha\rangle$ by $\langle O\rangle$. Show that the root mean square deviations are given by:

$$\Delta x = \sqrt{(\langle x^2\rangle - (\langle x\rangle)^2)} = \sqrt{\frac{\hbar}{2m\omega}}\,|\mu - \nu|, \quad \Delta p_x = \sqrt{(\langle p_x^2\rangle - (\langle p_x\rangle)^2)} = \sqrt{\frac{m\omega\hbar}{2}}\,|\mu + \nu|.$$

$$(1.125)$$

It then follows that

$$\Delta x \Delta p_x = \frac{\hbar}{2}|\mu^2 - \nu^2| \geq \frac{\hbar}{2}. \quad (1.126)$$

The last inequality follows because $|\mu^2 - \nu^2| \geq (|\mu|^2 - |\nu|^2)$. For the harmonic oscillator ground state $\Delta x = \sqrt{\hbar/2m\omega}$. We see from Eq. (1.125) that if $|\mu - \nu| < 1$, then the state is 'squeezed' in the sense that Δx is less than that of the ground state. However, Eq. (1.126) shows that the combined uncertainty $\Delta x \Delta p_x$ of a squeezed state is greater than that of the ground state.

(c) For a *coherent* state, set $\nu = 0$ and $|\mu| = 1$. The coherent state is in some sense like a particle, since its size in the phase space is localized in a minimum uncertainty wave packet, and it does not spread in time. Show that the normalized coherent state may be written as

$$|\alpha\rangle = e^{-|\alpha|^2/2}\, e^{\alpha a^\dagger}|0\rangle, \tag{1.127}$$

where $|0\rangle$ denotes the ground state of the harmonic oscillator.

(d) Writing the coherent state as

$$|\alpha\rangle = \sum_{n=0}^{\infty} f(n)|n\rangle,$$

show that $|f(n)|^2 = \exp(-|\alpha|^2)|\alpha|^{2n}/n!$, which is a Poisson distribution. Here $|n\rangle$ is the eigenstate of the harmonic oscillator with eigenvalue $(n + 1/2)\hbar\omega$.

(e) Prove that $\Delta x \Delta p_x = \hbar/2$ for the coherent state even at a later time $t > 0$.

(1.4) Charged particle in a magnetic field

The equation of motion of a particle with mass m and charge e in a static magnetic field $\mathbf{B}$ is given by

$$m\left(\frac{d^2\mathbf{r}}{dt^2}\right) = \frac{e}{c}\,(\mathbf{v} \times \mathbf{B}).$$

(a) Show that the Lagrangian

$$\mathcal{L} = \frac{1}{2}mv^2 + \frac{e}{c}\,\mathbf{v} \cdot \mathbf{A} \tag{1.128}$$

yields the above equation of motion, where $\nabla \times \mathbf{A} = \mathbf{B}$.

(b) Using the formula for the Hamiltonian $H = \sum_i p_i \dot{x}_i - \mathcal{L}$, show that

$$H = \frac{1}{2m}\left(\mathbf{p} - \frac{e}{c}\mathbf{A}\right)^2,$$

where $\mathbf{p}$ is the canonical momentum, $p_i = (\partial\mathcal{L}/\partial\dot{x}_i)$.

(c) Choose the case of a uniform magnetic field B along the z−direction, using the vector potential in the symmetric gauge $\mathbf{A} = \frac{1}{2}(\mathbf{B} \times \mathbf{r})$. Show that the Hamiltonian in this case is given by

$$H = \frac{p^2}{2m} + \frac{m}{2}\Omega_L^2\, r^2 - \Omega_L L_z, \tag{1.129}$$

where Ω_L is the Larmor frequency

$$\Omega_L = eB/2mc \tag{1.130}$$

which is equal to one-half of the cyclotron frequency defined by Eq. (1.87).

(d) Using the classical equations of motion, and the quantization condition $\oint \mathbf{p} \cdot \mathbf{r} = (n + 1/2)h$, show that the energy levels of the Hamiltonian (1.129) are given by

$$E_{n,p_z} = \frac{p_z^2}{2m} + \hbar\Omega_c \left(n + \frac{1}{2}\right), \qquad (1.131)$$

where $n = 0, 1, 2, \ldots$, and $\Omega_c = eB/mc$ is the cyclotron frequency.

(e) As noted in Sect. 1.1, the number of available quantum states is obtained by calculating the accessible phase space, and dividing it by h, $\mathcal{N} = \int dp\, dx/h$. Using this rule for the two-dimensional motion in the $x - y$ plane, and the result of part (d), show that the number of quantum states in each Landau level of a given n is eB/hc per unit area.

(1.5) The Garton-Tomkin resonances

The Garton-Tomkin resonances at spacings of $3/2\ \hbar\omega_c$ may be derived in a some-what over-simplified model [88]. Consider the motion of an electron governed by the Hamiltonian (1.88), but with $z = 0$ and also $l_z = 0$. Denoting $(x^2 + y^2) = \rho^2$,

$$H = \frac{p_\rho^2}{2m} - \frac{e^2}{\rho} + \frac{1}{2}m\Omega_L^2\rho^2 = \frac{p_\rho^2}{2m} + V_{eff}(\rho) \ .$$

(a) Plot the effective potential $V_{eff}(\rho)$ as a function of ρ, and determine the classical turning points (denoted by ρ_1, ρ_2) for a fixed energy E.

(b) The vibrational period of the classical orbit is given by

$$T = 2\int_{\rho_1}^{\rho_2} \frac{d\rho}{v(\rho)}\ , \qquad (1.132)$$

where $v(\rho)$ is the speed of the particle at ρ. Use the above relation to show that the vibrational period T for $E = 0$ is given by

$$T = \frac{2\pi}{\left(\frac{3}{2}\Omega_c\right)}\ ,$$

where Ω_c is the cyclotron frequency defined by Eq. (1.87). You will need the relations

$$\int_0^1 \frac{\sqrt{x}dx}{\sqrt{(1-x^3)}} = \frac{1}{3}\int_0^1 \frac{dy}{\sqrt{y(1-y)}} = \frac{\pi}{3}\ .$$

Bibliography

[1] A. Tonomura, J. Endo, T. Mastuda, T. Kawasaki, and H. Ezawa, Am. J. Phys. **57**, 117 (1989).

[2] E. Schrödinger, Ann. Phys. (Leipzig) **79**, 361 (1926) [English translation in *Collected Papers on Wave Mechanics*, translated by J. Shearer and W. Deans, (Blackie, Glasgow, 1928)]; Ann. Phys. (Leipzig) **81**, 109 (1926).

[3] W. Heisenberg, Z. Phys. **33**, 879 (1925); M. Born, W. Heisenberg, and P. Jordan, Z. Phys. **35**, 557 (1925); English translation in B. L. v. d. Waerden, *Sources in Quantum Mechanics* (Dover, New York, 1968).

[4] R. P. Feynman, Rev. Mod. Phys. **20**, 267 (1948).

[5] J. H. Van Vleck, Proc. Natl. Acad. Sci. USA **14**, 178 (1928).

[6] M. Gutzwiller, J. Math. Phys. **8**, 1979 (1967).

[7] R. Balian and C. Bloch, Ann. Phys. (N. Y.) **69**, 76 (1972).

[8] M. C. Gutzwiller, J. Math. Phys. **12**, 343 (1971); *Chaos in Classical and Quantum Mechanics* (Springer Verlag, New York, 1990).

[9] N. Bohr, Phil. Mag. (Series 6) **26**, 1, 857 (1913).

[10] For a fascinating historical account, see A. Pais: *Niels Bohr's Times* (Clarendon Press, Oxford, 1991).

[11] A. Einstein, Verh. Dtsch. Phys. Ges. **19**, 82 (1917).

[12] E. J. Heller and S. Tomsovic, Physics Today, July 1993, p. 38.

[13] For a popular account, see James Gleick, *Chaos* (Viking Penguin, Toronto, 1987).

[14] A readable collection of nontechnical articles on chaos will be found in *Exploring Chaos*, ed. Nina Hall (W. W. Norton, 1991).

[15] M. Gutzwiller: *Quantum Chaos*, The Scientific American, January 1992, p. 78.

[16] D. Kleppner: *Quantum Chaos and the bow-stern enigma*, Physics Today, August 1991, p. 9.

[17] R. V. Jensen, Nature **355**, 311 (1992).

[18] *Nobel Symposium on Quantum Chaos*, eds. K.-F. Berggren and S. Åberg; Physica Scripta **T90** (2001).

[19] H.-J. Stöckmann: *Quantum Chaos: An Introduction* (Cambridge University Press, 1999)

[20] P. M. Koch and K. A. H. van Leeuwen, Phys. Rep. **255**, 1 (1991).

[21] M. M. Sanders and R. V. Jensen, Am. J. Phys. **64**, 21 (1996).

[22] A. M. Chang, H. U. Baranger, L. N. Pfeiffer, and K. W. West, Phys. Rev. Lett. **73**, 2111 (1994).

[23] P. B. Wilkinson, T. M. Fromhold, L. Eaves, F. W. Sheard, N. Miura, and T. Takamasu, Nature **380**, 608 (1996).

[24] P. A. M. Dirac, Phys. Zeit. der Sowjetunion, Band **3**, Heft 1; reprinted in *Quantum Electrodynamics*, ed. J. Schwinger (Dover, New York, 1958), p. 321.

[25] P. A. M. Dirac, Proc. Roy. Soc. (London), **113A**, 621 (1927).

[26] P. A. M. Dirac: *The Principles of Quantum Mechanics* (Third edition, Clarendon Press, Oxford, 1958).

[27] R. P. Feynman and A. R. Hibbs: *Quantum Mechanics and Path Integrals* (McGraw-Hill, New York, 1964).

[28] L. S. Schulman: *Techniques and Applications of Path Integration* (John Wiley, New York, 1981).

[29] M. S. Swanson: *Path Integrals and Quantum Processes* (Academic Press, New York, 1992).

[30] P. Ehrenfest, Phil. Mag. **33**, 500 (1917).

[31] C. Gignoux and F. Brut, Am. J. Phys. **57**, 422 (1989).

[32] I. Percival and D. Richards: *Introduction to Dynamics* (Cambridge University Press, Cambridge, 1982) p. 142.

[33] Sin-ichiro Tomonaga: *Quantum Mechanics, Vol. I* (North- Holland, 1962) p. 21.

[34] O. Klein, reprinted in: *The Oscar Klein Memorial Lectures*, ed. G. Ekspong (World Scientific, Singapore, 1991), p. 108.

[35] L. Brillouin, J. Phys. Radium **7**, 353 (1926).

[36] J. B. Keller, Ann. Phys. (N. Y.) **4**, 180 (1958).

[37] G. Wentzel, Z. Phys. **38** 518 (1926); H. Kramers, Z. Phys. **39**, 828 (1926); L. Brillouin, Compt. Rend. **183**, 24 (1926). Other historical references may be found in L. Schiff: *Quantum Mechanics* (McGraw-Hill, New York, Third edition, 1968) p. 269.

[38] L. D. Landau and E. M. Lifshitz: *Quantum Mechanics* (Addison Wesley, Reading, 1958) p. 157, 491.

[39] A. B. Migdal and V. P. Krainov: *Approximation Methods in Quantum Mechanics* (Benjamin, New York, 1969) p. 111.

[40] V. M. Strutinsky, Nucl. Phys. **A 95**, 420 (1967); **A 122**, 1 (1968).

[41] R. K. Sheline, I. Ragnarsson, and S. G. Nilsson, Phys. Lett. **41 B**, 115 (1972).

[42] S. G. Nilsson, Mat.-Fys. Medd. Dan. Vid. Selsk. **29**, no. 16 (1955).

[43] P. J. Twin *et al.*, Phys. Rev. Lett. **57**, 811 (1986); B. Haas *et al.*, Phys. Rev. Lett. **60**, 503 (1988); A. Galindo-Uribarri *et al.*, Phys. Rev. Lett. **71**, 231 (1993).

[44] N. Bohr, Proc. Phys. Soc. London **35**, 275 (1923).

[45] M. Nauenberg and J. Yeazell, The Scientific American, June 1994, p. 44.

[46] R. Bluhm, V. A. Kostelecky, and J. A. Porter, Am. J. Phys. **64**, 944 (1996).

[47] W. Heisenberg, Z. Phys. **31**, 617 (1925); E. Schrödinger, Ann. Phys. (Leipzig) **79**, 361 (1926).

[48] E. Schrödinger, in: *Letters on Wave Mechanics*, ed. K. Prizibram (Philosophical Library, 1967).

[49] E. Schrödinger, Naturwissenschaften **14**, 664 (1926).

[50] For a readable account on the Rydberg atom, see D. Kleppner, M. G. Littman, and M. L. Zimmerman, The Scientific American, May 1981, p. 130.

[51] C. R. Stroud Jr., in: *Physics and Probability, Essays in honor of E. T. Jaynes*, eds. W. T. Grandy Jr. and Milonini (Cambridge University Press, 1993).

[52] Z. D. Gaeta and C. R. Stroud, Jr., Phys. Rev. **A 42**, 6308 (1990).

[53] L. S. Brown, Am. J. Phys. **41**, 525 (1973).

[54] R. Bluhm and V. A. Kostelecky, Phys. Rev. **A 51**, 4767 (1995).

[55] A. ten Wolde, L. D. Noordam, A. Lagendijk, and H. B. van Linden van den Heuvell, Phys. Rev. Lett. **61**, 2099 (1988).

[56] J. A. Yeazell, M. Mallalieu, and C. R. Stroud, Jr., Phys. Rev. Lett. **64**, 2007 (1990).

[57] Z. D. Gaeta, M. W. Noel, and C. R. Stroud Jr., Phys. Rev. Lett. **73**, 636 (1994).

[58] M. L. Du and J. B. Delos, Phys. Rev. **A 38**, 1896 (1988); *ibid.* 1913 (1988).

[59] H. Friedrich and D. Wintgen, Phys. Rep. **183**, 39 (1989).

[60] H. Hasegawa, M. Robnik, and G. Wunner, Prog. Th. Phys. Suppl. **98**, 198 (1989).

[61] T. S. Monteiro, Contemp. Phys. **35**, 311 (1994).

[62] R. A. Pullen, Ph.D. thesis, Imperial College, London (1981, unpublished).

[63] C. Iu, G. R. Welch, M. M. Kash, L. Hsu, and D. Kleppner, Phys. Rev. Lett. **63**, 1133 (1989).

[64] A. Holle, J. Main, G. Wiebusch, H. Rottke, and K. H. Welge, Phys. Rev. Lett. **61**, 161 (1988).

[65] W. R. S. Garton and F. S. Tomkins, Astrophys. J. **158**, 839 (1969).

[66] K. T. Lu, F. S. Tomkins, and W. R. S. Garton, Proc. Roy. Soc. London, Ser. A **362**, 421 (1978).

[67] J. Main, G. Wiebusch, and K. H. Welge, Comm. At. Mol. Phys. **25**, 233 (1991).

[68] W. P. Reinhardt, J. Phys. **B 16**, L635 (1983).

[69] E. J. Heller, J. Chem. Phys. **68**, 2066, 3891 (1978).

[70] D. Wintgen, Phys. Rev. Lett. **58**, 1589 (1987).

[71] Y. Aharonov and D. Bohm, Phys. Rev. **115**, 485 (1959)

[72] D. Weiss, M. L. Roukes, A. Menschig, P. Grambow, K. von Klitzing, and G. Weimann, Phys. Rev. Lett. **66**, 2790 (1991).

[73] D. Weiss, K. Richter, A. Menschig, R. Bergmann, H. Schweizer, K. von Klitzing, and G. Weimann, Phys. Rev. Lett. **70**, 4118 (1993).

[74] T. Ericson, Phys. Rev. Lett. **5**, 430 (1960); Advances in Physics **9**, 426 (1966).

[75] R. A. Jalabert, H. U. Baranger, and A. D. Stone, Phys. Rev. Lett. **65**, 2442 (1990); H. U. Baranger, R. A. Jalabert, and A. D. Stone, Chaos **3**, 665 (1993).

[76] R. Landauer, Phil. Mag. **21**, 863 (1970); M. Büttiker, Phys. Rev. Lett. **57**, 1761 (1986).

[77] P. A. Lee and T. V. Ramakrishnan, Rev. Mod. Phys. **57**, 287 (1985).

[78] R. V. Jensen, Nature **373**, 16 (1995).

[79] G. S. Ezra, K. Richter, G. Tanner and D. Wintgen, J. Phys. **B 24**, L413 (1991).

[80] S. W. McDonald, Lawrence Berkeley Lab. Rep. LBL 14837 (1983).

[81] E. J. Heller, Phys. Rev. Lett. **53**, 1515 (1984); E. J. Heller, P. W. O'Connor, and J. Gehlen, Physica Scripta **40**, 354 (1989).

[82] E. B. Bogomolny, Physica D **31**, 169 (1988).

[83] D. Weiss and K. Richter, Physica D **83**, 290 (1995).

[84] R. E. Prange, in: *The Quantum Hall Effect*, eds. R. E. Prange and S. M. Girvin (Springer-Verlag, N. Y., 1987).

[85] K. Richter, Europhys. Lett. **29**, 7 (1995).

[86] G. Hackenbroich and F. von Oppen, Europhys. Lett. **29**, 151 (1995).

[87] H. Silberbauer, J. Phys. C **4**, 7355 (1992); H. Silberbauer and U. Rössler, Phys. Rev. B **50**, 11911 (1994).

[88] A. R. Edmonds, J. Phys. Colloq. (Paris) **31**, (1970) C4-71; A. F. Strace, J. Phys B **6**, 585 (1973); R. F. O'Connell, Astrophys. J. **187**, 275 (1974); A. R. P. Rau, Phys. Rev. A **16**, 613 (1977).

Chapter 2

Quantization of integrable systems

2.1 Introduction

Our plan in this chapter is to discuss in some detail the connection between the classical and quantum mechanical equations of motion, largely using examples that are solvable. In Sect. 1.1, we had introduced Hamilton's principal function $R(q, q'; t)$ and the classical action $S(q, q'; E)$ in Eqs.(1.17) and (1.20), respectively, to show the roles they played in the path-integral formulation of quantum mechanics. We also recounted how the old quantum theory was formulated by quantizing the action. Here we go one step further by developing the Hamilton-Jacobi equations (both the time-dependent and the time-independent forms), and showing their formal similarity with Schrödinger's wave equation. This leads us directly to the WKB approximation for one-dimensional quantum problems, and its generalization to higher dimensions. We had noted earlier in Eq. (1.57) that there is on the average one quantum state in a cell of area h in the phase space (in one-dimensional motion). It turns out that even for multi-dimensional problems, the description of classical dynamics in phase space is naturally amenable to the semiclassical generalization. For this, it is the Hamiltonian (rather than the Lagrangian) formulation that is more suitable. A very readable article on the classical limit of quantum mechanics, with particular emphasis on the role of the generators of canonical transformations, has been written by Miller [1]. It is the Hamiltonian itself that is the generator of the canonical transformation that unfolds the motion of the particle in the phase space. Analogously, in quantum mechanics, the time evolution of the state vector is governed by a unitary transformation generated by the Hamiltonian.

Throughout this book, we deal with Hamiltonians that have no explicit time-dependence: such systems are called *autonomous*. After defining what is meant by a classically integrable system, we examine closely the so-called torus structure of the available phase space for such motion. Choosing irreducible cycles around the tori leads naturally to defining the action and angle variables of motion, and the quantization of action around each such cycle. This is explained, and some well-known solvable examples are given. Periodic orbits in the multi-dimensional torus structure have commensurate frequencies. We show that families of these periodic orbits contribute to the oscillating part of the density of states of an integrable system in the form of a trace formula (1.109). Finally, we discuss the transition from integrability to chaos in such systems when perturbations are added that couple resonantly to these natural frequencies.

While introducing the path integral method and the propagator in Sect. 1.1, we used only one space dimension for simplicity in notation. Since most of the theory and applications will be developed in two and higher dimensions, we begin by introducing the notation that will be used. This will entail repetition of some formulae already used in Chapter 1. The *Lagrangian* $\mathcal{L}$ of a system with N degrees of freedom is written as:

$$\mathcal{L}(\mathbf{q}, \dot{\mathbf{q}}, t); \qquad \mathbf{q} = \{q_i\}, \qquad \dot{\mathbf{q}} = \{\dot{q}_i\}, \qquad i = 1, 2, \ldots n. \tag{2.1}$$

Along a path $\mathbf{q}(t)$ in the configuration space, we denote

$$\mathbf{q}' = \mathbf{q}(t') ; \qquad \mathbf{q}'' = \mathbf{q}(t'') . \tag{2.2}$$

In some problems it may be convenient to eliminate time t using the equations of motion, and use some continuous parameter s to parameterize the trajectory $\mathbf{q}(s)$. We shall also denote this Lagrangian simply as $\mathcal{L}(q_i, \dot{q}_i, t)$. The canonically conjugate *momenta* are

$$\mathbf{p} = \{p_i\} ; \qquad p_i = \frac{\partial \mathcal{L}}{\partial \dot{q}_i} . \tag{2.3}$$

A periodic orbit (PO) has the property that the particle comes back to the same point with the same momentum after a time period T:

$$\mathbf{q}(t + T) = \mathbf{q}(t) , \quad \mathbf{p}(t + T) = \mathbf{p}(t) . \tag{2.4}$$

Hamilton's principal function, introduced in Eq. (1.17), is

$$R(\mathbf{q}, \mathbf{q}'; (t - t')) = \int_{t'}^{t} \mathcal{L}(\mathbf{q}'', \dot{\mathbf{q}}'', t'') \, dt'' . \tag{2.5}$$

Hamilton's *variational principle*

$$\delta R = 0 \quad \text{under} \quad \mathbf{q}''(t'') \to \mathbf{q}''(t'') + \delta \mathbf{q}''(t'') \quad \text{(with end points } \mathbf{q}', \mathbf{q} \text{ fixed)} \tag{2.6}$$

gives *Lagrange's equations of motion*:

$$\frac{\partial \mathcal{L}}{\partial q_i} - \frac{d}{dt} \frac{\partial \mathcal{L}}{\partial \dot{q}_i} = 0 . \tag{2.7}$$

Classical trajectories are solutions of Eq. (2.7) satisfying the appropriate boundary conditions. In Chapter 7, we shall also consider trajectories and periodic orbits that are diffractive, and do not obey Newtonian mechanics.

For an autonomous system, it is often more convenient to label the trajectory in the coordinate space for a given energy in terms of the action $S(E)$ rather than the principal function $R(t - t')$ given by (2.5). This change of variables is brought about by what is called a Legendre transformation

$$R(\mathbf{q}, \mathbf{q}', (t - t')) = S(\mathbf{q}, \mathbf{q}', E) - E(t - t') . \tag{2.8}$$

This relation was already used in Chapter 1 [see Eq. (1.20)], and has a deep significance in the Hamilton-Jacobi theory to be discussed presently. We have also emphasized in the last chapter the important role of the classical action, which may be written in either of the two following forms (we take $t'=0$):

$$S(\mathbf{q}(t), \mathbf{q}'(0); E) = \sum_i \int_0^t p_i \dot{q}_i \, d\tau = \sum_i \int_{q_i'}^{q_i} p_i(\mathbf{q}'', E) \, dq_i'' \,. \qquad (2.9)$$

For a free particle in Euclidean space one finds, analogously to (1.21), the expression

$$S_0(\mathbf{q}, \mathbf{q}', E) = \sqrt{2mE(\mathbf{q} - \mathbf{q}')^2} = p \,|\mathbf{q} - \mathbf{q}'| \,. \qquad (2.10)$$

When the particle is moving in a mean field, the corresponding action may be calculated by integrating the *local* momentum along the trajectory using a suitable parameter.

Note: In many modern textbooks on classical mechanics (e.g., Landau & Lifshitz [2], Vol. 1, or Goldstein [3]), Hamilton's principal function R (2.5) is called "action", whereas S (2.9) is sometimes called "abbreviated action". We shall follow Gutzwiller in calling S the action.

The Hamiltonian H is a function of the coordinates $\{q_i, p_i\}$, in contrast to the Lagrangian $\mathcal{L}$ that is expressed in the variables $\{q_i, \dot{q}_i\}$. Such a change of independent variables is brought about by a Legendre transformation:

$$H(\mathbf{p}, \mathbf{q}, t) = \sum_i \dot{q}_i \frac{\partial \mathcal{L}}{\partial \dot{q}_i} - \mathcal{L} \,, \qquad (2.11)$$

which includes the time-dependent situation. The equations of motion are given by

$$\frac{dp_i}{dt} = -\frac{\partial H}{\partial q_i}, \qquad \frac{dq_i}{dt} = \frac{\partial H}{\partial p_i} \,. \qquad (2.12)$$

We now go on to develop the Hamiltonian formalism and its connection with quantum mechanics in more detail.

2.2 Hamiltonian formalism and the classical limit

As pointed out already, the Hamiltonian formalism will be used often. In this section, after reviewing some standard formulae, we concentrate on the generating functions that link different canonical transformations. As Miller [1, 4] has emphasized, the classical limit of a quantum mechanical wave function, using the Dirac transformation theory, is naturally expressed in terms of these generating functions and their second derivatives. This is insightful, as well as important for the development of the semi-classical wave function later in this chapter. Using Hamilton's equations (2.12), we see that

$$\frac{d}{dt}H = \frac{\partial H}{\partial q}\dot{q} + \frac{\partial H}{\partial p}\dot{p} + \frac{\partial H}{\partial t} = \frac{\partial H}{\partial t} \,, \qquad (2.13)$$

so that $H = E$ is a constant of motion for an autonomous system. More generally, consider two dynamical variables A_1 and A_2, and define the *Poisson bracket* as

$$[A_1, A_2]_{q,p} = \sum_k \left(\frac{\partial A_1}{\partial q_k} \frac{\partial A_2}{\partial p_k} - \frac{\partial A_1}{\partial p_k} \frac{\partial A_2}{\partial q_k} \right). \tag{2.14}$$

Using the above definitions, it immediately follows that the total time derivative of the dynamical variable A_1 may be written as

$$\frac{dA_1}{dt} = [A_1, H]_{q,p} + \frac{\partial A_1}{\partial t}. \tag{2.15}$$

Thus a dynamical variable A_1 (that is not explicitly time-dependent) is a constant of motion if its Poisson bracket with the Hamiltonian vanishes. The constants of motion are also called *integrals* or *invariants*. Two integrals are said to be in *involution* if the Poisson bracket of the pair vanishes. *An integrable Hamiltonian with N degrees of freedom has N integrals in involution.* This is because a canonical transformation may be performed from the coordinates $\{q_i, p_i\} \rightarrow \{Q_i, P_i\}$, choosing the momenta P_i to be the constants I_i, and then

$$\dot{P_i} = -\frac{\partial H}{\partial Q_i} = 0. \tag{2.16}$$

This ensures that H is independent of the coordinates $\{Q_i\}$. Such coordinates are called "cyclic". The Hamiltonian is then only a function of the constants $\{I_i\}$, which are just the actions for the cyclic 'angle' variables $\{Q_i\}$. In some cases, there are more than N constants of motion, with still N of them in involution. Such systems are called *superintegrable*. More about superintegrable systems may be found in some recent papers [5].

A fundamental property of a canonical transformation is that it preserves the element of an area in the phase space. For simplicity, we now consider one-dimensional motion, but the results are valid generally. This is apparent by noting that the area integral $\int dq dp$ transforms to $\int [Q, P]_{q,p} dQ dP$, where the Poisson bracket $[Q, P]_{q,p}$, defined by Eq. (2.14), is the same as the Jacobian of the transformation. For a canonical transformation, it may easily be checked that this Jacobian is unity. For $H = E$, $\int_S dq dp$ over a surface area S may be written as $\int_C p(q, E) dq$ over a line integral C enclosing the area. So we may write, over an arbitrary closed contour $\mathcal{C}$,

$$\int_C (p \, dq - P dQ) = \int_C (p \, dq + Q dP) = 0.$$

In the last step, we have used the property of the perfect differential $\int_C d(PQ) = 0$. The last equation above implies that the quantity $(pdq + QdP)$ is itself expressible as the perfect differential of a function $F_2(q, P)$, obeying

$$\frac{\partial F_2}{\partial q} = p(q, P), \quad \frac{\partial F_2}{\partial P} = Q(q, P). \tag{2.17}$$

The function $F_2(q, P)$ is one of the standard generating functions of the canonical transformation, with the identity transformation being given by the particular choice $F_2 = qP$. In a canonical theory, one may take any two of the variables from (q, p, Q, P)

as the independent ones. In the above, $F_2(q, P)$ is the one with the choice of (q, P) as the independent variables, and the other two dependent variables (p, Q) are obtained from Eq. (2.17). This is a well known mathematical technique, known in the literature as Legendre transformation [3]. There are, of course, four generating functions corresponding to the four possible choices of picking one 'old' and one 'new' variable from (q, p) and (Q, P). The relations between the different generating functions may be easily derived [3], for example

$$F_1(q, Q) = F_2(q, P) - PQ. \qquad (2.18)$$

The function $F_1(q, Q)$ generates the other two dependent variables from the relations

$$\frac{\partial F_1}{\partial q} = p(q, Q), \quad -\frac{\partial F_1}{\partial Q} = P(q, Q). \qquad (2.19)$$

As Miller emphasizes in his beautiful article [1], the significance of the generating functions is revealed when the classical limit of quantum mechanics is taken using Dirac's transformation theory. In the Dirac notation, the momentum eigenvector $|p\rangle$ in the coordinate space representation q is given by

$$\langle q|p \rangle = \psi_p(q) = \frac{1}{\sqrt{2\pi\hbar}} \exp(ipq/\hbar), \qquad (2.20)$$

Classically, (q, p) are canonically conjugate variables. A similar equation may be written down for $\langle Q|P \rangle$ involving the other conjugate pair (Q, P):

$$\langle Q|P \rangle = \psi_P(Q) = \frac{1}{\sqrt{2\pi\hbar}} \exp(iPQ/\hbar), \qquad (2.21)$$

One may well ask about the form of a wave function $\langle q|Q \rangle$, or $\langle q|P \rangle$, in a mixed representation. This may be expressed quite generally as

$$\langle q|Q \rangle = \mathcal{A}_1(q, Q) \exp[if_1(q, Q)/\hbar], \qquad (2.22)$$
$$\langle q|P \rangle = \mathcal{A}_2(q, P) \exp[if_2(q, P)/\hbar]. \qquad (2.23)$$

The amplitude $\mathcal{A}_i$ and the phase f_i are real, and the subscripts match those of the classical generating functions with the same independent variables. Our objective is to find the semiclassical limit of the above wave functions as $\hbar \to 0$. This may be done by taking the completeness relation

$$\int_{-\infty}^{\infty} dq \, \langle P|q \rangle \, \langle q|P' \rangle = \delta(P - P'), \qquad (2.24)$$

and substituting for the wave functions from Eqs. (2.22, 2.23). We then get

$$\int_{-\infty}^{\infty} dq \, \mathcal{A}_2(q, P) \, \mathcal{A}_2(q, P') \exp\left\{i[f_2(q, P') - f_2(q, P)]/\hbar\right\} = \delta(P - P'). \qquad (2.25)$$

As $\hbar \to 0$, the exponent varies rapidly, and the dominant contribution to the integral comes only from P' in the neighborhood of P. For a fixed P, keeping only the leading term in the exponent, we have

$$[f_2(q, P') - f_2(q, P)] \simeq [\partial f_2(q, P)/\partial P](P' - P) = w(q, P)(P' - P), \qquad (2.26)$$

where

$$w(q, P) = [\partial f_2(q, P)/\partial P].$$ (2.27)

We may thus express Eq. (2.25) in the form

$$\int_{-\infty}^{\infty} dw \frac{[\mathcal{A}_2(q, P)]^2}{|\partial w/\partial q|} \exp\left[iw(P' - P)/\hbar\right] = \delta(P - P').$$ (2.28)

Recall that the Dirac delta function itself is of the form

$$\delta(P - P') = \frac{1}{2\pi\hbar} \int_{-\infty}^{\infty} dw \, \exp\left[iw(P' - P)/\hbar\right].$$ (2.29)

Combining this with Eq. (2.28), we get

$$\int_{-\infty}^{\infty} dw \, \exp\left[iw(P' - P)/\hbar\right] \left(\frac{[\mathcal{A}_2(q, P)]^2}{|\partial w/\partial q|} - (2\pi\hbar)^{-1}\right) = 0.$$ (2.30)

We therefore get the expression for the amplitude:

$$|\mathcal{A}_2(q, P)| = (2\pi\hbar)^{-1/2} \left|\frac{\partial w}{\partial q}\right|^{1/2} = (2\pi\hbar)^{-1/2} \left|\frac{\partial^2 f_2(q, P)}{\partial P \partial q}\right|^{1/2}.$$ (2.31)

In the last line, we have made use of Eq. (2.27). We still have to find the connection between $f_2(q, P)$, defined by Eq. (2.23) for the quantum mechanical wave function $\langle q|P\rangle$, and the generating function $F_2(q, P)$ of classical mechanics in the limit of $\hbar \to 0$. This may be found by expressing the state vector $\langle q|Q\rangle$ as

$$\langle q|Q\rangle = \int_{-\infty}^{\infty} dP \, \langle q|P\rangle \langle P|Q\rangle.$$ (2.32)

Substituting Eqs.(2.21, 2.23) above, we get

$$\langle q|Q\rangle = (2\pi\hbar)^{-1/2} \int_{-\infty}^{\infty} dP \, \mathcal{A}_2(q, P) \exp\left\{i[f_2(q, P) - PQ]/\hbar\right\}.$$ (2.33)

This is still an exact relation. We now go to the limit of $\hbar \to 0$, so that the exponent in the above integral oscillates rapidly. We now expand the exponent in a Taylor series about the stationary point $P = \tilde{P}$:

$$\left(\frac{\partial f_2(q, P)}{\partial P}\right)_{P=\tilde{P}} = Q.$$ (2.34)

Comparing with Eq. (2.17), we see that at the saddle point $\tilde{P}$, the function $f_2(q, \tilde{P})$ appearing in the quantum wave function obeys the same equation as the classical generator $F_2(q, P)$. This equivalence could be extended to the variable q also. We may therefore identify $f_2(q, P)$ with the generator $F_2(q, P)$ in the limit $\hbar \to 0$, and write

$$\mathcal{A}_2(q, P) = (2\pi\hbar)^{-1/2} \left|\frac{\partial^2 F_2(q, P)}{\partial P \partial q}\right|^{1/2}.$$ (2.35)

We thus get the classical limit of Eq. (2.23):

$$\langle q|P \rangle = (2\pi\hbar)^{-1/2} \left| \frac{\partial^2 F_2(q,P)}{\partial P \partial q} \right|^{1/2} \exp[iF_2(q,P)/\hbar], \qquad \hbar \to 0. \qquad (2.36)$$

This may be regarded as a statement of the *correspondence principle*. Since we know from Eq. (2.17) that

$$\frac{\partial F_2}{\partial q} = p(q,P),$$

and the local momentum in a mean potential $V(q)$ is

$$p(q,E) = \sqrt{2m[E - V(q)]}, \qquad (2.37)$$

it follows that in this case, if we choose P to be E, the amplitude $\mathcal{A}_2$ given by Eq. (2.35) is proportional to $[p(q)]^{-1/2}$. With this choice, we have

$$F_2(q,E) = \int^q p(q'', E)\, dq'', \quad F_1(q,t) = \int^q p(q'', E)\, dq'' - Et. \qquad (2.38)$$

The lower limit of the integrals is fixed at a value of q' for which $F_2(q', E) = 0$. Note that F_2 and F_1 may be identified with the action S and the principal function R with this choice [see Eq. (2.8)]. The wave function (2.36) is what is called the WKB approximation, and will be discussed in Sect. 2.4 in more detail. When Eq. (2.35) is generalized from one to N dimensions, the amplitude takes the form:

$$\mathcal{A}_2(\mathbf{q}, \mathbf{P}) = \left[(2\pi\hbar)^{-N} \det \left| \frac{\partial^2 F_2(\mathbf{q}, \mathbf{P})}{\partial P_i \partial q_j} \right| \right]^{1/2}. \qquad (2.39)$$

We shall use this form in Sect. 2.5 for the wave function in torus quantization. Miller goes on to evaluate the integral (2.33) by the saddle-point method, which we do not discuss here. Instead, we go on to develop the semiclassical approximations based on the Hamiltonian formalism.

2.3 Hamilton-Jacobi theory and wave mechanics

The advent of quantum mechanics brought a sharp change from classical physics. Nevertheless, the Hamilton-Jacobi formalism of classical physics already contained hints of the propagation of a "wave-front", whose phase is governed by the time-integral of the classical Lagrangian of the system, defined in Eq. (2.5) as Hamilton's principal function. Schrödinger [6] was guided by the Hamilton-Jacobi (HJ) equation in obtaining his wave-equation. We shall moreover see that the HJ-equation naturally leads to a semiclassical approximation, named after Wentzel, Kramers, and Brillouin (WKB) [7], that is very successful in integrable problems with a separable Hamiltonian. In its time-independent form, the HJ-equation involves the classical action S [see Eq. (2.9)], that plays a central role in this book.

To follow this connection, consider once again Hamilton's principal function R

$$R = \int_{t'}^{t} \mathcal{L}(q_i, \dot{q}_i, t'') \, dt'' \tag{2.40}$$

for the fixed time interval $(t - t')$, along an *actual* classical trajectory. In the above, q_i stands for the N generalized coordinates $q_1, q_2, \ldots q_N$, and similarly for the velocities $\dot{q}_i$. Since the equation of motion is of second order, each degree of freedom requires two initial conditions to specify the trajectory completely. Alternately, two points $\mathbf{q}^{(1)}$ and $\mathbf{q}^{(2)}$ on the trajectory covered in the fixed time interval specifies the trajectory uniquely. Thus R may be regarded as a function $R(\mathbf{q}^{(1)}, \mathbf{q}^{(2)}, (t - t'))$.

Consider now a family of classical paths, as shown in Fig. 2.1, each with the same starting point $\mathbf{q}^{(1)}$ at time t', but reaching different points $\mathbf{q}^{(2)}$ at the same time t. The time interval $(t - t')$ is the same for each trajectory, but the energy E of the particle along the different trajectories need not be the same. The difference δR between two such adjacent paths (we now take $\mathcal{L}$ to be time-independent) is

$$\delta R = \int_{t'}^{t} \left(\frac{\partial \mathcal{L}}{\partial q_i} \delta q_i + \frac{\partial \mathcal{L}}{\partial \dot{q}_i} \delta \dot{q}_i \right) dt'' . \tag{2.41}$$

Writing $\delta \dot{q}_i = d(\delta q_i)/dt''$, and integrating by parts, we get

$$\delta R = \int_{t'}^{t} \left(\frac{\partial \mathcal{L}}{\partial q_i} - \frac{d}{dt''} \frac{\partial \mathcal{L}}{\partial \dot{q}_i} \right) \delta q_i \, dt'' + \frac{\partial \mathcal{L}}{\partial \dot{q}_i} \delta q_i \big|_{t'}^{t} . \tag{2.42}$$

A classical trajectory obeys Lagrange's equation (2.7), so the first term on the right-hand side of Eq. (2.42) vanishes, giving

$$\delta R = p_i \, \delta q_i \big|_{t'}^{t} . \tag{2.43}$$

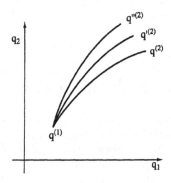

Figure 2.1: An ensemble of classical paths, starting at $\mathbf{q}^{(1)}$ at time t' and ending simultaneously at the time t at different points $\mathbf{q}^{(2)}$. For simplicity, the diagram is for two-dimensional motion, though the treatment in the text is general.

In the above, we have used the relation that the canonical momentum is defined by Eq. (2.3). To simplify the notation, let $q_i^{(1)} = 0$, $q_i^{(2)} = q_i$, and put the starting time $t' = 0$, so that $(t - t') = t$. Noting that it is only the final point q_i that is varied from one path to another, we denote $R(q_i, 0, t)$ by $R(q_i, t)$. We then have,

$$\delta R(q_i, t) = \sum_j p_j \, \delta q_j \,, \tag{2.44}$$

and

$$\frac{\partial R}{\partial q_i} = p_i \,. \tag{2.45}$$

Taking the total time-derivative of $R(q_i, t)$,

$$\frac{dR}{dt} = \frac{\partial R}{\partial q_i} \dot{q}_i + \frac{\partial R}{\partial t} \,. \tag{2.46}$$

Using Eq. (2.45), this reduces to

$$\frac{dR}{dt} = \sum_i p_i \dot{q}_i + \frac{\partial R}{\partial t} \,. \tag{2.47}$$

Note, from Eq. (2.40), that $\frac{dR}{dt} = \mathcal{L}$, and by definition the Lagrangian $\mathcal{L}$ is related to the Hamiltonian by

$$\mathcal{L} = \sum_i p_i \dot{q}_i - H \,. \tag{2.48}$$

Comparing this with Eq. (2.47), we obtain

$$-\frac{\partial R}{\partial t} = H \,. \tag{2.49}$$

Substituting this in Eq. (2.47),

$$\frac{dR}{dt} = \sum_i p_i \dot{q}_i - H. \tag{2.50}$$

Integrating between 0 and t, we thus obtain

$$R(q_i, t) = \sum_j \int_0^{q_j} p_j \, dq_j - Ht \,. \tag{2.51}$$

In an autonomous system with energy $H = E$, for a given trajectory traversing $q_i = 0$ to q_i in a time interval t, we thus get the important relation (2.8) written down earlier:

$$R(q_i, 0, t) = S(q_i, 0, E) - Et \,. \tag{2.52}$$

Eq. (2.49) is just the *Hamilton-Jacobi equation*

$$H\left(q_i, \frac{\partial R}{\partial q_i}\right) = -\frac{\partial R}{\partial t}. \tag{2.53}$$

This constitutes a set of partial differential equations whose significance will be discussed after we write down its time-independent form.

In its time-independent form, we use the action $S(q_i, 0, E)$ defined by Eq. (2.9) rather than Hamilton's principal function $R(q_i, 0, t)$. Again, we suppress the fixed starting coordinate at 0, and express the action as $S(q_i, E)$. We note from Eq. (1.24) that

$$p_i = \frac{\partial S}{\partial q_i} = (\nabla_q S)_i \,. \tag{2.54}$$

The Energy conservation equation may thus be written as

$$H\left(q_i, \frac{\partial S}{\partial q_i}\right) = E \,. \tag{2.55}$$

This is the time-independent form of the HJ equation. The HJ equation actually describes a whole family of orbits in the configuration space. Take, for example, a two-dimensional configuration space defined by the coordinates (q_1, q_2). In the diagram of Fig. 2.2, keeping E fixed, the equation $S(q_i, 0, E) = S_1$ is satisfied by a curve on the plane, and a neighboring value $S = S_2$ is given by an adjoining curve as shown. A given particle trajectory, starting on an initial point on S_1, is along a ray that is perpendicular to these curves of constant S, since its direction of motion is determined by (assuming a velocity-independent potential) $\dot{q}_i = p_i/m = (\nabla_q S)_i/m$. In this sense, there is an implicit wave picture in the family of particle trajectories, with the "wave fronts" defined by lines of constant phase S, and the propagating rays describing the particle trajectories as in geometrical optics. This picture may be developed more by considering the time-dependent function $R(q_i, t)$ given by Eq. (2.52). The curves of constant $S(q_i, E)$ are fixed in the configuration space, but those of constant $R(q_i, t)$ (obeying the condition $-\frac{\partial R}{\partial t} = E$) move from one surface of constant $S(q_i, E)$ to another with the progress of time [8]. The time-dependent phase of the "wave" is given by $R(q_i, t)$, and the phase velocity of the wave front is determined by demanding that

$$dS - E\, dt = 0 \,. \tag{2.56}$$

Along a trajectory l, $dS = |\nabla_q S|\, dl$, and so the phase velocity of the wave is

$$v_{phase} = \frac{dl}{dt} = \frac{E}{|\nabla_q S|} \,. \tag{2.57}$$

This phase velocity is different from the particle velocity. For example, for a free particle, it is easy to check that v_{phase} is only *half* of the particle velocity. The HJ equations (2.53, 2.55) not only hint towards a wave picture, but have a strong resemblance to the Schrödinger wave equation. Although the nature of the wave is not revealed by classical mechanics, it underlines the importance of Hamilton's principal function $R(q_i, t)$ and the associated action $S(q_i, E)$ that define the phase of the wave. In the next section this is elaborated with a view to develop the semiclassical approximation, and the associated quantization condition. The latter, in Bohr's atom, was the precursor to the development of quantum mechanics.

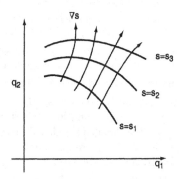

Figure 2.2: An ensemble of "rays" in the two-dimensional configuration space, each ray corresponding to a classical trajectory. The rays are transverse to the "wave fronts" of constant action $S(q_i, 0, E)$ for fixed E.

2.4 The WKB method

In this section, we first bring out the similarity between the Schrödinger equation of wave mechanics, and the HJ equation of the preceding section. This leads us to the WKB approximation directly, which already came up when we were developing the classical limit of a wave function in Sect. 2.2. Here we go into more details, particularly to show how a matching of the wave function across a turning point introduces an extra phase in it. (We also use the notation $\mathbf{r}$ instead of $\mathbf{q}$ for most of this section. The latter notation will often be used for developing the formalism, while the former is more convenient in concrete applications). In more general two-dimensional problems, similar phases will also be introduced at caustics. This will be discussed in more detail in Chapter 5. Consider the one-particle time-dependent equation

$$-\frac{\hbar^2}{2m}\nabla^2\psi + V(\mathbf{r})\psi(\mathbf{r}, t) = i\hbar\,\frac{\partial\psi}{\partial t}\,. \tag{2.58}$$

Following Madelung [9], make the substitution

$$\psi(\mathbf{r}, t) = A(\mathbf{r}, t)\,\exp\left[iR(\mathbf{r}, t)/\hbar\right], \tag{2.59}$$

where A and R are real. The phase R of the wave has the dimensions of $\hbar$. Equating the real and imaginary parts, and writing $\rho = A^2$, we obtain after some algebra

$$-\frac{\partial R}{\partial t} = \left[\frac{(\nabla R)^2}{2m} + V(\mathbf{r}) - \frac{\hbar^2}{2m}\frac{\nabla^2\sqrt{\rho}}{\sqrt{\rho}}\right], \tag{2.60}$$

and

$$\frac{\partial\rho}{\partial t} + \nabla\cdot\left(\rho\frac{\nabla R}{m}\right) = 0\,. \tag{2.61}$$

Using Eq. (2.45), $\nabla R/m$ is the velocity vector $\mathbf{v}$ of the particle, and Eq. (2.61) is the continuity equation corresponding to the conservation of the density ρ. Note that if the last term (called the Bohm, or quantum potential, see Ref. [10]) is neglected in Eq. (2.60), then it is precisely the HJ equation (2.49) of the previous section. Bohm [11], in fact, went on to give a different interpretation of quantum mechanics starting with the above equations (retaining the state-dependent "quantum potential" $-\frac{\hbar^2}{2m}\frac{\nabla^2\sqrt{\rho}}{\sqrt{\rho}}$). We would not pursue this path, but study the solution ψ first when the quantum potential is dropped. In such an approximation, the stationary solution is readily obtained for the one-dimensional problem. In higher dimensions the formalism is very similar if the potential depends only on the radial coordinate, since the angular kinetic effects may then be included by adding a centrifugal term. This will be taken up in the next subsection, with special emphasis on two-dimensional motion.

2.4.1   WKB in one dimension

For a given energy E, we may use Eq. (2.52) to relate R and the action S, and substitute this in Eq. (2.60). We then obtain

$$\frac{1}{2m}\left(\frac{\partial S}{\partial x}\right)^2 = E - V(x). \tag{2.62}$$

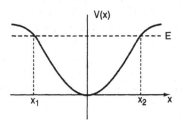

Figure 2.3:   Classical turning points are defined by x_1 and x_2, such that $V(x_i) = E$. A particle with energy E is confined classically in the region $x_1 \leq x \leq x_2$.

The action $S(x, x_1, E)$ is then given by

$$S(x, x_1, E) = \pm \int_{x_1}^{x} \sqrt{2m(E - V(x'))}\, dx', \tag{2.63}$$

The continuity equation (2.61) may also be integrated to yield

$$\rho = \frac{\text{const.}}{S'} = \frac{\text{const.}}{p(x)}, \tag{2.64}$$

where $S' = \left(\frac{\partial S}{\partial x}\right)$, and the local momentum $p(x)$ is identified with S':

$$p(x) = \sqrt{2m(E - V(x))}. \tag{2.65}$$

In Eq. (2.63), we have taken the starting point at x_1. Alternately, we could also define

$$S(x_2, x, E) = \pm \int_x^{x_2} \sqrt{2m(E - V(x'))} \, dx' \,, \tag{2.66}$$

where the starting point is the variable x. Consider first the classically allowed region for the particle, where $p(x)$ is real. This would be the range $x_1 \leq x \leq x_2$ shown in Fig. 2.3. The stationary wave function for a quantized energy E may be written as $\psi(x, t) = \psi(x) \exp(-iEt/\hbar)$. Using Eq. (2.59), and the form (2.66) for S, we may then write the WKB wave function in the form

$$\psi(x) = \frac{1}{\sqrt{p(x)}} \left\{ A \exp\left[\frac{i}{\hbar} \int_x^{x_2} p(x') \, dx'\right] + B \exp\left[-\frac{i}{\hbar} \int_x^{x_2} p(x') \, dx'\right] \right\}. \tag{2.67}$$

For $x < x_1$ or $x > x_2$, $p(x) = i|p(x)|$, and the above wave function may be analytically continued, retaining only the exponentially decaying part. For $x > x_2$, we then have

$$\psi(x) = \frac{1}{\sqrt{|p(x)|}} C \exp\left[-\frac{1}{\hbar} \int_{x_2}^x |p(x')| \, dx'\right]. \tag{2.68}$$

This semiclassical solution blows up at the turning points, where the local momentum $p(x)$ goes to zero. This in itself may be tolerated if the wave function is normalizable. The matching of the wavefunction at the turning points may still be done by examining the wave equation more closely in the vicinity of the turning point. Going past the classical turning point necessitates introducing some phase factor in the wave function. This is done carefully in several standard texts (see, e.g., Refs. [12, 13]). The time-independent Schrödinger wave equation in the vicinity of the turning point may be written by Taylor-expanding the potential $V(x)$ about $x = x_2$, and retaining only the leading term in $(E - V(x))$:

$$\frac{d^2\psi(x)}{dx^2} = \frac{2m}{\hbar^2} (x - x_2) V'(x_2) \psi(x). \tag{2.69}$$

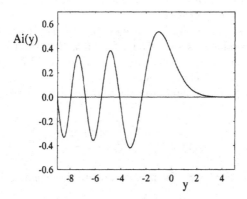

Figure 2.4: A plot of the Airy function $Ai(y)$.

Here $V'(x_2)$ is the first derivative of $V(x)$ with respect to x at $x = x_2$, where $E = V(x_2)$. This is Airy's differential equation, whose analytical solutions are known. These solutions may be used to connect the oscillatory part in the interior of the well with the exponentially decaying solution outside. To be more specific, we may shift the origin in Eq. (2.69) to x_2, thereby replacing $(x - x_2)$ by x. Letting

$$\frac{2m}{\hbar^2} V'(0) = \alpha^3, \qquad \alpha x = y, \tag{2.70}$$

Eq. (2.69) takes the form $\psi''(y) = y\psi(y)$. Excluding the exponentially growing solution, we focus on the solution $Ai(y)$, known as the Airy function, which is shown as a function of y in Fig. 2.4. For negative y, corresponding to the interior of the well ($x < x_2$), it is oscillatory, while for large positive y, outside the turning point x_2, it decays exponentially. Its asymptotic form for large negative y is given by [14]

$$Ai(y) \simeq |y|^{-1/4} \sin\left(\frac{2}{3} |y|^{3/2} + \frac{1}{4}\pi\right), \tag{2.71}$$

while for large positive y it is

$$Ai(y) \simeq \frac{1}{2} y^{-1/4} \exp\left(-\frac{2}{3} y^{3/2}\right). \tag{2.72}$$

By matching the asymptotic form of the oscillatory Airy function with Eq. (2.67), and the exponentially decaying form (2.72) with Eq. (2.68), we get $B = iC \exp(-i\pi/4)$, $A = -iC \exp(i\pi/4)$, where C is the same constant as in Eq. (2.68). The wave function in the interior of the potential well, Eq. (2.67), now takes the form

$$\psi(x) = \frac{2C}{\sqrt{p(x)}} \left\{ \sin\left[\frac{1}{\hbar} \int_x^{x_2} p(x')\,dx' + \frac{\pi}{4}\right] \right\}. \tag{2.73}$$

while the wavefunction for $x > x_2$ is given by Eq. (2.68). Inspecting these two forms, we conclude that *a phase factor of $\pi/4$ is introduced in the wave function at the turning point of a smoothly varying potential.*

The WKB quantization condition may now be obtained using the above result, and demanding single-valuedness of the wave function. Note from Fig. 2.3 that at the turning point x_1, $V'(x_1) < 0$, so on replacing x_2 by x_1 in Eq. (2.70), y is negative. The mirror image of Fig. 2.4, therefore, should be pictured at $x = x_1$. The semiclassical wave function for $x > x_1$ in the interior of the well may then be written as

$$\psi(x) = \frac{2\tilde{C}}{\sqrt{p(x)}} \left\{ \sin\left[\frac{1}{\hbar} \int_{x_1}^x p(x')\,dx' + \frac{\pi}{4}\right] \right\}, \tag{2.74}$$

where $\tilde{C}$ is a constant. For later reference, we rewrite this equation in the form

$$\psi(x) = \frac{2\tilde{C}}{\sqrt{p(x)}} \left\{ \cos\left[\frac{1}{\hbar} \int_{x_1}^x p(x')\,dx' - \frac{\pi}{4}\right] \right\}. \tag{2.75}$$

Similarly for $x < x_1$, the wave function is the same as Eq. (2.68) but with the limits of integration in the exponent going from x to x_1, and the constant C replaced by $\widetilde{C}$. The two phase factors in Eqs. (2.73) and (2.74) should be the same within a negative sign, as an extra minus sign may be absorbed in the overall constant. We may therefore add on one side an extra phase $(n+1)\pi$, where n is zero or a positive integer. We thus get, with $\widetilde{C} = (-1)^n C$,

$$-\frac{1}{\hbar} \int_x^{x_2} p(x') \, dx' - \pi/4 + (n+1)\pi = \frac{1}{\hbar} \int_{x_1}^x p(x') \, dx' + \pi/4 . \qquad (2.76)$$

This gives the quantization condition for a bound state in a potential well that is smoothly varying at both turning points,

$$\frac{1}{\hbar} \int_{x_1}^{x_2} p(x) \, dx = (n + 1/2)\pi \qquad n = 0, 1, 2, 3, \ldots . \qquad (2.77)$$

The derivation also makes it clear that n is the number of nodes in the wave function.

For a particle with energy E confined in a potential, a rough estimate of the normalization constant in Eq. (2.74) may be made by assuming that the contributions to normalization from regions $x < x_1$ and $x > x_2$ are negligible. It is easy to show that under this assumption, Eq. (2.74) reduces to the form

$$\psi(x) = \sqrt{\frac{2\omega}{\pi v(x)}} \left\{ \sin\left[\frac{1}{\hbar} \int_{x_1}^x p(x') \, dx' + \frac{\pi}{4} \right] \right\}, \qquad (2.78)$$

In the above, $\omega = 2\pi/T$ is the angular frequency, and $v(x) = p(x)/m$ is the speed of the particle. The quality of this approximation may be judged from Fig. 2.5, where the third and the fourth excited states are plotted for a harmonic oscillator potential, and compared with the exact solutions. Note that at the turning points, without the use of the connecting formulae, the wave function blows up. Nevertheless, in the interior region, the simple approximation (2.78) works remarkably well.

The quantization condition given by Eq. (2.77) changes if one of the walls of the potential is infinitely steep, so the wave function at that boundary is zero. For example, let there be an infinite wall at $x = 0$, but the potential at the turning point at x_2 be smooth (Fig. 2.6). Now the turning point $x_1 = 0$. We may still apply Eq. (2.73) with the condition that $\psi(x_1) = 0$. This yields

$$\sin\left[\frac{1}{\hbar} \int_{x_1}^{x_2} p(x') \, dx' + \frac{\pi}{4} \right] = 0 . \qquad (2.79)$$

We therefore have $\left[(1/\hbar) \int_{x_1}^{x_2} p(x') dx' + \pi/4 \right] = (n+1)\pi$, where n is zero or a positive integer. The quantization condition is then

$$\frac{1}{\hbar} \int_{x_1}^{x_2} p(x') \, dx' = (n + 3/4)\, \pi , \qquad n = 0, 1, 2, 3, \ldots . \qquad (2.80)$$

Since $\pi/4$ of the phase on the right-hand side of the above equation may be attributed as before to the turning point at x_2, we conclude that *the extra phase factor introduced at the steep wall is $\pi/2$.*

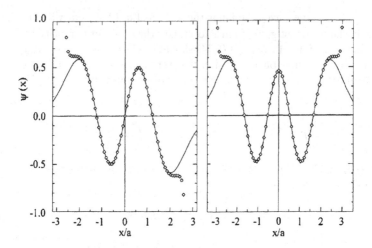

Figure 2.5:   Comparison of the exact wave functions (solid lines) of a one-dimensional harmonic oscillator for the excited states with $n = 3$ (left) and $n = 4$ (right) with those generated in the WKB approximation (diamond symbols) using Eq. (2.78). (Courtesy Carl Svensson.)

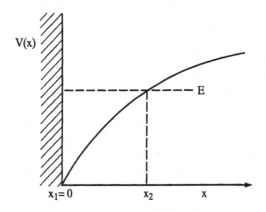

Figure 2.6: A potential with a hard wall on one side.

The WKB approximation that we have outlined above was made by neglecting the "Bohm potential"

$$V_B = \frac{\hbar^2}{2m} \frac{\nabla^2 \sqrt{\rho}}{\sqrt{\rho}}$$ (2.81)

in Eq. (2.60). This is justified only if it varies little over the de Broglie wave length $\lambda = \hbar/p$. Noting from Eq. (2.64) that the probability density ρ is proportional to p^{-1}, and the local momentum $p(x)$ is given by Eq. (2.65), it is easy to see that

$$\frac{V_B}{\lambda} \simeq \sqrt{\frac{\hbar^2}{2m}} \frac{(V')^2}{(E - V(x))^{3/2}}. \tag{2.82}$$

We have assumed above that the potential $V(x)$ is smooth enough for its second derivative to be disregarded. The condition $V_B \ll \lambda |V'(x)|$ then yields the condition of applicability of the lowest order WKB:

$$\sqrt{\frac{\hbar^2}{2m}} \frac{|V'(x)|}{(E - V(x))^{3/2}} \ll 1. \tag{2.83}$$

This is precisely the condition $|\partial \lambda / \partial x| \ll 1$. Neglecting V_B enabled us to write Eq. (2.62), that gives the leading term in an expansion of the quantum action in powers of $\hbar$. These higher order terms are developed in many texts, and will not be elaborated here.

For completeness we also mention that the WKB method is very useful for estimating the probability of quantum mechanical barrier penetration. Imagine Fig. 2.3 to be inverted, and a particle hitting the barrier from the left. One can again construct the appropriate WKB wave functions, and match them at the left and right turning points. It is now important to retain both the exponentially decreasing as well as the increasing solutions within the barrier, superposing them into a complex wave function. It is found that the transmission probability is approximately given by [15]

$$T \simeq \frac{1}{1 + \exp(2K)}, \tag{2.84}$$

where

$$K = \frac{1}{\hbar} \int_{x_1}^{x_2} [2m(V(x) - E)]^{1/2} \, dx. \tag{2.85}$$

It is known that Eq. (2.84) is exact for a parabolic barrier. Note that $T = 1/2$ at the top of the barrier when $E = V$. In case $K \gg 1$, Eq. (2.84) reduces to the simpler form

$$T \simeq \exp(-2K). \tag{2.86}$$

Obviously, this formula should not be used near the top of the barrier.

The semiclassical approach may also be used to calculate the level splitting in a double well with a barrier in between [16], [17]. In the case of a symmetric double well, for example, there is a lifting of the degeneracy in the the energy levels of the (isolated) left and the right wells, because of the coupling between them through tunneling. This is shown in Fig. 2.7, where the unperturbed energy level of each isolated well at E_0 is shown by dashed lines.

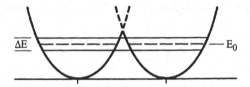

Figure 2.7: A double well with a barrier. Each isolated well has an energy level E_0 shown by the dashed lines. In this simplest version, other bound states are assumed to be absent, or far away in energy. The double well has two levels, with an energy splitting ΔE.

For the two-level problem, the quantum splitting may be estimated in a simple manner [14] using the WKB results developed in this section. It is the symmetric combination of the eigenstates of the isolated wells that goes down in energy by $\Delta E/2$, while the antisymmetric wave function goes up by the same amount. The WKB result is given by

$$\Delta E = \frac{\hbar \omega_0}{\pi} \, e^{-K} \,, \tag{2.87}$$

where $\omega_0 = 2\pi/T_0$ is the classical angular frequency of motion of the particle in *one* well, and K is the same as (2.85) for the barrier in the middle, both evaluated at the unperturbed energy E_0. In applying Eq. (2.85), take $x_1 = -a$, $x_2 = a$, where $\pm a$ are the two *inner* turning points of the double well potential at energy E_0. More generally, of course, there are many energy levels in each well, and there is a difference in the level density between the symmetric and the antisymmetric states brought about by tunneling. Creagh and Whelan [18] have used the Gutzwiller trace formula for chaotic orbits [see Eq. (1.109) in Chapter 1], and the technique of complex action given above to calculate this difference in a two-dimensional potential. The influence of chaotic dynamics in tunneling between states of orderly motion has been studied by Bohigas *et al.* [19], and by Doron and Frischat [20].

The algebra above was done for the one-dimensional problem. In higher-dimensions, the HJ equation (2.55) cannot be solved in general, unless the Hamiltonian is separable, i.e., written as a sum of uncoupled one-body terms in a suitably chosen coordinate frame. An obvious example of such a system is a three-dimensional harmonic oscillator. Leaving aside such simple cases, the method may still be generalized to many dimensions for integrable problems, where there are at least as many independent constants of motion as the number of (space) degrees of freedom. In that case, the important concept of "torus quantization" comes in, and the term EBK method (after Einstein, Brillouin and Keller [21]) is used. Before we develop the structure of the available phase space in multi-dimensional phase space and the EBK formalism, it is desirable to discuss how the one-dimensional WKB method is modified for the radial motion of a particle in a two or three-dimensional central potential. It is important to reexamine for such cases the phases that arise in the WKB wave function at the turning points. We mainly discuss two-dimensional motion in this context, and recommend the very readable work of Berry and Ozorio de Almeida [22] for a more rigorous treatment.

2.4.2 WKB for radial motion

Consider the Schrödinger equation for a particle moving in a two-dimensional *central* potential $V(r)$:

$$-\frac{\hbar^2}{2m}\left[\frac{\partial^2}{\partial r^2} + \frac{1}{r}\frac{\partial}{\partial r} + \frac{1}{r^2}\frac{\partial^2}{\partial \phi^2}\right]\Psi(r,\phi) + V(r)\Psi(r,\phi) = E\Psi(r,\phi).\qquad(2.88)$$

As is well-known, the radial and angular equations are separable in this case. Furthermore, by dividing the radial wave function by $\sqrt{r}$, one may obtain for it an equation in the radial variable that is of the same form (i.e., without the first derivative) as in the one-dimensional problem. Most of the formulae developed in the previous section may then be taken over, with the *proviso* that the domain of the radial variable is $0 \le r \le \infty$. By substituting

$$\Psi(r,\phi) = \sum_l \frac{u_l(r)}{\sqrt{r}}\, e^{il\phi},\qquad(2.89)$$

in Eq. (2.88), we obtain

$$\frac{d^2 u_l}{dr^2} + Q_l^2(r)\, u_l(r) = 0.\qquad(2.90)$$

In the above, $\hbar l$ is the eigenvalue of the angular momentum operator L_z, with integral values in the range $-\infty \le l \le \infty$, and

$$Q_l^2(r) = \frac{2m}{\hbar^2}\left[E - V(r) - \frac{\hbar^2}{2mr^2}\left(l^2 - \frac{1}{4}\right)\right].\qquad(2.91)$$

In our notation, $Q_l(r)$ is the local radial wave number in the lth partial wave. We note that Eq. (2.90) is of the same form as the one-dimensional Schrödinger equation, and we are tempted to write down a WKB solution in analogy with (2.75), with $p(x)/\hbar = k(x)$ replaced by $Q_l(r)$. To check if this procedure gives the correct asymptotic phase in the wave function, let us examine a solvable case. Take $V(r)$ to be a square-well of depth V_0, for example. Further, assume that $l \ne 0$. The exact regular solution of Eq. (2.90) is $J_l(kr)$, where the wave number

$$k^2 = 2m(E - V_0)/\hbar^2\qquad(2.92)$$

is a constant. For large r such that $kr \gg |l|$,

$$J_l(kr) \simeq \sqrt{\frac{2}{\pi kr}} \cos\left(kr - l\pi/2 - \pi/4\right).\qquad(2.93)$$

Now just as in Eq. (2.75), we construct the WKB wave function. To match the form (2.93), however, we replace $(l^2 - 1/4)$ in the expression for Q_l^2 above by a parameter s^2 to be determined shortly. Thus the WKB approximation for u_l is taken to be of the form [cf. Eq. (2.75)]:

$$u_l(r) \simeq \frac{2}{\sqrt{\tilde{Q}_l(r)}}\left\{\cos\left[\int_{r_1}^r \tilde{Q}_l(r')\, dr' - \frac{\pi}{4}\right]\right\},\qquad(2.94)$$

where

$$\tilde{Q}_l^2(r') = \frac{2m}{\hbar^2}\left[E - V_0 - \frac{\hbar^2}{2mr'^2}s^2\right].\qquad(2.95)$$

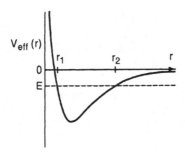

Figure 2.8: The effective potential for radial motion, ($l \neq 0$) with turning points at r_1 and r_2.

Here r_1 is the turning point closer to the origin, as shown in Fig. 2.8, at which $(k^2 r_1^2 - s^2) = 0$. The integral in Eq. (2.94) then yields

$$\int_{r_1}^r \left[k^2 r'^2 - s^2\right]^{1/2}\frac{dr'}{r'} = \sqrt{k^2 r^2 - s^2} - s\cos^{-1}\left(\frac{s}{kr}\right).\qquad(2.96)$$

For the asymptotic region $kr \gg s$, the right-hand side above is $(kr - s\pi/2)$. Substituting this in Eq. (2.94), and comparing with Eq. (2.93), we see that $s^2 = l^2$ for the phases to match. It turns out that this simple prescription is also adequate for the more general situation when the potential varies with r. Examining the behavior of the approximation more closely, it was shown in [22] that $V(r)$ should satisfy the condition $r^2 V(r) \to 0$ as $r \to 0$, and that Eq. (2.94) for the wave function is valid only when $l \neq 0$. This special case will be discussed shortly. Irrespective of this limitation, it is still valid to replace for *all* l in Q_l

$$\left(l^2 - \frac{1}{4}\right) \to l^2,\quad(2\text{ dimensions}).\qquad(2.97)$$

Additionally, we may construct the WKB wave function (2.94) as in one-dimension when $l \neq 0$. We may now derive the WKB quantization condition for the bound states in two dimensions by noting that the wave function is given by (2.94), where $\tilde{Q}_l$ is given by:

$$\tilde{Q}_l(r) = \left\{\frac{2m}{\hbar^2}\left[E - V(r) - \frac{\hbar^2}{2mr^2}l^2\right]\right\}^{1/2}.\qquad(2.98)$$

We shall often use the notation $p_r(r)$ for the radial momentum of the particle in the lth partial wave, where the subscript r to denote radial motion. This is given by

$$p_r(r) = \hbar\tilde{Q}_l(r) = \left\{2m\left[E - V(r) - \frac{\hbar^2}{2mr^2}l^2\right]\right\}^{1/2}.\qquad(2.99)$$

We may define the phase

$$
\phi_l(r_1, r) = \int_{r_1}^{r} \tilde{Q}_l(r')\, dr' ,
$$
$$
\phi_l(r, r_2) = \int_{r}^{r_2} \tilde{Q}_l(r')\, dr' . \tag{2.100}
$$

Since the WKB wave function at r may be written in terms of either one of these phases, it follows that

$$
\cos(\phi_l(r_1, r) - \pi/4) = \pm \cos\left(\phi_l(r, r_2) - \pi/4\right) . \tag{2.101}
$$

This implies the quantization condition [22]

$$
\phi_l(r_1, r_2) = \int_{r_1}^{r_2} \tilde{Q}_l(r')\, dr' = \left(n_r + \frac{1}{2}\right)\pi , \quad n_r = 0, 1, 2, \ldots \tag{2.102}
$$

Here n_r is the radial quantum number. Note that this is exactly of the same form as Eq. (2.77) in one dimension. Let us define, for later reference, the action variable by the relation

$$
I_r = \frac{1}{2\pi} S_r = \frac{1}{2\pi} \oint p_r dr . \tag{2.103}
$$

The quantization condition may then be written in the generalized form:

$$
I_r = \left(n_r + \frac{\mu}{4}\right)\hbar , \tag{2.104}
$$

where μ is called the Maslov index (see, e.g., Ref. [23]). In the example here, μ just counts the number of classical turning points, where the amplitude of the wave function diverges. Note that in this case the WKB wave function over a complete cycle acquires a factor $\pm 2\pi(n_r + \mu/4)$. In case the particle encounters a hard wall b times, the wave function goes to zero under Dirichlet boundary condition at every encounter with the wall, and picks up an extra phase of $b\pi$ for b reflections. The acquired phase may thus be written as $2\pi(n_r + \mu/4 + b/2)$. Under Neumann boundary condition, on the other hand, there is no change in the phase of the wave at the wall, and $b = 0$. Thus the quantization condition (2.104) now gets modified to

$$
I_r = \left(n_r + \frac{\mu}{4} + \frac{b}{2}\right)\hbar . \tag{2.105}
$$

We shall apply this condition below to find the eigenvalues of the circular disc. In Chapter 5, when non-integrable motion will be discussed, such quantization rules are not applicable. The Maslov index σ introduced by Gutzwiller also contains a number of conjugate points, called μ, but it contains additional contributions (cf. Sect. 5.1 and Appendix D).

For the corresponding three-dimensional problem, we have $l(l+1)$ instead of $(l^2 - 1/4)$ in Eq. (2.91) for Q_l, and the exact regular solution of Eq. (2.90) is proportional to $J_{(l+1/2)}(kr)$ for the square-well example. It is easy to verify, comparing the phase of this for large r with the model WKB phase as before, that for the correct behavior we should now replace

$$l(l+1) \rightarrow \left(l+\frac{1}{2}\right)^2, \quad (3 \text{ dimensions}). \tag{2.106}$$

This prescription, that works even for $l = 0$ in the three dimensional case, is originally due to Langer [24]. Langer realized that a careful treatment of WKB approximation for the radial motion must take into account the singular behavior of the centrifugal term at $r = 0$. Strictly speaking, this term dominates $Q_l^2(r)$ in Eq. (2.91), and the criterion (2.83) is no longer satisfied. For a rigorous discussion of this problem, the reader should consult the review article of Berry and Mount [25]. Langer made use of a transformation of the type $r = r_0 \ln x$, where r_0 has the dimension of length, and may be taken to be unity. The point $r = 0$ is now mapped on to $x = -\infty$, just as in one dimension. As already stated, this gave the same result as (2.106).

Berry and Ozorio de Almeida pointed out that Langer's method still fails for the $l = 0$ wave function in two dimensions. This case is very peculiar, since there is a centripetal potential instead of the centrifugal one, and the turning point r_1 for an attractive $V(r)$ is at $r = 0$. Using a formulation due to Miller and Good [26], they applied a further mapping to Langer's equation [22]. The important result was proved that the quantization condition given by Eq. (2.102) still holds for $l = 0$, with $r_1 = 0$. The WKB wave function in this case was shown to be given by

$$u_{l=0}(r) \simeq \left(\frac{\phi_1(0,r)}{k(r)}\right)^{1/2} J_0\left(\phi_1(0,r)\right), \tag{2.107}$$

where the wave number $k(r)$ is given by Eq. (2.92) with V_0 replaced by $V(r)$.

2.5 Torus quantization: from WKB to EBK

In the one-dimensional WKB problem of the preceding section, note from Eq. (2.63) that the action $S(x, x_1, E)$ (or $S(x_2, x, E)$) and its derivative S' (with respect to x) are double-valued functions of x. The two branches of S have the same value at the turning points $x = x_1$ and $x = x_2$. The consequence of this multiple-valuedness is directly reflected in the wave function $\psi(x)$ given by Eq. (2.67), showing two exponential terms of the form $\exp(\pm iS/\hbar)$. (From now on, we shall often refer to the action as $S(x, E)$, suppressing the starting point.)

There is an alternative way of expressing the same results, that is easily generalized to higher-dimensional integrable motion and that also yields the quantization condition given by Eq. (2.77). For the one-dimensional WKB, we draw the two branches of $p_x = S'(x, E)$ as a function of x in Fig. 2.9, showing that they join together at the turning points x_1 and x_2 to form a single closed curve. This plot is just the trajectory of the particle in the phase space, and every point on this curve has a unique value of the momentum p_x. On this curve, therefore, the semiclassical wave function is simply [see Eq. (2.36)]:

$$\psi(x, E) = \mathcal{A} \exp\left[iS(x, E)/\hbar\right] = \frac{1}{\sqrt{2\pi\hbar}} \left|\frac{\partial^2 S(x, E)}{\partial E \partial x}\right|^{1/2} \exp\left[iS(x, E)/\hbar\right]. \tag{2.108}$$

This wave function must be single-valued after a full cycle. Consider first the exponent containing the action S in the above equation. In one complete cycle, if the change in S is denoted by ΔS, then the phase change is $\Delta S/\hbar$. There are, additionally, extra phases introduced at each turning point through the amplitude factor $\mathcal{A}$. Note that $\mathcal{A}$ above is inversely proportional to the square root of the momentum, which itself changes sign at each turning point. A change in sign may be regarded as introducing a phase of π, since $\exp(i\pi) = -1$. So each turning point gives an additional phase of $-\pi/2$, and the single-valuedness of the wave function demands

$$\frac{1}{\hbar}\Delta S - 2\frac{\pi}{2} = 2n\pi\,, \tag{2.109}$$

where n is an integer. This is the same as Eq. (2.77) for one-dimensional motion, since $\Delta S = 2\int_{x_1}^{x_2} p(x)dx$ is the phase picked up over the complete period. We may write this as

$$\oint p\,dq = \left(n + \frac{\mu}{4}\right)2\pi\hbar\,, \qquad n = 0, 1, 2, \dots\,, \tag{2.110}$$

where $\mu = 2$ is the number of turning or "conjugate points" (a term borrowed from optics), at which the intensity $\mathcal{A}^2$ diverges. Exactly the same result was obtained in Eq. (2.104) for the radial integral. We may now generalize the above considerations to the multi-dimensional integrable problem.

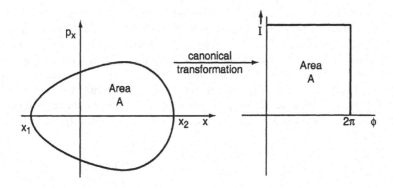

Figure 2.9: A plot of the two branches of momentum p_x in phase space. The two branches merge continuously at the turning points, producing a covering space that is topologically equivalent to a circle. After a canonical transformation that preserves the area in the phase space, an equivalent (I, ϕ) plot in action angle variables is also shown.

To this end, it is convenient to go over to the "action-angle" variables. We first do this for one-dimensional motion. Note from Eq. (2.16) that, in principle, we may canonically transform from the variables $\{p, x\} \to \{I, \phi\}$, where $\{\phi\}$ is a cyclic coordinate such that the Hamiltonian is only a function of I. This may be done by defining, for a given E,

$$\oint I\,d\phi = 2\pi I = \oint p\,dx\,, \tag{2.111}$$

so that

$$I = \frac{1}{2\pi} \, 2 \int_{x_1}^{x_2} \sqrt{2m \ (E - V(x'))} \, dx' . \tag{2.112}$$

By performing the integral on the right-hand side and inverting the equation, the energy E may be expressed in principle as a function of I only. This means that the Hamiltonian is expressible as $H(I)$. Since H is a constant of motion, so is I. Denoting the canonically conjugate variable to I by ϕ, we see from Hamilton's equation that the angular velocity

$$\omega = \dot{\phi} = \frac{\partial H}{\partial I} \tag{2.113}$$

is a constant in time. The time development of ϕ is given by

$$\phi(t) = \phi(0) + \omega t . \tag{2.114}$$

For one-dimensional motion, we may thus map the closed trajectory in the (x, p_x) phase space to a constant $I(E)$ versus ϕ graph, as shown in Fig. 2.9. The latter could alternately be drawn as a circle with radius $I(E)$ in polar coordinates. The quantization condition for the allowed energy levels, Eq. (2.110), is rewritten in terms of the action $I(E)$ as

$$I(E_n) = \left(n + \frac{\mu}{4} \right) \hbar , \quad n = 0, 1, 2, \ldots . \tag{2.115}$$

In the following, we shall use the Hamiltonian formalism to generalize these concepts for multi-dimensional motion. For this purpose, it is desirable to switch back to the standard notation (q, p), rather than (x, p_x). Throughout this book, we shall be using both notations interchangeably. Restricting still to one-dimensional motion, the angle variable ϕ may be expressed as a function of (q, I) by exploiting the property of the invariance of the phase space area in a canonical transformation, as shown in Fig. 2.9. In the (q, p) space, consider the area between two trajectories with energy E and $(E + \delta E)$. Since $E(I)$ is a single-valued function of I, each closed trajectory in the phase space may also be labeled by a unique I. The area in the (p, q) space is thus expressed as

$$\begin{aligned} \delta A &= \int_0^q \left[p(q', I + \delta I) - p(q', I) \right] dq' \\ &= \delta I \int_0^q \frac{\partial}{\partial I} p(q', I) \, dq' + O(\delta I^2) . \end{aligned}$$

In the (I, ϕ) plane, the same area is given by $\delta A = \delta I \, \phi(q, I)$. Equating the two expressions for δA, we obtain

$$\phi(q, I) = \int_0^q \frac{\partial}{\partial I} p(q', I) \, dq' . \tag{2.116}$$

One last point in one-dimensional motion need be emphasized before the generalization to N-dimensional integrable dynamics. (Note: we do *not* use curly brackets $\{..\}$ here to denote the Poisson brackets.) Let us denote a vector ξ with two components $\{q, p\}$:

$$\xi = \{\xi_1, \xi_2\} = \{q, p\} . \tag{2.117}$$

The phase space trajectory in the ξ-space for $H(p,q) = E$ is shown by a curve in Fig. 2.10. A tangent vector at any point on this curve is by construction normal to $\nabla_\xi H$. Note, however, that the velocity vector $\dot\xi$ with components $\{\dot q, \dot p\}$ is also perpendicular to $\nabla_\xi H$ by virtue of Hamilton's equations of motion Eq. (2.12):

$$\dot\xi \cdot (\nabla_\xi H) = \dot q \frac{\partial H}{\partial q} + \dot p \frac{\partial H}{\partial p} = 0 \, .$$

Thus the tangent vector at any point on the phase space trajectory must be in the direction of the flow-velocity vector $\dot\xi$, whose components are $\{\partial H/\partial p, \ -\partial H/\partial q\}$.

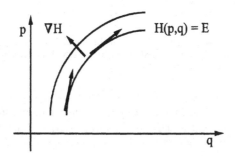

Figure 2.10: A tangent at any point on the trajectory with constant energy gives the flow vector in the phase space. The vector ∇H is perpendicular to the tangent.

Now consider a $2N$-dimensional phase space corresponding to N-dimensional (bounded) motion. For the integrable case, there are N independent constants of motion A_i ($i = 1, 2, \ldots, N$) obeying the pairwise Poisson-bracket relations given by Eq. (2.14). The trajectory in the phase space is thus confined to a N-dimensional "manifold". The term manifold is used to denote here a N-dimensional continuous surface of space embedded in the $2N$-dimensional phase space, defined by N equations $A_i = $ constant. We wish to know the topology of this manifold. Note that we may define in this manifold N independent "velocity vectors" ξ_i with components $\{\nabla_p A_i, -\nabla_q A_i\}$. (This is analogous to the velocity vector in the one-dimensional case with components $\partial H/\partial p$ and $-\partial H/\partial q$.) Moreover, these velocity vectors may be interpreted geometrically as tangential to the manifold, since they are orthogonal to the normals (with components $\{\nabla_q A_i, \nabla_p A_i\}$) by virtue of the Poisson bracket relations Eq. (2.14). There is a theorem in topology (the Poincaré-Hopf theorem) that states that if in a N-dimensional manifold one may construct N independent (and commuting) vector fields that are tangent to the manifold, then the manifold has the structure of an N-torus. It follows then that the manifold in phase-space that we are considering is an N-torus. For 2-dimensional motion, then, the topology is that of a doughnut, as illustrated in Fig. 2.11. A solvable example of this will be given below. At this point, we only need to recall that in an N-torus, there are N topologically independent closed curves, with N independent branches of the action S. A convenient basis is the choice of N closed curves, each wound around the angle variable ϕ_j. Any arbitrary closed curve in this topology may be expressed as a linear combination of these N windings in this basis.

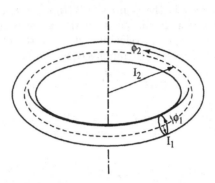

Figure 2.11: In two dimensions, integrable motion is confined within a two-dimensional manifold, called a two-torus, which is embedded in the four-dimensional phase space.

We now generalize the wave function $\psi(q, E)$ of the one-dimensional problem, given by Eq. (2.108), by expressing it in terms of the variables $(\mathbf{q}, \mathbf{I})$. The vector $\mathbf{I}$ denotes the N independent constants of motion $(I_1, I_2, ...I_N)$, and the action S is in the variables $(\mathbf{q}, \mathbf{I})$:

$$\psi(\mathbf{q}, \mathbf{I}) = (2\pi\hbar)^{-N/2} \left(\left| \det \frac{\partial^2 S}{\partial I_i \partial q_j} \right| \right)^{1/2} \exp\left[iS(\mathbf{q}, \mathbf{I})/\hbar \right] . \qquad (2.118)$$

Single-valuedness of the wave function then demands the generalization of Eq. (2.110) to yield N independent quantization conditions. This method, was applied by Keller and Rubinow [27] in some examples for the two- and three-torus topology. In the literature, this method is often called the EBK quantization.

Identifying the N independent actions I_j about the N closed curves in the torus is a nontrivial problem. The energy E of the system may then be expressed as $E = H(\mathbf{I})$, where $\mathbf{I}$ is a N-dimensional vector. The N conjugate angle variables ϕ_j do not come into the Hamiltonian. Take, for example, two-dimensional motion, with the two-torus manifold in the phase-space shown in Fig. 2.11. The quantization conditions are now given by

$$I_1 = \frac{1}{2\pi} \oint_{C_1} \mathbf{p} \cdot d\mathbf{q} = \left(n_1 + \frac{\mu_1}{4} \right) \hbar , \qquad I_2 = \frac{1}{2\pi} \oint_{C_2} \mathbf{p} \cdot d\mathbf{q} = \left(n_2 + \frac{\mu_2}{4} \right) \hbar . \qquad (2.119)$$

The quantized energy E_{n_1,n_2} is specified by two quantum numbers $\{n_1, n_2\}$, and is not separable in general in these quantum numbers. In the above, we assume for simplicity that the quantum numbers $\{n_1, n_2\}$ do not take negative values, although for rotational motion (angular momentum) this has to be corrected. The generalization of Eq. (2.119) from two- to N-dimensional motion is obvious. The appearance of the Maslov indices μ_i in trace formulae for integrable systems is discussed in Sects. 2.7 and 6.1.4 [see, in particular, Eq. (6.33)].

2.6 Examples

2.6.1 The two-dimensional hydrogen atom

The Coulomb problem that we consider below to illustrate the method of action-angle variables is superintegrable and chosen for its algebraic simplicity. The quantum spectrum in such a case can be found algebraically by using the extra symmetry of the Hamiltonian and group theoretical techniques (see Problem 2.1). Harmonic oscillator potentials in N dimensions are also superintegrable, having $SU(N)$ symmetry. Our approach here will be that used in the old quantum theory of Bohr and Sommerfeld. We shall also point out that the simple rules of the old theory are in this case consistent with the WKB rules based on quantum mechanics.

Consider the motion of an electron in the Coulomb field of an infinitely heavy nucleus, with the motion confined in the equatorial plane. To keep the discussion more general, first consider a central potential $V(r)$, and then specialize to the Coulomb case. The Lagrangian for the electron in the polar coordinates (r, ϕ) is given by

$$\mathcal{L} = \frac{1}{2} m \left(\dot{r}^2 + r^2 \dot{\phi}^2 \right) - V(r) \,. \tag{2.120}$$

Here the polar angle ϕ is a cyclic coordinate, which means that the Lagrangian does not depend on it explicitly. The canonical momenta corresponding to the coordinates r and ϕ are given by

$$p_r = \frac{\partial \mathcal{L}}{\partial \dot{r}} = m \dot{r} \,, \quad p_\phi = \frac{\partial \mathcal{L}}{\partial \dot{\phi}} = m r^2 \dot{\phi} \,. \tag{2.121}$$

Here p_r is the momentum for the radial motion, and p_ϕ is the angular momentum about an axis perpendicular to the plane of the motion. Since the equation of motion about the ϕ-axis is

$$\frac{d}{dt} \left(\frac{\partial \mathcal{L}}{\partial \dot{\phi}} \right) = 0 \,, \tag{2.122}$$

it follows that p_ϕ is a constant of motion. This is not the case for the radial momentum p_r, since r is not a cyclic variable. The Hamiltonian $H = (p_r \dot{r} + p_\phi \dot{\phi} - \mathcal{L})$ is given by

$$H = \frac{1}{2m} \left(p_r^2 + \frac{p_\phi^2}{r^2} \right) + V(r) \,. \tag{2.123}$$

There are two constants of motion in involution for this system, namely the energy $H = E$, and the angular momentum $p_\phi = L_z$. (For the additional constant of motion in this superintegrable example, see Problem 2.1). We may write, for given E and L_z (where the latter may have both signs),

$$p_r = \pm \sqrt{\left[2m \left(E - V(r) \right) - \frac{L_z^2}{r^2} \right]} \,. \tag{2.124}$$

From this expression, it is clear that for the radial motion there is an effective potential given by

$$V_{eff}(r) = V(r) + \frac{L_z^2}{2mr^2}, \tag{2.125}$$

and the radial momentum becomes

$$p_r = \pm\sqrt{2m(E - V_{eff})}. \tag{2.126}$$

For classical motion, $E \geq V_{eff}$ and p_r is necessarily real. The effective potential is plotted schematically in Fig. 2.8. For $E < 0$, it shows that the radial motion is confined within the classical "turning points" r_1 and r_2, defined by $V_{eff}(r_i) = E\ (i = 1, 2)$. For $E > 0$, the radial motion is unbound, and will not be discussed here. The radial equation of motion

$$\dot{p}_r = -\frac{d}{dr}V_{eff}(r) \tag{2.127}$$

is one-dimensional and may be integrated (numerically for any potential, but analytically in this case, see Problem 2.2) to yield a periodic solution. Since all one-dimensional motion is periodic, the radial solution alone repeats itself for a given energy between the turning points r_1 and r_2 (see Fig. 2.8). The periodicity in radial motion alone does not in general mean that the two-dimensional motion is also periodic. As we have noted before, the phase space of a system with two degrees of freedom is four-dimensional. But in the problem with a central potential, the energy E and the angular momentum p_ϕ are constants of motion and it is convenient to analyze the problem in terms of the action variables.

Since p_ϕ is a constant of motion, the action over a complete cycle is

$$S_\phi = \oint p_\phi\, d\phi = 2\pi L_z. \tag{2.128}$$

Using old quantum theory, this may be quantized by the rule:

$$I_\phi = \frac{1}{2\pi}S_\phi = L_z = l\hbar, \quad l = 0, \pm 1, \pm 2, \ldots. \tag{2.129}$$

This is also consistent with the EBK Eq. (2.119), since there is no turning point, and $\mu = 0$.

For radial motion, the action over the complete (radial) period for the attractive Coulomb potential $V(r) = -e^2/r$ and $E = -|E|$ is

$$\begin{aligned} S_r = 2\int_{r_1}^{r_2} p_r\, dr &= 2\int_{r_1}^{r_2}\sqrt{\left[2m\left(E + \frac{e^2}{r}\right) - \frac{L_z^2}{r^2}\right]}\, dr \\ &= -2\pi\left[|L_z| - \frac{me^2}{\sqrt{-2mE}}\right]. \end{aligned} \tag{2.130}$$

The Coulomb potential is special in the sense that not only can the radial integral above be performed analytically, but that Eq. (2.130) may be inverted to express E in terms of S_r and S_ϕ. (This is also true for the harmonic oscillator potential.) Combining Eqs. (2.128) and (2.130), a little algebra yields

$$E = -\frac{2\pi^2 m e^4}{(S_r + |S_\phi|)^2}. \tag{2.131}$$

We have noted in Sect 2.4.2 that WKB quantization should be done by replacing $(l^2 - 1/4)$ by l^2 for all l [see Eq. (2.97)]. This is quite consistent with the classical Bohr-Sommerfeld quantization we are doing here. The quantization condition (2.102) is the same as

$$I_r = \frac{1}{2\pi} S_r = \left[\frac{m e^2}{\sqrt{-2mE}} - |L_z|\right] = \left(n_r + \frac{1}{2}\right)\hbar, \quad n_r = 0, 1, 2, \ldots. \tag{2.132}$$

The quantized energy of the atom follows from these relations. For $L_z = 0$, special care should be taken, because the turning point at $r = 0$ coincides with the singularity of the equation. This will be taken up for the three dimensional case that we discuss next. Since the problem is separable in the radial and angular coordinates, the associated periods T_r and T_ϕ, corresponding to the two independent modes may be easily obtained. Given the form of p_r in Eq. (2.126), it follows that

$$T_r = \sqrt{2m} \int_{r_1}^{r_2} \frac{dr}{\sqrt{(E - V_{eff}(r))}}. \tag{2.133}$$

Using the definition of S_r above, we may then write $T_r = (\partial S_r/\partial E)$. Similarly, the period for the angular motion, T_ϕ, is given by $(\partial S_\phi/\partial E)$, where the action S_ϕ is given by Eq. (2.128). For separable motion, we may write in general

$$T_i = \frac{\partial S_i}{\partial E}, \quad \omega_i = \dot\phi_i = \frac{\partial H}{\partial I_i}. \tag{2.134}$$

In the above, $\omega_i = (2\pi/T_i)$, and (ϕ_i, I_i) are the standard action-angle variables with $I_i = S_i/(2\pi)$. Note that the Hamiltonian H is expressible in terms of S_i or I_i as in Eq. (2.131), and the angles ϕ_i are therefore all cyclic. The time-dependence of ϕ_i is obtained immediately by integrating Eq. (2.134) and given by $\phi_i(t) = \phi_i(0) + \omega_i t$. The motion may be conveniently depicted in the action-angle variables in the form of a torus. As shown in Fig. 2.11, I_1 and I_2 are the two radii of the torus, going around angles ϕ_1 and ϕ_2. The motion of the particle may be pictured as a combination of the two angular motions. Periodic motion in two-dimensional space is given by the condition that

$$\frac{\omega_r}{\omega_\phi} = \frac{T_\phi}{T_r} = \frac{n_1}{n_2}. \tag{2.135}$$

where n_1 and n_2 are positive integers. From Eqs. (2.131) and (2.134), we note that the above equation is always satisfied for the Coulomb potential at any energy. For a periodic orbit, the winding on the torus closes on itself, and retraces the trajectory. For other potentials, the nonperiodic orbits give rise to trajectories that continue to wind indefinitely on the torus without retracing the same path. In such a case, the ratio of the period for angular motion and the radial motion is an irrational number, and the motion is called "quasi-periodic". For periodic orbits, however, as the energy E is varied, a continuous family of trajectories on the tori are generated for the (integrable) motion under discussion.

Separability of the variables (r, ϕ) is ensured in a central potential $V(r)$, as shown above. However, separable equations also result if the two-dimensional potential is of the form [28]

$$V(r, \phi) = \tilde{U}(r) + \frac{U(\phi)}{r^2}, \qquad (2.136)$$

so that the motion remains integrable (see Problem 2.3).

2.6.2 The three-dimensional hydrogen atom

The action-angle formulation given above for a central potential may be generalized in a straightforward manner to higher dimensions. As an illustration, we consider the case of the hydrogen atom in three dimensions, using spherical polar coordinates (r, θ, ϕ). The Lagrangian is now given by

$$\mathcal{L} = \frac{1}{2} m \left(\dot{r}^2 + r^2 \dot{\theta}^2 + r^2 \sin^2\theta \, \dot{\phi}^2 + \frac{e^2}{r} \right). \qquad (2.137)$$

The canonical momenta are

$$p_r = \frac{\partial \mathcal{L}}{\partial \dot{r}} = m\dot{r}, \quad p_\theta = \frac{\partial \mathcal{L}}{\partial \dot{\theta}} = m r^2 \dot{\theta}, \quad p_\phi = \frac{\partial \mathcal{L}}{\partial \dot{\phi}} = m r^2 \sin^2\theta \, \dot{\phi}. \qquad (2.138)$$

The corresponding Hamiltonian is

$$H = \frac{1}{2m} \left(p_r^2 + \frac{p_\theta^2}{r^2} + \frac{p_\phi^2}{r^2 \sin^2\theta} \right) - \frac{e^2}{r}. \qquad (2.139)$$

Again, ϕ is a cyclic coordinate, so $p_\phi = L_z$ is a constant of motion. Since this again is an integrable problem in three dimensions, there should at least be three constants of motion in involution. Two of these are the energy E and the z-component of the angular momentum L_z. One can show from the equations of motion that a third constant of motion is given by

$$p_\theta^2 + \frac{p_\phi^2}{\sin^2\theta} = L^2. \qquad (2.140)$$

For given E, L^2, and L_z, we can calculate the respective actions by integrating over the r, θ, and ϕ variables. As before, for $E < 0$, only periodic orbits are obtained for the Coulomb problem, and the integrations can be done analytically. For the radial variable, we have

$$S_r = \oint p_r \, dr = 2\sqrt{2m} \int_{r_1}^{r_2} \left[E - \frac{L^2}{2mr^2} + \frac{e^2}{r} \right]^{\frac{1}{2}} dr, \qquad (2.141)$$

where r_1 and r_2 are the turning points of the radial motion, at which p_r is zero. The above integral gives

$$S_r = -2\pi \left[|L| - \frac{me^2}{\sqrt{-2mE}} \right] . \tag{2.142}$$

Similarly,

$$S_\theta = 2 \int_{\theta_1}^{\theta_2} p_\theta \, d\theta = 2 \int_{\theta_1}^{\theta_2} \left[L^2 - \frac{L_z^2}{\sin^2\theta} \right]^{\frac{1}{2}} d\theta , \tag{2.143}$$

where the angles θ_1 and θ_2 are the angular turning points which obey the relation

$$\sin^2\theta_{1,2} = \frac{L_z^2}{L^2} \tag{2.144}$$

requiring that $-L \le L_z \le L$. From Eq. (2.143), we obtain

$$S_\theta = 2\pi \left(|L| - |L_z| \right) . \tag{2.145}$$

As before, for the cyclic angle ϕ, we have

$$S_\phi = \oint p_\phi \, d\phi = 2\pi \, L_z . \tag{2.146}$$

Using the above equations, the energy E may now be expressed in terms of the actions of the periodic orbits:

$$E = -\frac{2\pi^2 \, m \, e^4}{(S_r + S_\theta + |S_\phi|)^2} . \tag{2.147}$$

The expression for E given by Eq. (2.147) is particularly simple, and shows that the individual periods for separable motion, T_r, T_θ, and T_ϕ, are all equal. More generally, for a periodic orbit in a given potential, their ratios are rational fractions, so that a closed curve is traced in the configuration space over a common period T. Over this time interval, the radial motion may execute k_r complete cycles, the θ motion k_θ cycles, and the ϕ motion k_ϕ cycles, where the k_i are integers. The total action S, computed over the period T, is given by

$$S = \left(k_r \, S_r + k_\theta \, S_\theta + k_\phi \, |S_\phi| \right) . \tag{2.148}$$

Quantization of the three actions separately gives the result obtained by Sommerfeld [29], generalizing the circular Bohr orbits to elliptical orbits (see Problem 2.2). In the modern version of WKB quantization, we apply Eq. (2.104):

$$I_r = \frac{1}{2\pi} \oint p_r \, dr = \frac{1}{2\pi} 2\sqrt{2m} \int_{r_1}^{r_2} \left[E - \frac{(l+1/2)^2\hbar^2}{2mr^2} + \frac{e^2}{r} \right]^{\frac{1}{2}} dr = \left(n_r + \frac{1}{2} \right) \hbar , \ n_r = 0, 1, 2 \ldots \tag{2.149}$$

The replacement of $L^2 \to (l+1/2)^2\hbar^2$, instead of by $l(l+1)\hbar^2$, is due to Langer [24]. It has already been discussed after Eq. (2.106), and is applicable for all l. We thus get, from Eq. (2.149),

$$I_r = \left[\frac{me^2}{\sqrt{-2mE}} - \left(l + \frac{1}{2} \right) \hbar \right] = \left(n_r + \frac{1}{2} \right) \hbar . \tag{2.150}$$

Simplifying, we obtain the exact Rydberg spectrum of the hydrogen atom in terms of the principal quantum number n:

$$E_n = -\frac{1}{2}\frac{me^4}{\hbar^2}\frac{1}{n^2} \quad \text{with} \quad n = (n_r + l + 1) = 1, 2, 3, \ldots . \quad (2.151)$$

2.6.3   The two-dimensional disk billiard

We shall now give an example of an integrable system in which the WKB quantization does not give an analytical expression for the energy and, furthermore, does not reproduce exactly the quantum-mechanical energy spectrum. This example shall be coming up again in this book, when we discuss different schemes of quantization.

Consider a particle confined to a two-dimensional circular domain of radius R with infinitely steep walls. Classically, the particle is reflected at the boundaries of the circular billiard. We apply the quantization condition given by Eq. (2.105) for radial WKB. Note that $V(r) = 0$ inside the billiard, so using the definition (2.99) for p_r, we may write

$$S_r = 2\int_{r_1}^R p_r dr = 2\hbar \int_{r_1}^R \left[k^2 - \frac{l^2}{r^2}\right]^{1/2} dr . \quad (2.152)$$

Here r_1 is the inner, and R the outer turning points for radial motion, and $k^2 = 2mE/\hbar^2$. Using Eq. (2.96) for the relevant integral, and the quantization condition (2.105), a little algebra yields

$$(k^2R^2 - l^2)^{1/2} - l\cos^{-1}\left(\frac{l}{kR}\right) = \left(n_r + \frac{\mu}{4} + \frac{b}{2}\right)\hbar , \quad n_r = 0, 1, 2, \ldots \quad (2.153)$$

For the problem under consideration, $\mu = 1$ and $b = 1$, there being one smooth and one hard turning point (Dirichlet boundary condition). Therefore

$$(k^2R^2 - l^2)^{1/2} - l\cos^{-1}\left(\frac{l}{kR}\right) = \left(n_r + \frac{3}{4}\right)\hbar , \quad n_r = 0, 1, 2, \ldots \quad (2.154)$$

The factor 3/4 should be replaced by 1/4 for the Neumann boundary condition. These results check with the formulae given by Keller and Rubinow [27]. Although this system is integrable, it is not possible to give a closed form for the energy $E(S_r, S_\phi)$ as in the harmonic-oscillator and Coulomb cases. However, the quantization of the radial action gives an implicit quantum condition for the energy levels $E(l, n_r)$ of the system.

Quantum mechanically, the eigenfunctions of this problem are given by the cylindrical Bessel functions $J_l(kr)$, where $\hbar l$ is an eigenvalue of the $\hat{L}_z$ operator, and $k = \sqrt{2mE}/\hbar$ is the wave number. The condition that the wavefunctions be zero at the boundary, $J_l(kR) = 0$, gives quantized energies of the form

$$E_{l,n} = \left(\frac{\hbar^2}{2mR^2}\right)x_{ln}^2 , \quad l = 0, \pm 1, \pm 2, \ldots \quad n = 1, 2, 3, \ldots , \quad (2.155)$$

where x_{ln} is the n-th zero of $J_l(x)$. Unlike in the Coulomb (and harmonic-oscillator) potentials, the spectrum obtained here from the semiclassical quantization via Eq. (2.154) is not exact, i.e., it does not reproduce the quantum-mechanical values (2.155). This is actually the rule, and the Coulomb and harmonic-oscillator cases must be considered as exceptions due to their special symmetries. We give in Table 2.1 a list of the lowest eigenvalues, obtained both exactly and by WKB quantization.

$|l|$	n	x_{ln} (exact)	x_{ln} (WKB)
0	1	2.40483	2.35619
	2	5.52008	5.49779
	3	8.65373	8.63938
	4	11.79153	11.78097
1	1	3.83171	3.79444
	2	7.01559	6.99700
	3	10.17347	10.16093
2	1	5.13562	5.10039
	2	8.41724	8.40014
	3	11.61984	11.60825
3	1	6.38016	6.34519
	2	9.76102	9.74463

Table 2.1: Roots x_{ln} of the energy eigenvalues of a two-dimensional disk billiard (in units of $\hbar^2/2mR^2$), calculated both quantum mechanically ("exact") and by semiclassical quantization ("WKB").

2.7 Connection to classical periodic orbits

We have in Chapter 1 repeatedly pointed at the close connection between semiclassical quantization and the periodic orbits of the classical system. The quantitative study of this connection is the object of the periodic orbit theory which we present in Chapters 5 and 6. For integrable systems, Berry and Tabor [30, 31] have shown that it is always possible to derive a "trace formula" of the form of Eq. (1.109) for the level density $g(E)$, whose oscillating part $\delta g(E)$ is expressed in terms of classical periodic orbits. The starting point of the first derivation of their trace formula [30] was the EBK quantization which we have discussed in the previous sections. In a second derivation [31], they started from the semiclassical Green's function which we will discuss and use extensively in Chapter 5 in the context of Gutzwiller's trace formula for systems with isolated periodic orbits.

We shall not present here the general trace formula of Berry and Tabor which has been widely used. Our incentive here is to point out the essential ideas and then to demonstrate them for a very simple system, the rectangular billiard. For the general case, we refer the reader to the original papers. Comparisons of results obtained by Berry and Tabor's method with those of other trace formulae will be presented in Chapter 6 for various model examples.

One starts from the fact that for integrable systems the torus quantization rule (2.119) can be used to obtain an (approximate) eigenvalue spectrum. We consider a two-dimensional system whose energy spectrum is specified by two non-negative integer quantum numbers (n_1, n_2). Its density of states is given by

$$g\left(E\right) = \sum_{n_1=0}^{\infty} \sum_{n_2=0}^{\infty} \delta(E - E(n_1, n_2)) \,. \tag{2.156}$$

The semiclassical form (1.109) is obtained in two steps: (a) using the *Poisson summation formula* to express the above as a double integral, and then (b) evaluating the integrals by the *saddle-point method*. For a single sum, the Poisson formula reads [32]

$$\sum_{n=0}^{\infty} f(n) = \sum_{M=-\infty}^{\infty} \int_0^{\infty} f(n) \, \exp\left(2\pi i M n\right) dn + \frac{1}{2} f(0) \,, \tag{2.157}$$

where it is assumed that the function $f(n)$ and its derivatives with respect to n vanish at infinity. The last term, which corresponds to the boundary correction in the trapezoidal rule or – more generally – in the Euler-MacLaurin summation formula (cf. Sect. 4.6.1), is of no interest when one only wants to know the function $f(n)$ for arguments $n > 0$. Generalizing (2.157) to two quantum numbers, however, this term gives rise to nontrivial contributions. Applying Eq. (2.157) twice to the double sum in (2.156), we obtain

$$\begin{aligned}
g\left(E\right) \;=\; & \sum_{M_1=-\infty}^{\infty} \sum_{M_2=-\infty}^{\infty} \int_0^{\infty} dn_1 \int_0^{\infty} dn_2 \, \delta(E - E(n_1, n_2)) \, \exp\left[2\pi i(M_1 n_1 + M_2 n_2)\right] \\
+ \;& \frac{1}{2} \sum_{M=-\infty}^{\infty} \int_0^{\infty} dn_1 \, \delta(E - E(n_1, 0)) \, \exp\left(2\pi i M n_1\right) \\
+ \;& \frac{1}{2} \sum_{M=-\infty}^{\infty} \int_0^{\infty} dn_2 \, \delta(E - E(0, n_2)) \, \exp\left(2\pi i M n_2\right) + \frac{1}{4} \delta(E) \,. \tag{2.158}
\end{aligned}$$

The double sum over the two-dimensional array of points (M_1, M_2) is called the "topological sum" by Berry and Tabor, for a reason which will become clear soon. The single sums, arising from the boundary correction in (2.157), lead to semiclassical corrections of higher order in $\hbar$ and are usually neglected. In the application below we shall, however, consider them explicitly, and in Sect. 6.1.2 we will discuss a case where they are not negligible at all. The last term, a delta function peaked at $E = 0$, is again uninteresting when applied to spectra with energies $E > 0$. (It contributes to the first integral of the level density, however, as discussed more generally in Sect. 4.6.)

We shall first concentrate on the first term in Eq. (2.158) which we call $g^{(2)}(E)$. Using the relations (2.119), we may change the integration variables from (n_1, n_2) to the torus action variables (I_1, I_2) and write $E(n_1, n_2) = H(I_1, I_2)$ to get

$$
g^{(2)}(E) = \frac{1}{\hbar^2} \sum_{M_1=-\infty}^{\infty} \sum_{M_2=-\infty}^{\infty} \exp\left[-i(M_1\mu_1 + M_2\mu_2)\frac{\pi}{2}\right]
$$
$$
\times \int_{\hbar\frac{\mu_1}{4}}^{\infty} dI_1 \int_{\hbar\frac{\mu_2}{4}}^{\infty} dI_2 \, \delta(E - H(I_1, I_2)) \exp\left[\frac{2\pi i}{\hbar}(M_1 I_1 + M_2 I_2)\right]. \quad (2.159)
$$

The double sum is taken over a lattice of points with integer values of M_1, M_2. Consider now the particular term with $M_1 = M_2 = 0$. For this point, the oscillating terms in Eq. (2.159) are absent, giving a smooth density of states

$$
\tilde{g}^{(2)}(E) = \frac{1}{\hbar^2} \int_{\hbar\frac{\mu_1}{4}}^{\infty} dI_1 \int_{\hbar\frac{\mu_2}{4}}^{\infty} dI_2 \, \delta(E - H(I_1, I_2)) \,. \quad (2.160)
$$

Replacing the lower limits $\hbar\mu_i/4$ by zero, the double integral on the right-hand side above may be expressed as going over the total available phase space. Using Eq. (2.111), this gives

$$
\tilde{g}^{(2)}(E) = \frac{1}{(2\pi\hbar)^2} \int_0^\infty dI_1 \int_0^{2\pi} d\phi_1 \int_0^\infty dI_2 \int_0^{2\pi} d\phi_2 \, \delta(E - H(I_1, I_2)) \,.
$$
$$
= \frac{1}{(2\pi\hbar)^2} \int\int d\mathbf{p}\, d\mathbf{q}\, \delta(E - H(\mathbf{p}, \mathbf{q})) \,. \quad (2.161)
$$

This is just the Thomas-Fermi expression for the density of states, where the sum over the quantum states is replaced by an integral over the available phase space, divided by h^2 (in two dimensions). The neglected boundary contributions from the lower limits $\hbar\mu_i/4$ in (2.160) give corrections of higher order in $\hbar$ and combine with those coming from the single-sum terms in Eq. (2.158) for $M = 0$. These corrections to the Thomas-Fermi model will be studied systematically in Chapter 4 and are included in the smoothly varying density of states $\tilde{g}(E)$. We shall in the following leave out the lattice point $M_1 = M_2 = 0$ and thus deal with the oscillating part $\delta g(E)$ of the density of states.

To proceed further, Berry and Tabor [30] get rid of the delta function in Eq. (2.159) by performing one integration exactly. This is done quite generally in the multi-dimensional action space (I_i) by introducing curvilinear coordinates on the "energy shell" defined by $E = H(I_1, I_2, \ldots)$. At this point, we go a slightly different route by using the Fourier representation of the delta function

$$
\delta(E - H(I_1, I_2)) = \frac{1}{2\pi\hbar} \int_{-\infty}^{\infty} \exp\left[i\,\tau(E - H(I_1, I_2))/\hbar\right] d\tau \,. \quad (2.162)
$$

For compactness, we also use the vector notation $\mathbf{M} \cdot \mathbf{I} = (M_1 I_1 + M_2 I_2)$ and $\mathbf{M} \cdot \boldsymbol{\mu} = (M_1\mu_1 + M_2\mu_2)$, obtaining

$$
\delta g^{(2)}(E) = \frac{1}{2\pi\hbar^2} {\sum_{M_1,M_2}}' \exp\left(-i\,\mathbf{M} \cdot \boldsymbol{\mu}\,\frac{\pi}{2}\right) \quad (2.163)
$$
$$
\times \frac{1}{\hbar} \int_{-\infty}^{\infty} d\tau \int_{\hbar\frac{\mu_1}{4}}^{\infty} dI_1 \int_{\hbar\frac{\mu_2}{4}}^{\infty} dI_2 \exp\left\{\frac{i}{\hbar}[2\pi\mathbf{M} \cdot \mathbf{I} + \tau(E - H(I_1, I_2))]\right\},
$$

where the prime on the sum sign indicates that $M_1 = M_2 = 0$ has to be left out. The next step is to perform the double integration over I_1, I_2 in the *stationary phase approximation*. The integrand is written as $\exp[i\Phi_{\mathbf{M}}(I_1, I_2, \tau)]$, where

$$\Phi_{\mathbf{M}}(I_1, I_2, \tau) = \frac{1}{\hbar}[2\pi\mathbf{M}\cdot\mathbf{I} + \tau(E - H(I_1, I_2))] . \qquad (2.164)$$

For each lattice point $\mathbf{M}$, the dominant contribution to the above double integral over I_1 and I_2 comes from the point in the (I_1, I_2) plane where the phase $\Phi_{\mathbf{M}}(I_1, I_2, \tau)$ is stationary. Taylor expanding about the saddle-point and retaining only up to the quadratic terms in the expansion, the complex Gaussian integrals may be performed analytically. The stationary point $(\tilde{I}_1, \tilde{I}_2)$ is such that (we drop the subscript 'M' of Φ)

$$\left.\frac{\partial\Phi}{\partial I_1}\right|_{\tilde{I}_1, \tilde{I}_2} = 0, \qquad \left.\frac{\partial\Phi}{\partial I_2}\right|_{\tilde{I}_1, \tilde{I}_2} = 0 . \qquad (2.165)$$

We immediately obtain

$$2\pi M_i = \tau\,\omega_i(\tilde{I}_1, \tilde{I}_2), \qquad (i = 1, 2) \qquad (2.166)$$

where we have used Eq. (2.113) for the angular velocity ω_i, with the derivatives taken at the saddle point $(\tilde{I}_1, \tilde{I}_2)$. Note that in general the solutions of Eq. (2.166) result in the actions $\tilde{I}_i$ being nonlinear functions of (M_1, M_2) and of τ:

$$\tilde{I}_i = \tilde{I}_i(M_1, M_2, \tau) . \qquad (i = 1, 2) \qquad (2.167)$$

Eq. (2.166) implies the important relation for the "resonant tori"

$$M_1 : M_2 = \omega_1 : \omega_2 . \qquad (2.168)$$

Since M_1 and M_2 are integers, Eq. (2.168) shows that in the saddle-point approximation, only those orbits on the torus with *commensurate frequencies* (ω_1, ω_2), i.e., only *closed periodic orbits* contribute to the density of states. This establishes the connection between $\delta g\,(E)$ and the periodic orbits, as one of the central results of Berry and Tabor in Ref. [30]. When the frequencies are not commensurate, the orbits do not close although the motion is still confined to the torus. Such orbits are termed multiply periodic or quasi-periodic. A periodic orbit is specified by (M_1, M_2); it closes after M_1 turns around the angle ϕ_1 and M_2 turns around ϕ_2. The topology of the periodic orbit is thus determined by the lattice indices $\mathbf{M}$. This is why the sum over (M_1, M_2) in Eq. (2.158) was called a "topological sum". In passing, it is to be noted that Eq. (2.168) also implies that the topological sum be restricted such that M_1, M_2 are both of the same sign.

The double integral over $dI_1\,dI_2$ in Eq. (2.163) may now be performed after pulling out the overall factor $\exp[i\Phi(\tilde{I}_1, \tilde{I}_2, \tau)]$ and substituting the variables as $I_i \to I_i' = (I_i - \tilde{I}_i)$. This transforms the lower limits in the relevant integrals to $\hbar\mu_i/4 - \tilde{I}_i$. For positive $\tilde{I}_i$ and large quantum numbers in Eq. (2.119), the lower limits may be replaced by $-\infty$; for negative $\tilde{I}_i$, the integrals will be exponentially small and may thus be neglected. We now complete the squares in the exponents and use for each of the I_i' integrations the following *Fresnel integral*:

$$\int_{-\infty}^{\infty}\exp\left(iay^2\right)dy = \sqrt{\frac{\pi}{a}}\,\exp\left(i\pi/4\right) = \sqrt{\frac{\pi}{|a|}}\,\exp\left[i\,\mathrm{sign}(a)\pi/4\right]. \qquad (2.169)$$

The constant a has been taken to be real here. If a lower limit $\hbar\mu_i/4 - \tilde{I}_i$ happens to be exactly zero, one has to take one half of this integral. If the stationary points lie close to the lower limits, the saddle-point approximation will in general be bad. After performing both integrations, the oscillating part $\delta g^{(2)}(E)$ of the density of states reduces (approximately) to

$$\delta g^{(2)}(E) \approx \frac{1}{2\hbar^2} \sum_{M_1,M_2}{}' \exp\left(-i\,\mathbf{M}\cdot\boldsymbol{\mu}\,\frac{\pi}{2}\right) \exp\left[-i\frac{\pi}{4}(N_+ - N_-)\right]$$
$$\times \int_{-\infty}^{\infty} \frac{1}{\tau|\mathcal{K}|^{1/2}} \exp\left\{\frac{i}{\hbar}\left[2\pi\mathbf{M}\cdot\tilde{\mathbf{I}} + \tau(E - H(\tilde{I}_1, \tilde{I}_2))\right]\right\} d\tau. \quad (2.170)$$

In the above, $\mathcal{K}$ is the following 2×2 determinant, which in general also depends on the energy E and on τ and is the combined result of the denominators under the root in Eq. (2.169) coming from the two integrations:

$$\mathcal{K}(E,\tau) = \det\left(\frac{\partial^2 H}{\partial I_i \partial I_j}\right)_{\tilde{I}_1,\tilde{I}_2} = \begin{vmatrix} \dfrac{\partial^2 H}{\partial I_1 \partial I_1} & \dfrac{\partial^2 H}{\partial I_1 \partial I_2} \\[2ex] \dfrac{\partial^2 H}{\partial I_2 \partial I_1} & \dfrac{\partial^2 H}{\partial I_2 \partial I_2} \end{vmatrix}_{\tilde{I}_1,\tilde{I}_2}. \quad (2.171)$$

In Eq. (2.170), N_+ and N_- are the numbers of positive and negative eigenvalues of $\mathcal{K}$, respectively. In order to obtain $\delta g\,(E)$ from Eq. (2.170), a final integration over τ has to be performed, again using the stationary phase method.[1] This step is not trivial since the $\tilde{I}_i$ by virtue of Eq. (2.167) also depend on τ. We shall not write out here the formal result. It will be given in Sect. 6.1.4 for the general D-dimensional case [see Eq. (6.31)]. In the following, we discuss an illustrative example for which all the steps outlined above can be analytically evaluated, including the neglected boundary contributions, and where we will find an exact trace formula for the level density.

The idea of Poisson summation and saddle-point integration for determining the oscillating part of the density of states has already been used in 1972 by Bogacheck and Gogadze [33] for a metallic cylinder in a weak longitudinal magnetic field. The authors mention neither the periodic orbit theory nor EBK quantization; the interest was in the flux quantization in normal metals and their magnetic properties. To get an approximate energy spectrum, they used the Debye expansion of the Bessel functions $J_l(x)$ (see Sect. 7.5), from whose zeros one finds the same eigenvalues (2.153) as from the EBK quantization. They found in this way the trace formula for the disk billiard discussed in Sect. 6.1.3, including a modulation factor due to weak magnetic fields, which is described in Sect. 6.2.3.

---

[1] Note that the energy conservation $E = H(I_1, I_2)$ must be *imposed* in this case, since it would in general only be approximately fulfilled when the τ integration coming from Eq.(2.162) is not done exactly.

2.7.1 Example: The two-dimensional rectangular billiard

Consider the motion of a particle in a two-dimensional rectangular billiard with sides of length a_1 and a_2 along the x and y axes. The quantum spectrum with Dirichlet boundary condition is given by

$$E_{n_1,n_2} = \frac{\hbar^2 \pi^2}{2m} \left(\frac{n_1^2}{a_1^2} + \frac{n_2^2}{a_2^2} \right), \qquad n_1, n_2 = 1, 2, \dots . \qquad (2.172)$$

Note that the above quantum spectrum is exactly reproduced by the EBK quantization rule (2.119). As we have seen in Sect. 2.4, the μ_i (including the contributions from b) in each degree of freedom equal 4 when a particle is reflected by infinitely steep walls on both sides. We will simplify the following by absorbing the $\mu_i = 4$ into the quantum numbers n_i which we therefore start counting from 1 instead of 0:

$$I_1 = \frac{1}{2\pi} \oint p_x \, dx = n_1 \hbar , \quad I_2 = \frac{1}{2\pi} \oint p_y \, dy = n_2 \hbar , \qquad n_1, n_2 = 1, 2, \dots . \qquad (2.173)$$

The classical Hamiltonian may be expressed as

$$H(I_1, I_2) = \frac{\pi^2}{2m} \left(\frac{I_1^2}{a_1^2} + \frac{I_2^2}{a_2^2} \right) . \qquad (2.174)$$

Since we sum the spectrum from $n_i=1$, we have to change the sign of the correction term in the Poisson formula (2.157) and the corresponding single sums in (2.158). The integrations over the I_i then go from zero to infinity, and the leading term of the level density becomes exactly

$$g^{(2)}(E) = \frac{1}{2\pi \hbar^2} \sum_{M_1, M_2 = -\infty}^{\infty} \frac{1}{\hbar} \int_{-\infty}^{\infty} d\tau \int_0^{\infty} dI_1 \int_0^{\infty} dI_2 \, \exp[i \Phi_{\mathbf{M}}(I_1, I_2, \tau)] . \qquad (2.175)$$

Since $H(I_1, I_2)$ is a quadratic expression in the I_i, the integrals over I_i in (2.175) can be performed exactly. Completing the squares in the exponent of (2.175), the shifted I_i' integrations go from the lower limits $-\tilde{I}_i$, where

$$\tilde{I}_i = \left(\frac{2m}{\tau \pi} \right) a_i^2 M_i , \qquad (i = 1, 2) \qquad (2.176)$$

correspond to the saddle points. The approximation discussed in the above treatment after Eq. (2.167), completing the lower limit to $-\infty$ when $\tilde{I}_i$ is positive and neglecting the integral when $\tilde{I}_i$ is negative, can in the present case seen to be exact. Since we have to sum over all M_i with both signs, we may use the simple relation

$$\int_{-cM_i}^{\infty} e^{-iaI_i'^2} \, dI_i' + \int_{+cM_i}^{\infty} e^{-iaI_i'^2} \, dI_i' = \int_{-\infty}^{\infty} e^{-iaI_i'^2} \, dI_i' = \sqrt{\frac{\pi}{a}} \, e^{-i\pi/4} , \qquad (2.177)$$

simply exploiting the fact that the integrand is even in I_i'. Thus, the two integrals coming from the opposite signs of each pair of nonzero M_i combine exactly to give one Fresnel integral (2.169). In the case where one of the M_i is zero, the corresponding integral gives one-half of a Fresnel integral (2.169). Altogether we can write the result of the I_i' integrals in the following compact form, which is exact:

$$g^{(2)}(E) = \frac{(-i)}{2\pi\hbar^2} \frac{ma_1a_2}{2\pi} \sum_{M_1,M_2=-\infty}^{\infty} \int_{-\infty}^{\infty} \frac{d\tau}{\tau} \exp\left[i\tau E/\hbar + \frac{i\hbar}{\tau}\frac{2m}{\hbar^2}(M_1^2a_1^2 + M_2^2a_2^2)\right].$$

(2.178)

Note that the point $M_1 = M_2 = 0$ is included here for convenience, so that $g^{(2)}(E)$ includes also the smooth Thomas-Fermi term.

The final τ-integration can also be done exactly by putting $i\tau/\hbar = \beta$ and noting that it then becomes an inverse Laplace transform (see Appendix B), yielding a cylindrical Bessel function of order zero:

$$\frac{1}{2\pi i} \int_{-i\infty}^{i\infty} \frac{d\beta}{\beta} \exp(\beta E - C/\beta) = \mathcal{L}_E^{-1}\left[\frac{1}{\beta}\exp(-C/\beta)\right] = J_0(2\sqrt{CE}),$$

(2.179)

where C is the coefficient of $i\hbar/\tau$ in the exponent of (2.178). Before writing down the final expression for $g^{(2)}(E)$, we now recognize that the argument of the Bessel function in (2.179) is just the classical action $S_{M_1M_2}$, corresponding to the rational torus given by the lattice point M_1, M_2, divided by $\hbar$

$$2\sqrt{CE} = \frac{\sqrt{2mE}}{\hbar}L_{M_1M_2} = \frac{p}{\hbar}L_{M_1M_2} = \frac{1}{\hbar}S_{M_1M_2},$$

(2.180)

where $L_{M_1M_2}$ is the length of the periodic orbit on this torus

$$L_{M_1M_2} = 2\sqrt{M_1^2a_1^2 + M_2^2a_2^2}.$$

(2.181)

We can thus write the leading part of the level density in the following exact form

$$g^{(2)}(E) = \frac{ma_1a_2}{2\pi\hbar^2} \sum_{M_1,M_2=-\infty}^{\infty} J_0\left(\frac{S_{M_1M_2}}{\hbar}\right),$$

(2.182)

where the topological sum over all lattice points **M** (except $M_1 = M_2 = 0$) corresponds exactly to the sum over all periodic orbits. Hereby, points on the axes (where one of the M_i is zero) occur only once: these are the straight-line bouncing orbits which are mapped onto themselves under time reversal. All other points give pairwise equal contributions, due to the time-reversal symmetry of the problem and the fact that all orbits with nonzero (M_1, M_2) change their orientation and thus are mapped onto topologically distinct orbits when time is reversed. As noted above, the smooth Thomas-Fermi part corresponding to $M_1 = M_2 = 0$ is included in (2.182).

When M_1 and M_2 are not relatively prime, one may write $(M_1, M_2) = n(m_1, m_2)$ with relatively prime m_1 and m_2. Then, (m_1, m_2) is usually referred to as a "primitive periodic orbit", and n is its repetition number. Evidently, the action is then just n times that of the primitive orbit.

Since the correct counting of the periodic orbits and their degeneracies is crucial, let us write out the double sum in Eq. (2.182) so that this becomes clear by separating the terms with $M_i = 0$ and restricting the remaining sums over positive values of the M_i:

$$g^{(2)}(E) = \frac{ma_1 a_2}{2\pi\hbar^2} \left\{ 1 + 2 \sum_{M_1=1}^{\infty} J_0\left(\frac{S_{M_1 0}}{\hbar}\right) + 2 \sum_{M_2=1}^{\infty} J_0\left(\frac{S_{0 M_2}}{\hbar}\right) + 4 \sum_{M_1,M_2=1}^{\infty} J_0\left(\frac{S_{M_1 M_2}}{\hbar}\right) \right\}.$$ (2.183)

If the τ integration were done by the stationary phase method, the Bessel function J_0 would be replaced by its asymptotic form:

$$J_0(x) \simeq \sqrt{\frac{2}{\pi x}} \cos\left(x - \frac{\pi}{4}\right).$$ (2.184)

In that approximation, the leading term in the trace formula for the rectangular billiard becomes

$$\delta g^{(2)}(E) \approx \frac{ma_1 a_2}{2\hbar^2 \pi^{3/2}} \sum_{PO} \sqrt{\frac{2\hbar}{S_{PO}}} \cos\left(\frac{1}{\hbar} S_{PO} - \frac{\pi}{4}\right),$$ (2.185)

where the sum over periodic orbits (PO) is to be taken as specified above. Equation (2.185) is just a special case of the general trace formula derived by Berry and Tabor [30], or that by Creagh and Littlejohn [34] for integrable systems, given in Eq. (6.31) of Chapter 6.

The boundary corrections, i.e., the single-sum terms in Eq. (2.158), can also be worked out analytically in this example. With the explicit form (2.172) of the energy, the integrations over the delta functions are performed in a straight-forward manner, leading to the corrections

$$-\frac{1}{2} g_i^{(1)}(E) = -\frac{a_i}{4\pi\hbar} \sqrt{\frac{2m}{E}} \sum_{M=-\infty}^{\infty} \cos\left(2M a_i \sqrt{2mE}/\hbar\right). \qquad (i = 1,2)$$ (2.186)

Hereby $g_i^{(1)}(E)$ represents the exact level density for the spectrum of a particle in a *one-dimensional* box of length a_i, which will be derived in Sect. 3.2.3 [cf. Eq. (3.63)]. Note that its amplitude is of order $\hbar^{1/2}$ with respect to that in Eq. (2.185), or alternately, of order $\hbar$ with respect to that in Eq. (2.182). (The counting of the powers of $\hbar$ in the amplitudes will be discussed in Sect. 6.1.4.)

Summing all contributions (2.182) and (2.186), we get the following exact trace formula for the two-dimensional rectangular billiard

$$\begin{aligned} g(E) &= \frac{ma_1 a_2}{2\pi\hbar^2} \sum_{M_1,M_2=-\infty}^{\infty} J_0\left(\frac{S_{M_1 M_2}}{\hbar}\right) \\ &\quad - \sum_{i=1,2} \frac{a_i}{4\pi\hbar} \sqrt{\frac{2m}{E}} \sum_{M=-\infty}^{\infty} \cos\left(2M a_i \sqrt{2mE}/\hbar\right). \end{aligned} \qquad (E > 0)$$ (2.187)

The last term in (2.158), containing a delta function, does not contribute to the level density in the physical region $E > 0$, but it contributes a constant $1/4$ to the number staircase function $N(E)$, defined by

$$N(E) = \int_{-\infty}^{E} g(E')\, dE',$$ (2.188)

which counts the number of levels up to the energy E.

Singling out the smooth parts coming from the contributions $M_1 = M_2 = 0$ in (2.182) and $M = 0$ in (2.186), we obtain the expression for the average level density

$$\tilde{g}(E) = g_{ETF}(E) = \frac{m}{2\pi^2\hbar^2} \left[\pi a_1 a_2 - \sqrt{\frac{\pi^2\hbar^2}{2mE}}(a_1 + a_2) \right], \qquad (2.189)$$

which includes a correction to the Thomas-Fermi term of relative order $\hbar$. It can also be obtained from the extended Thomas-Fermi approach discussed in Chapter 4.

In Fig. 2.12, we illustrate the convergence of the periodic orbit sums in the trace formula (2.187) for the square box with $a_1 = a_2 = a$. In order to be able to plot the sum of delta functions, we have convoluted $g(E)$ over the wave number $k = p/\hbar$ by a Gaussian with range $\gamma = 0.02/a$ and normalized it such that each delta function for a

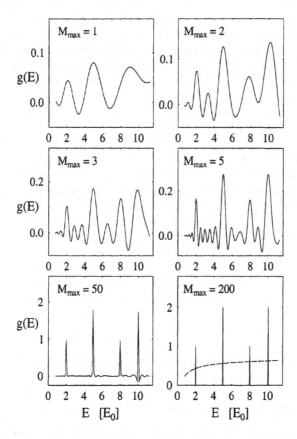

Figure 2.12: Level density of the square billiard $(a_1 = a_2 = a)$, calculated from the trace formula (2.187) using different numbers of periodic orbits and plotted versus the energy E (see text for normalization). In each panel, M_{max} is the largest number used for the orbit indices M_1, M_2 and M. The dashed line in the lower right panel is the ETF density (2.189).

singly degenerate level will have the height 1 in the converged sum (see Sect. 5.5 for the details of this procedure.) In each panel of Fig. 2.12, only orbits with labels M_1, M_2 or M up to a maximum value of M_{max} are included besides the average term (2.189), and the result is plotted versus the energy E in units of $E_0 = \hbar^2 \pi^2 / (2ma^2)$. Thus, in the upper left panel, only the two shortest primitive orbits (0,1) – which is identical to (1,0) – and (1,1) are included; with increasing M_{max}, longer and longer orbits will contribute. For $M_{max}=2$, the same two primitive orbits are included together with their second repetitions. Here, the highest peaks appear already at the positions of the exact energy levels $E/E_0 = 2, 5, 8$, and 10. By the time M_{max} has reached 200 (lower right), the exact quantum density has been obtained with the correct normalized peak heights which are equal to the quantum degeneracies. Also shown by the dashed line in the lower right panel is the average density (2.189).

We see in this example that the periodic orbit sum converges to the correct quantum level density, and that the inclusion of only a few of the shortest orbits already yields the rough positions of the quantum energy levels. The convergence to a sum of delta functions is rather an exception which only happens in integrable systems (and not always at the exact positions). But the fact that the few shortest periodic orbits yield an averaged level density which contains the main shell structure is also true for non-integrable systems, and will be a recurring theme in this book. We note in passing that partial summations of the periodic orbits sum may sometimes be possible. In fact, in the present example the sum over the repetitions of each given primitive periodic orbit can be performed analytically. But, as mentioned by Keating and Berry [35], these partial sums have singularities at false positions of the energy, and only after summing over all topologically different orbits the correct sum of delta functions is obtained. Thus, partial resummations of the periodic orbit sum may yield useless results unless all orbits are summed over. The physical way to use the periodic orbit sum is to order the contributions according to the lengths of the orbits, as shown in the above figure.

We will come back to the rectangular billiard in Sect. 6.1.5, where we will treat it in the context of the extended Gutzwiller theory. There we shall also give a physical interpretation of the correction terms (2.186). The trace formula (2.187) is obtained again in Sect. 3.2.6 where we investigate the three-dimensional rectangular box. We note that the exact trace formula given here has also been derived by Balian and Bloch [36] from a multiple-reflection expansion of the single-particle Green's function.

2.8   Transition from integrability to chaos

2.8.1   Destruction of resonant tori

We have pointed out in Sect. 2.5 that for integrable motion, the number of constants of motion in involution equals the number of independent degrees of freedom. In this case, the Hamiltonian may be expressed, in principle, in terms of the action variables $(I_1, I_2, ..)$ alone, and the corresponding canonical angle variables $(\phi_1, \phi_2, ..)$ are cyclic. Therefore the angle variables increase linearly with time, and all motion may be described as quasi-periodic or periodic. In these coordinates, the phase-space takes on a multi-dimensional torus structure. This is easy to visualize in the two-dimensional problem, as shown in Fig. 2.11. For simplicity, let us concentrate on this two-torus. Suppose one adds to the original Hamiltonian $H_0(I_1, I_2)$ a nonintegrable, small pertur- bation denoted by a potential $\epsilon V(I_1, I_2, \phi_1, \phi_2)$, where $\epsilon \ll 1$. One may then ask if the torus structure is destroyed, or preserved approximately. We could try to answer the question in a perturbative sense first. Let us assume that what the perturbation does is to only distort the torus in a continuous manner, and that approximate constants of motion $\mathcal{I}_1, \mathcal{I}_2$ (to order ϵ^2) may still be found by choosing an appropriate canonical transformation. If this is true, then the transformed Hamiltonian could be written as $H(\mathcal{I}_1, \mathcal{I}_2)$ neglecting terms of order ϵ^2. This procedure could then be repeated, re- garding $H(\mathcal{I}_1, \mathcal{I}_2)$ as the unperturbed Hamiltonian in the second step. Actually this procedure runs into problems due to vanishing denominators in the perturbation cal- culation for those periodic orbits of H_0 that are in resonance with the frequencies of the perturbing term. Nevertheless, it has been demonstrated [37] that with certain restrictions on the "smallness" parameter, the torus structure is indeed retained for those pseudo-periodic orbits that do *not* obey the resonance condition. On the other hand, the periodic orbits for which the ratio of the frequencies is a rational number, or very close to a rational number, may get grossly distorted or destroyed. We shall first illustrate this instability by taking an explicit example that may be worked out analytically, following closely a very readable paper of Walker and Ford [38]. More general statements regarding the transition to ergodic motion will then be made.

We are now ready to continue on our main theme, the stability of the periodic orbits under a perturbation. The perturbing potential $V(I_1, I_2, \phi_1, \phi_2)$ should remain identical if the angles ϕ_1, ϕ_2 change by a multiple of 2π. This periodicity ensures that V may be expressed in terms of a double Fourier series:

$$V(I_1, I_2, \phi_1, \phi_2) = \sum_m \sum_n f_{mn}(I_1, I_2) \cos(m\phi_1 + n\phi_2). \qquad (2.190)$$

To analyze the effect of this perturbation in a tractable way, we shall first assume that there is only one nonzero term in the above series. The Hamiltonian is then given by

$$H = H_0(I_1, I_2) + \epsilon f_{mn}(I_1, I_2) \cos(m\phi_1 + n\phi_2). \qquad (2.191)$$

To transform away the last term by a canonical transformation that takes the coordi- nates $(I_1, \phi_1, I_2, \phi_2) \rightarrow (\mathcal{I}_1, \theta_1, \mathcal{I}_2, \theta_2)$, let us choose a generating function

$$F_2(\mathcal{I}_i, \phi_i) = \mathcal{I}_1\phi_1 + \mathcal{I}_2\phi_2 + \epsilon B_{mn}(\mathcal{I}_1, \mathcal{I}_2) \sin(m\phi_1 + n\phi_2), \qquad (2.192)$$

that differs from the identity transformation by order ϵ. The coefficient B_{mn} has to be chosen as follows. We note that

$$I_1 = \frac{\partial F_2}{\partial \phi_1} = \mathcal{I}_1 + \epsilon B_{mn} m \cos(m\phi_1 + n\phi_2),$$

and

$$\theta_1 = \frac{\partial F_2}{\partial \mathcal{I}_1} = \phi_1 + \epsilon \frac{\partial B_{mn}}{\partial \mathcal{I}_1} \sin(m\phi_1 + n\phi_2).$$

Similar equations hold for I_2, θ_2. We may substitute these expressions for I_1, I_2 in the equation for H above. Taylor expanding about the new action variables $\mathcal{I}_1, \mathcal{I}_2$, and retaining only terms up to order ϵ, we then get

$$
\begin{aligned}
H &= H_0(\mathcal{I}_1, \mathcal{I}_2) \\
&\quad + \epsilon \left[(m\omega_1 + n\omega_2) B_{mn}(\mathcal{I}_1, \mathcal{I}_2) + f_{mn}(\mathcal{I}_1, \mathcal{I}_2) \right] \cos \left(m\theta_1 + n\theta_2 \right).
\end{aligned} \tag{2.193}
$$

In the above, $\omega_i = \partial H_0(\mathcal{I}_1, \mathcal{I}_2)/\partial \mathcal{I}_i$ are the unperturbed natural frequencies of the system. We see from the above that the θ_i-dependent term, of order ϵ, may be eliminated from the transformed Hamiltonian by choosing

$$
B_{mn}(\mathcal{I}_1, \mathcal{I}_2) = -\frac{f_{mn}(\mathcal{I}_1, \mathcal{I}_2)}{(m\omega_1 + n\omega_2)}. \tag{2.194}
$$

This is only valid, however, if the denominator in the above expression is not zero or, at least, not very close to zero. The condition that

$$
m\omega_1 + n\omega_2 = 0 \tag{2.195}
$$

is just that for a quasi-periodic orbit to turn into a periodic one. Such orbits may be grossly distorted and shoot out of the torus geometry. More quantitatively, the band of frequencies for which

$$
|m\omega_1 + n\omega_2| \ll |f_{mn}(I_1, I_2)|, \tag{2.196}
$$

are unstable to the angle-dependent perturbation. The external perturbation is coupling *resonantly* to the natural frequencies of the unperturbed system to bring about this amplitude instability. When the perturbation V has a number of angle-dependent terms on the right-hand side of Eq. (2.190) that may couple to a band of frequencies resonantly satisfying Eq. (2.195), *overlapping resonances* may destroy the torus structure, and bring about ergodicity in the sense of statistical mechanics.

2.8.2   The model of Walker and Ford

To illustrate the distortions in the torus by the resonances, Walker and Ford [38] solve some model Hamiltonians exactly, rather than by perturbation approximation. First they consider the Hamiltonian (2.191) with the single perturbing term. In this case, it is easy to verify that the Poisson bracket $[H, I]$ equals zero, where $I = (nI_1 - mI_2)$. Thus, even in the presence of the angle-dependent perturbation (in this specific form), two constants of motion, E and I, are present in this two-dimensional problem which is therefore still integrable. First consider

$$
H = H_0(I_1, I_2) + \alpha I_1 I_2 \cos \left(2\phi_1 - 2\phi_2 \right), \tag{2.197}
$$

which has the constant of motion $I = I_1 + I_2$. For H_0, choose

$$
H_0 = I_1 + I_2 - I_1^2 - 3I_1 I_2 + I_2^2. \tag{2.198}
$$

The unperturbed frequencies of the system are

$$
\omega_1 = \frac{\partial H_0}{\partial I_1} = 1 - 2I_1 - 3I_2, \qquad \omega_2 = \frac{\partial H_0}{\partial I_2} = 1 - 3I_1 + 2I_2. \tag{2.199}
$$

Note that the two frequencies are equal for $I_1 = 5I_2$. Putting this relation back in $H_0 = E$ yields real solutions for I_1, I_2 only in the energy range $0 \leq E \leq 3/13$. It is instructive to see first how the trajectories in the phase space get distorted by the simple one-term perturbation $\alpha I_1 I_2 \cos(2\phi_1 - 2\phi_2)$. For this purpose, note that the coordinate I_2 may be eliminated in Eq. (2.197) by writing $I_2 = I - I_1$, so a given energy E equals

$$E = H_0(I_1, I - I_1) + \alpha I_1 (I - I_1) \cos(2\phi_1 - 2\phi_2). \qquad (2.200)$$

Rather than solving the actual equations of motion, the trajectory of a particle in this simple problem may be obtained algebraically from the above equation. To see this, keep E fixed on the left-hand-side of Eq. (2.200). A trajectory is then labeled by a constant I and is obtained by changing the coordinates I_1 and $\phi = 2(\phi_1 - \phi_2)$ in the right-hand side to equal E. It is more convenient to fix ϕ_2 at a particular value $(3\pi/2$ in the following) and to track the motion of the system in the (I_1, ϕ_1)-plane. Now Eq. (2.200) may be written as a quadratic equation for I_1:

$$I_1^2 [3 + \alpha \cos(2\phi_1)] - I_1 [5I + \alpha I \cos(2\phi_1)] + I + I^2 - E = 0. \qquad (2.201)$$

This defines a Poincaré surface of section which is presented in Fig. 2.13. Following Walker and Ford, the plot is shown in the cartesian coordinates

$$q_1 = \sqrt{2I_1} \cos\phi_1, \qquad p_1 = -\sqrt{2I_1} \sin\phi_1. \qquad (2.202)$$

In this figure, the energy E is the same for every trajectory, but each trajectory has a different I. In the absence of the angle-dependent perturbation term, only the first term H_0 in Eq. (2.197) survives, giving rise to concentric circles for the plot. When the perturbing term is present in Eq. (2.201), the curves get distorted. The distortions suffered by the tori are principally in the region of the polar caps and at the shoulders of the equator. These are precisely the points where the $(2, 2)$ resonance condition given by Eq. (2.195) is satisfied with $m = 2, n = -2$. This may be recognized easily if one takes the perturbing potential V in (2.197) as a function of the angle $\phi = 2(\phi_1 - \phi_2)$. The points of zero slope of $V(\phi)$, i.e., the maxima and minima of the cosine function, tell us immediately that the motion is unstable about $\phi = 0, 2\pi, ..$ and stable at $\phi = \pi, 3\pi, ...$ At these values of ϕ, it is easy to check from the equations of motion that $\dot{I}_i = 0$. Furthermore, for a periodic orbit satisfying the resonance condition $2\omega_1 - 2\omega_2 = 0$, we get $\partial H/\partial I_1 = \partial H/\partial I_2$. This condition now yields the modified relation

$$I_1 = \frac{(5 \pm \alpha)}{(1 \pm \alpha)} I_2 \qquad (2.203)$$

at the stable (upper sign) and the unstable (lower sign) fixed points. In Fig. 2.13 on the next page, the central point of each crescent represents one of the two distinct stable periodic orbits. Note from Eq. (2.202) that ϕ_1 increases clockwise, starting from zero, having the values $\pi/2$ and $3\pi/2$ at the lower and the upper polar caps, respectively. The two polar caps represent the two distinct unstable periodic orbits, about which the unperturbed circular plots are substantially distorted for $E < 3/13$.

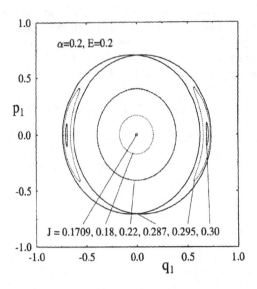

Figure 2.13: Poincaré surface of section, defined by Eqs. (2.201) and
(2.202), for the analytically soluble model Hamiltonian (2.197).

A somewhat different pattern of the polar plot emerges when the perturbing potential, again a single periodic term, has an $m - n$ resonance with $m \neq n$. Consider,
for example,

$$H = H_0 + \beta I_1^{3/2} I_2 \cos\left(3\phi_1 - 2\phi_2\right), \qquad (2.204)$$

where H_0 is the same as before. This is again an integrable problem with the constant
of motion $I = 2I_1 + 3I_2$, and the phase plots may be obtained algebraically. Some
typical polar plots in the (I_1, ϕ_1) plane are shown in Fig. 2.14. The main difference
from the previous case (where $m = n$) is that instead of four distinct periodic orbits,
now there are only two. The perturbing potential is still a periodic function of one angle
ϕ, but now $\phi = (3\phi_1 - 2\phi_2)$. As before, ϕ_2 is kept fixed at $3\pi/2$, so that $\phi_1 = (\pi + \phi/3)$.

In Fig. 2.14, therefore, $\phi_1 = \pi$ corresponds to the unstable point $\phi = 0$, and
it again increases clockwise. The other unstable points occur at $\phi_1 = 5\pi/3$, $7\pi/3$,
corresponding to $\phi = 2\pi$ and 4π, respectively. Similarly, the centers of the three
crescents at $\phi_1 = 4\pi/3$, 2π, and $8\pi/3$ represent stable points, but all belonging to the
same periodic orbit. As such, the three crescents are termed *island chains*. By contrast,
the two crescents in Fig. 2.13 are *not* island chains. Another important characteristic
is that the resonance condition $3\omega_1 = 2\omega_2$ may only be satisfied for a certain threshold
energy and beyond, for which $I_1 \geq 0$. This may be seen directly by substituting
$3\omega_1 = 2\omega_2$ in the Eq. (2.199), which gives $I_2 = 1/13$. Using this value in Eq. (2.198)
for $H_0 = E$ yields

$$I_1 = \frac{5}{13} - \left(\frac{3}{13} - E\right)^{1/2}.$$

For $I_1 \geq 0$, we must then have $E \geq 14/169$. The distortion of the tori and the appear-

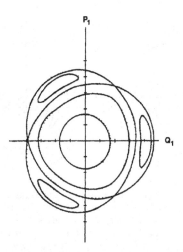

Figure 2.14: Poincaré surface of section for the $(3-2)$ resonance perturbation Eq. (2.204), for fixed energy E and $\phi_2 = 3\pi/2$. The coordinates Q_1, P_1 are defined as in Eq. (2.202); the angle ϕ_1 increases in the clock-wise orientation. (After [38].)

ance of the island chains in the (I_1, ϕ_1) plane by such a perturbing potential takes into effect abruptly only beyond this threshold energy.

Till now the analysis was for an isolated periodic term in the potential (2.190), and the problem remained integrable. The resonant condition (2.195) yielded pairs of periodic orbits, for which the perturbation formula (2.194) broke down. The *resonance zones* were regions of the phase space for which Eq. (2.196) is approximately satisfied. We have seen that the tori in the resonance zones are substantially distorted even for this integrable problem. The important question is: what happens if there is more than one term in the potential (2.190)? In this situation, there is no strict constant of motion other than E. Can we predict under what condition the torus structure may break down and chaos may set in? To answer this question, let us consider two terms in the perturbing potential (2.190), and examine the situation at large enough energy for which the resonance zones of the two resonances begins to overlap. The trajectories have to be calculated numerically. Following Walker and Ford, we take

$$H = H_0(I_1, I_2) + \alpha I_1 I_2 \cos(2\phi_1 - 2\phi_2) + \beta I_1 I_2^{3/2} \cos(2\phi_1 - 3\phi_2). \qquad (2.205)$$

In the following, $\alpha = \beta = 0.02$. Note that in the last term, in comparison to Eq. (2.204), ϕ_1 and ϕ_2 are interchanged. For $\alpha = 0$ in the above equation, the $2-3$ unperturbed resonance does not exist for $E \leq 0.16$. Therefore, in this energy range, this term may be eliminated by a suitable canonical transformation, as described by Eqs. (2.192) – (2.194), and there is no complication of a vanishing energy denominator.

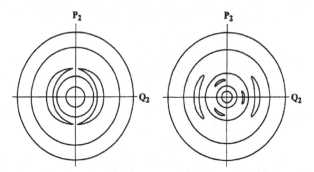

Figure 2.15:   Poincaré surface of section for the Hamiltonian with the double resonance given by Eq. (2.205). The notation is the same as in the previous figure. In the left part, the energy has the value $E = 0.0561$ for which the $(2-3)$ resonance does not exist. In the right part, $E = 0.18$; the $(2-3)$ resonance zones have now appeared but are well separated from the $(2-2)$ resonances. (After [38].)

Actual numerical calculations show that for E substantially less than 0.16, the phase space is only distorted by the $2-2$ resonance, as shown in the left part of Fig. 2.15. This is now plotted in the (I_2, ϕ_2) plane for $E = 0.0561$. The system behaves essentially like an integrable one, with $(I_1 + I_2)$ and $(3I_1 + 2I_2)$ remaining approximate constants of motion. On the right side of Fig. 2.15 now, the energy is increased to $E = 0.18$, and the $2-3$ island chains make their appearance. Nevertheless, these remain well separated from the $2-2$ resonances at this energy. The appearance of an island chain signals the distortion of a torus with commensurate frequencies [see Eq. (2.195)] due to the perturbation. The width of this chain grows with increasing perturbation, and the overlapping of two resonances destroys the corresponding periodic orbits.

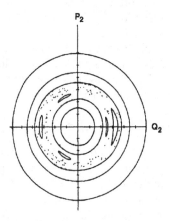

Figure 2.16:   The same as in Fig. 2.15, but for $E = 0.2095$, for which the $(2-2)$ and $(2-3)$ resonances begin to overlap. Narrow zones of instability appear abruptly in the vicinity of the overlapping resonances. (After [38].)

Numerical integration of the equations show that the intercepts of the $2 - 2$ and the $2 - 3$ zone boundaries overlap at $E = 0.2095$, and for this value of the energy, Fig. 2.16 depicts considerable filling-up of the phase space. In this way, we begin to have a quantitative criterion for the breakdown of the torus structure in the phase space, namely the overlapping of different resonance zones with increasing perturbation and the appearance of more disorderly motion in the phase space. This is possible for energies beyond a threshold value for sufficiently large-amplitude motion. We should warn the reader that even in this simple example, the situation is more complicated by the appearance of secondary resonance zones, the details of which will be found in the paper by Walker and Ford [38].

The above example shows that even when nontrivial perturbations are introduced, up to a certain strength of the perturbation much of the orderly mode of motion within the torus structure persists. As stated in Eq. (2.196), only the periodic orbits in a narrow band of frequencies are affected; the others continue to coexist with the more disorderly motion. This observation was made into a quantitative statement by Kolmogorov; it was proved for flows by Arnol'd and for maps by Moser and has become known in the literature as the "KAM theorem" [37]. We quote MacKay and Meiss [39]: *The KAM theorem states that those invariant tori with sufficiently incommensurate frequencies (the frequencies ω must obey a relation of the form $|\mathbf{m} \cdot \omega| > C |\mathbf{m}|^{-\tau}$ for all integer vectors $\mathbf{m} \neq 0$) persist for sufficiently small perturbations of an integrable system. Such orbits form a set of positive measure, and as the system approaches being integrable, the measure of the complement approaches zero.* In the above, C and τ are (positive) constants. The vector $\mathbf{m}$ refers to the set of resonances $\{m_1, m_2, ...\}$. For the theorem to hold, the N frequencies ω_i of the unperturbed Hamiltonian should be linearly independent of each other, i.e.,

$$\det \left| \frac{\partial \omega_i}{\partial I_j} \right| = \det \left| \frac{\partial^2 H(I)}{\partial I_i \partial I_j} \right| \neq 0 \,.$$

In the above, N is the spatial dimension of the system.

An example of the destruction of periodic orbits that are resonant (or commensurate) with the frequency of the perturbing potential is manifested in the so-called Kirkwood gaps in the asteroid belt. There are more than four thousand asteroids that circle the sun, their orbits mainly lying between Mars and Jupiter. The dominant perturbation in the asteroids' central attraction to the sun is Jupiter, which has roughly a period of 12 years. The distribution of the asteroid periods (that depend on the distance from the sun) are shown in Fig. 2.17 on the next page. The absence of asteroids with periods commensurate with Jupiter's period leads to the so-called Kirkwood gaps, named after their discoverer. Non-technical articles on this subject will be found in Refs. [40] and [41].

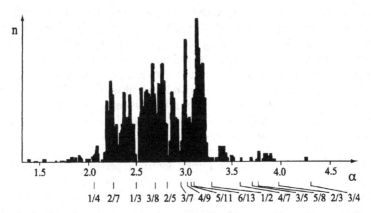

Figure 2.17: Gaps in the asteroid belt occur at locations corresponding to an asteroid period being in a commensurate ratio with the period of Jupiter around the sun. The vertical axis gives a measure of the number n of asteroids as a function of the major semiaxis α of the asteroid orbit in astronomical units (1$a.u.$ is the mean distance of the earth from the sun). A 1/3 fraction with Jupiter's period corresponds to $\alpha = 2.5$, showing pronounced gaps. (After [41].)

2.9   Problems

(2.1) Constants of motion in two-dimensional Coulomb potential
Consider the Hamiltonian of the two-dimensional H-atom given by Eq. (2.123) and use cartesian coordinates.

(a) In addition to the constants of motion H and $L = (xp_y - yp_x)$, show that there are the following two additional classical constants of motion:

$$L_x = \frac{1}{m} p_y L - \frac{\alpha}{r} x \, , \quad L_y = -\frac{1}{m} p_x L - \frac{\alpha}{r} y \, ,$$

such that the following Poisson brackets are zero

$$[L_x, H] = [L_y, H] = 0 \, .$$

You may further verify that

$$[L_x, L_y] = -\frac{2i\hbar}{m} HL \, , \quad [L, L_x] = i\hbar L_y \, , \quad \text{and} \quad [L_y, L] = i\hbar L_x \, .$$

For a bound state, H is replaced by $-|E|$. By redefining

$$L_x \to \left(\frac{m}{2|E|}\right)^{\frac{1}{2}} L_x \, , \quad L_y \to \left(\frac{m}{2|E|}\right)^{\frac{1}{2}} L_y \, ,$$

check that L, L_x, L_y generate the rotation group $SO(3)$.

(b) In quantum mechanics, the Poisson brackets are replaced by the commutators of the corresponding operators, symmetrized to make them hermitian:

$$L_x = \frac{1}{2m}(p_y L + L p_y) - \frac{\alpha}{r}x\,,$$

and likewise for L_y. Check that the scaled operators obey the Lie algebra of the rotation group. For more details about obtaining the quantum spectrum algebraically, see the very readable article by S. Moszkowski [42].

(2.2) Bound-state Kepler orbits
The equation of an ellipse in polar coordinates (r, ϕ) on a plane is given by

$$r = \frac{a(1 - \epsilon^2)}{(1 - \epsilon \cos\phi)}\,,$$

where ϵ is the eccentricity of the ellipse, and a its semi-major axis.
(a) It is well-known that the classical bound state solution of Eq. (2.120) is an ellipse. Noting that the turning points r_1 and r_2 are solutions of $p_r = 0$, where p_r is given by Eq. (2.124), show that

$$a = \frac{1}{2}(r_1 + r_2) = -\frac{k}{2E}\,, \qquad b = \sqrt{r_1 r_2} = \sqrt{-\frac{L_z^2}{2mE}}\,,$$

where b is the semi-minor axis and $V(r) = -k/r$.
(b) Using Eqs. (2.128), (2.131), show that

$$\frac{a}{b} = \frac{(S_r + |S_\phi|)}{|S_\phi|}\,.$$

With the quantization conditions

$$S_r = n_r h, \quad S_\phi = kh, \quad n = n_r + |k|\,,$$

where $n_r = 0, 1, 2, \ldots$ and $k = \pm 1, \pm 2 \ldots$, we see that $a/b = n/|k|$. A circular orbit results when $n_r = 0$, so $n = |k|$. The quantum number $k = 0$ is excluded to keep the energy E finite. The correspondence with modern quantum notation is given by $|k| = |l| + 1$.

(2.3) Motion in a non-central potential
Consider the classical equations of motion for a non-central potential of the form given in Eq. (2.136). Choose

$$\tilde{U}(r) = \frac{3}{4}m\omega^2 r^2\,, \qquad U(\phi) = \frac{a^2}{2m}f(\phi)\,.$$

Show that one may write the energy E as

$$E = \frac{p_r^2}{2m} + \frac{1}{2mr^2}\left(p_\phi^2 + a^2 f(\phi)\right) + \frac{3}{4}m\omega^2 r^2\,.$$

Generalizing the procedure of Sect. 1.2.1, take the angular constant of motion as $p_\phi^2 + a^2 f(\phi) = \alpha_\phi^2$. Now calculate S_ϕ and S_r as defined by Eqs. (2.128) and (2.130) in the text for the special case of $f(\phi) = 1/\sin^2(3\phi)$. Show that for this case,

$$E = \frac{1}{2\pi}\sqrt{\frac{3}{2}}\,\omega\left(2S_r + 3|S_\phi| + 2\pi a\right)\,.$$

Show, using Eq. (2.135), that $\omega_r/\omega_\phi = 2/3$. For more details on this problem, and its connection to a one-dimensional three-body problem, see the paper by Khandekar and Lawande [43].

Consider now the problem in quantum mechanics, for a noncentral potential given by Eq. (2.136). By assuming that the wave function may be written in the separable form $\Psi(r,\phi) = \frac{1}{\sqrt{r}}u(r)K(\phi)$, show that the Schrödinger equation takes the form ($2m = \hbar = 1$)

$$\left(-\frac{d^2}{dr^2} + \tilde{U}(r) + \frac{(M^2 - 1/4)}{r^2}\right)u(r) = E\,u(r),$$

where M^2 is the eigenvalue of the equation in the angle variable:

$$\left(-\frac{d^2}{d\phi^2} + U(\phi)\right)K(\phi) = M^2 K(\phi).$$

Note that M^2 need not in general be an integer, even though the energies are quantized. For an explicit example with $f(\phi) = 1/\sin^2(3\phi)$, see Ref. [28].

(2.4) One-dimensional Coulomb problem

Consider a one-dimensional Coulomb potential

$$V(x) = -\frac{e^2}{x}, \quad x \geq 0,$$
$$= \infty, \quad x < 0.$$

The motion is confined to the positive half-axis only. Show that

$$H(I) = -\frac{me^4}{2I^2}.$$

Using Eq. (2.116), show that

$$x = \frac{2I^2}{me^2}\sin^2\alpha, \qquad p_x = \frac{me^2}{I\tan\alpha},$$

where the action-angle variable is $\phi = (2\alpha - \sin 2\alpha)$.

It is interesting that the one-dimensional Coulomb potential, defined in the entire domain,

$$V(x) = -\frac{e^2}{|x|}, \quad -\infty \leq x \leq \infty, \tag{2.206}$$

still has a classical turning point at the origin for the bound-state problem. It is as if the motion of the electron in this case is the limit of a two-dimensional Kepler orbit with the eccentricity going to unity, the electron going around the nucleus at the origin. This problem has a long history (Euler already knew about it in 1675). The interested reader should look up the books by Szebehely [44] and Stiefel and Scheifele [45], as well as the paper by Biechele et al. [46].

Bibliography

[1] W. H. Miller, Adv. Chem. Phys., **25**, 69 (1974).

[2] L. D. Landau and E. M. Lifshitz: *Mechanics* (Third edition, Pergamon Press Ltd., 1976).

[3] H. Goldstein: *Classical Mechanics* (Addison-Wesley, Reading, Mass., 1950). Chapters 7 - 9 contain relevant material.

[4] W. H. Miller, J. Chem. Phys. **53**, 1949 (1970).

[5] J. Hietarinta, Phys. Rep. **147**, 87 (1987); D. Bonatsos, C. Daskaloyannis, and K. Kokkotas, Phys. Rev. A **50**, 3700 (1994).

[6] E. Schrödinger, Ann. Phys. (Leipzig) **79**, 361 (1926).

[7] G. Wentzel, Z. Phys. **38**, 518 (1926); H. Kramers, Z. Phys. **39**, 828 (1926); L. Brillouin, Compt. Rend. **183**, 24 (1926). Other historical references may be found in L. Schiff: *Quantum Mechanics* (Third edition, McGraw-Hill, New York, 1968) p. 269.

[8] H. Goldstein: *Classical Mechanics* (Addison-Wesley, Reading, Mass., 1950) p. 308.

[9] E. Madelung, Z. Phys. **40**, 322 (1926).

[10] P. R. Holland: *The Quantum Theory of Motion* (Cambridge University Press, Cambridge, England, 1993).

[11] D. Bohm, Phys. Rev. **85**, 166, 180 (1952).

[12] A. S. Davydov: *Quantum Mechanics* (Pergamon Press, Oxford, 1965).

[13] D. J. Griffiths: *Introduction to Quantum Mechanics* (Prentice Hall, Englewood Cliffs, New Jersey, 1995).

[14] L. D. Landau and F. M. Lifshitz: *Quantum Mechanics* (Addison-Wesley, Reading, Mass., 1958).

[15] N. Fröman and P. O. Fröman: *JWKB Approximation* (North Holland, Amsterdam, 1965).

[16] K. W. Ford, D. L. Hill, M. Wakano and J. A. Wheeler, Ann. Phys. (N. Y.) **7**, 239 (1959).

[17] W. H. Miller, J. Phys. Chem. **83**, 960 (1979).

[18] S. C. Creagh and N. D. Whelan, Phys. Rev. Lett. **77**, 4975 (1996); Ann. Phys. (N. Y.) **272**, 196 (1999).

[19] O. Bohigas, S. Tomsovic, and D. Ullmo, Phys. Rep. **223**, 43 (1993).

[20] E. Doron and S. D. Frischat, Phys. Rev. Lett. **75**, 3661 (1995).

[21] A. Einstein, Verh. Dtsch. Phys. Ges. **19**, 82 (1917); L. Brillouin, J. Phys. Radium **7**, 353 (1926); J. B. Keller, Ann. Phys. (N. Y.) **4**, 180 (1958).

[22] M. V. Berry and A. M. Ozorio de Almeida, J. Phys. A **6**, 1451 (1973).

[23] M. Tabor: *Chaos and integrability in nonlinear dynamics* (John Wiley & Sons, New York, 1989).

[24] R. E. Langer, Phys. Rev. **51**, 669 (1937).

[25] M. V. Berry and K. E. Mount, Rep. Prog. Phys. **35**, 315 (1972).

[26] S. C. Miller and R. H. Good, Phys. Rev. **91**, 174 (1953).

[27] J. B. Keller and S. I. Rubinow, Ann. Phys. (N. Y.) **9**, 24 (1960).

[28] A. Khare and R. K. Bhaduri, Am. J. Phys. **62**, 1008 (1994).

[29] A. Sommerfeld: *Wave Mechanics* (Dutton, New York 1911).

[30] M. V. Berry and M. Tabor, Proc. R. Soc. Lond. **A 349**, 101 (1976); *ibid.* **A 356**, 375 (1977).

[31] M. V. Berry and M. Tabor, J. Phys. **A 10**, 371 (1977).

[32] see, e.g., E. C. Titchmarsh: *Introduction to the theory of Fourier Integrals* (Second edition, Clarendon Press, Oxford, 1948), p. 60.

[33] E. N. Bogachek and G. A. Gogadze, Sov. Phys. JETP **36**, 973 (1973) [Zh. Eksp. Teor. Fiz. **63**, 1839 (1973)].

[34] S. C. Creagh and R. G. Littlejohn, Phys. Rev. **A 44**, 836 (1991).

[35] J. P. Keating and M. V. Berry, J. Phys. **A 20**, L1139 (1987).

[36] R. Balian and C. Bloch, Ann. Phys. (N. Y.) **69**, 76 (1972).

[37] A. N. Kolmogorov, Dokl. Akad. Nauk. SSSR **98**, 527 (1954); V. I. Arnol'd, Russian Math. Surveys **18**, 9 (1963); J. Moser, Nachr. Akad. Wiss. Göttingen, II Math. Physik Kl. **1** (1962).

[38] G. H. Walker and J. Ford, Phys. Rev. **188**, 416 (1969).

[39] R. S. Mackay and J. D. Meiss: "Survey of Hamiltonian Dynamics", in: *Hamiltonian Dynamical Systems*, eds. R. S. MacKay and J. D. Meiss (Adam Hilger, Bristol and Philadelphia, 1987).

[40] Carl Murray, "Is the solar system stable?", in: *Exploring Chaos*, ed. Nina Hall, (W. W. Norton, New York, 1991).

[41] J. Moser, Mathematical Intelligencer **1**, 65 (1978).

[42] S. Moszkowski, Phys. Rev. **110**, 403 (1958).

[43] D. C. Khandekar and S. V. Lawande, Am. J. Phys. **40**, 458 (1972).

[44] V. G. Szebehely: *Theory of Orbits, the Restricted Problem of Three Bodies* (Academic Press, New York, 1967).

[45] E. L. Stiefel and G. Scheifele: *Linear and Regular Celestial Mechanics* (Springer-Verlag, New York, 1971).

[46] P. Biechele, D. A. Goodings, and J. H. Lefèbvre, Phys. Rev. **E 53**, 3198 (1996).

Chapter 3

The single-particle level density

3.1 Introduction

In Chapter 2 we have reviewed some basic elements of quantization and its close connection to classical motion. We have seen that for integrable systems with N degrees of freedom, where the classical motion takes place on a N-torus in the phase space of action-angle variables (I_i, ϕ_i), one can quantize the actions $S_i = 2\pi I_i$ independently of each other. This is the EBK quantization (or "torus quantization") scheme, discussed in more detail in Sect. 2.5, representing the generalization of the WKB method into several dimensions. It is possible because in integrable systems, the classical motion always separates in the action-angle variables. Furthermore, when the basic frequencies $\omega_i = \partial H / \partial I_i$ have rational ratios, the motion is periodic.

In Chapters 5 and 6 we will develop the periodic orbit theory which quite generally establishes a link between the density of states of a given quantum-mechanical system with the periodic orbits of the corresponding classical system through so-called "trace formulae". As a preparation to this development, we present in Sect. 3.1.1 of this chapter some basic quantum-mechanical tools needed to calculate the density of states, or – as we will call it often in this book – the level density. In Sect. 3.2, a number of examples of integrable systems are given for which exact trace formulae may be obtained analytically.

For applications to many-fermion systems, we are using the mean-field approximation (see Appendix A). Thus the formulae that we develop here for the density of states and other quantities are specialized to the single-particle approach. In a more general formulation of the many-body problem one has to start from a density matrix with correlations. For the N-particle system, the single-particle density matrix is then obtained by integrating over the coordinates of $N - 1$ particles, see Eq. (A.2).

One fundamental feature of the level density is that it can always be decomposed into a smooth and an oscillating part. The smooth part, which in the periodic orbit theory may formally be attributed to "orbits of length zero", will be discussed in Chapter 4 where we derive it from the extended Thomas-Fermi model.

3.1.1 Level density and other basic tools

We start from the Hamiltonian for a particle with mass m in a local potential $V(\mathbf{r})$

$$\hat{H} = \hat{T} + V(\mathbf{r}), \qquad \hat{T} = -\frac{\hbar^2}{2m}\,\nabla^2. \tag{3.1}$$

We shall treat $V(\mathbf{r})$ throughout as a fixed external one-body potential. When dealing with realistic interacting fermion systems, $V(\mathbf{r})$ will be a self-consistent Hartree or Kohn-Sham potential. In Appendix A we summarize some methods by which such a local mean field can be obtained.

The stationary Schrödinger equation for bound states gives the spectrum E_n:

$$\hat{H}|n\rangle = E_n|n\rangle, \qquad E_n > 0. \tag{3.2}$$

For convenience, we have adjusted the energy scale such that all eigenvalues are positive. The eigenstates $|n\rangle$ form a complete, orthonormal basis set of wavefunctions:

$$\psi_n(\mathbf{r}) = \langle\mathbf{r}|n\rangle; \quad \langle n|m\rangle = \delta_{nm}; \quad \sum_n \psi_n^*(\mathbf{r}')\,\psi_n(\mathbf{r}) = \delta(\mathbf{r}' - \mathbf{r}). \tag{3.3}$$

Here and in the following, sums over n always run over the *complete* spectrum, whereby in more than one dimension n stands for a complete set of quantum numbers, and degeneracies are explicitly summed over. The inclusion of a continuum causes no principal difficulty; if not explicitly mentioned otherwise, we shall therefore assume the spectrum to be discrete.

The single-particle *level density*, or density of states, $g(E)$ is defined as the sum of delta functions:

$$g(E) = \sum_n \delta(E - E_n). \tag{3.4}$$

The first integral of $g(E)$ is often called the "number staircase function" $N(E)$ which counts the number of levels (including their degeneracies) up to a given energy E:

$$N(E) = \int_0^E g(E')\,dE'. \tag{3.5}$$

When dealing with many-body systems in the mean-field approximation (see Appendix A), we are considering a system of N independent particles with Fermi statistics at temperature $T = 0$ to fill the states of the lowest energy. Using the occupation numbers

$$\begin{aligned}\nu_n &= 1 \quad \text{for } E_n < \mu, \\ \nu_n &= 0 \quad \text{for } E_n > \mu,\end{aligned} \tag{3.6}$$

where μ is the Fermi energy, we may write the particle number N as

$$N = \sum_n \nu_n = N(\mu) \tag{3.7}$$

in terms of the staircase function (3.5).

For particles bound in a box with reflecting walls, the level density arises more naturally as a function of the wave number k defined by

$$k = \sqrt{2mE}/\hbar. \tag{3.8}$$

It is therefore sometimes useful to define a level density $g(k)$ by

$$g(k) = \sum_n \delta(k - \sqrt{2mE_n}/\hbar) \tag{3.9}$$

which measures the number of eigenvalues found per unit interval on the k scale. Obviously, using the differential of Eq. (3.8), the two level densities (3.4) and (3.9) are related by

$$g(k) = \frac{\hbar^2}{2m} 2k\, g(E). \tag{3.10}$$

Note that, if k and E are related by Eq. (3.8), the number staircase function may also be written as

$$N(E) = N(k) = \int_0^k g(k')\, dk'. \tag{3.11}$$

The two level densities may therefore also be simply defined as the respective derivatives of the number staircase function:

$$g(E) = \frac{d}{dE} N(E), \qquad g(k) = \frac{d}{dk} N(k). \tag{3.12}$$

The single-particle *density matrix* $\rho(\mathbf{r}, \mathbf{r}')$ is defined as

$$\rho(\mathbf{r}, \mathbf{r}') = \sum_n \nu_n \psi_n^*(\mathbf{r}') \psi_n(\mathbf{r}) = \sum_{E_n < \mu} \psi_n^*(\mathbf{r}') \psi_n(\mathbf{r}). \tag{3.13}$$

Note that this quantity depends on μ and thus on the statistics, whereas $g(E)$, as well as the partition function, the Bloch density and the propagator to be discussed below, do not. This holds, in particular, also for the *local density* function $\rho(\mathbf{r})$ given as the diagonal of (3.13):

$$\rho(\mathbf{r}) = \rho(\mathbf{r}, \mathbf{r}) = \sum_{E_n < \mu} \psi_n^*(\mathbf{r}) \psi_n(\mathbf{r}). \tag{3.14}$$

When used for fermions, the right-hand side above will acquire an extra spin degeneracy factor of 2 (see the note at the end of this section).

A Laplace transform of $g(E)$ yields the single-particle *canonical partition function* $Z(\beta)$:

$$Z(\beta) = \mathcal{L}_\beta [g(E)] = \int_0^\infty e^{-\beta E} g(E)\, dE = \sum_n e^{-\beta E_n}. \tag{3.15}$$

In quantum statistics and thermodynamics, β has the physical significance of an inverse temperature: $\beta = 1/k_B T$ (k_B is the Boltzmann constant). In our present context, however, β is just a mathematical variable which in general is taken to be complex after an appropriate analytical continuation of $Z(\beta)$.

We define the *Bloch density* $C(\mathbf{r}, \mathbf{r}'; \beta)$ by (see also Sect. 4.2)

$$C(\mathbf{r}, \mathbf{r}'; \beta) = \sum_n \psi_n^*(\mathbf{r}') \psi_n(\mathbf{r})\, e^{-\beta E_n} = \langle \mathbf{r} | e^{-\beta \hat{H}} | \mathbf{r}' \rangle. \tag{3.16}$$

It obeys the Bloch equation

$$-\frac{\partial}{\partial\beta} C(\mathbf{r}, \mathbf{r}'; \beta) = \hat{H}_{\mathbf{r}} C(\mathbf{r}, \mathbf{r}'; \beta), \qquad (3.17)$$

where the index $\mathbf{r}$ on the Hamiltonian $\hat{H}_{\mathbf{r}}$ indicates that it acts on the variable $\mathbf{r}$, subject to the boundary condition

$$C(\mathbf{r}, \mathbf{r}'; \beta = 0) = \delta(\mathbf{r} - \mathbf{r}') \qquad (3.18)$$

which follows directly from the completeness of the wavefunctions. Due to the orthogonality of the states $|n\rangle$, we can obtain $Z(\beta)$ as the trace of $C(\mathbf{r}, \mathbf{r}'; \beta)$ in coordinate space:

$$Z(\beta) = \operatorname{tr} C = \int C(\mathbf{r}, \mathbf{r}; \beta)\, d^3r. \qquad (3.19)$$

It is straight-forward to calculate the Bloch density matrix for a free particle, using a plane-wave basis. For the one-dimensional case, Eq. (3.16) with $V = 0$ gives

$$C(x, x'; \beta) = \frac{1}{2\pi} \int_{-\infty}^{\infty} dk \, \exp\left[ik(x - x') - \beta\frac{\hbar^2 k^2}{2m}\right]. \qquad (3.20)$$

The integral may be done analytically by completing the square in the exponent, and the final result is

$$C(x, x'; \beta) = \left(\frac{m}{2\pi\hbar^2\beta}\right)^{1/2} \exp\left[-\frac{m}{2\hbar^2\beta}(x - x')^2\right]. \qquad (3.21)$$

For harmonic oscillator potentials in any dimension, the Bloch density matrix can also be obtained exactly (see Problem 3.6). For later reference, we give here the "classical" Bloch density matrix in D dimensions for an arbitrary local potential $V(\mathbf{r})$. It is obtained by including the potential in the exponent of (3.21) (cf. Sect. 4.3):

$$C_{cl}(\mathbf{r}, \mathbf{r}'; \beta) = \left(\frac{m}{2\pi\hbar^2\beta}\right)^{D/2} \exp\left[-\beta V(\mathbf{r})\right] \exp\left[-\frac{m}{2\hbar^2\beta}(\mathbf{r} - \mathbf{r}')^2\right]. \qquad (3.22)$$

If we replace β in (3.16) above by an imaginary time interval: $\beta \to i(t - t')/\hbar$, the Bloch density transforms to the *single-particle propagator* $K(\mathbf{r}, \mathbf{r}'; t - t')$ that describes the propagation of the particle from $\mathbf{r}' \to \mathbf{r}$ in a time interval $t - t' > 0$:

$$K(\mathbf{r}, \mathbf{r}'; t - t') = \sum_n \psi_n^*(\mathbf{r}')\psi_n(\mathbf{r})\, e^{-\frac{i}{\hbar}E_n(t - t')} = \langle\mathbf{r}|e^{-\frac{i}{\hbar}\hat{H}(t - t')}|\mathbf{r}'\rangle. \qquad (t - t' > 0) \quad (3.23)$$

From the time evolution of the single-particle state $|n(t)\rangle$

$$|n(t)\rangle = e^{-\frac{i}{\hbar}\hat{H}(t - t')}|n(t')\rangle \qquad (3.24)$$

it follows that $K(\mathbf{r}, \mathbf{r}'; t - t')$ is the kernel of an integral operator $\hat{K}$ which propagates the particle from $\mathbf{r}'$ at time t' to $\mathbf{r}$ at time t:

$$\psi_n(\mathbf{r}, t) = \hat{K}\psi_n(\mathbf{r}', t') = \int d^3r'\, K(\mathbf{r}, \mathbf{r}'; t - t')\, \psi_n(\mathbf{r}', t'). \qquad (3.25)$$

$K(\mathbf{r}, \mathbf{r}'; t-t')$ is also identical to the Green's function of the time dependent Schrödinger equation:

$$\left(-i\hbar\frac{\partial}{\partial t} + \hat{H}_{\mathbf{r}}\right) K(\mathbf{r}, \mathbf{r}'; t - t') = -i\hbar\,\delta(\mathbf{r} - \mathbf{r}')\,\delta(t - t').$$ (3.26)

Using the completeness relation (3.3), it is easy to prove that K fulfills the following important property:

$$K(\mathbf{r}, \mathbf{r}'; t - t') = \int K(\mathbf{r}, \mathbf{r}''; t - t'')\, K(\mathbf{r}'', \mathbf{r}'; t'' - t')\, d^3r''.$$ (3.27)

This has the following meaning: the probability to propagate a particle from $\mathbf{r}'$ at time t' directly to $\mathbf{r}$ at time t is the same as the integrated probability for propagating it in two steps through all possible intermediate points $\mathbf{r}''$ at the time t''.

Since the time *difference* $(t - t')$ for causality reasons always has to be positive definite, as indicated in (3.23), we shall for the sake of simplicity set $t' = 0$ from now on and assume $t \geq 0$. Taking the Fourier integral of K, we get

$$-\frac{i}{\hbar}\int_0^\infty K(\mathbf{r}, \mathbf{r}'; t)\, e^{\frac{i}{\hbar}Et}dt = -\frac{i}{\hbar}\sum_n \psi_n^*(\mathbf{r}')\,\psi_n(\mathbf{r})\int_0^\infty e^{\frac{i}{\hbar}(E-E_n)t}dt.$$ (3.28)

Since the integral on the r.h.s. above oscillates and thus does not converge, we give the energy a small positive imaginary part ϵ: $E \to E + i\epsilon$ ($\epsilon > 0$), thereby keeping E real. Then the integral becomes proportional to $1/(E + i\epsilon - E_n)$, and in the limit $\epsilon \to 0$, the Fourier integral of K yields the *Green's function* in energy representation:

$$G(\mathbf{r}, \mathbf{r}'; E) = -\frac{i}{\hbar}\lim_{\epsilon\to 0}\int_0^\infty K(\mathbf{r}, \mathbf{r}'; t)\, e^{\frac{i}{\hbar}(E+i\epsilon)t}dt = \sum_n \psi_n^*(\mathbf{r}')\,\psi_n(\mathbf{r})\frac{1}{(E - E_n)}.$$ (3.29)

Thus the Green's function obeys the equation

$$(E - \hat{H}_{\mathbf{r}})G(\mathbf{r}, \mathbf{r}'; E) = \delta(\mathbf{r} - \mathbf{r}').$$ (3.30)

For a free particle, the Green's function $G_0(\mathbf{r}, \mathbf{r}'; E)$ can be derived analytically from Eq.(3.29). After replacing β by $it/\hbar$ in Eq.(3.21) and performing the Fourier transform, we obtain for the one-dimensional problem

$$G_0(x, x'; E) = -\left(\frac{2m}{\hbar^2}\right)\frac{i}{2k}\exp\left(-ik|x - x'|\right). \quad \text{(D=1)}$$ (3.31)

We also give the free Green's functions for $D = 2$ and $D = 3$ dimensions for later use:

$$G_0(\mathbf{r}, \mathbf{r}'; E) = -\left(\frac{2m}{\hbar^2}\right)\frac{i}{4}H_0^+(k|\mathbf{r} - \mathbf{r}'|), \quad \text{(D=2)}$$ (3.32)

$$= -\left(\frac{2m}{\hbar^2}\right)\frac{\exp(ik|\mathbf{r} - \mathbf{r}'|)}{4\pi|\mathbf{r} - \mathbf{r}'|}. \quad \text{(D=3)}$$ (3.33)

In the above, H_0^+ is the Hankel function in standard notation. Note that the Green's function is singular for $D = 2$ and $D = 3$ as $|\mathbf{r} - \mathbf{r}'| \to 0$.

Another important quantity is the *trace* of the Green's function, defined by

$$\text{tr}\, G = \int G(\mathbf{r}, \mathbf{r}; E)\, d^3r = G(E) = \sum_n \frac{1}{(E - E_n)}. \tag{3.34}$$

It has its poles along the real energy axis at the position of the eigenvalues E_n and is often also called a *resolvent*.[1] Now, using the general identity

$$\frac{1}{(E + i\epsilon - E_n)} = \mathcal{P}\frac{1}{(E - E_n)} - i\pi\delta(E - E_n), \qquad (\epsilon > 0) \tag{3.35}$$

where the first term on the r.h.s. is a Cauchy principal value and thus real, we find the level density $g(E)$, defined in (3.4), from the imaginary part of $G(E + i\epsilon)$:

$$g(E) = -\frac{1}{\pi}\Im m\, G(E + i\epsilon) = -\frac{1}{\pi}\Im m \int G(\mathbf{r}, \mathbf{r}; E + i\epsilon)\, d^3r. \qquad (\epsilon > 0) \tag{3.36}$$

Rather simple relations are also found if one starts in β space from $Z(\beta)$ or $C(\mathbf{r}, \mathbf{r}'; \beta)$ and performs inverse Laplace transforms. The level density is, due to Eq. (3.15), simply given by the inverse Laplace transform of $Z(\beta)$:

$$g(E) = \mathcal{L}_E^{-1}[Z(\beta)] = \frac{1}{2\pi i}\int_{\epsilon - i\infty}^{\epsilon + i\infty} e^{\beta E} Z(\beta)\, d\beta. \qquad (\epsilon > 0) \tag{3.37}$$

Hereby, the integral is to be taken in the complex β plane along a contour $\mathbf{C}$ which is parallel to the imaginary axis, as illustrated in Fig. 3.1. The positive distance ϵ of the contour from the imaginary axis has to be chosen large enough so that *all* poles of the integrand lie to the left of the contour. If the integrand has only isolated poles, one can easily evaluate Eq. (3.37) using the residue calculus (see Sect. 3.2.1 for a specific example). Note that the same formula (3.37) holds also for the inverse of the two-sided Laplace transform, in which the integration over E in Eq. (3.15) starts at $-\infty$, and which we shall encounter in Chapter 4.

From the formulae on inverse Laplace transforms given in the Appendix B, we find the following relations for some of the properties of a system of N independent particles described by the single-particle Hamiltonian (3.1). The particle number becomes, using the integral (3.5) with the Fermi energy μ as the upper limit,

$$N = N(\mu) = \int_0^\mu g(E)\, dE = \mathcal{L}_\mu^{-1}\left[\frac{1}{\beta}Z(\beta)\right]. \tag{3.38}$$

The sum of occupied single-particle energies in the ground state is

$$E_{sp} = \sum_{E_n < \mu} E_n = \int_0^\mu E\, g(E)\, dE = -\mathcal{L}_\mu^{-1}\left[\frac{1}{\beta}\frac{\partial}{\partial\beta}Z(\beta)\right] = \mu N - \mathcal{L}_\mu^{-1}\left[\frac{1}{\beta^2}Z(\beta)\right]. \tag{3.39}$$

The density matrix (3.13) is obtained directly from the Bloch density by

$$\rho(\mathbf{r}, \mathbf{r}') = \mathcal{L}_\mu^{-1}\left[\frac{1}{\beta}C(\mathbf{r}, \mathbf{r}'; \beta)\right]. \tag{3.40}$$

---

[1] In the literature on periodic orbit theory, the quantity $G(E)$ is often denoted by $g(E)$. In this book, we reserve this symbol for the level density as defined by Eq. (3.4) or (3.36).

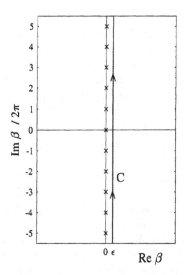

Figure 3.1: Complex contour integration used for the inverse Laplace transform. Crosses indicate the poles of Eq. (3.44).

Note: In this chapter, we have not treated explicitly the spin degrees of freedom, although most of our physical applications of the semiclassical methods will be concerned with fermionic systems. When one includes the Pauli spinors in the wavefunctions for fermions, the density matrix $\rho(\mathbf{r}, \mathbf{r}')$, and similarly the Bloch density C, the propagator K and the Green's function G, becomes a (2×2) matrix in spin space. All traces in coordinate space will then have to be accompanied by a trace in spin space, when one is evaluating expectation values of observables like energy or particle number of a many-body system. As long as the Hamiltonian $\hat{H}$ is not explicitly spin dependent, this will result in a simple spin degeneracy factor of 2 in all the sums over the single-particle spectrum, or the corresponding integrals over the level density, such as Eqs. (3.38,3.39) above. This factor is *not* included in any of the formulae given in the remainder of this chapter, and we will also keep the convention *never* to include it in the level density $g(E)$. The spin factor 2 will, however, be explicitly included in the local densities $\rho(\mathbf{r})$ and $\tau(\mathbf{r})$ to be discussed and used in Chapter 4.

3.1.2 Separation of $g(E)$ into smooth and oscillating parts

The level density $g(E)$ contains all the information about the irregularities of the quantum spectrum. On the average, it has a smooth energy dependence which is determined by the number of degrees of freedom and the degeneracies of the single-particle levels, given by the symmetries of the Hamiltonian (3.1). One very basic feature is that $g(E)$ can always be written as a sum of an average level density $\tilde{g}(E)$ and an oscillating part $\delta g(E)$:

$$g(E) = \tilde{g}(E) + \delta g(E) . \tag{3.41}$$

There are several ways of obtaining either part. In principle, for a given system (3.1), both parts are determined uniquely. This will be illustrated in a number of analytically soluble cases in Sect. 3.2. We shall discuss the smooth part $\tilde{g}(E)$, which is most commonly known from the (extended) Thomas-Fermi model, in Chapter 4. The oscillating part is the object of the semiclassical periodic orbit theory to be developed in Chapters 5 and 6. The main outcome of this semiclassical theory is the so-called "trace formula" for the oscillating part $\delta g(E)$:

$$\delta g(E) = \sum_{\Gamma \epsilon \{ppo\}} \sum_{k=1}^{\infty} \mathcal{A}_{\Gamma k}(E) \cos \left[\frac{k}{\hbar} S_{\Gamma}(E) - \sigma_{\Gamma k} \frac{\pi}{2} \right]. \tag{3.42}$$

Here Γ counts classes of topologically distinct primitive periodic orbits (ppo) (with energy E as a parameter); k counts the repeated revolutions around each primitive orbit which yield a series of harmonics; $S_{\Gamma}(E) = \oint \mathbf{p}_{\Gamma} \cdot d\mathbf{q}_{\Gamma}$ is the classical action integral along the orbit Γ; and the "Maslov index" $\sigma_{\Gamma k}$ is a phase factor depending on its topology. The amplitudes $\mathcal{A}_{\Gamma k}$ depend on energy, time period and stability of the orbit and of its nature being isolated or non-isolated. In the latter case, one is summing in Eq.(3.42) over distinct families of degenerate orbits.

The trace formula (3.42) represents a Fourier decomposition of the oscillating part of the level density. It gives the basis for many interesting interpretations of quantum phenomena in terms of classical orbits (see Sect. 1.4 and the examples in Chapter 8). Before deriving (3.42) in Chapter 5 for general cases, we will consider in the remaining part of this section a few examples for which the trace formula can be found exactly from the quantum-mechanical energy spectrum. These cases will allow us to illustrate the separation (3.41) of the level density.

3.2 Some exact trace formulae

3.2.1 The linear harmonic oscillator

As a model study, we start from the following schematic non-degenerate spectrum:

$$E_n = n, \qquad n = 0, 1, 2, 3, \dots \tag{3.43}$$

For this spectrum, we can easily sum the partition function to give

$$Z(\beta) = \sum_n e^{-n\beta} = \frac{1}{1 - e^{-\beta}} = \frac{e^{\beta/2}}{2 \operatorname{Sinh}(\beta/2)}. \tag{3.44}$$

Now we Laplace invert this using Eq. (3.37). This is most easily done by realizing that the inverse Laplace integral is tantamount to taking the sum of residues of all poles of the integrand lying to the left of the contour. Since E always can be chosen to be positive, the contour can be closed by an infinite semicircle around the left half plane. The function $1/\operatorname{Sinh}(z)$ has only simple poles, all lying on the imaginary axis at the values $z_k = ik\pi$ with $k = 0, \pm 1, \pm 2, \dots$ with residues $(-1)^k$:

$$\frac{1}{\operatorname{Sinh}(z)} = \sum_{k=-\infty}^{k=+\infty} \frac{(-1)^k}{(z - k\pi i)}. \tag{3.45}$$

Thus, the poles of (3.44) lie at $\beta_k = 2k\pi i$, as shown in Fig. 3.1. Summing up all residues, taking into account the extra factors $\exp(-\beta_k/2)$ and $\exp(E\beta_k)$ appearing under the inverse Laplace integral, we obtain

$$g\left(E\right) = \sum_{n=0}^{\infty} \delta(E - n) = \sum_{k=-\infty}^{k=+\infty} e^{2\pi i k E}. \qquad (E \geq 0) \qquad (3.46)$$

This formula is equivalent to the Poisson summation formula, presented in a more general form in Eq. (2.157) of Sect. 2.7, after carrying out the integration over n on the right-hand side of Eq. (2.157). Since the spectrum is bounded at a lower value $E = 0$, only non-negative energies E are relevant. These are automatically selected by the lower integration limit in the definition of the Laplace transform (3.15). The imaginary parts in (3.46) cancel so that we can rewrite $g\left(E\right)$ as

$$g\left(E\right) = \left\{1 + 2\sum_{k=1}^{\infty} \cos(2\pi k E)\right\}. \qquad (E \geq 0) \qquad (3.47)$$

Note that the constant term in (3.47), which corresponds to the average level density, comes from the pole $\beta_{k=0} = 0$, whereas the poles along the imaginary β axis combine pairwise ($\beta_{\pm k} \neq 0$) to make up all Fourier components of the oscillating part.

It is now a trivial step to apply this result to the linear harmonic-oscillator potential $V(x) = (m/2)\,\omega^2 x^2$ with the spectrum

$$E_n = \hbar\omega(n + 1/2). \qquad n = 0, 1, 2, \ldots \qquad (3.48)$$

All we have to do is to introduce the energy unit $\hbar\omega$ and to shift the energy scale by the zero-point energy $\hbar\omega/2$. Thus, the level density becomes

$$g\left(E\right) = \frac{1}{\hbar\omega}\left\{1 + 2\sum_{k=1}^{\infty} \cos\left[2\pi k\left(\frac{E}{\hbar\omega} - \frac{1}{2}\right)\right]\right\} = \frac{1}{\hbar\omega}\left\{1 + 2\sum_{k=1}^{\infty}(-1)^k \cos\left(2\pi k\frac{E}{\hbar\omega}\right)\right\}.$$
$$(3.49)$$

This is our first exact trace formula which has, indeed, the form of (3.41). The smooth part is just the constant $1/\hbar\omega$, namely one energy level per unit $\hbar\omega$, as it is obtained also in the Thomas-Fermi model (see Sect. 4.1). The oscillating part is exactly of the form (3.42). It contains cosine functions with constant amplitudes; their arguments are multiples of $2\pi E/\hbar\omega$. This latter quantity is easily recognized as the classical action: taking the classical momentum $p(x) = \sqrt{2m[E - V(x)]}$ and integrating it between the turning points $x_0 = \pm\sqrt{(2E/m)}/\omega$ gives

$$S(E) = \oint p\,dx = 2\int_{-|x_0|}^{|x_0|} p(x)\,dx = 2\pi E/\omega. \qquad (3.50)$$

3.2.2   General spectrum depending on one quantum number

Next we assume the spectrum E_n to be given by a function $f(n)$ and each level to have a degeneracy $d_n = D(n)$

$$E_n = f(n), \quad d_n = D(n), \qquad n = 0, 1, 2, 3, \dots \tag{3.51}$$

where we assume $f(n)$ to be an arbitrary monotonous function with differentiable inverse $f^{-1}(x) = F(x)$, such that $n = F(E_n)$, and $D(n)$ to be another arbitrary function. Using general properties of the delta function, we write

$$\delta(E - E_n) = \delta(E - f(n)) = \delta(n - F(E))|F'(E)|. \tag{3.52}$$

Further defining $D(E) = D(F(E))$, we can insert all this into Eq. (3.4) to find

$$g(E) = D(E)|F'(E)| \sum_{n=0}^{\infty} \delta(n - F(E)). \tag{3.53}$$

Now we can apply Eq. (3.46) without problems and find the following important general formula for the level density [1]

$$g(E) = D(E)|F'(E)| \left\{ 1 + 2 \sum_{k=1}^{\infty} \cos[k2\pi F(E)] \right\}. \qquad (E \geq 0) \tag{3.54}$$

This formula can be applied to any system where the E_n and d_n are known explicitly. This is not restricted to one-dimensional problems: there exist higher-dimensional potentials for which the spectrum can be written analytically in terms of one single quantum number n with known degeneracies d_n. For such systems, the general formula (3.54) can also be applied and leads to an exact trace formula (see several examples below).

From now on, we shall not write down the condition $E \geq 0$ explicitly; the lower boundary of the level spectrum at $E = 0$ shall always be understood. We can see from (3.54) that the average level density is given by

$$\tilde{g}(E) = D(E)|F'(E)| \tag{3.55}$$

and the oscillating part by

$$\delta g(E) = 2D(E)|F'(E)| \sum_{k=1}^{\infty} \cos[k2\pi F(E)]. \tag{3.56}$$

Within periodic orbit theory, we will recognize $2\pi\hbar F(E)$ quite generally as the *classical action* S, integrated along an elementary closed orbit of the classical motion of the particle. The number k corresponds to the number of revolutions around the orbit. (Note that in the one-dimensional case, of course, every orbit is a periodic orbit.) If several distinct orbits exist, these have to be summed over separately. The amplitudes of the oscillating terms in $\delta g(E)$ then depend on the orbit and are in general much more difficult to obtain. The smooth level density (3.55) is in many analytical cases found to be identical with the extended Thomas-Fermi (ETF) result (see Chapter 4).

That $F(E)$ is proportional to the classical action may intuitively be understood at this point for a one-dimensional system with $d_n = 1$. The Bohr-Sommerfeld quantization rule in one dimension reads

$$S(E_n) = \oint p\, dx = 2 \int_{x_1}^{x_2} p(x)\, dx = 2\pi\hbar\, n \tag{3.57}$$

(up to a possible constant to be added to n, if one wants to use the more accurate WKB method – but that is not relevant here), where x_1, x_2 are the classical turning points. Now we just have to remember that, by construction, $n = F(E_n)$. Indeed, inserting this into (3.57) gives directly

$$F(E) = \frac{1}{2\pi\hbar} S(E). \qquad \text{(in 1 dimension)} \tag{3.58}$$

On the other hand, we can also derive the Bohr-Sommerfeld rule (3.57) from the Thomas-Fermi theory in one dimension. Integrating the level density (3.4) from 0 (i.e., from below the lowest level) up to a Fermi energy $E_n < \mu < E_{n+1}$, we pick up one unity for each delta function peak and thus get the number n of levels up to μ:

$$\int_0^\mu g(E)\, dE = \int_0^\mu \sum_m \delta(E - E_m)\, dE = n. \tag{3.59}$$

Let us now replace $g(E)$ under the integral on the left-hand side of Eq. (3.59) by its Thomas-Fermi value which we can relate to that of the (diagonal) classical Bloch function [cf. Eqs. (3.38) and (3.19)]:

$$\int_0^\mu g_{TF}(E)\, dE = \mathcal{L}_\mu^{-1} \left[\frac{1}{\beta} Z_{cl}(\beta) \right] = \int_{x_1}^{x_2} \mathcal{L}_\mu^{-1} \left[\frac{1}{\beta} C_{cl}(x, x; \beta) \right] dx. \tag{3.60}$$

We use the classical Bloch density (3.22) in one dimension

$$C_{cl}(x, x; \beta) = \left(\frac{m}{2\pi\hbar^2\beta} \right)^{1/2} e^{-\beta V(x)}, \tag{3.61}$$

and do the Laplace inversion using the relations in Appendix B to get

$$\int_0^\mu g_{TF}(E)\, dE = \frac{1}{\pi\hbar} \int_{x_1}^{x_2} \sqrt{2m[\mu - V(x)]}\, dx = \frac{1}{2\pi\hbar} \oint p(x)\, dx = n, \tag{3.62}$$

since $\sqrt{2m[\mu - V(x)]} = p(x)$ is the momentum of the particle at the Fermi energy (in the classically allowed region). We have thus, indeed, found Eq. (3.57).

We shall in the following apply the general formula (3.54) to some specific model potentials.

3.2.3   One-dimensional box

We take a one-dimensional box of length L. The spectrum then is

$$E_n = E_0 \left(n + 1\right)^2, \quad E_0 = \frac{\hbar^2 \pi^2}{2mL^2}. \quad (n = 0, 1, 2, 3...) \qquad (3.63)$$

In the above notation, we have $F(E) = \sqrt{E/E_0} - 1$ and thus find

$$g(E) = \frac{1}{2\sqrt{E_0 E}} \left\{ 1 + 2 \sum_{k=1}^{\infty} \cos\left(2k\pi\sqrt{E/E_0}\right) \right\}. \qquad (3.64)$$

The average part is again identical to the Thomas-Fermi result.

   To interpret this result as a trace formula, we remember that the momentum of the particle is $p = \sqrt{2mE} = \hbar\pi\sqrt{(E/E_0)}/L$. The classical trajectory of the particle in this 1-dimensional box is a straight line of length $2L$ with constant momentum, so that the action integral simply is

$$S(E) = 2Lp = 2\hbar\pi\sqrt{E/E_0}. \qquad (3.65)$$

We thus recognize that the oscillating part of (3.64) is identical to Eq. (3.42) with one term in the sum over Γ since there can only be one orbit in one dimension.

3.2.4   More-dimensional spherical harmonic oscillators

For a harmonic oscillator with spherical symmetry in D dimensions, the spectrum can be written in terms of one single principal quantum number n

$$E_n = \hbar\omega(n + D/2), \qquad n = 0, 1, 2, 3, ... \qquad (3.66)$$

and the degeneracy d_n is a polynomial of order $D - 1$ ($d_n = n + 1$ for $D = 2$ and $d_n = \frac{1}{2}(n+1)(n+2)$ for $D = 3$). The partition function can in the general case easily be summed to give

$$Z(\beta) = \frac{1}{[2\operatorname{Sinh}(\beta\hbar\omega/2)]^D}. \qquad (3.67)$$

(This is obtained most easily in cartesian coordinates; due to the separability of the potential, Z is just the D-th power of the 1-dimensional result.)

   To obtain the exact $g(E)$, we do not need $Z(\beta)$ but proceed as shown above using Eq. (3.54). This leads rather easily in 2 dimensions to

$$g(E) = \frac{E}{(\hbar\omega)^2} \left\{ 1 + 2 \sum_{k=1}^{\infty} \cos\left(2\pi k \frac{E}{\hbar\omega}\right) \right\} \quad (D = 2) \qquad (3.68)$$

and in 3 dimensions to

$$g(E) = \frac{1}{2(\hbar\omega)^3} \left[E^2 - \frac{1}{4}(\hbar\omega)^2 \right] \left\{ 1 + 2 \sum_{k=1}^{\infty} (-1)^k \cos\left(2\pi k \frac{E}{\hbar\omega}\right) \right\}. \quad (D = 3) \qquad (3.69)$$

Actually, from our derivation it becomes clear that the exact level density of the D-dimensional harmonic oscillator is given by

$$g\left(E\right) = g_{ETF}^{(D)}(E) \left\{ 1 + 2 \sum_{k=1}^{\infty} (-1)^{kD} \cos\left(2\pi k \frac{E}{\hbar\omega}\right) \right\}. \qquad (3.70)$$

The factor in front is identical to the average level density $\tilde{g}\left(E\right)$ obtained analytically in the ETF model discussed in Chapter 4 (cf. Problem 4.2). The relation of the oscillating part to the trace formula (3.42) is also clear: the classical action equals $S(E) = 2\pi E/\omega$ for all primitive orbits in any dimension. Note that the average part of the level density not only emerges separately, according to Eq. (3.41), but it multiplies also the oscillating part $\delta g\left(E\right)$ as a factor – as is the case for the one-dimensional box, too, see Eq. (3.64). This is not always so, as we shall see in the following examples.

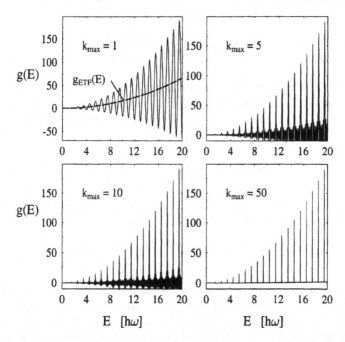

Figure 3.2: Level density of the three-dimensional isotropic harmonic oscillator, calculated according to Eq. (3.69) with different numbers k_{max} of harmonics k included. $g_{ETF}(E)$ (dashed line) is the average part of the level density which appears as the factor in front of (3.69).

In Fig. 3.2 we demonstrate the convergence of the sum over harmonics k in the trace formula (3.69), plotting $g\left(E\right)$ versus the energy (in units of $\hbar\omega$) for different maximum values k_{max} of the index k, after normalizing it by a factor $1 + 2k_{max}$. The dashed line in the upper left panel is the average part $g_{ETF}(E)$ given by the factor in front of (3.69), which is identical to that derived in the extended Thomas-Fermi model in Chapter 4 (see Eq. (4.199) in Problem 4.2). We see how with increasing k_{max}, the delta function peaks come out sharper and sharper at the correct positions of the quantum spectrum (3.66). $k_{max} = 50$ harmonics are enough to practically reproduce the exact sum of delta functions (normalized here to their quantum degeneracies). On the other hand, the use of the lowest harmonic $k = 1$ alone already gives the rough information about the locations and averaged strengths of the main shells.

3.2.5   Harmonic oscillators at finite temperature

Here we give an example of a level density describing a fermion system at finite temperature, which allows us to see analytically how the shell effects are smoothed out by the temperature. We refer to Sect. 4.4.3 for a general formalism used to describe single-particle properties of a fermion system at $T > 0$. As we show there, the level density $g_T(E)$ can be obtained from the inverse Laplace transform of the zero-temperature partition function $Z_0(\beta)$, multiplied by a universal factor:

$$g_T(E) = \mathcal{L}_E^{-1}\left[Z_0(\beta)\frac{\pi\beta T}{\sin(\pi\beta T)}\right]. \qquad (3.71)$$

(We put $k_B = 1$ and measure T in energy units.)

We shall apply this formalism here for spherical harmonic oscillators in one and three dimensions. Let us first take the one-dimensional case. Knowing already its partition function at $T = 0$ (see Eq. (3.67)) we have

$$Z_T(\beta) = \frac{1}{2\operatorname{Sinh}(\beta\hbar\omega/2)}\frac{\pi\beta T}{\sin(\pi\beta T)}. \qquad (3.72)$$

We now Laplace invert this back by summing over all residues. Compared to the situation at $T = 0$, we have new poles along the real axis at $\beta_p = p/T$ for $p = \pm 1, \pm 2, \dots$ due to

$$\frac{\pi\beta T}{\sin\pi\beta T} = \beta\sum_{p=-\infty}^{+\infty}\frac{(-1)^p}{(\beta - p/T)}. \qquad (3.73)$$

The inversion formula for the two-sided Laplace transform is the same as that of the one-sided transform; the presence of poles to the right of the integration contour is normal in this case and need not disturb us. For $E \geq 0$ we close the contour on an infinite circle around the left half plane and pick up the residues of the new poles with negative p; for $E < 0$ we close it around the right half plane and pick up those with positive p. Writing the level density again as $g_T(E) = \tilde{g}_T(E) + \delta g_T(E)$, the result for the smooth part is

$$\tilde{g}_T(E) = \frac{1}{\hbar\omega}\left\{\Theta(E) + \operatorname{sign}(E)\sum_{p=1}^{\infty}(-1)^p e^{-p|E|/T}\frac{(p\hbar\omega/2T)}{\operatorname{Sinh}(p\hbar\omega/2T)}\right\}, \qquad (3.74)$$

and that for the oscillating part is

$$\delta g_T(E) = \frac{2}{\hbar\omega}\sum_{k=1}^{\infty}(-1)^k \cos\left(\frac{2\pi kE}{\hbar\omega}\right)\frac{\tau_k}{\operatorname{Sinh}\tau_k}\Theta(E), \quad \text{with} \quad \tau_k = \frac{2\pi^2 kT}{\hbar\omega}. \qquad (3.75)$$

Proceeding similarly for the three-dimensional harmonic oscillator, we obtain for the smooth part

$$\tilde{g}_T(E) = \frac{1}{2(\hbar\omega)^3}\left\{\left[E^2 - \frac{1}{4}(\hbar\omega)^2 + \frac{\pi^2}{3}T^2\right]\Theta(E)\right.$$
$$\left. + \text{sign}(E)\, T^2 \sum_{p=1}^{\infty}\frac{(-1)^p}{p^2}e^{-p|E|/T}\left[\frac{(p\hbar\omega/2T)}{\text{Sinh}(p\hbar\omega/2T)}\right]^3\right\}, \quad (3.76)$$

and for the oscillating part

$$\delta g_T(E) = \frac{1}{(\hbar\omega)^3}\sum_{k=1}^{\infty}(-1)^k\frac{\tau_k}{\text{Sinh}\,\tau_k}$$
$$\times\left\{\left[E^2 - \frac{(\hbar\omega)^2}{4} + 2\pi^2T^2\left(\frac{1}{2} + \frac{\text{Coth}\,\tau_k}{\tau_k} - \text{Coth}^2\tau_k\right)\right]\cos\left(\frac{2\pi kE}{\hbar\omega}\right)\right.$$
$$\left. + \frac{E\hbar\omega}{k\pi}\left(1 - \tau_k\,\text{Coth}\,\tau_k\right)\sin\left(\frac{2\pi kE}{\hbar\omega}\right)\right\}\Theta(E). \quad (3.77)$$

An alternative form of the smooth part is derived in the framework of the extended Thomas-Fermi model in Chapter 4, Problem 4.6; Eq. (3.76) may be gained from an asymptotic expansion of the result (4.217) given there. Note that the above results converge exactly to Eqs. (3.49),(3.69) in the limit $T \to 0$. The interesting thing is that we see how a finite temperature suppresses the oscillating part $\delta g_T(E)$. All oscillating terms are damped by the factor $(\tau_k/\text{Sinh}\,\tau_k)$. At a temperature of $T \simeq \hbar\omega/2$, this already gives a suppression by a factor $\sim 10^{-3}$ and there are practically no shell effects left in $g_T(E)$. This is well illustrated in Figure 3.3, where we show $g_T(E)$ for the three-

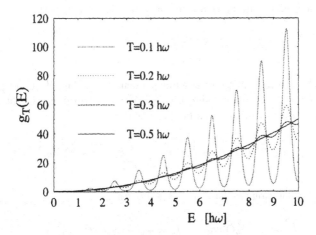

Figure 3.3: Level density $g_T(E)$ of the three-dimensional isotropic harmonic oscillator, calculated as the sum of Eqs. (3.76) and (3.77) for four different temperatures.

dimensional case at four different finite temperatures. The solid line, corresponding to $T = 0.5\,\hbar\omega$, is perfectly smooth and coincides with $\tilde{g}_T(E)$ whose temperature dependence is negligible in the range shown. Thus, a temperature of half the level spacing $\hbar\omega$ is sufficient to completely smooth out the shell effects.

3.2.6   Three-dimensional rectangular box

For box potentials in more than one dimension, it is not possible to write down the single-particle energies E_n in terms of one single quantum number. However, using the Poisson summation formula repeatedly, it is nevertheless possible to give exact trace formulae due to the separability of these systems in cartesian coordinates. We have actually given an exact result for the two-dimensional rectangular box in Sect. 2.7.1. We shall presently derive the analogous result for a rectangular box in three dimensions, thereby retrieving the results in one and two dimensions as byproducts. These results have also been derived by Balian and Bloch [2].

We consider a particle with mass m, confined by a rectangular box in three dimensions with side lengths a_1, a_2 and a_3 and ideally reflecting walls. The exact quantum-mechanical eigenenergies, obtained by solving the Schrödinger equation with Dirichlet boundary conditions, are given by

$$E_{n_1,n_2,n_3} = E_0 \left(n_1^2 \frac{L^2}{a_1^2} + n_2^2 \frac{L^2}{a_2^2} + n_3^2 \frac{L^2}{a_3^2} \right), \qquad n_1, n_2, n_3 = 1, 2, \dots. \qquad (3.78)$$

where we again use the quantity E_0 as energy unit:

$$E_0 = \frac{\hbar^2 \pi^2}{2mL^2}. \qquad (3.79)$$

We apply now the Poisson summation formula to the partition function

$$Z(\beta) = \sum_{n_1=1}^{\infty} \sum_{n_2=1}^{\infty} \sum_{n_3=1}^{\infty} \exp\left(-\beta E_{n_1,n_2,n_3}\right). \qquad (3.80)$$

Note that $Z(\beta)$ above is a product of three independent sums. Therefore, it is sufficient to treat the one-dimensional case. Using Eq. (2.157) and performing the integral over n_i, one obtains for the one-dimensional box with length a_i

$$Z_i^{(1)}(\beta) = \sum_{n_i=1}^{\infty} \exp\left(-\beta E_0 n_i^2 L^2/a_i^2\right) = \frac{1}{2}\sqrt{\frac{\pi}{\beta E_0}} \frac{a_i}{L} \sum_{M=-\infty}^{\infty} \exp\left[-\pi^2 M^2 a_i^2/(\beta E_0 L^2)\right] - \frac{1}{2}. \qquad (3.81)$$

In order to find this solution, we have used the integral

$$\int_0^{\infty} \exp(-cx^2)\,\cos(bx)\,dx = \frac{1}{2}\sqrt{\frac{\pi}{c}}\exp(-b^2/4c). \qquad (3.82)$$

We now take the product of three expressions of the form (3.81), with the three respective side lengths a_i inserted. The inverse Laplace transforms necessary to determine the level density are given in Appendix B. The final result is the following exact trace formula for the level density of the three-dimensional rectangular box:

$$g(E) = g^{(3)}(E) - \frac{1}{2}\left[g_{12}^{(2)}(E) + g_{23}^{(2)}(E) + g_{31}^{(2)}(E)\right]$$
$$+ \frac{1}{4}\left[g_1^{(1)}(E) + g_2^{(1)}(E) + g_3^{(1)}(E)\right] - \frac{1}{8}\delta(E). \tag{3.83}$$

The leading term, coming from the three-fold summations, is given by

$$g^{(3)}(E) = \frac{1}{E_0}\frac{\pi}{4}\sqrt{\frac{E}{E_0}}\frac{a_1 a_2 a_3}{L^3} \sum_{M_1,M_2,M_3=-\infty}^{\infty} j_0\left[\frac{1}{\hbar}S_{M_1,M_2,M_3}^{(3)}\right], \tag{3.84}$$

where $j_0(x)$ is the spherical Bessel function of order zero, and the quantities $S_{M_1,M_2,M_3}^{(3)}$ are given by

$$S_{M_1,M_2,M_3}^{(3)} = pL_{M_1,M_2,M_3} = \sqrt{2mE}\, 2\sqrt{M_1^2 a_1^2 + M_2^2 a_2^2 + M_3^2 a_3^2} \tag{3.85}$$

and correspond to the actions of the classical periodic orbits with lengths L_{M_1,M_2,M_3}. The absolute values of the M_i count the numbers of reflections from one of the walls in the i-th direction. (We refer to Sect. 6.1.5 for a more explicit discussion of these periodic orbits in two dimensions.) Using $j_0(x) = \sin(x)/x$, the oscillating part of (3.84) has the usual form (3.42) of the trace formula.

The result (3.83) for the three-dimensional box contains also combinations of the level densities for the two- and the one-dimensional box. These extra terms emerge from the edge correction in the Poisson formula (2.157). Their physical interpretation within the periodic orbit theory will be given in Sect. 6.1.5. The leading term for a two-dimensional box with side lengths a_i and a_j is, in agreement with the result (2.178) found in Sect. 2.7.1,

$$g_{ij}^{(2)}(E) = \frac{1}{E_0}\frac{\pi}{4}\frac{a_i a_j}{L^2} \sum_{M_1,M_2=-\infty}^{\infty} J_0\left[\frac{1}{\hbar}S_{M_1,M_2}^{(2)}(a_i,a_j)\right], \tag{3.86}$$

with an obvious expression, analogous to (3.85), for the actions $S_{M_1,M_2}^{(2)}$ of the two-dimensional orbits [see also Eq. (2.181)]. The oscillating part in (3.86) is not exactly of the usual form (3.42) of the trace formula. This is, however, retrieved when using the asymptotic expansion of the Bessel function $J_0(x)$ (see Sect. 2.7.1). The level density for a one-dimensional box with length a_i is finally

$$g_i^{(1)}(E) = \frac{1}{\sqrt{E_0 E}}\frac{1}{2}\frac{a_i}{L} \sum_{M=-\infty}^{\infty} \cos\left[\frac{1}{\hbar}S_M^{(1)}(a_i)\right], \tag{3.87}$$

where $S_M^{(1)}(a_i) = 2Ma_i\sqrt{2mE}$, in agreement with Eq. (3.64).

In the result (3.83), we have kept the delta function peaked at $E = 0$. It does not contribute to the level density at $E > 0$, where all the eigenvalues lie, but when integrating $g(E)$ in order to obtain the number N of particles according to Eq. (3.38), it contributes a constant -1/8 to N and thus affects the position of the Fermi energy μ.

The average part of the level density (3.83) is found from those contributions in all three terms when all the M_i are zero, and reads

$$\tilde{g}(E) = \frac{1}{E_0} \left\{ \frac{\pi}{4} \sqrt{\frac{E}{E_0}} \frac{\mathcal{V}}{L^3} - \frac{\pi}{8} \frac{\mathcal{S}}{2L^2} + \frac{1}{8} \sqrt{\frac{E_0}{E}} \frac{\mathcal{L}}{L} \right\} , \qquad (3.88)$$

where $\mathcal{V} = a_1 a_2 a_3$ is the volume and $\mathcal{S} = 2(a_1 a_2 + a_2 a_3 + a_3 a_1)$ the surface of the box, and $\mathcal{L} = a_1 + a_2 + a_3$ is the sum of the side lengths. As in the earlier examples given above, this smooth level density is identical with that found from the extended Thomas-Fermi model or the Weyl expansion (see Chapter 4). In Sect. 4.6.2, we will see that a corresponding expression holds actually for the average density of states in a cavity of arbitrary smooth shape [see Eq. (4.143)]. The first two terms, containing the volume $\mathcal{V}$ and the surface $\mathcal{S}$, have exactly the same coefficients. The next term, however, is in general proportional to the mean curvature $\mathcal{C}$ which cannot be defined for cavities with edges such as the rectangular box.

When one uses the Neumann boundary conditions in solving the Schrödinger Equation for the rectangular box, demanding that not the wave functions but their normal derivatives be zero along the boundary, the eigenenergies are still given by Eq. (3.78); the quantum numbers n_i may, however, also be zero (but not all three at the same time). The derivation of the trace formula then proceeds in exactly the same way, except that now the edge correction for the Poisson summation formula in each dimension changes sign. As a result, the second and fourth terms in the final result (3.83) get a positive instead of the negative sign. This is also in agreement with the opposite signs of the surface term in the general Weyl expansion for the two types of boundary conditions (see Sect. 4.6.2).

3.2.7 Equilateral triangular billiard

All the examples given so far in this chapter represent separable systems and are therefore particularly simple to solve exactly, even when their spectra are given in terms of several quantum numbers. A non-separable but still integrable system which is exactly solvable shall be treated in the present example.[2] We will go through it in some detail because its treatment involves some nontrivial steps and the result exhibits several interesting features; in Chapters 5 and 6 we shall apply the semiclassical periodic orbit theory to reproduce the resulting trace formula.

We consider a particle with mass μ in a two-dimensional billiard whose boundary forms an equilateral triangle with side length L. We take the boundary to be given by the three points $(0,0)$, $(-L/2, \sqrt{3}L/2)$ and $(L/2, \sqrt{3}L/2)$ in the (x,y) plane. The quantum spectrum of this system is found by solving the Schrödinger equation

$$-\frac{\hbar^2}{2\mu} \left(\frac{\partial^2}{\partial x^2} + \frac{\partial^2}{\partial y^2} \right) \Psi(x,y) = E\Psi(x,y) \qquad (3.89)$$

---

[2] We note in passing that for a corresponding three-dimensional system, a tetrahedral box with reflecting walls, the exact quantum spectrum has been obtained in Ref. [3] in an elegant way, along with that of the triangular billiard discussed here.

with some boundary conditions. We first discuss the case with Dirichlet boundary conditions, demanding that the wavefunctions be zero along the boundary. The following two sets of wavefunctions $\Psi_{mn}^{(c,s)}(x,y)$ fulfill this condition [4, 5] for arbitrary integers m, n:

$$
\begin{aligned}
\Psi_{mn}^{(c,s)}(x,y) = \; & (\cos, \sin) \left[(2m-n)\frac{2\pi}{3L}x \right] \sin\left(n\frac{2\pi}{\sqrt{3}L}y \right) \\
& - (\cos, \sin) \left[(2n-m)\frac{2\pi}{3L}x \right] \sin\left(m\frac{2\pi}{\sqrt{3}L}y \right) \\
& + (\cos, \sin) \left[-(m+n)\frac{2\pi}{3L}x \right] \sin\left[(m-n)\frac{2\pi}{\sqrt{3}L}y \right].
\end{aligned}
\tag{3.90}
$$

Hereby the x-dependent part may be chosen to be either the cos or the sin function, simultaneously in all three terms, leading to the wavefunctions denoted by $\Psi_{mn}^{(c)}$ and $\Psi_{mn}^{(s)}$, respectively. These wavefunctions have the following symmetry properties:

$$
\Psi_{mn}^{(c,s)} = -\Psi_{nm}^{(c,s)}; \qquad \Psi_{m,m-n}^{(c)} = \Psi_{mn}^{(c)}; \qquad \Psi_{m,m-n}^{(s)} = -\Psi_{mn}^{(s)}; \tag{3.91}
$$

$$
\Psi_{-m,-n}^{(c)} = -\Psi_{mn}^{(c)}; \qquad \Psi_{-m,-n}^{(s)} = \Psi_{mn}^{(s)}; \qquad \Psi_{m,-n}^{(c,s)} = -\Psi_{m+n,n}^{(c,s)}. \tag{3.92}
$$

Note that $\Psi_{mn}^{(c,s)}$ vanishes identically if either n or m is zero, but also if $n = m$. Furthermore, due to Eqs. (3.92), states with any negative quantum number are identical with states having only positive quantum numbers. Thus, the selection rules for allowed quantum states are

$$
m = 1, 2, 3, \ldots; \qquad n = 1, 2, 3, \ldots; \qquad m \geq 2n. \tag{3.93}
$$

Inserting the wavefunction (3.90) into the Schrödinger equation (3.89) leads to the energy spectrum

$$
E_{mn} = E_\Delta \left(n^2 + m^2 - mn \right), \qquad E_\Delta = \frac{16}{9}\frac{\hbar^2\pi^2}{2\mu L^2}, \tag{3.94}
$$

where the allowed values of m and n are given by the selection rules (3.93). All states are doubly degenerate due to the two independent wavefunctions $\Psi_{mn}^{(c)}$ and $\Psi_{mn}^{(s)}$, except for the states with $m = 2n$ which are singly degenerate because $\Psi_{2n,n}^{(s)}$ vanishes identically. Note that the eigenfrequency spectrum ω_{mn} of an equilateral triangular membrane, which is the same as that in (3.94) apart from the constants, has been derived already in 1852 by Lamé [6] in a course on the theory of elasticity.[3]

The exact level density for the Dirichlet boundary conditions is thus given by

$$
g^{(D)}(E) = 2 \sum_{n=1}^{\infty} \sum_{m=2n+1}^{\infty} \delta\left(E - E_{mn} \right) + \sum_{n=1}^{\infty} \delta\left(E - E_{2n,n} \right). \tag{3.95}
$$

---

[3]Lamé introduces hereby a set of three coordinates perpendicular to the three sides of the boundary. What he writes after presenting this result appeared so interesting to us that we quote it in the concluding remarks at the end of this book.

For the calculations below, it is advantageous to rewrite the summations in the above equation by exploiting the symmetries of the wavefunctions. First, using the second and the third identity in Eq. (3.91), we can extend the summation over one octant of the (n, m) plane, including once all points lying between (but not on) the borders defined by $n = 0$ and $n = m$, so that the level density may be written as

$$g^{(D)}(E) = \sum_{n=1}^{\infty} \sum_{m=n+1}^{\infty} \delta\left(E - E_{mn}\right). \tag{3.96}$$

Next, we use the first identity in Eq. (3.91) to extend the summation over the whole first quadrant of the (n, m) plane, excluding the diagonal $n = m$ and the m and n axes, and dividing the result by two:

$$g^{(D)}(E) = \frac{1}{2} \sum_{n=1}^{\infty} \sum_{m=1}^{\infty} \delta\left(E - E_{mn}\right) - \frac{1}{2} \sum_{n=1}^{\infty} \delta\left(E - E_{nn}\right). \tag{3.97}$$

Finally, we use the third identity in Eq. (3.92) and the obvious identity $E_{nn} = E_{n0}$ to write the level density as

$$g^{(D)}(E) = \frac{1}{3} \sum_{n=1}^{\infty} \sum_{m=-\infty}^{\infty} \delta\left(E - E_{mn}\right) - \frac{2}{3} \sum_{n=1}^{\infty} \delta\left(E - E_{n0}\right). \tag{3.98}$$

Correspondingly, the exact partition function is given by

$$Z^{(D)}(\beta) = \frac{1}{3} \sum_{n=1}^{\infty} \sum_{m=-\infty}^{\infty} \exp(-\beta E_{mn}) - \frac{2}{3} \sum_{n=1}^{\infty} \exp(-\beta E_{n0}). \tag{3.99}$$

Now we perform the summations in Eq. (3.99) using the one-dimensional and two-dimensional Poisson formulae Eqs. (2.157) and (2.158), respectively. In the resulting two-dimensional integral of the leading term, the m integration can be done using the substitution $m' = m - n/2$, leading to a simple Gaussian integral. The remaining integrals can also be done exactly using the identity (3.82). After Laplace inversion using the relations given in Appendix B and a careful regrouping of terms, we readily find the following exact trace formula for the level density:

$$g^{(D)}(E) = \frac{\pi}{3\sqrt{3}E_{\Delta}} \left\{ 1 + 12 \sum_{M_2 > M_1 > 0} J_0(kL_{M_1 M_2}) + 6 \sum_{M > 0} J_0(kL_{MM}) + 6 \sum_{M > 0} J_0(kL_{M0}) \right\}$$

$$- \frac{1}{2\sqrt{EE_{\Delta}}} \left\{ 1 + 2 \sum_{M > 0} \cos(kL_M) \right\} + \frac{1}{3}\delta(E), \tag{3.100}$$

where $k = \sqrt{2\mu E}/\hbar$ is the wave number and

$$L_{M_1 M_2} = \sqrt{3(M_1^2 + M_2^2 + M_1 M_2)}\, L, \qquad L_M = \frac{3}{2}ML. \tag{3.101}$$

In order to arrive at Eq. (3.100), we have exploited the fact that the quantities $L_{M_1 M_2}$ have the same symmetry properties under sign changes of the M_i as the E_{mn} (3.94) under sign changes of m and n due to Eqs. (3.91) and (3.92). A more compact form of this result, again exploiting these symmetry properties, is

$$g^{(D)}(E) = \frac{\pi}{3\sqrt{3}E_\Delta} \sum_{M_1,M_2=-\infty}^{\infty} J_0(kL_{M_1 M_2}) - \frac{1}{2\sqrt{EE_\Delta}} \sum_{M=-\infty}^{\infty} \cos(kL_M) + \frac{1}{3}\delta(E). \quad (3.102)$$

If the Bessel function J_0 is expanded asymptotically for large values of its argument, $J_0(x) \sim (2/\pi x)^{1/2} \cos(x - \pi/4)$, both terms in Eq. (3.102) have the form of Eq. (3.42). The terms with $L_{M_1 M_2} = L_M = 0$ yield the average part of the level density which agrees with the results of the ETF model and the Weyl expansion discussed in Sect. 4.6.2. Note, in particular, that the coefficient $1/3$ of the delta function term agrees with the general formula (4.152) for the corner correction. We have kept this term, which comes here from the edge correction in the Poisson formula. It does not contribute to the level density in the physical domain $E > 0$, but gives an additive constant to its first integral.

We now do the same for the Neumann boundary condition, demanding that the normal derivative of the wave function be zero along the boundary, i.e., $\partial\Phi/\partial n = 0$, where n is the unit vector normal to the boundary. This condition is fulfilled for the following wavefunctions

$$\begin{aligned}
\Phi_{mn}^{(c,s)}(x,y) &= (\cos, \sin)\left[(2m-n)\frac{2\pi}{3L}x\right]\cos\left(n\frac{2\pi}{\sqrt{3}L}y\right) \\
&+ (\cos,\sin)\left[(2n-m)\frac{2\pi}{3L}x\right]\cos\left(m\frac{2\pi}{\sqrt{3}L}y\right) \\
&+ (\cos,\sin)\left[-(m+n)\frac{2\pi}{3L}x\right]\cos\left[(m-n)\frac{2\pi}{\sqrt{3}L}y\right]. \quad (3.103)
\end{aligned}$$

and leads to the same eigenvalues (3.94) given above. However, either of the quantum numbers m and n (but not both!) may now be zero. These wavefunctions have the same symmetries as the Ψ_{mn} in Eqs. (3.91) and (3.91) (up to irrelevant phases), and the additional symmetry

$$\Phi_{0n}^{(c)} = \Phi_{n0}^{(c)} = \Phi_{nn}^{(c)}; \quad \Phi_{0n}^{(s)} = \Phi_{n0}^{(s)} = -\Phi_{nn}^{(s)}. \quad (3.104)$$

We can therefore write (using $E_{nn} = E_{0n}$)

$$g^{(N)}(E) = 2\sum_{n=1}^{\infty}\sum_{m=2n+1}^{\infty} \delta(E - E_{mn}) + \sum_{n=1}^{\infty} \delta(E - E_{2n,n}) + 2\sum_{n=1}^{\infty} \delta(E - E_{0n}). \quad (3.105)$$

Performing similar mappings of the eigenvalues in the (n, m) plane as above, this leads to the same trace formula as in Eq. (3.102) but with a plus sign in front of the second term and a different constant in front of the delta function term:

$$g^{(N)}(E) = \frac{\pi}{3\sqrt{3}E_\Delta} \sum_{M_1,M_2=-\infty}^{\infty} J_0(kL_{M_1 M_2}) + \frac{1}{2\sqrt{EE_\Delta}} \sum_{M=-\infty}^{\infty} \cos(kL_M) - \frac{2}{3}\delta(E). \quad (3.106)$$

The sign change of the second term is in agreement with the general Weyl expansion for the smooth level density, obtained by retaining only the contributions from $M_1 =$

$M_2 = M = 0$, as discussed in Sect. 4.6.2. For later reference, we give here its explicit form

$$g_{ETF}(E) = \frac{\pi}{3\sqrt{3}E_\Delta} \mp \frac{1}{2\sqrt{EE_\Delta}} \; ; \tag{3.107}$$

the upper and lower signs of the second term holds for Dirichlet and Neumann boundary conditions, respectively.

The derivation of the above trace formulae from the periodic orbit theory will be discussed in Sect. 6.1.2 and in Problem 5.2. The first sum in (3.102) comes from degenerate families of orbits, characterized by pairs $(M_1 M_2)$ of integers, with lengths $L_{M_1 M_2}$. The two shortest orbit families (10) and (11) are illustrated in Fig. 6.1 (center and right side). The second sum in (3.102) represents corrections of relative order $\hbar$, coming from an isolated orbit with triangular shape of length $L_1 = 3L/2$ (left side of Fig. 6.1) and its odd repetitions (odd values of M); the terms with even values of M in this sum come from those orbits of the family (MM) with length $L_{MM} = 3ML$ which graze the sides of the triangle.

3.2.8 Cranked or anisotropic harmonic oscillator

Consider the two-dimensional motion of a particle in the (x, y)-plane, governed by the Hamiltonian (see, e.g., Ref. [7])

$$\hat{H} = \hat{T} + \frac{m}{2}\omega^2 r^2 + \Omega\,\hat{L}_z \,. \tag{3.108}$$

The particle is bound by a harmonic oscillator with frequency ω, and the oscillator itself is rotating about the z-axis with a frequency Ω. Since the last term in Eq. (3.108), $\Omega \hat{L}_z$, commutes with the harmonic oscillator Hamiltonian, the eigenspectrum of $\hat{H}$ may be immediately written down in terms of a radial quantum number n_r and the eigenvalues $\hbar l$ of the angular momentum operator $\hat{L}_z$:

$$E_{n_r,l} = \hbar\omega\,(2n_r + |l| + 1) + \hbar\Omega l, \quad n_r = 0, 1, 2, ..., \quad l = 0, \pm 1, \pm 2, ... \tag{3.109}$$

The partition function can be easily summed for this spectrum to give

$$Z_\Omega(\beta) = \frac{1}{4\,\mathrm{Sinh}[\beta\hbar(\omega + \Omega)/2]\,\mathrm{Sinh}[\beta\hbar(\omega - \Omega)/2]}. \tag{3.110}$$

Note that this partition function is strictly identical to that of a *deformed* two-dimensional harmonic oscillator with different frequencies in x and y direction:

$$V(x, y) = \frac{m}{2}\left(\omega_x^2 x^2 + \omega_y^2 y^2\right), \tag{3.111}$$

if we identify the frequencies with

$$\omega_x = \omega + \Omega, \qquad \omega_y = \omega - \Omega, \tag{3.112}$$

and restrict the cranking frequency to values $-\omega < \Omega < +\omega$. This is so in spite of the fact that the two systems and their dynamics are different. It just so happens that their spectra (with the correct degeneracies) can be mapped onto each other and that the partition sums become identical (see also Problem 3.3). There is actually a third system that has the same partition function and thus the same level density: that of a particle in two dimensions moving in a homogeneous transverse magnetic field. This system will be discussed further below.

The exact form of the level density (and that of the classical orbits) depends on the frequency ratio $\omega_x : \omega_y$ being rational or irrational. $Z(\beta)$ has poles of both first and second order. The poles of first order lead to terms with constant amplitudes and those of second order to terms proportional to the energy E (cf. Appendix B).

For the case of *irrational* Ω/ω, i.e. two *incommensurate frequencies* ω_x, ω_y, the only double pole is at $\beta = 0$ and yields the average part of $g(E)$ which is linear in E; all other poles are single and yield the oscillating part of $g(E)$ with constant amplitudes. The result of summing all residues is [8]

$$g(E, \Omega) = \frac{1}{\hbar\omega_x \hbar\omega_y} E + \frac{1}{\hbar\omega_x} \sum_{k=1}^{\infty} \frac{(-1)^k}{\sin\left(k\pi\frac{\omega_y}{\omega_x}\right)} \sin\left(k\frac{2\pi E}{\hbar\omega_x}\right)$$
$$+ \frac{1}{\hbar\omega_y} \sum_{k=1}^{\infty} \frac{(-1)^k}{\sin\left(k\pi\frac{\omega_x}{\omega_y}\right)} \sin\left(k\frac{2\pi E}{\hbar\omega_y}\right). \qquad (3.113)$$

The explicit dependence on the cranking frequency Ω is given through Eq. (3.112).

For the case of a *rational* cranking frequency Ω, i.e. two *commensurate frequencies* ω_x, ω_y, the situation is more complicated. There are double poles whenever integer multiples of the two frequencies coincide. Let us assume that this ratio is rational:

$$\omega_x : \omega_y = (\omega + \Omega) : (\omega - \Omega) = n : p \qquad (n, p \in \mathbf{N}) \qquad (3.114)$$

with relatively prime integers n and p. We write

$$\omega_x = (\omega + \Omega) = n\omega_0, \quad \omega_y = (\omega - \Omega) = p\omega_0. \qquad (3.115)$$

The final result for the exact quantum level density then becomes [8]

$$g(E) = \frac{E}{\hbar\omega_x \hbar\omega_y}\left[1 + 2\sum_{k=1}^{\infty} (-1)^{k(p+n)} \cos\left(k\frac{2\pi E}{\hbar\omega_0}\right)\right]$$
$$+ \frac{1}{\hbar\omega_x} \sum_{k \neq ln}^{\infty} \frac{(-1)^k}{\sin\left(k\pi\frac{p}{n}\right)} \sin\left(k\frac{2\pi E}{\hbar\omega_x}\right)$$
$$+ \frac{1}{\hbar\omega_y} \sum_{k \neq lp}^{\infty} \frac{(-1)^k}{\sin\left(k\pi\frac{n}{p}\right)} \sin\left(k\frac{2\pi E}{\hbar\omega_y}\right). \qquad (l \in \mathbf{N}) \qquad (3.116)$$

For n or p equal to unity, the corresponding sum(s) on the second and third lines vanish (since then no term fulfills the condition $k \neq ln$ or $k \neq lp$, respectively). In particular, for the uncranked isotropic harmonic oscillator we get back the result (3.68).

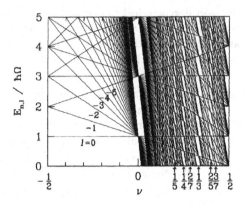

Figure 3.4: The energy spectrum of a cranked two-dimensional harmonic oscillator as given by Eq. (3.109), as a function of ν defined in Eq. (3.117). The Landau levels are formed at $\nu = 0$. (From [7].)

Some intriguing points emerge from the study of the spectrum (3.109). It generates the so-called "Farey fan" pattern [9] and exhibits clearly the intimate connection between quantum gaps and classical periodic orbits [2]. These quantum gaps shall be illustrated in the following. In Fig. 3.4, the pattern of the energy levels (3.109) is shown by varying the cranking frequency Ω in the range $0 \leq \Omega \leq 2\omega$, keeping the oscillator frequency ω fixed. The energy levels are plotted versus the dimensionless quantity ν, defined by

$$\nu = \frac{(\Omega - \omega)}{2\omega}, \tag{3.117}$$

in the range $-\frac{1}{2} \leq \nu \leq \frac{1}{2}$. The collapse of the single-particle states for $\Omega = \omega$ into highly degenerate equally spaced levels is clearly seen. These degenerate levels are called *Landau levels* [10] in the context of the motion of a charged particle in a uniform magnetic field. For such a particle (without spin), the Hamiltonian is

$$\hat{H}_B = \frac{1}{2m}(\mathbf{p} - \frac{e}{c}\mathbf{A})^2, \tag{3.118}$$

where e is its charge and $\mathbf{p}$ its momentum in the (x, y)-plane (cf. Problem 2.7). The vector potential $\mathbf{A} = \frac{1}{2}(\mathbf{B} \times \mathbf{r})$ generates a transverse homogeneous magnetic field $\mathbf{B}$ in the z direction. Taking $e = -|e|$ and the symmetric gauge, $\mathbf{A} = (-\frac{1}{2}By, \frac{1}{2}Bx, 0)$, $\hat{H}_B$ reduces to $\hat{H}$ in Eq. (3.108) for $\omega = \Omega = \Omega_c/2$, where

$$\Omega_c = \frac{|e|B}{mc} \tag{3.119}$$

is called the cyclotron frequency.

As is clear from Fig. 3.4, the collapsing single-particle states in the lowest Landau level have all aligned (but different) angular momenta, and originate from different shells of the harmonic oscillator. Even for the range $-\frac{1}{2} \leq \nu < 0$, the convergence of the states to the Landau levels is preceded by a repeating zig-zag pattern of gaps. This is shown vividly for the higher excited states in Fig. 3.5. Note that the quantum gaps appear at those values of the cranking frequency for which Ω/ω is a rational number, a fact that links these gaps to the occurrence of classical periodic orbits. These may be easily seen by examining the solutions of the classical equations of motion. Using the variable $z = x + iy$, the two equations of motion may be written compactly as

$$\ddot{z} = \left(\Omega^2 - \omega^2\right) z + 2i\,\Omega\,\dot{z}. \tag{3.120}$$

The first term on the right is an attractive harmonic force for $\Omega < \omega$, but it becomes repulsive for cranking frequencies beyond the Landau levels. The general solution of Eq. (3.120) is

$$z = A\,e^{i(\Omega-\omega)t} + B\,e^{i(\Omega+\omega)t}, \tag{3.121}$$

where A and B are constants. The two normal mode frequencies (for $\Omega \neq \omega$) are $|\omega - \Omega|$ and $(\omega + \Omega)$. When the ratio of these frequencies is a rational fraction, a closed periodic orbit in the x-y plane is obtained. Some of these periodic orbits are shown in Fig. 3.6 on the next page for various values of ν. For the special case of $\Omega = \omega$, the solutions of Eq. (3.120) are given by circles in the (x, y)-plane with arbitrary centres and are nothing but the cyclotron orbits.

The accidental degeneracy of the anisotropic two-dimensional harmonic oscillator has been discussed by Louck *et al.* [11]. The quantum analogue of this motion with coherent states is analyzed in Refs. [12, 13]. Apart from this case, the most prominent gaps occur in Figs. 3.4 and 3.5 for the simplest fractions. For example, the large gaps

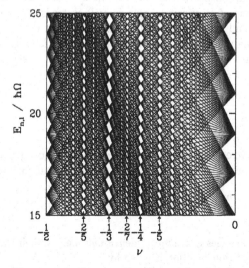

Figure 3.5: The same as the preceding figure, showing the pattern of the quantum gaps for higher excitations in the range $-1/2 \leq \nu \leq 0$. (From [7].)

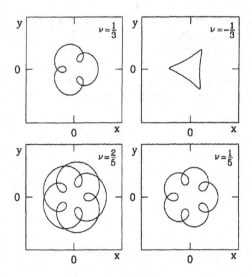

Figure 3.6:   Some periodic orbits in the $x - y$ plane for a few values of ν. These are solutions of Eq. (3.120). (From [7].)

at $\nu = \pm 1/3$ correspond to the situation when one normal mode frequency is twice the other. It is interesting to study the degeneracy of the levels for the situation when the cranking frequency Ω is greater than the oscillator frequency ω. The Farey fan pattern [9] has also been studied in connection with number theory and continued fractions. At the energy $E = \hbar\omega$, inspection of the level at $\nu = 1/k$ (with an integer $k > 1$) reveals that the number of converging single-particle states is exactly a fraction $1/k$ of the Landau level. Finally, it may be worth mentioning that the sequence of the gaps generated by this dynamical model for $\nu > 0$ are the same, for the odd denominators, as the Haldane hierarchy [14] in the fractional quantum Hall effect [15, 16].

3.3   Problems

(3.1) Anisotropic three-dimensional harmonic-oscillator potential
Consider the Hamiltonian of an anisotropic three-dimensional harmonic oscillator

$$H = \frac{m}{2}(p_x^2 + p_y^2 + p_z^2) + \frac{m}{2}(\omega_x^2 x^2 + \omega_y^2 y^2 + \omega_z^2 z^2), \qquad (3.122)$$

assuming that the frequencies ω_i are *incommensurable*, i.e., all their ratios are irrational.
(a) Show that the partition function is given by

$$Z(\beta) = \frac{1}{8\,\mathrm{Sinh}[\beta\hbar\omega_x/2]\,\mathrm{Sinh}[\beta\hbar\omega_y/2]\,\mathrm{Sinh}[\beta\hbar\omega_z/2]}. \qquad (3.123)$$

(b) Using the residue technique described in Sect. 3.2.1, show that the exact level density has the following form:

$$
g(E) = \frac{1}{2\,\hbar\omega_x\hbar\omega_y\hbar\omega_z} \left\{ E^2 - \frac{1}{12}\left[(\hbar\omega_x)^2 + (\hbar\omega_y)^2 + (\hbar\omega_z)^2\right] \right\} +
$$
$$
+ \frac{1}{2\hbar\omega_x} \sum_{k=1}^{\infty} \frac{(-1)^{k+1}}{\sin(k\pi\omega_y/\omega_x)\sin(k\pi\omega_z/\omega_x)} \cos\left(k\frac{2\pi E}{\hbar\omega_x}\right)
$$
$$
+ \frac{1}{2\hbar\omega_y} \sum_{k=1}^{\infty} \frac{(-1)^{k+1}}{\sin(k\pi\omega_x/\omega_y)\sin(k\pi\omega_z/\omega_y)} \cos\left(k\frac{2\pi E}{\hbar\omega_y}\right)
$$
$$
+ \frac{1}{2\hbar\omega_z} \sum_{k=1}^{\infty} \frac{(-1)^{k+1}}{\sin(k\pi\omega_x/\omega_z)\sin(k\pi\omega_y/\omega_z)} \cos\left(k\frac{2\pi E}{\hbar\omega_z}\right). \tag{3.124}
$$

Note that the first term above is the smooth part which can also be derived in the extended Thomas-Fermi model (see Sect. 4.3), and the remaining terms give the oscillatory part which can be found semiclassically as described in Chapter 5 (see, in particular, Problem 5.1). The analogous expressions for the cases where some or all of the frequencies have rational ratios may be found in Ref. [8].

(3.2) Landau orbits of a charged particle in a uniform magnetic field
From Fig. 3.4, it is clear that each electron in the lowest Landau level (for $\omega = \Omega = \Omega_c/2$ and $n = 0$) is in a *stretched* state of the harmonic oscillator with the lowest possible angular momentum projection $L_z = -l$. The wave function of a particle in such a state is given by (with $z = x - iy$)

$$
\phi_{n,-l}(z) = \frac{1}{\sqrt{\pi\,l!}}\frac{1}{a^{l+1}} z^l \exp(-|z|^2/2a^2) ,
$$

where $a = \sqrt{\hbar/m\Omega} = \sqrt{2}l_B$ and l_B is the magnetic length defined by Eq. (1.91).

(a) Construct a determinantal N-electron state ($N \to \infty$) in the lowest Landau level ($n = 0$), and show that it is given by

$$
\Psi = \prod_{l=0}^{\infty} \left(\frac{1}{\sqrt{\pi\,l!}}\frac{1}{a^{l+1}}\right) \prod_{i<j}(z_i - z_j) \exp(-|z|^2/2a^2) .
$$

(b) Hence show that the single-particle density of the filled Landau level is

$$
\rho_0 = \sum_{l=0}^{\infty} |\phi_l(z)|^2 = \frac{1}{\pi a^2} = \frac{|e|B}{hc}
$$

per unit area. This agrees with the semiclassical estimate made previously.

(3.3) Isotropic two-dimensional harmonic oscillator with homogeneous magnetic field
Consider the motion of a particle in an isotropic two-dimensional harmonic-oscillator potential with a transverse homogeneous magnetic field [17]. Start from the Hamiltonian (3.118) and add the potential $V(x,y) = \frac{m}{2}\omega^2(x^2 + y^2)$, so that the total Hamiltonian has the form

$$
\hat{H} = \frac{1}{2m}(\mathbf{p} - \frac{e}{c}\mathbf{A})^2 + \frac{m}{2}\omega^2 r^2 = \frac{p^2}{2m} + \frac{m}{2}\left(\omega^2 + \omega_L^2\right)r^2 - \omega_L\,\hat{L}_z , \tag{3.125}
$$

where $\omega_L = eB/2mc$ is the Larmor frequency (1.130) which is half the cyclotron frequency (3.119).

(a) Show that the quantum spectrum of the Hamiltonian (3.125) is given by

$$E_{n_r l} = \hbar \sqrt{\omega_L^2 + \omega^2} \, (2n_r + |l| + 1) - \hbar \omega_L \, l \,, \qquad (3.126)$$

with $n_r = 0, 1, 2, \dots$ and $l = 0, \pm 1, \pm 2, \dots$.

(b) Show that the spectrum (3.126), including the degeneracies, can be mapped onto the energy spectrum of an anisotropic two-dimensional harmonic oscillator potential (3.111)

$$E_{n_x, n_y} = \hbar \omega_x (n_x + 1/2) + \hbar \omega_y (n_y + 1/2) \qquad (3.127)$$

with $n_x, n_y = 0, 1, 2, \dots$, if the frequencies ω_x, ω_y are chosen as

$$\omega_x = \sqrt{\omega_L^2 + \omega^2} + \omega_L \,, \qquad \omega_y = \sqrt{\omega_L^2 + \omega^2} - \omega_L \,. \qquad (3.128)$$

(Note that $\omega_x \omega_y = \omega^2$.) Consequently, the level density corresponding to the spectrum (3.126) is given by the trace formula in Eq. (3.113), where the dependence on the magnetic field is given through Eqs. (3.128) and (1.130).

(c) Expand the trace formula (3.113) for small magnetic fields B and show that the resulting trace formula is

$$g\left(E, B\right) = \frac{E}{(\hbar \omega)^2} \left[1 + 2 \sum_{n=1}^{n_{max}} \cos\left(n \frac{2\pi E}{\hbar \omega} \right) j_0\left(\beta n \frac{2\pi E}{\hbar \omega} \right) \right] \,, \qquad (3.129)$$

where β is the dimensionless magnetic field strength

$$\beta = \frac{\omega_L}{\omega} = \frac{eB}{2mc\omega} \,. \qquad (3.130)$$

Hereby, n_{max} must be chosen small enough so that $2\pi \beta \, n_{max} \ll 1$, but will be a large number for small field strengths B. (Attention: you will actually have to use a Laurent expansion, since the leading amplitudes in (3.113) go like $1/B$; care must be taken to get these diverging terms to cancel exactly.) Note that Eq. (3.129) is just the trace formula (3.68) for the isotropic two-dimensional harmonic oscillator with, however, an extra modulating amplitude factor involving the spherical Bessel function $j_0(x)$. In Problem 6.2, this modulation factor is derived from a semiclassical perturbative viewpoint.

(3.4) Two anyons with harmonic-oscillator confinement (After [18, 19])
Consider a particle with mass $m=1$ and charge e in a two-dimensional harmonic-oscillator potential with an additional vector potential $\mathbf{A}(\mathbf{r})$ defined by

$$A_x(\mathbf{r}) = -\frac{\phi}{2\pi} \frac{y}{r^2} \,, \qquad A_y(\mathbf{r}) = \frac{\phi}{2\pi} \frac{x}{r^2} \,, \qquad (r > 0) \qquad (3.131)$$

where ϕ is a constant parameter defining the strength of the magnetic flux. The Lagrangian of this system is given by

$$\mathcal{L} = \frac{1}{2} \left(\dot{x}^2 + \dot{y}^2 \right) - \frac{1}{2} \omega^2 (x^2 + y^2) + \frac{e}{c} \mathbf{v} \cdot \mathbf{A} \,, \qquad (3.132)$$

where $\mathbf{v} = (\dot{x}, \dot{y})$ is the velocity of the particle. The magnetic field defined by the vector potential (3.131) is oriented in the z-direction; its $\mathbf{r}$-dependence in the x, y-plane is a two-dimensional delta function with strength ϕ

$$B_z(\mathbf{r}) = \phi \, \delta^{(2)}(\mathbf{r}), \tag{3.133}$$

so that the magnetic flux through any finite area surrounding the origin is given by ϕ. This system of a harmonic oscillator with magnetic flux tube can actually be understood as that of two anyons with statistical interaction [20], bound by a harmonic attractive force and described in their relative coordinates.

(a) Show that the classical equations of motion in r-space are independent of ϕ:

$$\ddot{r} = -\omega^2 r, \qquad \frac{d}{dt}(r^2 \dot{\theta}) = 0. \tag{3.134}$$

This is easily understood once one realizes that the term $\mathbf{v} \cdot \mathbf{A}$ in the Lagrangian (3.132) is proportional to the time derivative of the polar angle θ

$$\mathbf{v} \cdot \mathbf{A} = \frac{\phi}{2\pi} \frac{1}{r^2} (x\dot{y} - y\dot{x}) = \frac{\phi}{2\pi} \dot{\theta}, \tag{3.135}$$

and thus cannot contribute to the equations of motion.

(b) Show that the Hamiltonian in phase-space, using polar coordinates (r, θ), is given by

$$H(r, p_r, \theta, p_\theta) = \frac{1}{2} p_r^2 + \frac{1}{2r^2} \left(p_\theta - \frac{e}{c} \frac{\phi}{2\pi} \right)^2 + \frac{1}{2} \omega^2 r^2, \tag{3.136}$$

where the canonical momenta are

$$p_r = \dot{r}, \qquad p_\theta = L = r^2 \dot{\theta} + \frac{e}{c} \frac{\phi}{2\pi}. \tag{3.137}$$

The phase-space dynamics will therefore depend on α. Since both energy and angular momentum L are conserved, we have two constants of motion. The linear canonical momenta p_x and p_y diverge, however, at the origin; therefore the origin of the phase space must be omitted from the classical dynamics. Such a system is called *pseudo-integrable* (see, e.g., Ref. [18]).

(c) Quantize Eq. (3.136) and show that the quantum spectrum is given by

$$E = \hbar\omega(2n_r + |l - \alpha| + 1), \quad n_r = 0, 1, 2, ..., \quad l = 0, \pm 1, \pm 2, ..., \tag{3.138}$$

where l and n_r are the angular momentum and radial quantum numbers, respectively. Hereby we have introduced the dimensionless parameter

$$\alpha = e\phi/hc \tag{3.139}$$

which is the magnetic flux ϕ in units of the elementary flux quantum hc/e.

(d) Show that for the spectrum (3.138), the partition function can be summed to give

$$Z(\beta, \alpha) = \frac{\mathrm{Cosh}[\beta\hbar\omega(\alpha - 1)] + \mathrm{Cosh}(\beta\hbar\omega\alpha)}{2\,\mathrm{Sinh}^2(\beta\hbar\omega)}. \tag{3.140}$$

(e) Laplace invert $Z(\beta, \alpha)$ by the residue technique described in Sect. 3.2.1 in order to derive the following exact trace formula for the level density:

$$
g(E, \alpha) = \frac{E}{(\hbar\omega)^2} \left\{ 1 + \sum_{k=1}^{\infty} \left[\cos\left(2\pi k \frac{E}{\hbar\omega} - 2\pi k\alpha \right) + \cos\left(2\pi k \frac{E}{\hbar\omega} + 2\pi k\alpha \right) \right] \right\}
$$
$$
+ \frac{1}{\hbar\omega} \sum_{k=1}^{\infty} \left\{ (-1)^k \sin(\pi k\alpha) \sin\left(\pi k \frac{E}{\hbar\omega} \right) \right.
$$
$$
\left. -2\alpha \sin(2\pi k\alpha) \sin\left(2\pi k \frac{E}{\hbar\omega} \right) \right\}. \tag{3.141}
$$

Note that for any integer value of α, this reduces to the level density (3.68) of the isotropic two-dimensional harmonic oscillator.

(f) Give a semiclassical interpretation of the result (3.141). First show, using the canonical momenta (3.137), that the action of any classical orbit that surrounds the origin once is given by

$$
S = \oint \mathbf{p} \cdot d\mathbf{q} = S_0 \pm 2\pi\hbar\alpha, \tag{3.142}
$$

where $S_0 = 2\pi E/\omega$ is the action of the unperturbed harmonic oscillator and the second term is due to the vector potential in (3.132), its sign depending on the time orientation (clockwise or anticlockwise) of the orbit. Next, note that $\pm 2\pi\alpha$ is just the Aharonov-Bohm phase [21] that a charged particle picks up when it surrounds a solenoid with dimensionless flux α. It is this phase that appears on the first line of the trace formula (3.141). We can therefore interpret these terms in (3.141), which are the leading ones, as a simple inclusion of the Aharonov-Bohm effect into the classical periodic orbit theory. Note, however, that the exact quantum result (3.141) contains two more terms which are of order $\hbar$ relative to the leading ones; they represent quantum corrections which are less straightforward to interpret classically. We observe [19] that the first of these terms for $k=1$ corresponds to a contribution with double frequency or half the fundamental period of the classical harmonic-oscillator orbits; they may be understood as due to orbits that are reflected by the flux tube. Indeed, the factor $\sin(\pi\alpha)$ appearing in this term for $k=1$ is proportional to the quantum-mechanical scattering amplitude, given by Aharonov and Bohm [21], for a free particle that is scattered backwards by a magnetic flux tube (cf. also Refs. [19, 22]).

(3.5) Trace formula for the hydrogen atom

Use the method of Sect. 3.2.2 to derive a trace formula for the bound spectrum of the Coulomb potential.

(a) Write the hydrogen spectrum (2.151) in Rydberg units (Ry $\simeq 13.606$ eV) as

$$
E_n = -\frac{\mathrm{Ry}}{n^2}, \qquad n = 1, 2, 3, \ldots, \qquad \mathrm{Ry} = \frac{me^4}{2\hbar^2} \tag{3.143}
$$

and evaluate the correct degeneracy d_n of the n-th level.

(b) Show that the application of Eq. (3.54) leads directly to the trace formula

$$
g(E) = \frac{(\mathrm{Ry})^{3/2}}{2(-E)^{5/2}} \left\{ 1 + 2 \sum_{k=1}^{\infty} \cos\left(2\pi k \sqrt{-\mathrm{Ry}/E} \right) \right\}. \qquad E < 0 \tag{3.144}
$$

(3.6) Bloch density matrix for harmonic oscillator potentials
Consider the Bloch density matrix as defined by Eq. (3.16). For a one-dimensional harmonic oscillator potential, the definition yields (with $\hbar = m = \omega = 1$)

$$C(x, x'; \beta) = \frac{e^{-(x^2+x'^2)/2}}{\sqrt{\pi}} \sum_{n=0}^{\infty} \frac{H_n(x)H_n(x')}{2^n \, n!} \, e^{-\beta(n+1/2)}. \tag{3.145}$$

Making use of the identity (for $z < 1$)

$$e^{-(x^2+x'^2)/2} \sum_{n=0}^{\infty} \frac{H_n(x)H_n(x')}{2^n \, n!} z^n = e^{(x^2+x'^2)/2} \frac{e^{-(x^2+x'^2-2xx'z)/(1-z^2)}}{\sqrt{(1-z^2)}}, \tag{3.146}$$

show that

$$C(x, x'; \beta) = \frac{1}{\sqrt{2\pi \, \mathrm{Sinh}(\beta)}} \exp\left[-\frac{1}{2}\left(x^2 + x'^2 \right) \mathrm{Coth}(\beta) + \frac{xx'}{\mathrm{Sinh}(\beta)} \right]. \tag{3.147}$$

This expression may be simplified to

$$C(x, x'; \beta) = \frac{1}{\sqrt{2\pi \, \mathrm{Sinh}(\beta)}} \exp\left[-\frac{1}{4}\left(x + x' \right)^2 \mathrm{Tanh}(\beta/2) - \frac{1}{4}\left(x - x' \right)^2 \mathrm{Coth}(\beta/2) \right]. \tag{3.148}$$

Show further that it is straightforward to generalize this to higher dimensions, and in particular for $D = 3$, that

$$C(\mathbf{R} + \mathbf{s}/2, \mathbf{R} - \mathbf{s}/2; \beta) = \frac{1}{[2\pi \, \mathrm{Sinh}(\beta)]^{3/2}} \exp\left[-R^2 \, \mathrm{Tanh}(\beta/2) - \frac{1}{4} s^2 \, \mathrm{Coth}(\beta/2) \right], \tag{3.149}$$

where we have put $\mathbf{r} = \mathbf{R} + \mathbf{s}/2$, $\mathbf{r}' = \mathbf{R} - \mathbf{s}/2$.

(3.7) Two-dimensional electron gas with spin-orbit interaction in a homogeneous magnetic field (After [23])
The Hamiltonian for an electron gas confined to the (x, y) plane with a spin-orbit interaction of Rashba type [24] in a magnetic field $\mathbf{B} = (0, 0, B_0)$ has the form:

$$\hat{H} = \hat{\boldsymbol{\pi}}^2/2m^* + \hbar\kappa \left(\hat{\pi}_x \, \sigma_y - \hat{\pi}_y \, \sigma_x \right), \tag{3.150}$$

where m^* is the effective mass, κ a coupling constant, and σ_x, σ_y are Pauli matrices. $\hat{\boldsymbol{\pi}} = (\hat{p}_x, \hat{p}_y, 0) - (eB_0/2c)(-y, x, 0)$ is the mechanical momentum and e the charge of the electron. The spin Zeeman interaction $-\mu \, B_0 \, \sigma_z$ has been neglected here. The exact eigenvalues of the Hamiltonian (3.150) are known analytically [24]:

$$E_0 = \hbar\omega_c/2, \quad E_n^{\pm} = \hbar\omega_c \left(n \pm \sqrt{1/4 + 2n\,\hbar\,\tilde{\kappa}^2} \right), \quad n = 1, 2, 3, \ldots \tag{3.151}$$

where $\omega_c = eB_0/m^*c$ is the cyclotron frequency and $\tilde{\kappa} = \kappa \sqrt{m^*/\omega_c}$.

Using the Poisson summation formula (2.157), show that the quantum-mechanical level density can be written in the form of the following exact trace formula:

$$\begin{aligned} g(E) = \frac{2}{\hbar\omega_c}\Bigg\{ & 1 + \sum_{\pm}\left(1 \pm \frac{\hbar\tilde{\kappa}^2}{\sqrt{1/4 + 2E\,\tilde{\kappa}^2/\omega_c + \hbar^2\tilde{\kappa}^4}} \right) \\ & \times \sum_{k=1}^{\infty} \cos\left[k\,2\pi\left(\frac{E}{\hbar\omega_c} + \hbar\,\tilde{\kappa}^2 \pm \sqrt{1/4 + 2E\,\tilde{\kappa}^2/\omega_c + \hbar^2\tilde{\kappa}^4} \right) \right] \Bigg\}. \end{aligned} \tag{3.152}$$

Bibliography

[1] B. K. Jennings, Ph.D. thesis, McMaster University (1974, unpublished).

[2] R. Balian and C. Bloch, Ann. Phys. (N. Y.) **69**, 76 (1972).

[3] H. R. Krishnamurthy, H. S. Mani, and H. C. Verma, J. Phys. **A 15**, 2131 (1982).

[4] F. E. Borgnis and C. H. Papas, in: *Encyclopedia of Physics*, ed. S. Flügge (Springer-Verlag, Berlin, 1958) Vol. XVI, p. 285.

[5] P. J. Richens and M. V. Berry, Physica **2D**, 495 (1981).

[6] G. Lamé: *Leçons sur la théorie mathématique de l'élasticité des corps solides* (Bachelier, Paris, 1852), §57, p. 131.

[7] R. K. Bhaduri, Shuxi Li, K. Tanaka, and J. C. Waddington, J. Phys. **A 27**, L553 (1994).

[8] M. Brack and S. R. Jain, Phys. Rev. **A 51**, 3462 (1995).

[9] M. Douglas McIlroy, Proc. Symp. Appl. Math. **46**, 105 (1992); J. C. Lagarias, *ibid.*, p. 35.

[10] L. D. Landau and E. M. Lifshitz: *Quantum Mechanics* (Pergamon Press, London, 1958) p. 474.

[11] J. D. Louck, M. Moshinski, and K. B. Wolf, J. Math. Phys. **14**, 692 (1973).

[12] I. A. Malkin and V. I. Man'ko, Soviet Phys. JETP **28**, 527 (1969).

[13] A. Feldman and A. H. Kahn, Phys. Rev. **B 1**, 4584 (1970).

[14] F. D. M. Haldane, Phys. Rev. Lett. **51**, 605 (1983).

[15] D. C. Tsui, H. L. Störmer, and A. C. Gossard, Phys. Rev. Lett. **48**, 1559 (1982); R. R. Du, H. L. Störmer, D. C. Tsui, L. N. Pfeiffer, and K. W. West, Phys. Rev. Lett. **70**, 2944 (1993).

[16] R. B. Laughlin, Phys. Rev. Lett. **50**, 1395 (1983).

[17] V. Fock, Z. Phys. **47**, 446 (1929).

[18] G. Date and M. V. N. Murthy, Phys. Rev. **A 48**, 105 (1993).

[19] M. Brack, R. K. Bhaduri, J. Law, M. V. N. Murthy, and Ch. Maier, Chaos **5**, 317 (1995); Erratum: Chaos **5**, 707 (1995).

[20] A. Lerda: *Anyons*, Lecture Notes in Physics, Monographs **m4** (Springer, Berlin-Heidelberg, 1992).

[21] Y. Aharonov and D. Bohm, Phys. Rev. **115**, 485 (1959).

[22] S. M. Reimann, M. Brack, A. G. Magner, J. Blaschke, and M. V. N. Murthy, Phys. Rev. **A 53**, 39 (1996).

[23] Ch. Amann and M. Brack, J. Phys. **A 35**, 6009 (2002).

[24] Y. Bychkov and E. Rashba, J. Phys. **C 17**, 6039 (1984); and earlier references quoted therein.

Chapter 4

The extended Thomas-Fermi model

4.1 Introduction

In the previous chapters, we have seen in some detail how the quantum density of states may be resolved into a part that varies smoothly with energy, and another part that oscillates with energy. For large principal quantum numbers, the classical action $S(E)$ for the corresponding energy E is much larger than $\hbar$, and one may then neglect the fast varying oscillatory part which is governed by the frequency $S(E)/\hbar$ (see Eq. (3.42)). In this chapter we systematically develop a semiclassical expansion for the smooth part of the single-particle quantum density of states in powers of $\hbar$. The leading term of this series is the well-known Thomas-Fermi (TF) level density. As already noted in Sect. 2.7, this is obtained by calculating the volume of the accessible phase space and dividing it by the volume h^D of the elementary quantum cell, where D is the spatial dimension of the system. When higher-order correction terms in $\hbar$ are added to the TF term, this is called the extended Thomas-Fermi (ETF) approximation. In the application of the ETF method to a many-fermion system, only those terms in the $\hbar$ expansion will be included that give finite contribution to the energy. This point will be discussed later.

As we saw in Sect. 3.1, the level density $g(E)$ is the inverse Laplace transform of the canonical partition function. A direct method of deriving the ETF approximation is to use the Wigner-Kirkwood expansion [1, 2] of the partition function in powers of $\hbar$, involving the derivatives of the potential. By taking the inverse Laplace transform of the successive terms of this series, one may generate the ETF approximation systematically [3]. The method is only applicable when the potential in which the particle is moving is smooth, in the sense that its derivatives exist, and when the higher derivatives may be neglected. We follow Kirkwood's method [2] of obtaining the semiclassical series for the partition function. It is possible to derive the same result using Wigner's distribution function in the mixed $(\mathbf{r}, \mathbf{p})$ representation of the quantum density matrix in the Hilbert space. This idea will be discussed in the next section before the derivation due to Kirkwood is presented in Sect. 4.3.

The ETF model, as pointed out already, yields the smooth part of the level density and other quantities. It will be derived from the Wigner-Kirkwood expansion in Sect.

4.4.1. In Sect. 4.4.2 we discuss the ETF density variational method and its use for obtaining in a self-consistent way the average parts of the energy and other observables of a many-fermion system in its ground state. For the description of the excited states of such a system, the concept of (non-zero) temperature is often useful. The TF method and its extended version may therefore be adapted for finite-temperature calculations, which is the subject of Sect. 4.4.3. For a billiard (or cavity) with sharp boundaries the Wigner-Kirkwood expansion cannot be used, but other methods for finding $\hbar$ expansions may be applied. The Euler-MacLaurin formula requires the knowledge of the quantum spectrum, but is otherwise applicable to any system. We discuss it in Sect. 4.6 and show that the first three terms in the level density of a three-dimensional cavity go like the volume, the surface, and the curvature of the enclosure, as is well-known from the so-called Weyl expansion. (The harmonic oscillator is an exception and has no surface term.) Finally, a numerical smoothing of the level density may also be done, by replacing each delta function spike by an appropriate smoothing function, like a Gaussian or a Lorentzian. For mathematical consistency in the global variation with energy, a requisite number of the lower moments of the distribution must be preserved. Strutinsky [4, 5] developed a powerful smoothing method of this kind that has been very successful in nuclear physics and has recently also been introduced to the physics of metal clusters. The Strutinsky averaging method and its relationship to the ETF model will be the subject of Sect. 4.7.

The details of the above topics will unfold in the following sections, but it may still be useful to have an overview of the TF approximation from the partition function point of view. As stated in Eq. (3.37) in the previous chapter, the single-particle density of states may be obtained from the inverse Laplace transform of the quantum canonical partition function. Since the TF approximation is obtained by replacing the sum over the quantum states by the integral over the classical phase space, it is not surprising that it amounts to taking the inverse Laplace transform of the *classical* canonical partition function. To see this, take the simple example of the three-dimensional isotropic harmonic oscillator [6]. The quantum single-particle partition function is (see also Sect. 3.2.4, with $\beta = 1/k_B T$)

$$Z(\beta) = \left[\sum_{n=0}^{\infty} e^{-\beta\hbar\omega(n+1/2)} \right]^3 = \frac{1}{8} \frac{1}{\mathrm{Sinh}^3(\beta\hbar\omega/2)}. \tag{4.1}$$

The semiclassical approximation for the partition function is obtained for the situation $\hbar\omega < k_B T$, for which the hyperbolic function may be expanded in powers of $\beta\hbar\omega$ (see, e.g., Ref. [7], p. 85). We then obtain

$$Z_{WK} = \frac{1}{(\beta\hbar\omega)^3} - \frac{1}{8(\beta\hbar\omega)} + \frac{17}{1920}(\beta\hbar\omega) + \dots. \tag{4.2}$$

The first term on the right-hand side of the above expansion is just the classical partition function:

$$Z_{cl}(\beta) = \frac{1}{h^3} \int d^3p\, d^3r\, e^{-\beta[p^2/2m + V(r)]} = \frac{1}{(\beta\hbar\omega)^3}, \tag{4.3}$$

where we have taken $V(r) = \frac{1}{2}m\omega^2 r^2$. The TF level density is obtained by taking the inverse Laplace transform of $Z_{cl}(\beta)$ above, yielding $g_{TF}(E) = E^2/2(\hbar\omega)^3$. The higher order terms in $(\beta\hbar\omega)$ in Eq. (4.2) give systematic corrections to the TF result. This is termed the ETF approximation. Note that one does not recover the oscillating part $\delta g(E)$ of the level density by making this $\hbar$ expansion.

Instead of specializing to the harmonic oscillator, we could as well have obtained the inverse Laplace transform of $Z_{cl}(\beta)$ given in Eq. (4.3) for any given $V(\mathbf{r})$. For this purpose, it is convenient to first perform the Gaussian p integration, and then to use the standard results of Laplace inversion. This immediately gives the TF result for the level density [3]

$$g_{TF}(E) = \frac{1}{4\pi^2}\left(\frac{2m}{\hbar^2}\right)^{3/2} \int d^3r\, [E - V(\mathbf{r})]^{1/2}\, \Theta(E - V(\mathbf{r})). \qquad (4.4)$$

In the above, $\Theta(x)$ is the unit step-function. In the following section, higher-order corrections to this expression will be obtained using the Wigner-Kirkwood expansion.

Although one of the main topics of this chapter is the development of the ETF approximation using the Wigner-Kirkwood series, it is important to appreciate that the TF method, in its self-consistent form, is a very useful tool in studying an interacting many-fermion system. The TF method was in this connection first applied to the electrons in an atom [8, 9]. Since the derivatives of the Coulomb potential become more and more singular near the origin, the ETF expansion is not valid for this problem. Nevertheless, we give a brief account of TF in the atomic problem in this introduction, in order to emphasize the self-consistency requirement that the spatial density has to fulfill. A more general account of the ETF method in the many-body context will be given in Sect. 4.4.2, and the formalism of the self-consistent mean-field approach is given in Appendix A. Here we follow the classic paper of Thomas [8] to illustrate how the TF self-consistent mean field is obtained in an atom.

An electron in an atom is bound to the nucleus by its attractive Coulomb field, and it also interacts repulsively with all the other bound electrons in the atom. Let us assume that as a result of all this, an electron moves in an average central potential $V(r)$, where r is the distance of the electron from the nucleus. Ignoring the size of the nucleus, we expect $V(r)$ to behave like the bare Coulomb potential of the nucleus, $-Ze^2/r$, for $r \to 0$. For large r, on the other hand, the mean field should decrease faster than $1/r$ due to screening of the nucleus by the other electrons. Since the electrons are identical, each electron moves in the same mean field, and in the simplest picture, they move independently of each other. For a neutral atom, there are Z electrons, and in the ground state, these have to be filled into the lowest energy states of $V(r)$ respecting the Pauli principle. The most energetic electron has an energy μ, which is called the Fermi energy. Had the electron been in a translationally invariant system, the linear momentum would be a good quantum number, and one could associate a Fermi momentum p_F to it, such that $p_F^2/2m = \mu$. This is not the case, however, since the electrons are confined in the potential $V(r)$ with a fixed origin. Unlike for a translationally invariant system, the density distribution is not constant either. In the TF method, this density is directly related to the volume of the available phase space, i.e., to p_F^3. In order to apply the method, it is assumed that the Fermi momentum p_F itself is a function of the spatial distance r, but that $p_F(r)$ varies slowly over the spatial size of the atom. We may then express the energy of the most energetic electron to be

$$\mu = \frac{p_F^2(r)}{2m} + V(r). \qquad (4.5)$$

Note that although both terms on the right-hand side vary with r, the Fermi energy μ is a constant. The next step is to obtain an expression for the density distribution of the electrons in the atom. This is easily done by recalling that each quantum state, according to the Pauli principle, may be occupied by two electrons, one with spin up and the other with spin down. In the ground state, only the lowest energy single-particle states are occupied using this rule. We then get

$$N = \sum_{i < E_F} 2 \simeq 2 \frac{\Omega}{h^3} \frac{4\pi}{3} p_F^3, \tag{4.6}$$

where we denote the number of electrons in the atom by N (which only equals Z in a neutral atom), and the atomic volume by Ω. Defining the number density $\rho = N/\Omega$, we immediately get $\rho = k_F^3/3\pi^2$, with $p_F = \hbar k_F$. Substituting for p_F from the relation (4.5), we obtain the TF expression for the density:

$$\rho(r) = \frac{1}{3\pi^2} \left(\frac{2m}{\hbar^2}\right)^{3/2} (\mu - V(r))^{3/2} \quad (r \leq r_0); \qquad \rho(r) = 0 \quad (r \geq r_0). \tag{4.7}$$

Here r_0 is the classical turning point defined by $V(r_0) = \mu$. From the above relation, we see that ρ depends on the average one-body potential $V(r)$, which itself should depend on the density distribution of the electrons. Neglecting the exchange contribution to the potential, this dependence is given by

$$V(r) = -\frac{Ze^2}{r} + e^2 \int \frac{\rho(\mathbf{r}')d^3r'}{|\mathbf{r} - \mathbf{r}'|}. \tag{4.8}$$

We see that Eqs. (4.7, 4.8) are coupled and have to be solved self-consistently. This self-consistency requirement is common to all mean-field theories, and the TF atom is one of the simplest examples of it. The Poisson equation of electrostatics follows from Eq. (4.8):

$$\nabla^2 V(r) = 4\pi Ze^2\delta(\mathbf{r}) - 4\pi e^2\rho(r). \tag{4.9}$$

For completeness, we indicate how the self-consistency problem is solved here. Following Fermi, it is best to define a screening function $\phi(r)$:

$$\frac{Ze^2}{r}\phi(r) = \mu - V(r). \tag{4.10}$$

Obviously, $\phi(0) = 1$. Combining Eqs. (4.7,4.10), self-consistency now reduces to the nonlinear differential equation for the screening function

$$\frac{d^2\phi(x)}{dx^2} = \frac{\phi^{3/2}(x)}{\sqrt{x}}. \tag{4.11}$$

In the above, x is a dimensionless variable in terms of a length scale b (proportional to the Bohr radius $a_0 = \hbar^2/me^2$), such that $x = r/b$, with

$$b = \left(\frac{3\pi}{4}\right)^{2/3} \frac{\hbar^2}{2me^2} \frac{1}{Z^{1/3}} = \frac{0.855a_0}{Z^{1/3}}. \tag{4.12}$$

The remarkable point about Eq. (4.11) is that it is universal to all atoms, with the scale factor b decreasing for larger atoms. For a neutral atom, the number of electrons, N, must equal the atomic charge Z. Since $\int \rho(r)d^3r = N = Z$, it follows, by taking the large r limit of Eq. (4.8) that $V(r) \to \infty$ as $r \to 0$. Demanding for large r that $\rho(r) \to 0$, it then follows from Eq. (4.7) that the chemical potential $\mu = 0$, and r_0 is infinity in this case. For a positive ion, on the other hand, r_0 is finite, and $V(r) = -(Z-N)e^2/r$ for $r > r_0$. The self-consistent screening function $\phi(x)$ has to be found numerically by solving Eq. (4.11) with the appropriate boundary condition. We do not give further details, since our main aim here was to present the self-consistency requirement of the mean field. A good discussion of the TF atom will be found in the book by March [10].

Although the above description of the TF method is only applicable to electrons interacting with the Coulomb potential, similar equations may be obtained in a nucleus, where the nucleons (protons and neutrons) are dominantly interacting with short-range forces. The atomic nucleus is a dense many-nucleon system bound together by strong, short-range forces. Just as the Coulomb potential may be obtained by the exchange of a photon between two charged particles, the short-range nuclear potential may be derived by the exchange of massive mesons between two nucleons. Yukawa [11] proposed this picture first, and derived the form of the meson-exchange potential. This was later identified as the pion, the lightest of many mesons. The one-pion exchange potential gives rise to a strong tensor potential, and a relatively long-range spin-dependent central potential. The two-nucleon potential due to a meson exchange falls off exponentially with the inter-particle distance, and the range parameter in the exponent is inversely proportional to the mass of the exchanged meson [12]. In a spherical spin-zero nucleus, the tensor as well as the spin-dependent potentials average out to zero in the first approximation, and the effective attractive force may be regarded as coming from the exchange of a scalar meson. In addition, there is a repulsive potential at very short range due to vector meson exchange. With this simplified picture in mind, Serber [13] writes the mean potential $V(r)$ experienced by a nucleon just as in Eq. (4.8) for the atomic case:

$$V(r) = \frac{g}{4\pi} \int \frac{\exp(-|\mathbf{r} - \mathbf{r}'|/b)}{|\mathbf{r} - \mathbf{r}'|} \, \rho(r') \, d^3r' \,. \tag{4.13}$$

Comparing this with Eq. (4.8), we do not have anything similar to the first term on the right, $-Ze^2/r$, since there is no center of attraction in the nucleus. The mean field arises solely from the inter-particle interaction. The coupling constant g is analogous to e^2, and the range parameter b is the Compton wave length of the exchanged scalar meson, $b = \frac{\hbar}{Mc}$. From Eq. (4.13), we immediately obtain the Poisson-like equation

$$\nabla^2 V(r) - \frac{1}{b^2} V(r) = -g\rho(r) \,. \tag{4.14}$$

Serber also modified the kinetic energy contribution to account for the very short-range nuclear repulsion between the nucleons. This helps the nucleus in not collapsing to a point. He went on to perform, mostly analytically, a TF-like calculation for the ground-state properties of the nucleus (including the profile of the surface). This toy-model is instructive for its simplicity, and for its formal similarity with the atomic problem. A more realistic approach, however, is to construct an energy density functional and to obtain a differential equation for the density profile from a variational principle. This approach was pioneered by Skyrme [14] in nuclear physics, and developed and refined very fruitfully by Kohn and Sham [15] in atomic and condensed-matter systems. The energy density functional treats the kinetic energy term more carefully and includes the exchange as well as the correlation energy between electrons. The density functional approach is discussed in appendix A.2, and some applications are presented later.

4.2   The Wigner distribution function

In classical statistical mechanics, the distribution factor (at thermal equilibrium) of a configuration of particles in the phase space is given by the usual Boltzmann factor

$$f_{cl}(x_1, x_2, \ldots, x_D; p_1, p_2, \ldots, p_D; \beta) = \frac{1}{h^D} \exp\left[-\beta \left(p_1^2 + p_2^2 + \ldots + p_D^2\right)/2m\right]$$
$$\times \exp\left[-\beta V(x_1, x_2, \ldots, x_D)\right]. \qquad (4.15)$$

The probability of finding particle 1 between (x_1, p_1) and $(x_1 + dx_1, p_1 + dp_1)$, and likewise for particles $2, 3, \ldots, D$, is then given by $f_{cl}\, dx_1 dx_2..dx_D dp_1 dp_2 \ldots dp_D$, divided by the integral of the same quantity over all of phase space. In Eq. (4.15), $V(x_1, x_2, \ldots, x_D)$ is the potential energy of the configuration, and we have used a one-dimensional notation for the coordinates x_n and momenta p_n of the nth particle. The simplicity arises because one works in the classical phase space, and the momentum distribution factor $\exp[-\beta(p_1^2 + p_2^2 + \ldots + p_D^2)/2m]$ is decoupled from the coordinate part.

Wigner [1] sought the analogue of the above expression in quantum statistical mechanics, with a view to obtain the semiclassical corrections in powers of $\hbar$. This has two complications: (a) one cannot specify the coordinate and momentum of a particle simultaneously in quantum mechanics, so the phase space concept is ambiguous, and (b) correlations due to quantum statistics are involved. Point (b) is tractable, and does not concern us at the moment. Moreover, to avoid writing a long string of coordinates, let us focus on the one-particle distribution function, assuming that each particle is moving independently of the others in a mean-field potential $V(x)$. This way we may focus on the central issue (a). For the classical situation, Eq. (4.15) for one particle now reduces to the form

$$f_{cl}(x, p, \beta) = \frac{1}{h} \exp(-\beta p^2/2m) \exp[-\beta V(x)]. \qquad (4.16)$$

We may regard the Planck constant h (or h^D in the D-dimensional case) as the elementary volume of a quantum cell in the phase space. Then $f_{cl}(x, p)/h$, when normalized to unity over the accessible phase space volume, is the classical probability density of the particle being in the dynamical state specified by (x, p). The choice of h as the elementary volume of phase space ensures correspondence with quantum mechanics for high quantum numbers.

Now consider the situation in quantum mechanics. For a particle in a pure state $\psi(x)$, the probability of its being between x and $x + dx$ is simply $|\psi(x)|^2 dx$. Alternately, it could also be described in momentum space using the wave function $\phi(p)$, the Fourier transform of $\psi(x)$. Then the probability of the particle to have a momentum between

p and $p + dp$ is $|\phi(p)|^2 dp$. Unlike the classical probability $\rho_{cl}(x, p)$ given by Eq. (4.16), the quantum probability is only a function of x or p, not both. Wigner transformed this function of one variable to that of two to make it more analogous to the classical quantity. Consider, for the pure state $|\psi\rangle$, the density matrix operator $\hat{\rho} = |\psi\rangle\langle\psi|$. This contains all the quantum information for the state. The quantum probability factor in x space is then the diagonal element $\rho(x, x)$ of the density matrix

$$\rho(x, x') = \langle x|\hat{\rho}|x'\rangle = \psi^*(x')\psi(x). \qquad (4.17)$$

We are using the standard Dirac representation $\psi(x) = \langle x|\psi\rangle$. The density matrix formalism is very useful in the many-body problem also. For a many-fermion system at zero temperature, the single-particle density matrix is defined in Appendix A in Eqs. (A.2,A.6). Here we shall continue with the simple form (4.17) to define the Wigner transform of the density matrix $\hat{\rho}$. This is denoted by $\rho_W(x, p)$, and is given by (x, p are *not* operators)

$$
\begin{aligned}
\rho_W(x, p) &= \frac{1}{2\pi\hbar} \int_{-\infty}^{\infty} dy \, \langle x - y/2|\hat{\rho}|x + y/2\rangle \exp(ipy/\hbar) \\
&= \frac{1}{2\pi\hbar} \int_{-\infty}^{\infty} dy \, \psi^*(x + y/2)\psi(x - y/2) \exp(ipy/\hbar). \qquad (4.18)
\end{aligned}
$$

The quantity $\rho_W(x, p)$ is often called the Wigner distribution function, and its form is due to Wigner and Szilard [1]. It is easily checked that

$$\int_{-\infty}^{\infty} \rho_W(x, p) dp = \rho(x, x) = |\psi(x)|^2, \qquad (4.19)$$

$$\int_{-\infty}^{\infty} \rho_W(x, p) dx = \rho(p, p) = |\phi(p)|^2. \qquad (4.20)$$

We have taken the momentum space wave function $\phi(p)$ to be

$$\phi(p) = \frac{1}{\sqrt{2\pi\hbar}} \int_{-\infty}^{\infty} dx \, \psi(x) \exp(ipx/\hbar). \qquad (4.21)$$

The relations (4.19) and (4.20) imply that one obtains the correct quantum mechanical expectation values of any function of the coordinate or momentum using the Wigner transform ρ_W:

$$\int_{-\infty}^{\infty} dx \int_{-\infty}^{\infty} dp \, [f(p) + F(x)] \, \rho_W(x, p) = \int_{-\infty}^{\infty} dx \, \psi^*(x) \, [f(-i\hbar \partial/\partial x) + F(x)] \, \psi(x).$$
$$(4.22)$$

We again emphasize that in the left-hand side of the above equation, x and p are regarded as classical variables, and in this sense $\rho_W(x, p)$ is analogous to the classical phase space density of the dynamical ensemble. This should not be taken literally, since $\rho_W(x, p)$ is a fully quantum-mechanical object and can take on negative values, as is clear from its definition. Moreover, there is a great deal of freedom in the choice of the transform that may still satisfy Eqs. (4.19) and (4.20). The particular Wigner form $\rho_W(x, p)$ happens to be one of the simplest.

The usefulness of this approach in semiclassical physics becomes apparent when one goes over to statistical mechanics and constructs the quantum analogue of Eq. (4.15).

The quantum density matrix operator for a mixed state after statistical averaging is given by [16]

$$\hat{\rho} = \sum_n |n\rangle P_n \langle n|, \tag{4.23}$$

where P_n is the probability that the particle is in the nth quantum state. This follows by writing, for a mixed state $|\psi\rangle$, the series $\sum_n c_n |n\rangle$, so that $P_n = |c_n|^2$, and neglecting all cross terms in the expression for $\hat{\rho} = |\psi\rangle\langle\psi|$. In equilibrium quantum statistical mechanics, the density matrix in the Gibbs ensemble (denoted by the letter C_β) is of the form [16]

$$\hat{C}_\beta = e^{-\beta\hat{H}}. \tag{4.24}$$

We shall assume here for simplicity that the potential V in coordinate space is local, so that

$$\langle x|\hat{H}|x'\rangle = \left(-\hbar^2/2m\, \partial^2/\partial x^2 + V(x)\right)\delta(x - x'). \tag{4.25}$$

We may then write

$$\begin{aligned}
C(x, x'; \beta) &= \langle x|e^{-\beta\hat{H}}|x'\rangle = e^{-\beta\hat{H}_x}\delta(x - x') \\
&= \frac{1}{2\pi\hbar}\int_{-\infty}^{\infty} dp\, \exp(-ipx'/\hbar)\, e^{-\beta\hat{H}_x}\, \exp(ipx/\hbar),
\end{aligned} \tag{4.26}$$

where we have used the relation $\delta(x - x') = (1/2\pi\hbar)\int_{-\infty}^{\infty}\exp[ip(x - x')/\hbar]dp$. The ensemble average of an operator $\hat{Q}$ is given by

$$\langle\hat{Q}\rangle = Tr\left(\hat{C}_\beta\hat{Q}\right)/Tr\hat{C}_\beta. \tag{4.27}$$

The Gibbs density matrix operator $\hat{C}_\beta$ may be put in the form given by Eq. (4.23) by inserting a unit operator $\sum_n |n\rangle\langle n|$, where $|n\rangle$ is an eigenstate of $\hat{H}$. We then get

$$\hat{C}_\beta = \sum_n |n\rangle e^{-\beta E_n}\langle n|. \tag{4.28}$$

Thus the quantum probability P_n has in the Gibbs ensemble been replaced by the Boltzmann factor $\exp(-\beta E_n)$. In the x-representation, the density matrix is given by

$$C(x, x'; \beta) = \langle x|\hat{C}_\beta|x'\rangle = \sum_n \psi_n^\star(x')e^{-\beta E_n}\psi_n(x). \tag{4.29}$$

We shall often refer to $C(x, x'; \beta)$ as the "Bloch density" matrix. Wigner's work forged a systematic link between the quantum and classical distribution functions of statistical mechanics. The Wigner transform of the Bloch density matrix may be expanded in a power series of the Planck's constant h, and its leading term is just the classical distribution function. To be more precise, the Wigner transform $C_W(x, p; \beta)$ of the statistical density matrix is given by

$$\begin{aligned}
C_W(x, p, \beta) &= \frac{1}{2\pi\hbar}\int_{-\infty}^{\infty} dy\, C(x - y/2, x + y/2; \beta)\, \exp(ipy/\hbar) \\
&= \frac{1}{2\pi\hbar}\sum_n e^{-\beta E_n}\int_{-\infty}^{\infty} dy\, \psi_n^\star(x + y/2)\psi_n(x - y/2)\, e^{ipy/\hbar}.
\end{aligned} \tag{4.30}$$

Wigner [1] made a semiclassical $\hbar$-expansion of this, whose leading term is $f_{cl}(x,p)/h$, and the corrections are given in even powers of h involving derivatives of the potential. The generalization to the many-particle case (4.15) is easily done and may be found in Wigner's paper. The Wigner transform method has been used for a general one-body Hamiltonian to obtain semiclassical expansions of physical quantities like the one-body spatial density and the kinetic energy density by Grammaticos and Voros [17]. In passing, it may be mentioned that in the non-equilibrium case, $f_{cl}(x,p,t)$ is the dynamical distribution function which is governed by the Boltzmann equation [18]. In quantum mechanics, when the time-dependent Hartree-Fock equation is obtained in the mean-field approximation, its Wigner transform to leading order in $\hbar$ again gives the (collisionless) Boltzmann equation. We do not pursue this here, and the interested reader may consult Ref. [19].

We have noted that the Wigner transform method is fruitful in developing semiclassical expansions. It is less successful, however, in generating the quantum analogue of the classical distribution function in phase space, since $\rho_W(x,p)$, given by Eq. (4.18), may oscillate rapidly and take on negative values. In Problem (4.1) we introduce the so-called Husimi distribution function [20, 21] which is always non-negative and therefore more successful in this respect. The general topic of quantum phase space distribution functions (of which the Wigner function is only one example) is discussed in detail in review articles by Balasz and Jennings [22] and Lee [21].

Rather than following Wigner's derivation of the semiclassical expansion of the function $C_W(x,p;\beta)$ in phase space, we shall follow Kirkwood [2] and directly deal with $C(x,x';\beta)$ in coordinate space. Since the integral of $C(x,x;\beta)$ over all x yields the canonical partition function $Z(\beta)$, this expansion immediately leads to the ETF approximation.

4.3 The Wigner-Kirkwood expansion

In this section, we make a semiclassical $\hbar$-expansion of the Bloch density matrix using Kirkwood's method [2], and then Laplace invert it term by term to obtain the ETF approximation at zero temperature [23, 24]. Since the applications to physical problems will be in two or three dimensions, we denote the Bloch matrix elements in the vector notation as in Eq. (3.16) by

$$C(\mathbf{r},\mathbf{r}';\beta) = \langle \mathbf{r}|\hat{C}_\beta|\mathbf{r}'\rangle = \sum_n \psi_n^\star(\mathbf{r}')\psi_n(\mathbf{r})\exp(-\beta E_n)\,. \qquad (4.31)$$

We follow the presentation given by Jennings [23]. The canonical partition function is obtained by integrating over all space the diagonal Bloch matrix $C(\mathbf{r},\mathbf{r};\beta) = C(\mathbf{r},\beta)$, and is given by (in D dimensions)

$$Z(\beta) = \int d^D r\, C(\mathbf{r},\beta) = \sum_n \exp(-\beta E_n) = Tr\exp(-\beta\hat{H})\,. \qquad (4.32)$$

The trace of the operator may be taken with respect to any complete set of states. For the semiclassical expansion involving the integral over the phase space, it is most convenient to take plane waves as the complete set. We may then write

$$Z(\beta) = \frac{1}{h^D} \int d^D r \int d^D p \, e^{-i\mathbf{p}\cdot\mathbf{r}/\hbar} e^{-\beta\hat{H}} e^{i\mathbf{p}\cdot\mathbf{r}/\hbar} . \tag{4.33}$$

As in the Wigner transform of the previous section, $\mathbf{p}$ and $\mathbf{r}$ are not to be taken as operators. In $\hat{H}$, however, the kinetic energy operator $\hat{T}$ does not commute with $V(\mathbf{r})$, so that $\exp(-\beta\hat{H}) \neq \exp(-\beta\hat{T})\exp(-\beta V(r))$. In Eq. (4.33), therefore, we put

$$e^{-\beta\hat{H}} e^{i\mathbf{p}\cdot\mathbf{r}/\hbar} = e^{-\beta H_{cl}} e^{i\mathbf{p}\cdot\mathbf{r}/\hbar} w(\mathbf{r},\mathbf{p};\beta) = u(\mathbf{r},\mathbf{p};\beta) , \tag{4.34}$$

where $w(\mathbf{r},\mathbf{p};\beta)$ has to be determined using the rules of quantum mechanics. This may be done by noting first that $u(\mathbf{r},\mathbf{p};\beta)$ obeys the Bloch equation

$$\frac{\partial u}{\partial \beta} + \hat{H}u = 0 \tag{4.35}$$

with the boundary condition that

$$\lim_{\beta\to 0} u = e^{i\mathbf{p}\cdot\mathbf{r}/\hbar} . \tag{4.36}$$

It is now straightforward to substitute for u from Eq. (4.34) in the differential equation (4.35), and to show that $w(\mathbf{r},\mathbf{p};\beta)$ obeys the equation

$$\begin{aligned}
\frac{\partial w}{\partial \beta} &= -i\hbar\left[\frac{\beta}{m}(\mathbf{p}\cdot\nabla V)w - \frac{1}{m}(\mathbf{p}\cdot\nabla w)\right] \\
&+ \frac{\hbar^2}{2m}\left[\beta^2(\nabla V)^2 w - \beta(\nabla^2 V)w + \nabla^2 w - 2\beta(\nabla V\cdot\nabla w)\right].
\end{aligned} \tag{4.37}$$

The boundary condition (4.36) reduces to $\lim_{\beta\to 0} w(\mathbf{r},\mathbf{p};\beta) = 1$. Till now no semiclassical approximation has been made. We do so at this point, by assuming that $w(\mathbf{r},\mathbf{p};\beta)$ may be expanded in a power series of $\hbar$:

$$w = 1 + \hbar w_1 + \hbar^2 w_2 + \dots . \tag{4.38}$$

Note that w_1, w_2 etc. are not dimensionless. The semiclassical series for the partition function takes the form

$$Z_{WK}(\beta) = \frac{1}{h^D} \int d^D r \int d^D p \, e^{-\beta H_{cl}} \left(1 + \hbar w_1 + \hbar^2 w_2 + \dots \right). \tag{4.39}$$

With this substitution, Eq. (4.37) may be solved in each order of $\hbar$ for w_i. For example, by equating terms of first order in $\hbar$ on both sides, we obtain

$$w_1 = -\frac{i\beta^2}{2m}\mathbf{p}\cdot\nabla V . \tag{4.40}$$

This expression for w_1 may now be substituted in the equation for w_2 that is obtained by equating all terms of order $\hbar^2$. This procedure yields

$$w_2 = -\frac{\beta^2}{4m}\nabla^2 V + \frac{\beta^3}{6m}(\nabla V)^2 - \frac{\beta^4}{8m^2}(\mathbf{p}\cdot\nabla V)^2 + \frac{\beta^3}{6m^2}(\mathbf{p}\cdot\nabla)^2 V . \tag{4.41}$$

The algebra is straightforward, but escalates rapidly. From the above expressions of w_1 and w_2, some characteristics of w_i are already apparent, and hold generally. Only odd powers of p occur in w_i with odd i, and similarly for the even powers. This results in only the w_i with even i contributing to $Z_{WK}(\beta)$ in Eq. (4.39). We also note that w_1 involves only ∇V, while w_2 has $(\nabla \cdot \nabla)V$ and $\nabla^2 V$. This is also a general feature of the expansion, with the ∇ operators occurring i times in every term of w_i. Performing the p integrations in Eq. (4.39) up to fourth order in $\hbar$, we obtain

$$
Z_{WK}(\beta) = \left(\frac{1}{\lambda}\right)^D \int d^D r \, e^{-\beta V(\mathbf{r})} \left\{ 1 - \frac{\beta^2}{12} \frac{\hbar^2}{2m} \nabla^2 V \right.
$$
$$
\left. + \frac{\beta^3}{1440} \left(\frac{\hbar^2}{2m}\right)^2 \left[-7\nabla^4 V + 5\beta(\nabla^2 V)^2 + \beta \nabla^2 (\nabla V)^2 \right] \right\}. \quad (4.42)
$$

In the above,

$$
\lambda = (2\pi \hbar^2 \beta / m)^{1/2} \quad (4.43)
$$

is the thermal wave length [25]. It should be mentioned that in deriving Eq. (4.42), use has been made of the identity

$$
\int d^D r \, e^{-\beta V} (\nabla V)^2 = \frac{1}{\beta} \int d^D r \, e^{-\beta V} \nabla^2 V . \quad (4.44)
$$

A similar technique may be used to obtain the semiclassical expansion of the Bloch density $C(\mathbf{r}, \mathbf{r}'; \beta)$. We again follow the presentation due to Jennings [23]. From Eq. (4.31), it is clear that C satisfies the equation

$$
\hat{H}_\mathbf{r} \, C(\mathbf{r}, \mathbf{r}'; \beta) + \frac{\partial C(\mathbf{r}, \mathbf{r}'; \beta)}{\partial \beta} = 0 , \quad (4.45)
$$

where the Hamiltonian in the $\mathbf{r}$-representation is denoted by $\hat{H}_\mathbf{r}$. The boundary condition satisfied by the Bloch density is

$$
\lim_{\beta \to 0} C(\mathbf{r}, \mathbf{r}'; \beta) = \delta(\mathbf{r} - \mathbf{r}') . \quad (4.46)
$$

Recall from Eq. (4.26) that $C(\mathbf{r}, \mathbf{r}'; \beta)$ is given by

$$
C(\mathbf{r}, \mathbf{r}'; \beta) = \frac{1}{h^D} \int d^D p \, e^{-i\mathbf{p} \cdot \mathbf{r}'/\hbar} \, e^{-\beta \hat{H}_\mathbf{r}} \, e^{i\mathbf{p} \cdot \mathbf{r}/\hbar} . \quad (4.47)
$$

We may now use Eq. (4.34) again to introduce $w(\mathbf{r}, \mathbf{p}; \beta)$ and its semiclassical expansion given by Eq. (4.38). At this point we have to worry about the spin components of the wave functions for fermions. As mentioned at the end of Sect. 3.1.1, the density matrix $\rho(\mathbf{r}, \mathbf{r}'; \beta)$ and the Bloch density matrix $C(\mathbf{r}, \mathbf{r}'; \beta)$, and hence also the $w_i(\mathbf{r}, \mathbf{p}; \beta)$ introduced above, become (2×2) matrices in spin space when the Pauli spinors are included in the wave functions. When taking traces in $\mathbf{r}$ space, this entails also taking a trace in spin space. In the mean-field treatment of a many fermion system, one will in general introduce separate spin densities (one for "spin up" and one for "spin down"), as discussed in Appendix A.2. As long as the Hamiltonian does not depend on the spin explicitly, however, the result of taking the traces is a simple spin degeneracy factor of 2 in the quantities obtained after taking the traces, and the local density $\rho(\mathbf{r})$ (3.14) is just the sum of the two spin densities. Since most applications of the ETF model discussed later on will be for fermionic systems, we shall in the remainder of this section include this spin degeneracy factor in all *local* densities such as $\rho(\mathbf{r})$ or the kinetic energy density $\tau(\mathbf{r})$ introduced below. We refer to the literature for the Wigner-Kirkwood expansion for Hamiltonians including spin-orbit or momentum-dependent terms (e.g., the cranking term used to describe rotating systems) [17, 24, 26, 27].

Denoting the diagonal Bloch density by $C(\mathbf{r}, \beta)$, as before, making use of Eqs. (4.34), (4.38) and taking the spin trace (assuming a spin-independent Hamiltonian $\hat{H}$), we obtain the expansion

$$C_{WK}(\mathbf{r}, \beta) = \frac{2}{h^D} \int d^D p \, e^{-\beta H_{cl}(\mathbf{r}, \mathbf{p})} \left(1 + \hbar w_1 + \hbar^2 w_2 + \dots\right). \qquad (4.48)$$

As before, w_1 and w_2 are given by Eqs. (4.40,4.41). Because there is no $\mathbf{r}$ integration, we cannot combine various terms by using Eq. (4.44), and the expression for $C_{WK}(\mathbf{r}, \beta)$ is not as compact as $Z_{WK}(\beta)$ in Eq. (4.42). We shall therefore only give the expression for C_{WK} to second order:

$$C_{WK}(\mathbf{r}, \beta) = 2\left(\frac{1}{\lambda}\right)^D e^{-\beta V(\mathbf{r})} \left(1 - \frac{\hbar^2 \beta^2}{12m}\left[\nabla^2 V - \frac{\beta}{2}(\nabla V)^2\right] + \dots\right). \qquad (4.49)$$

This expansion will be used to obtain the spatial density $\rho(\mathbf{r})$ of a noninteracting Fermi system using the inverse Laplace transform

$$\rho(\mathbf{r}) = \mathcal{L}_\mu^{-1}\left[\frac{1}{\beta}C(\mathbf{r}, \beta)\right]. \qquad (4.50)$$

We also need a similar expansion for the kinetic energy density. Note that a particle in a state $\psi_n(\mathbf{r})$ has a kinetic energy density of $\frac{\hbar^2}{2m}\nabla\psi_n(\mathbf{r})\cdot\nabla\psi_n^\star(\mathbf{r})$. For a noninteracting many-fermion system, the total kinetic energy density of the ground state is thus

$$\tau(\mathbf{r}) = \frac{\hbar^2}{2m}\sum_{E_n < \mu} \nabla\psi_n^\star(\mathbf{r}) \cdot \nabla\psi_n(\mathbf{r}). \qquad (4.51)$$

To obtain a semiclassical expansion of this, we define a quantity $\mathcal{T}(\mathbf{r}, \beta)$ analogous to the local Bloch density by

$$\mathcal{T}(\mathbf{r}, \beta) = \frac{\hbar^2}{2m}\sum_n \nabla\psi_n^\star(\mathbf{r}) \cdot \nabla e^{-\beta\hat{H}}\psi_n(\mathbf{r}), \qquad (4.52)$$

from which the kinetic energy density is obtained via an inverse Laplace transform analogous to (4.50):

$$\tau(\mathbf{r}) = \mathcal{L}_\mu^{-1}\left[\frac{1}{\beta}\mathcal{T}(\mathbf{r}, \beta)\right]. \qquad (4.53)$$

The next step is to express $\psi(\mathbf{r})$ in the plane wave basis

$$\psi_n(\mathbf{r}) = \frac{1}{h^{D/2}}\int d^D p \, \phi_n(\mathbf{p}) \, e^{i\mathbf{p}\cdot\mathbf{r}/\hbar},$$

and substitute it into Eq. (4.52). We then obtain, including the spin factor of 2,

$$\mathcal{T}(\mathbf{r}, \beta) = \frac{\hbar^2}{2m} \frac{2}{h^D} \int d^D p\, e^{-i\mathbf{p}\cdot\mathbf{r}/\hbar} \left(-\frac{i}{\hbar}\mathbf{p}\cdot\nabla\right) e^{-\beta\hat{H}_r}\, e^{i\mathbf{p}\cdot\mathbf{r}/\hbar}. \tag{4.54}$$

Using the same method as before, the semiclassical expansion of this quantity is now given by

$$\mathcal{T}_{WK}(\mathbf{r}, \beta) = \frac{2}{h^D} \int d^D p \left(\frac{p^2}{2m} - \frac{i\hbar}{2m}\mathbf{p}\cdot\nabla\right) e^{-\beta H_{cl}} \left(1 + \hbar w_1 + \hbar^2 w_2 + \ldots\right). \tag{4.55}$$

4.4 The extended Thomas-Fermi model

4.4.1 The ETF model at zero temperature

In this section, we take advantage of the semiclassical expansions derived above to obtain the higher order corrections to the TF expressions of the level density, the sum of the occupied single-particle energies, and the spatial and kinetic energy densities. We assume, as before, that the fermions are moving in a one-body mean potential $V(\mathbf{r})$. The self-consistency condition does not concern us at the moment since we only want to write down the basic equations, relating the above global properties of the system to the given potential $V(\mathbf{r})$.

The ETF level density is easily obtained by taking the inverse Laplace transform of $Z_{WK}(\beta)$ given by Eq. (4.42). This approach, which treats β merely as an intermediate integration variable, has been explained earlier in Sect. 3.2. We make use of the standard formulae (cf. Appendix B)

$$\mathcal{L}_E^{-1}\left[\frac{e^{-\beta V(\mathbf{r})}}{\beta^\nu}\right] = \frac{(E - V(\mathbf{r}))^{\nu-1}}{\Gamma(\nu)}\Theta(E - V(\mathbf{r})), \tag{4.56}$$

and

$$\mathcal{L}_E^{-1}\left[\beta^n Z(\beta)\right] = \frac{\partial}{\partial E^n}\mathcal{L}_E^{-1}[Z(\beta)]. \tag{4.57}$$

The TF level density in 3 dimensions was given by Eq. (4.4), and now it may be given in any spatial dimension D by Laplace inverting the leading classical term in Eq. (4.42):

$$g_{TF}(E) = \left(\frac{m}{2\pi\hbar^2}\right)^{D/2} \int d^D r\, \frac{[E - V(\mathbf{r})]^{\frac{D}{2}-1}}{\Gamma(D/2)}\,\Theta\left(E - V(\mathbf{r})\right). \tag{4.58}$$

Note that in two dimensions ($D=2$), the energy dependence comes in only through the integrations limits in (4.58) which are given by the classical turning points.

Since we often deal with billiard systems, i.e., with particles confined to a finite spatial domain by infinitely steep potential walls (see, e.g., Sect. 4.6), we give the TF level density for such cases. Here $V \equiv 0$ inside the system and E is always positive. The spatial integral in (4.58) is then limited by the boundary and gives simply

$$g_{TF}(E) = \left(\frac{m}{2\pi\hbar^2}\right)^{D/2} \frac{E^{D/2-1}}{\Gamma(D/2)}\, \mathcal{V}_D, \tag{4.59}$$

where $\mathcal{V}_D$ is the volume of the finite D-dimensional domain. (Note that here, the level density will be independent of E in two dimensions.) It is important to realize, however, that the higher-order $\hbar$-corrections given below cannot be used for billiard systems, since they rely on the potential V being differentiable. Ways of deriving the ETF level density for these systems will be discussed in Sect. 4.6.

For the higher-order corrections to the TF level density, it is more convenient to write down the expressions separately for $D = 2$ and 3. These are given by

$$g_{ETF}(E) = g_{TF}(E) - \frac{1}{48\pi} \frac{\partial}{\partial E} \int d^2r \ \nabla^2 V(\mathbf{r}) \delta \left(E - V(\mathbf{r})\right), \qquad (D=2) \qquad (4.60)$$

and

$$g_{ETF}(E) = g_{TF}(E) - \frac{1}{96\pi^2} \left(\frac{2m}{\hbar^2}\right)^{1/2} \frac{\partial}{\partial E} \int d^3r \ \Theta \left(E - V(\mathbf{r})\right) \frac{\nabla^2 V}{(E - V)^{1/2}}. \quad (D=3)$$
$$(4.61)$$

The second term on the right-hand side of Eq. (4.60) contributes to the particle number N that is obtained by integrating the level density.

Similarly, the spatial density $\rho(\mathbf{r})$ is obtained by taking the inverse Laplace transform of $C(\mathbf{r}, \beta)$ according to Eq. (4.50). We obtain for $D=2$

$$\begin{aligned}
\rho_{ETF}(\mathbf{r}) &= \frac{1}{2\pi} \frac{2m}{\hbar^2} (\mu - V(\mathbf{r})) \\
&\quad - \frac{1}{12\pi} \nabla^2 V \ \delta(\mu - V(\mathbf{r})) + \frac{1}{24\pi} (\nabla V)^2 \frac{\partial}{\partial \mu} \delta(\mu - V(\mathbf{r})). \quad (4.62)
\end{aligned}$$

For $D=3$, the semiclassical density up to second order in $\hbar$ is given by

$$\begin{aligned}
\rho_{ETF}(\mathbf{r}) &= \frac{1}{3\pi^2} \left(\frac{2m}{\hbar^2}\right)^{3/2} (\mu - V(\mathbf{r}))^{3/2} \\
&\quad - \frac{1}{24\pi^2} \left(\frac{2m}{\hbar^2}\right)^{1/2} \left[\frac{\nabla^2 V}{(\mu - V(\mathbf{r}))^{1/2}} + \frac{(\nabla V)^2}{4(\mu - V(\mathbf{r}))^{3/2}} \right]. \quad (4.63)
\end{aligned}$$

Formally, the right-hand side of (4.63) has to be multiplied by a step function $\Theta(\mu - V(\mathbf{r}))$, like also the first term in (4.62), but we omit this overall factor since all the ETF densities are anyhow only defined in the classically allowed region where $E \geq V(\mathbf{r})$. The corresponding expressions up to second order for the kinetic energy density, obtained from Eqs. (4.53,4.55) are, for $D=2$:

$$\begin{aligned}
\tau_{ETF}(\mathbf{r}) &= \frac{1}{4\pi} \frac{2m}{\hbar^2} (\mu - V(\mathbf{r}))^2 \\
&\quad - \frac{1}{6\pi} \nabla^2 V \ \Theta(\mu - V(\mathbf{r})) + \frac{1}{6\pi} (\nabla V)^2 \ \delta(\mu - V(\mathbf{r})), \quad (4.64)
\end{aligned}$$

and for $D=3$:

$$\begin{aligned}
\tau_{ETF}(\mathbf{r}) &= \frac{1}{5\pi^2} \left(\frac{2m}{\hbar^2}\right)^{3/2} (\mu - V(\mathbf{r}))^{5/2} \\
&\quad - \frac{1}{8\pi^2} \left(\frac{2m}{\hbar^2}\right)^{1/2} \left[\frac{5}{3} \nabla^2 V \ (\mu - V(\mathbf{r}))^{1/2} - \frac{3}{4} \frac{(\nabla V)^2}{(\mu - V(\mathbf{r}))^{1/2}} \right]. \quad (4.65)
\end{aligned}$$

Higher-order corrections in $\hbar$ have been worked out for $D=3$, but they are rather lengthy and inelegant. Nevertheless, the fourth-order term is important for obtaining a sufficient convergence of the total energy in realistic three-dimensional problems [24]. One deficiency of the ETF model is apparent from the above expressions. The $\hbar$-corrections of both the density $\rho_{ETF}(\mathbf{r})$ and the kinetic energy density $\tau_{ETF}(\mathbf{r})$ are singular at the classical turning points $\mathbf{r}_\mu$ where $V(\mathbf{r}_\mu) = \mu$. This problem becomes more acute for the higher-order corrections. It does not occur at finite temperatures, as will be discussed in Sect. 4.4.3.

Nevertheless, the integrated particle number N and the total energy E_{sp}, obtained via Eqs. (3.39), (3.38), stay finite to sufficiently high order in $\hbar$, if suitable integrations by parts are used. We do not give here their lengthy expressions which may be found in the literature [3, 17, 24]. As a result, one obtains the average part of the single-particle energy sum (3.39) in the form

$$E_{ETF}(\mu) = \int_0^\mu E\, g_{ETF}(E)\, dE = E_{TF} + E_2 + E_4 + \ldots , \qquad (4.66)$$

where the indices on the correction terms give the power of $\hbar$ relative to the TF term. Remember that, as explained in the previous section, only terms with even powers in $\hbar$ appear for systems with differentiable potentials. (This is different for billiard systems, see Sect. 4.6.) Each term in the series (4.66) depends on the Fermi energy μ which is determined by the particle number integral (3.38) which will similarly result in a series of the form

$$N = N_{ETF}(\mu) = \int_0^\mu g_{ETF}(E)\, dE = N_{TF} + N_2 + \ldots . \qquad (4.67)$$

If one succeeds in finding explicit analytical expressions in the above expansions, which is possible, e.g., for harmonic-oscillator potentials (see Problem 4.2), one may attempt to eliminate the Fermi energy from both quantities and to arrive at an explicit expression for the energy E_{ETF} as a function of the particle number N. This will then have the form:

$$E_{ETF}(N) = E_{TF}(N)\left[1 + c_s N^{-1/3} + c_c N^{-2/3} + \ldots\right]. \qquad (4.68)$$

For saturated interacting fermion systems, the total TF energy is proportional to the volume and to the particle number, and the expansion takes the form well-known from the liquid drop model (LDM) [28, 29]

$$E_{LDM}(N) = a_v N + a_s N^{2/3} + a_c N^{1/3} + \ldots , \qquad (4.69)$$

where a_v, a_s and a_c are called the volume, surface and curvature energies, respectively. A general systematic scheme for deriving the LDM expression for the energy within the framework of the density functional theory makes use of the so-called "leptodermous expansion" [26, 30]. Note that for three-dimensional harmonic-oscillator potentials, for which we evaluate the expansion (4.68) explicitly in Problem 4.2, the leading term is going like $N^{4/3}$ and there is no surface correction of relative order $N^{-1/3}$.

A word must be said about the convergence of the series (4.66), (4.67). These are *asymptotic* in nature and thus do not converge in the strict mathematical sense. With the exception of billiard systems with piecewise straight (for $D=2$) or plane ($D=3$) boundaries, they contain, in fact, always infinitely many terms. In the case of harmonic-oscillator potentials, the correction terms are all infinite starting from a certain order in $\hbar$ (which is different for E_{ETF} and N_{ETF}, see also Problem 4.2). It is then obvious to ignore the divergent terms and to include only the finite contributions in the sum. This partial sum is then usually referred to as the "ETF value" of the corresponding quantity. In the general case, however, the series may be semi-convergent in the sense that the correction terms start to increase in absolute value after reaching a certain order in $\hbar$. The place where to truncate the ETF series is then a matter of convenience and thus not unique. We return to this formal problem in Sect. 4.6.1. May it suffice at this place to mention that in practical applications of the ETF model to realistic three-dimensional systems, the fourth-order contribution to the energy, E_4, was found to be sufficiently small so that the expansion could be stopped at this place [24]. Moreover, the agreement of the result so obtained with the Strutinsky-averaged single-particle energy sum gave a numerical confirmation of this truncation (see Sect. 4.7).

The divergence problems at the classical turning points at $T = 0$ are circumvented when one expresses the kinetic energy density $\tau(\mathbf{r})$ as a functional of the local density $\rho(\mathbf{r})$, by eliminating the Fermi energy μ and the potential V and its derivatives from the expressions above. This cannot be done in a closed form but iteratively, including terms of increasing order in $\hbar$, leading to a functional series of the form

$$\tau_{ETF}[\rho] = \tau_{TF}[\rho] + \tau_2[\rho] + \tau_4[\rho] + \dots . \tag{4.70}$$

Hereby ρ is always the ETF density (4.63) to the same order in $\hbar$, up to which the correction terms in (4.70) are included. In order to illustrate this procedure, we take the 3-dimensional case. The lowest-order TF expressions for $\rho(\mathbf{r})$ in (4.63) and $\tau(\mathbf{r})$ in (4.65). It is easily seen that they obey the functional relation (for $D=3$)

$$\tau_{TF}[\rho] = \frac{\hbar^2}{2m}\kappa\,\rho^{5/3}\,, \qquad \text{with} \quad \kappa = \frac{3}{5}(3\pi^2)^{2/3}. \tag{4.71}$$

(Note that this quantity depends on the spin degeneracy factor 2 which is included here.) In order to obtain the correction functional $\tau_2[\rho]$, we now write down the right-hand side of (4.71), insert the ETF density up to the second term in $\hbar$, as given in Eq. (4.63), and Taylor expand the 5/3-th power up to relative order $\hbar^2$ to obtain

$$\frac{\hbar^2}{2m}\kappa\,\rho_{ETF}^{5/3}(\mathbf{r}) = \frac{1}{5\pi^2}\left(\frac{2m}{\hbar^2}\right)^{3/2}(\mu - V(\mathbf{r}))^{5/2} \times$$
$$\left\{1 - \frac{5}{24}\frac{\hbar^2}{2m}\left[\frac{\nabla^2 V}{(\mu - V(\mathbf{r}))^2} + \frac{(\nabla V)^2}{4(\mu - V(\mathbf{r}))^3}\right] + \mathcal{O}(\hbar^4)\right\}.$$
$$= \tau_{TF}[\rho_{ETF}] - \frac{1}{24\pi^2}\left(\frac{2m}{\hbar^2}\right)^{1/2}\left[\nabla^2 V\,(\mu - V(\mathbf{r}))^{1/2} + \frac{1}{4}\frac{(\nabla V)^2}{(\mu - V(\mathbf{r}))^{1/2}}\right]$$
$$+ \mathcal{O}(\hbar). \tag{4.72}$$

The correction terms of order $1/\hbar$ in square brackets on the right-hand side above are now exactly of the same form as those in Eq. (4.65) for $\tau_{ETF}(\mathbf{r})$. Both these terms can be eliminated taking gradients of the ETF density

$$\nabla \rho_{ETF} = -\frac{1}{2\pi^2}\left(\frac{2m}{\hbar^2}\right)^{3/2}\nabla V(\mu - V)^{1/2} + \mathcal{O}(1/\hbar)\,,$$
$$\nabla^2 \rho_{ETF} = \frac{1}{2\pi^2}\left(\frac{2m}{\hbar^2}\right)^{3/2}\left[-\nabla^2 V(\mu - V)^{1/2} + \frac{1}{2}\frac{(\nabla V)^2}{(\mu - V)^{1/2}}\right] + \mathcal{O}(1/\hbar)\,, \tag{4.73}$$

and noting that, to the given order in $\hbar$, we need only insert the leading terms of (4.73) into Eqs. (4.65) and (4.72). This leads finally to the second-order correction functional for $D=3$

$$\tau_2[\rho] = \frac{\hbar^2}{2m}\left[\frac{1}{36}\frac{(\nabla\rho)^2}{\rho} + \frac{1}{3}\nabla^2\rho\right]. \tag{4.74}$$

Whereas the TF functional (4.71) is exact for a system with constant density ρ, the term (4.74) corrects the functional, taking into account contributions to the kinetic energy coming from spatial variations of the density. Note that the second term in (4.74) does not contribute to the total kinetic energy, obtained by integrating τ over the whole space, if the density $\rho(\mathbf{r})$ goes to zero at infinity which will always be the case for a finite fermion system. The form of the first term in (4.74) was for the first time evaluated by Weizsäcker [28], in what he called the "inhomogeneity correction" to the TF functional (4.71), but he obtained a 9 times larger prefactor. There has been quite some discussion of the correct coefficient of this so-called "Weizsäcker correction" in the literature (see, e.g., Refs. [31, 32]). By now it is well established that the factor $1/36$ is correct for smoothly varying densities, whereas the original coefficient $1/4$ derived by Weizsäcker is correct in the limit of rapid density oscillations of small amplitude (see Jones and Gunnarsson [33] for a detailed discussion). Since the ETF model by default always yields smoothly varying densities (in particular, no quantum shell oscillations can be included), the coefficient $1/36$ given in (4.74) is the correct one to be used in the ETF model.

The fourth-order correction term $\tau_4[\rho]$ for $D=3$ is much harder to evaluate, but essentially found in a completely analogous manner as shown above. It has been given for the first time by Hodges [34] for a Fermi system governed by a local mean field $V(\mathbf{r})$. Since in many applications only the spatial integral of $\tau(\mathbf{r})$, i.e., the kinetic energy, is needed, we give here only its integrated form which, after some suitable integrations by parts, contains three terms (the explicit form of the seven original terms before integration may be found in Refs. [35, 36]):

$$\begin{aligned} T_4[\rho] &= \int \tau_4[\rho(\mathbf{r})]d^3r \\ &= \frac{\hbar^2}{2m}\frac{(3\pi^2)^{-2/3}}{6480}\int\rho^{1/3}\left[8\left(\frac{\nabla\rho}{\rho}\right)^4 - 27\left(\frac{\nabla\rho}{\rho}\right)^2\frac{\nabla^2\rho}{\rho} + 24\left(\frac{\nabla^2\rho}{\rho}\right)^2\right]d^3r. \end{aligned} \tag{4.75}$$

This procedure can, in principle, be continued *ad infinitum*. Practically, there is hardly any need for going to higher correction terms.

For two-dimensional systems, one obtains analogously

$$\tau_{ETF}[\rho] = \frac{\hbar^2}{2m}\left[\pi\,\rho^2 + \frac{1}{3}\nabla^2\rho\right]. \qquad (D=2) \tag{4.76}$$

The coefficient of the Weizsäcker-like term in two dimensions turns out to be proportional to $(\mu - V)\delta(\mu - V)$ which is zero everywhere, so that for the total kinetic energy, the TF functional is correct including all terms of order $\hbar^2$.

It is easily recognized from the above results that the functionals $\tau_{ETF}[\rho]$ yield finite results anywhere in space, as long as the density $\rho(\mathbf{r})$ is finite and falls to zero slowly enough as r tends towards infinity (see the following section). Thus, it may be used also outside the classically allowed region and, in particular, at the classical turning points without any problems. In fact, the functional $\tau_{ETF}[\rho]$ plays an important role in the density variational method and its applications discussed in Sect. 4.4.2 below.

The derivation given above is only valid inside the classically allowed region, since the ETF expressions for the densities ρ and τ are only defined there. One may therefore question the validity of the functional relation $\tau_{ETF}[\rho]$ at the turning points where they diverge, and beyond the turning points where they do not even exist. For a long time it has, in fact, been taken as a miraculous coincidence that the functional can be used everywhere in space for reasonably realistic densities. Only after the successful generalization of the ETF method to finite temperatures has it been shown that the generalized functional $\tau_{ETF}[\rho]$ at $T > 0$ not only is finite and well-defined everywhere in space, but also leads to its form given above when the limit $T \to 0$ is taken carefully. We shall discuss this in Sect. 4.4.3 below, after we have applied the zero-temperature variational ETF method to finite fermion systems.

The convergence of the functional series (4.70) in the 3-dimensional case, up to the fourth-order term $\tau_4[\rho]$, has been checked numerically for typical nuclear potentials in Refs. [35, 36] and for atoms in Ref. [37]. To be consistent with the ETF model, one should use *average* densities $\rho(\mathbf{r})$, since quantum shell oscillations cannot be correctly described by this model. In Refs. [35, 36], therefore, the quantum-mechanical densities and kinetic energies were numerically averaged using the Strutinsky procedure which we will discuss extensively in Sect. 4.7 [see, in particular, Eq. (4.179)].

In Fig. 4.1, we reproduce some results obtained in nuclear applications. In the left part of the figure, a Woods-Saxon potential containing 126 nucleons has been used, with deformations $(c, h=0)$ which will be explained in Sect. 8.1.2 (see Fig. 8.2). In the right part, an axially symmetric harmonic-oscillator potential containing 112 particles is used; the ratio of the frequencies is $q = \omega_\perp : \omega_z$, where z is the symmetry axis. In both parts of the figure $T_m[\tilde{\rho}]$ denotes the total integrated kinetic energy using the functional (4.70), including *all* terms up to order m, in terms of the Strutinsky-averaged densities $\tilde{\rho}(\mathbf{r})$. The curve denoted by $\tilde{T}$ is the Strutinsky-averaged quantum-mechanical kinetic energy which in the right panel cannot be distinguished from the result for $T_4[\tilde{\rho}]$. The agreement reached up to fourth order in the ETF functional is thus perfect for the harmonic oscillator potential, and within about one part in thousand for the Woods-Saxon potential, independently of deformation (which is quite extreme for $q=3$). Note that if only the Weizsäcker term is included ($T_2[\tilde{\rho}]$), the error is of the order of a percent and the deformation dependence is not quite correct; to lowest order (TF), the results are completely wrong. In the right part of this figure, the exact quantum-mechanical kinetic energy T, which contains the typical shell effects, is also shown. This is to demonstrate that the ETF functional only applies to the averaged quantities. (A completely wrong deformation dependence was shown to result [36] when the exact quantum-mechanical density $\rho(\mathbf{r})$ is used in the functional $\tau_{ETF}[\rho]$.)

For nonlocal potentials, like a spin-orbit or any other momentum-dependent potential, or for the case of an $\mathbf{r}$-dependent effective mass of the particle (all of which is used in nuclear physics), the kinetic-energy density functional becomes much more complicated and additional densities (such as a spin-orbit density) and their ETF expansions and functionals have to be derived. Relevant expressions for these cases may be found in Refs. [17, 26, 27].

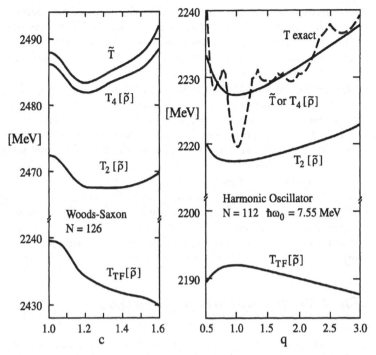

Figure 4.1: Tests of the kinetic energy functional $\tau_{ETF}[\rho]$. *Left:* Woods-Saxon potential with $N = 126$ nucleons at typical deformations for nuclear fission (see the c, h shapes in Fig. 8.2), taken along $h = 0$. *Right:* Axially symmetric harmonic-oscillator potential with frequency ratio q and $N = 112$ particles. (After [35, 36].)

The extension of the ETF model to relativistic systems has also been studied and extensively used, both for atomic physics (see the text book by Dreizler and Gross [38] for references) and nuclear physics (see two recent review articles [39, 40]).

The Wigner-Kirkwood expansion of the single-particle density matrix $\rho(\mathbf{r}, \mathbf{r}')$ may also be obtained by direct Laplace inversion of the Bloch density $C(\mathbf{r}, \mathbf{r}'; \beta)$, according to Eq. (3.40). The $\hbar$ expansion of the latter is found by inserting (4.34) with the expanded form (4.38) of $w(\mathbf{r}, \mathbf{p}; \beta)$ into Eq. (4.47) and integrating over $\mathbf{p}$. The results are most conveniently expressed in center of mass and relative coordinates, defined by $\mathbf{R} = (\mathbf{r} + \mathbf{r}')/2$ and $\mathbf{s} = \mathbf{r} - \mathbf{r}'$, respectively, and have been discussed by Jennings [41]. The TF term of the density matrix in three dimensions ($D=3$) is the familiar form already given by Slater [42]:

$$\rho_{TF}(\mathbf{R}, \mathbf{s}; \beta) = \rho_{TF}(\mathbf{R}) \, \frac{3}{sk_F} \, j_1(sk_F), \qquad s = |\mathbf{s}|, \qquad (4.77)$$

where

$$k_F = \left[3\pi^2 \rho_{TF}(\mathbf{R})\right]^{1/3} \qquad (4.78)$$

is the Fermi wave number and j_1 is the spherical Bessel function of order one. The higher-order terms involve Bessel functions of higher order. A very similar expansion of the density matrix can also derived quantum-mechanically. This "density matrix expansion" (DME) has played an important role in establishing the connection of the Skyrme force [14] to more general nuclear interactions with a finite range [43]. It is generally being used for the calculation of two-body matrix elements and expectation values of two-body operators (see, e.g., Ref. [26]).

Finally we mention very briefly that another way of solving the divergence problem at the classical turning points consists in a *partial resummation* of the Wigner-Kirkwood series (4.49) for the Bloch density. Indeed, all terms in this infinite series that contain arbitrary powers of the first derivatives of the potential $V(\mathbf{r})$ can be summed up analytically [44]. After Laplace inversion of the Bloch density by saddle-point integration, one finds a density that varies smoothly through the classical turning points and falls off to zero further outside. This method has been extended to resum also all terms with second derivatives of V and applied to nuclear shell-model potentials [45, 46]. A related method of partial resummation, which is suited to avoid the divergence occurring at the origin in a Coulomb problem, has been proposed for atomic physics [47] and was applied self-consistently to calculate semiclassical binding energies and densities of atoms [48].

4.4.2 The ETF density variational method

In the introduction 4.1 to this chapter, we have outlined the variational treatment of the classical Thomas-Fermi atom, which yields good average densities and energies of an atom. With the use of the ETF kinetic energy functional $\tau_{ETF}[\rho]$ derived above, this self-consistent treatment can be extended to explicitly include quantum corrections to the average energy of a system, taking in particular into account the spatial variation of the average density $\rho(\mathbf{r})$. In Appendix A.2 we discuss the density variational method in general terms and show how the total energy of an interacting many-fermion system can be expressed through the local density $\rho(\mathbf{r})$. Without specifying here the nature of the two-body interaction, we use the fact [32] that the total energy of the system is uniquely given as a functional of the local density $\rho(\mathbf{r})$ and write it as the sum of kinetic and potential parts:

$$E_{tot} = E_{ETF}[\rho] = T_{ETF}[\rho] + E_{pot}[\rho] = \int \left\{ \tau_{ETF}[\rho] + \mathcal{E}_{pot}[\rho] \right\} d^3r \qquad (4.79)$$

where $\mathcal{E}_{pot}[\rho]$ also contains the contribution from any external one-body potential $V_{ext}(\mathbf{r})$ which just gives a term $V_{ext}(\mathbf{r})\rho(\mathbf{r})$ under the integral above.

We now want to apply the variational principle, making the energy $E[\rho]$ stationary with respect to an arbitrary local variation $\delta\rho(\mathbf{r})$ of the density. In order to keep the number of particles $N = \int \rho(\mathbf{r})d^3r$ constant, we make use of a Lagrange multiplier which here is identical to the Fermi energy μ, so that the variational equation is

$$\frac{\delta}{\delta\rho(\mathbf{r})}\left[E_{ETF}[\rho(\mathbf{r})] - \mu \int \rho(\mathbf{r})d^3r \right] = 0. \qquad (4.80)$$

The variation of the potential energy gives, quite generally, the local mean-field potential V which depends in a self-consistent way on the density itself:

$$\frac{\delta}{\delta\rho(\mathbf{r})}E_{pot}[\rho(\mathbf{r})] = V[\rho(\mathbf{r})]. \qquad (4.81)$$

The exact form of $V[\rho]$ depends, of course, on the particular system under consideration and on the two-body interaction. The variation of the kinetic energy part, however, is universal and can be calculated directly from the explicit forms of the ETF functionals given above. To avoid lengthy formulae, we give here explicitly only the result obtained up to the Weizsäcker term. Assuming the density variation $\delta\rho(\mathbf{r})$ to vanish on the boundary (which usually is at infinity) and using the standard techniques of variational calculus, one gets the following Euler-Lagrange equation

$$\frac{\hbar^2}{2m}\left\{\frac{5}{3}\kappa\rho^{2/3} + \frac{1}{36}\left[\frac{(\nabla\rho)^2}{\rho^2} - 2\frac{\nabla^2\rho}{\rho}\right]\right\} + D_4[\rho] + V[\rho] = \mu\,, \qquad (4.82)$$

where D_4 is the contribution from the fourth-order gradient term, $D_4[\rho] = \delta T_4[\rho]/\delta\rho$, whose explicit form (containing seven terms with up to fourth-order gradients of ρ) may be found in Ref. [26]. Equation (4.82) is a partial differential equation which in general can only be solved numerically. To lowest TF order, neglecting all gradient terms, the solution of (4.82) for the density $\rho(\mathbf{r})$ is algebraic and gives, as it should, exactly the TF expression (4.7). The variational equation (4.82) has played an important role in early studies of the nuclear surface energy, which are nicely summarized in the review by Wilets [31]. (For a simple model with an analytical solution of (4.82) up to second order gradients, see also Problem 4.5.)

If the second-order gradient terms are kept in Eq. (4.82) and the potential $V[\rho]$ is assumed to vanish at large distances, the solutions for the density are found to decay exponentially, as has been noted already long ago [14, 49]. This is easily recognized from the fact that the Fermi energy μ must be a constant which can only be cancelled by the highest gradient term in Eq. (4.82). The rate of the exponential fall-off turns, however, out to be too fast in comparison to that of exact quantum-mechanical densities. One pragmatic way out of this problem is to artificially increase the coefficient of the Weizsäcker term, adjusting it by optimizing the results. The dilemma hereby is that the same constant cannot be used for densities and for integrated energies. Nevertheless, this procedure has been widely used in physics and physical chemistry (see Ref. [38] for details and applications). It violates, however, the aspect of universality which is one of the nice features of the ETF model, and can easily be avoided by including the higher-order gradient terms in Eq. (4.82). For atomic systems which are governed by a $1/r$ divergence at the center due to the external nuclear potential, special problems arise for which we refer to the book by Dreizler and Gross [38], where also many applications of the ETF variational method to atomic systems can be found.

For the following discussions and applications, we shall assume that the total potential is finite in the interior of the system and goes to zero at large distances. Then the asymptotic fall-off of the density is always governed by the highest gradient term included in the variational equation (4.82). Including the fourth-order contribution $D_4[\rho]$, it has been noted [36] that the density of a spherical system falls off only with the sixth inverse power of r, i.e., $\rho(r) \sim r^{-6}$. More generally, the highest derivative term in $\tau_{2n}[\rho]$ goes radially like

$$\tau_{2n}[\rho(r)] \sim \rho^{(5-2n)/3}(r)\left(\frac{1}{\rho}\frac{d^{2n}\rho(r)}{dr^{2n}}\right)\,, \qquad (4.83)$$

and the solution of Eq. (4.80) is found [36] to fall off like

$$\rho(r) \sim r^{-3n/(n-1)} \quad \text{for} \quad r \to \infty. \quad (n \geq 2) \tag{4.84}$$

This is not the correct asymptotic behavior for most systems, as the quantum-mechanical density usually falls of exponentially. It has turned out, however, that this asymptotic solution is reached only mathematically very far away from the actual surface of a realistic system, and that the inclusion of the fourth-order term definitely improves the variational results. This was shown in numerical solutions of the ETF variational equation (4.82) up to fourth order, both for metallic clusters within the jellium model [50] and for atomic nuclei [51]. Spherical symmetry was assumed in both cases so that the differential equation was only one-dimensional. In the latter work, realistic effective nucleon-nucleon interactions including spin-orbit terms were used and the results were compared to those of fully microscopical Hartree-Fock calculations. Note also that in nuclei, two coupled variational equations for the individual densities of protons and neutrons have to be solved.

The numerical solution of a nonlinear fourth-order differential equation is not quite easy, and very difficult if not practically impossible for deformed systems without symmetry. For that case one may perform the variation in (4.80) not exactly in **r** space, but in a restricted space of parameterized variational densities. The choice of the variational density is then a matter of physical intuition and, of course, not free of a certain bias. A spherical density profile that has proved useful for both nuclei [26] and metal clusters [52, 53] is a generalized Fermi function:

$$\rho(r) = \frac{\rho_0}{\left[1 + \exp\left(\frac{r-R}{a}\right)\right]^\gamma}. \tag{4.85}$$

Of the four parameters above, only three are independent due to the constraint of the particle number conservation. The independent parameters are then determined variationally by minimizing the total energy of the system under consideration, and no adjustable parameter is used anywhere. With this restricted variational procedure up to fourth order in the ETF functionals, density profiles and energies were obtained that agreed very closely with the fully variational ETF results [50, 51] on one hand, and with the averaged results of fully microscopic Kohn-Sham calculations for metal clusters [55, 56] and Hartree-Fock calculations for nuclei [26, 51] on the other hand. In particular, the profiles of the variationally optimized densities (4.85) match the self-consistent quantum-mechanical ones over the entire region of the surface, demonstrating that the mathematically inadequate asymptotic inverse-power decays of the exact solutions of (4.82) have no practical significance.

Note that the sixth- and higher-order gradient corrections to $\tau_{ETF}[\rho]$ cannot be used in connection with densities that decay exponentially like $\rho(r) \sim e^{-ar}$. It has repeatedly been pointed out in the literature that this leads to divergences. Indeed, the general form (4.83) of $\tau_{2n}[\rho](r)$ is immediately seen to diverge for $n \geq 3$ in this case, due to the factor $\rho^{(5-2n)/3}$. This is, however, not a shortcoming of the ETF variational method itself. As has been pointed out (see the note added in proof to Ref. [36]), when one uses the correct variational solution with the asymptotic behavior given in Eq. (4.84) and inserts this into the general form (4.83) of τ_{2n}, the latter is found to go to zero with the same power law as the density (4.84) and thus can be integrated to give a finite kinetic energy at any finite order $2n$.

One basic problem of the ETF variational method is that there is effectively no lower bound to the total energy, different from the Hartree-Fock method where this is granted by the Ritz principle (see Appendix A.1). *A priori*, density functional theory (see Appendix A.2) tells us that, as a consequence of the Hohenberg-Kohn theorem [32], the total energy (4.79) should have its variational minimum at the value of the exact energy of the system which therefore gives the lower bound. But this is only true if the *exact* density functional $E[\rho]$ is used which, however, is unknown. Due to the approximations necessary to put the density functional method into use, and here in particular due to the ETF approximation used to derive the kinetic energy functional, this lower bound is no longer given. As a consequence, the variational ETF results turn out to overbind systematically, i.e., the negative energies are too large in absolute value [26, 51]. The amount of overbinding is, however, much less than a percent of the total energy and very little dependent on the deformation for the systems that have been investigated.

We should like to emphasize again that due to the very nature of the ETF model, the Euler equation (4.82) yields only *average* densities and energies, and can thus never reproduce any of the quantum shell oscillations typical of finite fermion systems. The ETF variational method is thus limited to yield average properties of finite fermion systems. A perturbative way to include the shell effects, without going through the fully self-consistent microscopic Hartree-Fock or Kohn-Sham schemes (see Appendix A), makes use of Strutinsky's shell-correction method and will be discussed at the end of Sect. 4.7.3 below.

As an example of a variational ETF calculation for a complex deformed system, we show in Fig. 4.2 on the next page the deformation energy of the nucleus ^{240}Pu which exhibits the famous double-humped fission barrier, to which we shall return in Sect. 8.1.2. For the details of this calculation, which made use of an effective nucleon-nucleon interaction of the Skyrme type [14, 57], including spin-orbit and effective mass terms, and the variational densities (4.85) (independently for neutrons and protons), we must refer the reader to the original papers [26, 57]. For the deformed shapes of the nucleus, a suitable parametrization [58] obtained within the liquid drop model was used (see Fig. 8.2). The density profiles were assumed to vary along the normal distance d from the liquid-drop surface according to Eq. (4.85), with $(r - R)$ replaced by d. The resulting total binding energy, plotted versus the quadrupole moment Q of the nucleus, is shown in Fig. 4.2 by the heavy dashed line, denoted as E_{ETF}. The thin solid line shows the exact energy E_{HF} obtained in a self-consistent, fully microscopic Hartree-Fock (HF) calculation [57] using the same effective interaction, as a function of the constrained quadrupole moment Q. The HF curve has the typical features of a double-humped fission barrier which is due to the quantum shell effects (see Sect. 8.1.2 for a more explicit discussion of nuclear fission barriers). The ETF result can therefore not directly be compared to it, apart from the fact that it represents a smooth average of the deformation energy, with the ground-state minimum at the spherical shape ($Q = 0$) as is to be expected from the ETF model.

A direct comparison is, however possible, if one averages the Hartree-Fock results in a self-consistent way using the Strutinsky method [59]. This was done in Refs. [26, 57],

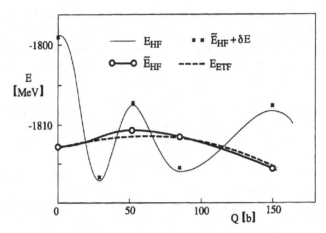

Figure 4.2:   Comparison of different deformation energies of the ^{240}Pu nucleus.
Heavy dashed line: variational ETF energy; *heavy solid line:* self-consistently
Strutinsky-averaged Hartree-Fock (HF) energy; *thin solid line:* exact microscopic
HF energy; *crosses:* sum of average HF energy plus shell-correction energy (cf. Sect.
4.7). (After [26, 57].)

and the result is given in Fig. 4.2 by the circles connected with a heavy solid line,
denoted by $\tilde{E}_{HF}$. It should be mentioned that due to the systematic overbinding of
the ETF variational method, the energy E_{ETF} was adjusted in absolute position to
agree with $\tilde{E}_{HF}$ at the spherical point. The close agreement of the two curves within
~ 1 MeV shows that the amount of overbinding (here ~ 8 MeV) is roughly constant
and thus does not affect the deformation energies. The crosses in Fig. 4.2 indicate the
results achieved when adding the Strutinsky shell-correction energy (see Sect. 4.7) to
the smooth part of the energy, yielding an excellent approximation to the exact HF
energy. (The fact that the heights of the barriers are lower than in experiment here is
due to a deficiency of the effective force used, as discussed in Refs. [26, 57], and does
not affect the present comparison.)

An interesting way of including the shell effects perturbatively, which at the same
time effectively cancels the above mentioned overbinding errors, has been proposed
by Pearson and collaborators [60]. They have performed extensive ETF variational
calculations for nuclear binding energies, employing realistic effective nucleon-nucleon
interactions of the type originally proposed by Skyrme [14]. In order to include the
shell effects, they write the total energy as[1]

$$E_{tot} = E_{ETF}[\rho] + \sum_{E_n < \mu} E_n - \int \{\tau_{ETF}[\rho] + \rho V[\rho]\} d^3 r \, , \qquad (4.86)$$

where ρ are the variational densities, $V[\rho]$ is the ETF mean field given by Eq. (4.81),
E_n are its eigenvalues, and μ is the Fermi energy. The integral in the third term on
the right-hand side above represents an average of the sum of occupied single-particle

---

[1]We do not exhibit here the modifications necessary to include the effective mass and spin-orbit
components of the nuclear interaction, nor the fact that one is dealing with two kinds of nucleons.

energies in the second term. This way of including the shell effects into the total energy is inspired by the Strutinsky method, which we present in Sect. 4.7 below, and is formally justified from the self-consistent mean-field approach by the Strutinsky energy theorem discussed in Appendix A.3. In order to distinguish their approach from the standard Strutinsky method, in which the last term of Eq. (4.86) is given by a numerical energy averaging of the single-particle energy sum (see Sect. 4.7.1), they termed their method the "ETF plus Strutinsky integral" (ETFSI) method. In Ref. [61], the total energy defined in Eq. (4.86) was fitted to 1492 experimental nuclear masses, by varying the parameters of the effective nuclear force (including pairing interaction) such as to minimize the rms error. The ETF calculation for each set of parameters was done in a similar way as described above in connection with Fig. 4.2; the variational densities for neutrons and protons were of the form (4.85) with $\gamma = 1$.

To illustrate their results, we show in Fig. 4.3 the best mass fit obtained in Ref. [61] with the force ETFSI-1. The upper part of the figure gives the difference between the experimental nuclear masses (in MeV) and the ETFSI energies (4.86), plotted as a function of the neutron number N. It is seen to be roughly $\sim \pm 2$ MeV or less; the rms error was found to be 0.736 MeV. This fit is of the same quality as those obtained with the standard Strutinsky method, employing phenomenological liquid drop model energies and shell-model potentials with many more parameters (see Ref. [62] for the most recent example). To appreciate this achievement, one should bear in mind that the total nuclear energies, of which the difference here has been taken, are of the order of one to two thousand MeV for the larger nuclei. It gives an impressive example of the practical use of the variational ETF method.

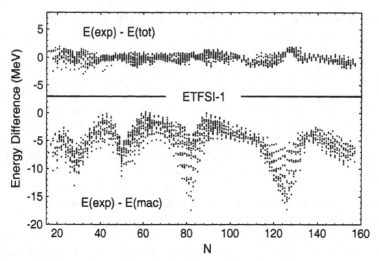

Figure 4.3: *Upper part:* Error in the total nuclear binding energy E(tot), defined in Eq. (4.86), as compared to the experimental binding energy E(exp). *Lower part:* Empirical energy shell-correction, given as the difference between E(exp) and the macroscopic part E(mac) defined as the ETF energy for spherical shape (see text). All ETF energies were evaluated by the self-consistent ETFSI method [60]. (Courtesy J. M. Pearson.)

In the lower part of Fig. 4.3, we show the difference between the experimental nuclear masses and the spherical part of the average ETF energies, denoted E(mac) in the figure. The use of spherical ETF energies is motivated by the argument that when fitting experimental masses by LDM type mass formula, no shell effects can be included. Since deformations are driven by the shell effects, one leaves their contributions in the empirical shell-correction, which is the quantity shown. Its oscillating behavior clearly reflects the shell structure. It has pronounced minima at the "magic" neutron numbers $N = 28, 50, 82$, and 126. These will be further discussed in Chapter 8.

Let us turn to an example from the physics of metal clusters to which we shall return in Sect. 8.2 in a different context. Seidl et al. [63] have used restricted variational ETF calculations to fourth order, with the density profiles (4.85), to calculate ionization potentials (IP) and electron affinities (EA) of various metal clusters, and compared them to experimental data. These quantities are defined as the energy required to remove an electron and the energy gained by adding an electron, respectively, according to

$$\text{IP}(N) = E(N, -1) - E(N, 0), \qquad \text{EA}(N) = E(N, 0) - E(N, +1). \tag{4.87}$$

Here $E(N, z)$ is the total energy of a cluster containing N atoms and z excess valence electrons. This means that $E(N, -1)$ is the energy of a singly ionized cluster and $E(N, 0)$ that of a neutral cluster, etc. Experimentally, these quantities exhibit strong shell effects, with the saw-tooth pattern typical also of separation energies in nuclei. Major steps occur at the so-called magic numbers that indicate an enhanced stability of the cluster and can be associated to a filled spherical shell of the valence electrons (see Sect. 8.2 for more details). The ETF model can, of course, not reproduce these oscillations. However, they occur at the same magic numbers for both IP and EA and are roughly in phase. Therefore, if one takes the differences $\text{IP}(N) - \text{EA}(N)$, the shell effects tend to cancel and the results become relatively smooth functions of the cluster size N.

In Fig. 4.4, the differences $\text{IP}(N) - \text{EA}(N)$ for aluminum clusters are plotted versus the quantity $N^{-1/3}$ which is proportional to the inverse cluster radius. The dots give the experimental values (see [63] for the references to the data), and the solid line is the ETF result for the difference. The largest cluster measured has $N=66$ atoms, the smallest is the dimer ($N=2$). Some of the structures seen in the experimental values are remainders of the shell effects, some of them are due to the ionic structure (e.g., for $N=13$, i.e., at $N^{-1/3} = 0.425$). In the ETF calculations, the jellium model [55, 64] was used which neglects the explicit structure of the ions (cf. Sect. 8.2). That the largest clusters are best described by the ETF model was to be expected, but that the ETF curve actually fits the average experimental values so well all the way down to $N=2$ is quite remarkable and came as a surprise.

A more theoretically oriented application of the ETF variational method is that it allows one to systematically obtain an asymptotic expansion of the total energy of a given system in powers of the inverse cube root $N^{-1/3}$ of the particle number, which is the characteristic expansion of a mass formula or the liquid drop model, leading to the general form (4.69) of the energy. This is done with the so-called "leptodermous expansion" that was developed by Myers and Swiatecki [30] in nuclear physics. It has been used with variational ETF results for nuclei in Ref. [26] and for metal clusters in Refs. [50, 54, 53].

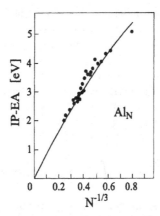

Figure 4.4: Difference between ionization potential IP and electron affinity EA of aluminum clusters, shown versus the inverse cluster radius $N^{-1/3}$ (in dimensionless units). Dots are experimental values, and the solid line is the result of a fourth-order ETF density variational calculation using the jellium model. (From [63].)

4.4.3 The finite-temperature ETF model

In the present section, we shall develop the extension of the ETF model to nonzero temperatures T. This is interesting from the physical point of view because it enables us to describe a system in an excited state. Mathematically, it is also of interest because the finite-temperature extension removes the divergence of the ETF densities $\rho(\mathbf{r})$ and $\tau(\mathbf{r})$ at the classical turning points, which plagued the $T = 0$ treatment. Before proceeding with the mathematical formalism, we should clarify what is meant by the temperature of a many-fermion system like an atomic nucleus, or a metal cluster. Although the number of fermions in the system may not be too large, the density of excited states of the system increases rapidly with the excitation energy. Of course, at very low energies, there may be collective excitations in the form of rotations, vibrations etc. These represent *ordered* motion of the particles with strong phase correlations, and are often best described by introducing collective degrees of freedom. But at higher excitation energies, the motion of the particles becomes increasingly *random* and, in classical language, chaotic. Then, the counting of the states of the total system becomes basically a combinatorial problem of distributing the excitation energy amongst the particles that are excited above the Fermi sea. This is the picture of a (noninteracting) Fermi gas which has absorbed a certain amount of excitation energy. The large number of ways of partitioning this energy is directly related to the density of states which in turn grows exponentially with the statistical entropy [65]. The inverse temperature of a system in thermodynamic equilibrium is the partial derivative of the entropy with respect to the energy. Thus, the inverse temperature may be related to the (logarithmic) derivative of the density of states with respect to the energy. To be more specific, we may consider a grand canonical ensemble for a Fermi gas and calculate its thermodynamic properties at a temperature T, which is precisely related to the excitation energy of the system under consideration. The reader is cautioned that the fluctuations from the ensemble average may be large for small particle numbers.

The generalization to the finite-temperature TF model is straightforward. We know that the ground state of an N-fermion system in the mean-field approximation is obtained by occupying the lowest N single-particle states fully. In the TF approximation, the sum over the quantum single-particle states is replaced by an integration over the available phase space, divided by the elementary cell volume h^D (see, for example, Eq. (2.161) in Sect. 2.7). The ground state is at $T = 0$, and we have a step-function occupancy, so that (including the spin factor of 2):

$$N = \frac{2}{h^D} \int d^D p \int d^D r \, \Theta \left(\mu - H_{cl}(\mathbf{p}, \mathbf{r}) \right) , \qquad (4.88)$$

where $H = p^2/2m + V(\mathbf{r})$ is the classical single-particle Hamiltonian. At a finite temperature[2] $T > 0$, the step-function occupancy is replaced by the Fermi occupancy factor, giving

$$N = \frac{2}{h^D} \int d^D p \int d^D r \, \frac{1}{[1 + \exp\left(H_{cl}(\mathbf{p}, \mathbf{r}) - \mu\right)/T]} . \qquad (4.89)$$

This shows that the spatial density ρ in the TF approximation is

$$\rho_{TF}(\mathbf{r}, T) = \frac{2}{h^D} \int d^D p \, \frac{1}{[1 + \exp\left(p^2/2m + V(\mathbf{r}) - \mu\right)/T]} . \qquad (4.90)$$

This integration may be done analytically only for $D = 2$. (For expressions of $\rho(\mathbf{r})$ and $\tau(\mathbf{r})$ in terms of the so-called Fermi integrals, see further below). In a physically interesting problem, the fermions interact with each other, and the mean-field potential V is itself a functional of ρ, just as in the $T = 0$ case. In that case, Eq. (4.90) must be solved self-consistently.

We now proceed to show that the TF approximation at finite temperature may be systematically improved by the same Wigner-Kirkwood expansion as developed above. The point to note is that a quantity like the sum of the occupied single-particle energies which we denoted by E_{sp} at zero temperature, is given by

$$E_{sp} = 2 \int_0^\mu g(E) E \, dE . \qquad (4.91)$$

Denoting the same quantity at finite temperature by U, in thermodynamics usually called the "internal energy", we have

$$U = 2 \int_0^\infty \frac{g(E) E}{[1 + \exp\left(E - \mu\right)/T]} \, dE . \qquad (4.92)$$

It is the same single-particle level density $g(E)$ that is used here, as in the zero-temperature expressions, given by the sum of delta functions in Eq. (3.4). In the TF approximation, the quantum-mechanical $g(E)$ is replaced by $g_{TF}(E)$, and the ETF corrections add on higher-order $\hbar$ terms to it, using the Wigner-Kirkwood series and

---

[2]T from now on stands for $k_B T$, where k_B is the Boltzmann constant, and is measured in suitable energy units.

the inverse Laplace transform. This may still be implemented at finite T as before. For computational ease, we shall define a finite-temperature level density $g_T(E)$ which will allow us to use the convolution theorem in Laplace transforms and compute the free energy F directly, following the formalism developed in Refs. [59, 66] (see also Ref. [26]). For the zero-temperature problem, the canonical partition function $Z(\beta)$ is calculated only for taking its inverse transform in order to find $g(E)$ to various orders in $\hbar$. In this context, β is only a mathematical complex-valued variable and *not* an inverse temperature. In the following expressions, the finite temperature T will appear explicitly; the quantity β still plays the same role as a complex mathematical variable and should not be taken as an inverse temperature.

In quantum statistics, a many-body system at finite temperature T is usually treated as a microcanonical, a canonical or a grand canonical ensemble, depending on the particular situation one is describing. We use here the grand canonical description of a system embedded in a heat bath at a fixed temperature T, whereby both its energy and its particle number are conserved only on the average. For a fermion system the Helmholtz free energy is, in the mean-field approximation, given in terms of the single-particle energies E_n by

$$F = U - TS = 2\sum_n E_n \nu_n - TS, \qquad (4.93)$$

where U is the internal energy and S the entropy, and ν_n are the Fermi-Dirac occupation numbers whose sum gives the average particle number:

$$\nu_n = \frac{1}{1 + \exp\left(\frac{E_n - \mu}{T}\right)}, \qquad N = \langle \hat{N} \rangle = 2\sum_n \nu_n. \qquad (4.94)$$

μ is the Fermi energy or chemical potential, which is determined by the right-hand side of Eq. (4.94). We put explicitly the spin degeneracy factor 2 for fermions before all sums over the quantum states n here and below, and do not include it in the level density. The explicit form of the entropy S (in units of k_B) is

$$S = -2\sum_n \left[\nu_n \log \nu_n + (1 - \nu_n) \log(1 - \nu_n)\right]. \qquad (4.95)$$

It can be shown [59] that the above quantities N, F and S may be expressed in terms of the convoluted *finite-temperature level density* $g_T(E)$ as

$$N = 2\int_{-\infty}^{\mu} g_T(E)\,dE, \quad F = 2\int_{-\infty}^{\mu} E\,g_T(E)\,dE, \quad S = -\left.\frac{\partial F}{\partial T}\right|_{\nu_n = const}. \qquad (4.96)$$

The function $g_T(E)$ is defined as the convolution of the normal ($T = 0$) level density (3.4) over the function $1/4T\mathrm{Cosh}^2(E/2T)$:

$$g_T(E) = \int_{-\infty}^{\infty} \frac{g(E')}{4T\mathrm{Cosh}^2[(E - E')/2T]}\,dE' = \sum_n \frac{1}{4T\mathrm{Cosh}^2[(E - E_n)/2T]}. \qquad (4.97)$$

Note that integrating the inverse Cosh^2 gives the Fermi function (see also Problem 4.6), so that the integral over $g_T(E)$ leads to (4.94). To show that the integral over $E g_T(E)$

gives the correct free energy including the "heat energy" $-TS$ needs some algebraic manipulations, but is straightforward (cf. Problem 4.7).

We now take the Laplace transform of $g_T(E)$, exploiting the folding theorem (cf. Appendix B) and the fact that the (*two-sided!*) Laplace transform of the inverse Cosh² is given by (see, e.g., [67])

$$\mathcal{L}_\beta^{(2)}\left[\frac{1}{4T\text{Cosh}^2(E/2T)}\right] = \int_{-\infty}^{\infty} e^{-\beta E} \frac{1}{4T\text{Cosh}^2(E/2T)}\, dE = \frac{\pi\beta T}{\sin(\pi\beta T)}. \qquad (4.98)$$

This gives us the following form of the partition function

$$Z_T(\beta) = \int_{-\infty}^{\infty} e^{-\beta E} g_T(E)\, dE = Z_0(\beta)\frac{\pi\beta T}{\sin(\pi\beta T)}, \qquad (4.99)$$

where $Z_0(\beta)$ is the partition function in its usual form (3.15):

$$Z_0(\beta) = \sum_n e^{-\beta E_n}. \qquad (4.100)$$

The average particle number and the Helmholtz free energy are now obtained by inverse Laplace transforms like in Eqs. (3.38,3.39)

$$N = 2\mathcal{L}_\mu^{-1}\left[\frac{1}{\beta}Z_T(\beta)\right], \qquad F = -2\mathcal{L}_\mu^{-1}\left[\frac{1}{\beta}\frac{\partial}{\partial\beta}Z_T(\beta)\right]; \qquad (4.101)$$

the entropy S is found from F by the partial derivative given in (4.96).

The local densities $\rho(\mathbf{r})$ and $\tau(\mathbf{r})$ at finite temperature are defined by

$$\rho(\mathbf{r}) = 2\sum_n |\psi_n(\mathbf{r})|^2 \nu_n, \qquad \tau(\mathbf{r}) = \frac{\hbar^2}{2m}2\sum_n |\nabla\psi_n(\mathbf{r})|^2 \nu_n. \qquad (4.102)$$

We also introduce an entropy density $\sigma(\mathbf{r})$ by

$$\sigma(\mathbf{r}) = -2\sum_n |\psi_n(\mathbf{r})|^2[\nu_n \log\nu_n + (1-\nu_n)\log(1-\nu_n)], \qquad (4.103)$$

and a free energy density $\mathcal{F}(\mathbf{r})$ by

$$\mathcal{F}(\mathbf{r}) = \tau(\mathbf{r}) + V(\mathbf{r})\rho(\mathbf{r}) - T\sigma(\mathbf{r}), \qquad (4.104)$$

so that

$$F = \int \mathcal{F}(\mathbf{r})\, d^3r, \qquad S = \int \sigma(\mathbf{r})\, d^3r. \qquad (4.105)$$

All these densities can be obtained by inverse Laplace transforms from the finite-temperature extension of the Bloch density C_T that carries the same temperature factor as that appearing in Z_T (4.99):

$$C_T(\mathbf{r}, \beta) = C_0(\mathbf{r}, \beta)\frac{\pi\beta T}{\sin(\pi\beta T)}, \qquad (4.106)$$

where $C_0(\mathbf{r}, \beta)$ is the $T = 0$ Bloch density given by

$$C_0(\mathbf{r}, \beta) = 2 \sum_n \psi_n^*(\mathbf{r})\, \psi_n(\mathbf{r})\, e^{-\beta E_n}. \tag{4.107}$$

The explicit inversion relations for the density and the free energy density are

$$\rho(\mathbf{r}) = \mathcal{L}_\mu^{-1}\left[\frac{1}{\beta} C_T(\mathbf{r}, \beta)\right], \qquad \mathcal{F}(\mathbf{r}) = \mu\rho(\mathbf{r}) - \mathcal{L}_\mu^{-1}\left[\frac{1}{\beta^2} C_T(\mathbf{r}, \beta)\right]. \tag{4.108}$$

The entropy density is similarly found from

$$\sigma(\mathbf{r}) = -\frac{\partial}{\partial T}\mathcal{F}(\mathbf{r})\bigg|_{\rho=const} = \mathcal{L}_\mu^{-1}\left[\frac{1}{\beta^2}\frac{\partial}{\partial T} C_T(\mathbf{r}, \beta)\right], \tag{4.109}$$

and the kinetic energy density, finally, is obtained in terms of the other densities simply from Eq. (4.104).

In order to derive the finite-temperature ETF expressions of all these quantities, it is now sufficient to replace $Z_0(\beta)$ and $C_0(\mathbf{r}, \beta)$ in the above relations by their Wigner-Kirkwood expansions given in Sect. 4.3. This leads, after Laplace inversion, inevitably to integrals over the Fermi function of the type[3]

$$F_\xi(\eta) = \int_0^\infty \frac{x^\xi}{1 + \exp\,(x - \eta)}\, dx\,, \tag{4.110}$$

the so-called "Fermi integrals", where η is a real variable. The F_ξ cannot be given analytically for non-integer ξ, but there exist several useful series expansions and numerical approximations [7, 69, 70, 71].

We do not give here the rather lengthy expressions of all the quantities of interest. They have been given by Bartel et al. [66] for a local potential $V(\mathbf{r})$ up to fourth order in the $\hbar$-expansion, and including the effective mass and spin-orbit terms used in nuclear physics up to second order (see also Ref. [26] for the latter). At this place we just quote the classical TF expressions in terms of the Fermi integrals for the three-dimensional case ($D=3$), the most important of which were given for the first time by Stoner [69]:

$$\rho_{TF}(\mathbf{r}, T) = A_T F_{1/2}(\eta_0)\,, \tag{4.111}$$

$$\mathcal{F}_{TF}(\mathbf{r}, T) = \mu\, \rho_{TF}(\mathbf{r}, T) - \frac{2}{3} A_T T F_{3/2}(\eta_0)\,, \tag{4.112}$$

$$\sigma_{TF}(\mathbf{r}, T) = -\eta_0\, \rho_{TF}(\mathbf{r}, T) + \frac{5}{3} A_T F_{3/2}(\eta_0)\,, \tag{4.113}$$

$$\tau_{TF}(\mathbf{r}, T) = A_T T F_{3/2}(\eta_0)\,. \tag{4.114}$$

Hereby the constant factor A_T and the variable quantity η_0 are given by

$$A_T = \frac{1}{2\pi^2}\left(\frac{2m}{\hbar^2}\right)^{3/2} T^{3/2}, \quad \eta_0(\mathbf{r}) = \frac{1}{T}[\mu - V(\mathbf{r})]\,. \tag{4.115}$$

---

[3]In the literature, there exist several conflicting notations for the Fermi integrals. Often one finds them referred to as $J_\xi(\eta)$ or $I_\xi(\eta)$, but that might lead to confusion with Bessel functions. We follow the notation of Landau and Lifshitz [68]. Note that the integral (4.110) does not exist for $\xi < -1$, but corresponding functions $F_\xi(\eta)$ may still be defined using the recursive relation $\xi F_{\xi-1}(\eta) = dF_\xi(\eta)/d\eta$.

The ETF correction terms involve gradients of the potential $V(\mathbf{r})$ as in Eqs. (4.63), (4.65) above, multiplied with Fermi integrals of lower orders. One important point to note is that they stay finite for $T > 0$ also at and outside the classical turning points where $\eta_0 = 0$.

Upon integration of the ETF densities, one obtains again series expansions of the forms given in Eq. (4.66) for the free Helmholtz energy

$$F_{ETF} = F_{TF} + F_2 + F_4 + \dots , \qquad (4.116)$$

and for the entropy

$$S_{ETF} = S_{TF} + S_2 + \dots . \qquad (4.117)$$

As an example of their convergence, it is instructive to study the case of a harmonic-oscillator potential for which these series can be obtained explicitly [66] in terms of the Fermi integrals (4.110). The densities (4.113,4.112) need not be derived in this case, but one may start directly from the partition function, whose $\hbar$-expansion is given in Eq. (4.1), and use the inverse Laplace transform in (4.101) and the last relation in Eq. (4.96) to get F_{ETF} and S_{ETF}. Their explicit expressions up to fourth order in $\hbar$ are worked out for the spherical harmonic oscillator in Problem 4.6.

In Table 4.1 we show the results obtained for the harmonic oscillator with N=70 particles for various temperatures. The individual contributions at the various orders in $\hbar$ are given in an obvious notation (the term S_4 is negligible); S_{ETF} and F_{ETF} denote their sums. The exact quantum-mechanical results, given by Eqs. (4.95) and (4.93), are denoted by S_{qm} and F_{qm}, respectively. We see that the convergence up to order 4 is excellent (for S already up to order 2). Furthermore, with increasing temperature the ETF values approach the quantum-mechanical ones. This is because the shell effects are wiped out by the finite temperature; we have demonstrated this in Sect. 3.2.5 for the level density of the same system. At the temperature $T = 5$ MeV, no shell effects are left and the two values agree up to six digits. This shows us that at sufficiently high temperatures, finite fermion systems may be described *exactly* by the ETF model, due to the fact that then the shell effects cease to exist.

T	S_{TF}	S_2	S_{ETF}	S_{qm}	F_{TF}	F_2	F_4	F_{ETF}	F_{qm}
0					2534.48	-35.37	-0.14	2498.97	2487.24
1	14.91	-0.10	14.81	6.67	2526.92	-35.21	-0.14	2491.57	2485.89
3	44.43	-0.31	44.12	43.86	2466.63	-33.96	-0.14	2432.53	2432.41
5	73.06	-0.52	72.54	72.54	2347.31	-31.49	-0.14	2315.68	2315.68

Table 4.1: Entropy S (in units of k_B) and free Helmholtz energy F (in units of MeV) of a system of N=70 fermions in a spherical harmonic-oscillator potential with $\hbar\omega$=7.896 MeV at different temperatures T (in MeV). Shown are the various terms of their ETF expansion; S_{qm} and F_{qm} are their exact quantum-mechanical values. (From [66].)

As the next step, we briefly indicate the nature of the density functionals at $T > 0$. Here the elimination of μ, the potential V, and its derivatives can no longer be done explicitly. Therefore the density functionals $\mathcal{F}[\rho]$, $\sigma[\rho]$ and $\tau[\rho]$ have no closed form at finite temperature. However, the relation between ρ and the other densities can easily be established numerically: since $F_{1/2}(\eta_0)$ is a monotonous function of η_0, Eq. (4.111) can be inverted numerically, which for monotonous densities always gives a unique result for $\eta_0(\rho)$ that can be inserted in the other TF densities above. The second-order gradient corrections to the TF functionals at $T > 0$ were derived for the first time by Perrot [72] for local potentials and by Brack [73] including effective mass and spin-orbit terms. The integrated fourth-order corrections have been given for local potentials by Bartel *et al.* [66]; we refer the interested reader to this paper for the explicit expressions and their derivations which are algebraically rather cumbersome. We just give here the results up to second order (leaving out terms whose spatial integral gives zero) for the free energy density

$$\mathcal{F}_{ETF}[\rho] = \mu\,\rho - \frac{2}{3}A_T T F_{3/2}(\eta) + \frac{\hbar^2}{2m}\frac{(\nabla\rho)^2}{\rho}\zeta(\eta)\,, \qquad (4.118)$$

and for the entropy density

$$\sigma_{ETF}[\rho] = -\eta\,\rho + \frac{5}{3}A_T F_{3/2}(\eta) - \frac{\hbar^2}{2m}\frac{(\nabla\rho)^2}{\rho}\left.\frac{\partial\zeta(\eta)}{\partial T}\right|_\rho\,; \qquad (4.119)$$

$\tau_{ETF}[\rho]$ is then easily found through Eq. (4.104). The coefficient $\zeta(\eta)$ is given in terms of the Fermi integrals by

$$\zeta(\eta) = -\frac{F_{1/2}(\eta)F_{-3/2}(\eta)}{12F_{-1/2}^2(\eta)} \simeq \frac{1}{36}\left[1 + \frac{2}{(1 + e^\eta)^{1/2}}\right]\,, \qquad (4.120)$$

where the right-hand side is a rough approximation that holds within less than 3 percent for all values of η. (A better numerical approximation is given in [72].) Note that in the above, η is no longer given by the simple relation in (4.115) due to the presence of the $\hbar$-corrections in the density, but through a numerical inversion of the relation

$$\rho = A_T F_{1/2}(\eta)\,. \qquad (4.121)$$

An important aspect now is the zero-temperature limit of the above results [26, 73]. Since we want to know the functionals at finite density, we have to keep ρ finite while letting T go to zero, also in the limit when ρ tends to zero at far distances from a finite system (the two limits are not interchangeable!). The key quantity here is η which turns out to go to infinity like $1/T$ *at finite density* ρ. Then, one may use the well-known asymptotic expansion of the Fermi integrals [69]

$$F_\xi(\eta) = \frac{1}{\xi + 1}\eta^{\xi+1}\left[1 + \xi(\xi + 1)\frac{\pi^2}{6}\eta^{-2} + \mathcal{O}(\eta^{-4})\ldots\right] \qquad \eta \gg 1 \qquad (4.122)$$

which, through the above expressions, yields back the zero-temperature functional $\tau_{ETF}[\rho]$ given in Eq. (4.74). (The same holds also for the fourth-order gradient terms derived in [66].) This gives a rigorous proof of the validity of the $T = 0$ functional at any point in space where the density is finite, including the classically forbidden region.

The finite-temperature functionals up to fourth order have been used in nuclear physics for calculations of the interface between the liquid and the gas phase of hot nuclear matter [26, 66]. As a practical result thereof, temperature-dependent liquid drop parameters were obtained [74] which can be used for phenomenological studies involving highly excited nuclei. Another recent application to astrophysics is concerned with the equation of state of inhomogeneous stellar nuclear matter and its role in the adiabatic collapse of stars [75].

4.5   Bose-Einstein condensation in a trap

Quantum effects in a gas (for atoms of mass M at equilibrium temperature T and a mean number density $\bar{\rho}$) should be manifest when the thermal de Broglie wave length, $\lambda_T = \sqrt{2\pi\hbar^2/mk_BT}$ is greater than, or of the same order as, the mean spacing $\bar{\ell} \simeq (\bar{\rho})^{-1/3}$ between the atoms. This means that

$$\lambda_T \, \bar{\rho}^{1/3} \gtrsim 1 , \tag{4.123}$$

i.e., quantum effects cannot be ignored at low temperatures and high density. To minimise interaction effects (and recombination of atoms), on the other hand, one requires a dilute gas, so that the mean spacing $\bar{\ell}$ is much larger than some measure of the interaction range. This makes the observation of BEC in dilute gases a challenging effort because of the requirement of very low temperatures. The first such observations of BEC in a gas of magnetically trapped neutral atoms were made only in the last decade [76, 77, 78]. BEC was observed in the alkali atoms ^{87}Rb, ^{23}Na, and ^{7}Li, which have respectively $(37 + 50), (11 + 12)$, and $(3 + 4)$ protons+neutrons in the atomic nucleus. Since the electron number is the same as the number of protons in a neutral atom, it is the (even) neutron number that is responsible for the bosonic nature of the atom. The magnetic trap is of harmonic oscillator shape, and typically there may be $10^4 - 10^8$ atoms in the trap. The trap is axially symmetric, but anisotropic, and the anisotropy parameter $\lambda = \omega_z/\omega_\perp$ may be varied over a wide range. We shall denote the geometric mean by $\omega_0 = (\omega_\perp^2 \omega_z)^{1/3}$. Typically, ω_0 is in the range of 100 to 200 radians/sec. Note that $\hbar\omega = 1$ nK corresponds to $\omega \simeq 131$ rads/sec. For ^{87}Rb, this corresponds to an oscillator length parameter $a_{ho} = \sqrt{\hbar/M\omega} = 2.3\,\mu$m. The critical temperature $T_c^{(0)}$ at which Bose condensation takes place is typically of the order of a few hundred nK, and for $T \geq T_c^{(0)}$ almost all of the atoms are still in the excited states of the trap. Consequently, the radius of the thermal cloud of atoms is much larger than a_{ho}. The average number density of atoms in the cloud is in the range $\bar{\rho} = 10^{11} - 10^{15}$ atoms/cc. For ^{87}Rb, taking $T = 100$ nK gives $\lambda_T \simeq 6 \times 10^{-5}$ cm, and for $\bar{\rho} = 10^{13}$/cc, we get $\lambda_T \, \bar{\rho}^{1/3} \simeq 1.3$, which satisfies the relation given by Eq. (4.123).

Although the gas is dilute so that interactions between the atoms do not cause them to recombine and change the nature of the system, the interactions do play an important role. Where as for a repulsive interaction between the atoms the condensate remains stable, an attractive interaction causes it to collapse if the number of atoms in the condensate exceeds a maximum value. More over, the density of states in the bulk of the condensate for the low-lying excitations are modified fundamentally. For example, for a bose gas in a large box, interactions cause the low-lying excitations to be (gap-less) sound waves. The interaction is also responsible for superfluidity. For the alkali atoms ^{87}Rb and ^{23}Na, the low-energy effective interaction between the atoms is repulsive, while for ^{7}Li, it is attractive. One of the interesting developments, however, is the use of the so-called Feshbach (molecular) resonance to tune the strength and even the sign of the interaction between the atoms [79].

In the next subsection, we first consider an ideal bose gas in a harmonic trap to examine the phenomenon of BEC semiclassically. Next, we introduce interactions between the atoms to see its effects. The reader is recommended the introductory article by Burnett [80], the review article by Dalfovo *et al.* [81], and the excellent book by Pethick and Smith on BEC [82] for more details.

4.5.1 BEC in an ideal trapped bose gas

At an equilibrium temperature T, the bosons are distributed in the states with energy levels E_n by the distribution function (we take all $E_n > 0$):

$$N = \sum_n \frac{1}{e^{(E_n - \mu)\beta} - 1}, \tag{4.124}$$

where the sum is over all states including their degenerecies, $\beta = 1/T$ here is the inverse temperature (we set $k_B = 1$), and the chemical potential μ is adjusted at every temperature T to satisfy the above equation. Note that every term on the r.h.s. above is positive, so we must have $\mu \leq E_0$, where E_0 is the lowest energy level. Since the ground state $n = 0$ (taken to be nondegenerate) plays a special role in BEC, it is best to isolate it from the rest:

$$N = \frac{z}{1 - z} + \sum_{n>0} \frac{1}{e^{(E_n - \mu)\beta} - 1}, \tag{4.125}$$

where $z = \exp[(\mu - E_0)\beta]$ is the fugacity, and the first term on the r.h.s. is the occupancy of the ground state. From this term, it is clear that only for $N \to \infty$ can $z \to 1$ for a macroscopic occupancy $N_0/N \neq 0$. In Fig. 4.5 on the next page, we show the behavior of $\mu/\hbar\omega$ (left-hand side) and N_0/N (right-hand side) versus $T/\hbar\omega$, when the trap is taken to be a spherical harmonic oscillator potential with $N = 10^4$ particles. It is remarkable that a sizable fraction of the particles start occupying the ground state rather abruptly at a tempurature $T \gg \hbar\omega$, which is not at all to be expected from classical Boltzmann statistics. This macroscopic occupancy of the ground state as a function of temperature becomes more and more sharp and tends to become discontinuous at $T = T_c^{(0)}$ in the limit $N \to \infty$. The "order parameter" N_0/N is then much greater than zero for $T < T_c^{(0)}$, and is zero for $T > T_c^{(0)}$. In this limit, the chemical potential μ also shows a discontinuity in its slope at $T = T_c^{(0)}$, and equals E_0.

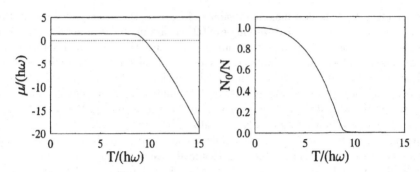

Figure 4.5: $N = 10^4$ ideal bosons in a three-dimensional isotropic harmonic oscillator. *Left:* Chemical potential μ versus temperature T. *Right:* Relative ground-state occupancy N_0/N versus temperature T. (Both μ and $k_B T$ are in units of $\hbar\omega$.)

We can make an estimate of the critical temperature $T_c^{(0)}$ by using the leading term in the semiclassical density of states $g_{TF}(E) = E^2/2(\hbar\omega)^3$, which is obtained from Eq. (4.3). We replace the sum over the excited states in the r.h.s. of Eq. (4.125) by an integral, and define a semiclassical $\tilde{\mu} \leq 0$ to conserve the particle number. We take the limit of $N \to \infty$, and note that in this limit

$$N = \frac{\tilde{z}}{1 - \tilde{z}} + \frac{1}{2(\hbar\omega)^3} \int_0^\infty dE \, \frac{E^2}{e^{(E-\tilde{\mu})\beta} - 1}, \qquad (4.126)$$

where $\tilde{z} = \exp(\beta\tilde{\mu})$ [for a more careful treatment, see Problem (4.8)]. Recalling that $\exp[(\tilde{\mu}-E)\beta] < 1$, we can multiply the numerator and the denominator of the integrand above by this factor, and then expand the denominator in an infinite geometric series. Performing the integrations term by term, we get

$$N = \frac{\tilde{z}}{1 - \tilde{z}} + \left(\frac{k_B T}{\hbar\omega}\right)^3 \sum_{l=1}^\infty \frac{e^{\tilde{\mu}l\beta}}{l^3}. \qquad (4.127)$$

Note that the sum above is convergent, since $\tilde{\mu} \leq 0$. It is convenient to rewrite the above equation as

$$1 = \frac{N_0}{N} + \frac{1}{N}\left(\frac{k_B T}{\hbar\omega}\right)^3 \sum_{l=1}^\infty \frac{e^{\tilde{\mu}l\beta}}{l^3}, \qquad (4.128)$$

where $N_0 = \tilde{z}/(1 - \tilde{z})$ is the ground state occupancy. We now take the limit $N \to \infty$, and note that for finite T, $\tilde{z} \neq 1$, so the first term on the r.h.s. of Eq. (4.128) vanishes. Still the equation can be satisfied by the second term on the r.h.s. if $N\omega^3$ is finite, even though $N \to \infty$. As T is reduced from above $T_c^{(0)}$, however, $\tilde{\mu}$ has to approach zero from the negative side to continue to satisfy Eq. (4.128). Finally, at $T = T_c^{(0)}$, $\tilde{\mu}$ hits zero, and then the sum becomes $\zeta(3) = \sum_{l=1}^\infty l^{-3}$, its maximum value. When $T < T_c^{(0)}$, N_0/N is a non-zero fraction, so that there is macroscopic occupancy of the ground state. The critical temperature $T_c^{(0)}$ can be found from Eq. (4.128) to be

$$T_c^{(0)} = \left(\frac{N}{\zeta(3)}\right)^{1/3} \hbar\omega, \qquad (4.129)$$

where $\zeta(3) \simeq 1.202$. This result is only in the limit of $N \to \infty$, with the condition that $\omega \to 0$ such that $N\omega^{1/3}$ is a constant. [For finite N, the transition to the condensed state is continuous, see Problem (4.8)]. For $T \le T_c^{(0)}$, since $\tilde{\mu} = 0$, we also deduce from Eqs. (4.128,4.129) that

$$\frac{N_0}{N} = 1 - \left(\frac{T}{T_c^{(0)}}\right)^3. \tag{4.130}$$

4.5.2 Inclusion of interactions in a dilute gas

The mean distance between the atoms in the trapped dilute gas is so large that the interaction effects between the atoms may be included by examining the asymptotic tail of the relative two-body wave function. More over, the relative momenta at the low temperatures under consideration are so low that only the scattering in the relative s-state between the atoms is relevant. This radial wave function, in the absence of the interaction, is proportional to $\sin(kr)/k$, which goes like r in the limit of the relative wave number $k \to 0$. This asymptotic wave function gets modified to $(r - a)$ by the interaction, independent of the shape of the potential. The constant a is called the scattering length; it is positive for repulsive and negative for attractive interactions. For ^{87}Rb atoms, for example, $a = 5.8$ nm, while for ^{7}Li atoms, $a = -1.45$ nm. When the mean distance between the atoms is much larger than the scattering length a, i.e., $\bar{\rho}|a|^3 \ll 1$, the effective interaction between two atoms may be be characterised by a one-parameter "pseudopotential" $V_{12}(\mathbf{r}) = g\,\delta(\mathbf{r})$, where $g = a\,(4\pi\hbar^2/M)$ [83]. Note that for $T < T_c$ (we have dropped the superscipt 0 for interacting bosons), a sizable fraction of the bosons are in the same quantum state. In the mean-field approximation, this is a single-particle wave function, $\Psi(\mathbf{r})$, but normalized to N_0 rather than unity. The system at $T = 0$ is described by the so-called "Gross-Pitaevskii energy functional" [84]:

$$E[\Psi] = \int d^3r \left[\frac{\hbar^2}{2M}|\nabla\Psi(\mathbf{r})|^2 + V_0(\mathbf{r})|\Psi(\mathbf{r})|^2 + \frac{g}{2}|\Psi(\mathbf{r})|^4 \right], \tag{4.131}$$

where V_0 is the external trap potential, and the last term is the interaction energy. Note that for N bosons, the interaction energy is is $g\,N(N-1)/2$, which, even for a very weak g, may be comparable to the trap potential energy term that is proportional to N. A nonlinear Schrödinger equation is obtained by varying $(E - \mu N)$ with respect to $\Psi^*(\mathbf{r})$:

$$\frac{\hbar^2}{2M}\nabla^2\Psi(\mathbf{r}) + V_0(\mathbf{r})\Psi(\mathbf{r}) + g\,|\Psi(\mathbf{r})|^2\Psi(\mathbf{r}) = \mu\,\Psi(\mathbf{r}) \tag{4.132}$$

For a repulsive interaction and large N, the atomic cloud is spread out, with the result that the kinetic energy term is negligible. By ignoring it, we immediately obtain the so-called Thomas-Fermi spatial density $\rho_{TF}(\mathbf{r}) = [\mu - V_0(\mathbf{r})]/g$, which goes to zero at the turning point and beyond it. The usage of the TF approximation (for large-N, rather than $\hbar \to 0$) is different from the fermionic case. We refer the reader to [82] to appreciate the richness of the many-body problem brought about by this "weak" interaction between the atoms.

4.6 $\hbar$ expansion for cavities and billiards

The Wigner-Kirkwood expansion described in the last section is implicitly an expansion in gradients of the one-body potential $V(\mathbf{r})$. It cannot be applied, therefore, to systems in which particles are confined to a finite spatial domain by reflecting walls. In these so-called 'cavities' in three or 'billiards' in two dimensions, the confining potential has infinite derivatives. In order to obtain $\hbar$ expansions of the average level density and other properties of such systems, one has to resort to alternative methods. In this section, we shall first present the Euler-MacLaurin expansion as a technical tool which can be used if the quantum spectrum is known analytically. We then discuss the general form of the Weyl expansion, which is older than quantum mechanics, and finally give the black-body radiation in a small cavity as an example.

4.6.1 The Euler-MacLaurin expansion

One particular method that has been proposed [6] for finding the $\hbar$ expansions of the level densities for a particle in a cavity is the so-called Euler-MacLaurin summation formula, which is usually stated as follows (see, e.g., Ref. [7], pp. 806 and 886). Let the real function $f(x)$ have its first $2K+2$ derivatives continuous on an interval (a, b) of the real x axis. Divide the interval into N equal parts of length $h = (b - a)/N$. Then, for some θ with $0 < \theta < 1$, depending on $f^{(2K+2)}(x)$ on (a, b), we have

$$
\begin{aligned}
\sum_{n=0}^{N} f(a + nh) &= \frac{1}{h} \int_a^b f(x)\, dx + \frac{1}{2}[f(a) + f(b)] \\
&+ \sum_{k=1}^{K} \frac{h^{2k-1}}{(2k)!} B_{2k} \left[f^{(2k-1)}(b) - f^{(2k-1)}(a) \right] \\
&+ \frac{h^{2K+2}}{(2K+2)!} B_{2K+2} \sum_{k=0}^{N-1} f^{(2K+2)}(a + kh + \theta h),
\end{aligned} \tag{4.133}
$$

where B_{2k} denotes Bernoulli numbers, the first of which are

$$
B_2 = \frac{1}{6}, \quad B_4 = -\frac{1}{30}, \quad B_6 = \frac{1}{42}, \ldots \tag{4.134}
$$

The formula (4.133) is often used for numerical integrations, after rearranging it so as to bring the integral to one side. The first line corresponds to the trapezoidal rule. Adding the correction terms containing odd derivatives of $f(x)$ at the end points a and b up to order $2K - 1$, the remaining error is of order h^{2K+2}. When the integrand $f(x)$ is known analytically, this method of numerical integration is actually very powerful and far superior to many others (including Simpson's rule and more fancy methods). Even going only to the first derivative corrections, it can give excellent results with a moderately small step size h.

In the present context, however, Eq. (4.133) may also be used to extract the smooth part from a given sum, such as the canonical partition function $Z(\beta)$ in (3.15) [6] or a Strutinsky-averaged level density [85] (see Sect. 4.7), when the spectrum is known analytically. Note that Eq. (4.133) yields an asymptotic expansion which may be semi-convergent. In practice, one therefore has to truncate the expansions, as will be discussed further below. When the summation index n on the left-hand side of Eq. (4.133) is a quantum number, it appears usually multiplied by Planck's constant h (or $\hbar$) – see, e.g., the harmonic-oscillator spectrum $E_n = \hbar\omega(n + 1/2)$ or that of an infinite box potential, $E_n = \pi^2\hbar^2 n^2/2mL^2$. (For integrable systems, this can be immediately seen from the torus quantization, see Sect. 2.5.) We may therefore readily identify the quantity h in Eq. (4.133) with Planck's constant. Then it becomes clear that this formula will lead to an $\hbar$ expansion of the sum on the left-hand side. The integral term yields the Thomas-Fermi approximation (see also Sect. 2.7), and the remaining terms constitute the ETF series of the expansion.

To illustrate this point, let us apply Eq. (4.141) to the partition function of the linear harmonic oscillator. We immediately get

$$\sum_{n=0}^{\infty} e^{-\beta\hbar\omega(n+1/2)} = \frac{1}{\beta\hbar\omega}e^{-\beta\hbar\omega/2} + \frac{1}{2}e^{-\beta\hbar\omega/2} + \frac{1}{12}(\beta\hbar\omega)\,e^{-\beta\hbar\omega/2} + \dots . \qquad (4.135)$$

On further expanding the exponential on the right-hand side, we get precisely the Wigner-Kirkwood expansion of the exact partition function of the one-dimensional harmonic oscillator, $Z(\beta) = [2\operatorname{Sinh}(\beta\hbar\omega/2)]^{-1}$ [cf. the three-dimensional case in (4.1)].

The advantage of the Euler-MacLaurin expansion is that it may also be applied to potentials with infinitely steep walls if their spectra are analytically known, or to any given quantum spectrum even if the corresponding potential is unknown. The average level density of a rectangular box potential in any dimension can be easily determined in this way [6]. In Sect. 3.2.6, we have already done this implicitly by applying the Poisson summation formula (2.157) to the partition function of these systems. In fact, isolating the term with $M = 0$, Eq. (2.157) may be rewritten as

$$\sum_{n=0}^{\infty} f(n) = \int_0^{\infty} f(x)\,dx + \frac{1}{2}f(0) + 2\sum_{M=1}^{\infty} \int_0^{\infty} f(x)\cos(2\pi M x)\,dx . \qquad (4.136)$$

It has been assumed here that the function $f(x)$ and all its derivatives at infinity are zero: $d^n f/dx^n(\infty) = 0$ ($n = 0, 1, 2, \dots$). This is the case for the partition function of a rectangular box, which is just a Gaussian: $f(x) = \exp(-x^2)$. Now, the first two terms of the Poisson formula (4.136), which are the only non-oscillating parts, are identical to the first two terms of the Euler-MacLaurin formula (4.133) (using $a = 0$ and $b = \infty$). Moreover, since *all* odd derivatives of the Gaussian are zero also at $x = 0$, these first two terms yield the entire smooth part of the level density [see Eq. (3.88), and also Eq. (2.189)]. We repeat here the result for the three-dimensional rectangular box, which has the form

$$\tilde{g}^{(3)}(E) = \frac{1}{E_0}\left\{ \frac{\pi}{4}\sqrt{\frac{E}{E_0}}\frac{\mathcal{V}}{L^3} - \frac{\pi}{8}\frac{\mathcal{S}}{2L^2} + \frac{1}{8}\sqrt{\frac{E_0}{E}}\frac{\mathcal{L}}{L} \right\} \qquad (4.137)$$

in terms of its volume $\mathcal{V}$, its surface $\mathcal{S}$, and the sum of the side lengths $\mathcal{L}$. Remembering that $E_0 = \hbar^2\pi^2/2mL^2$ (with L being an arbitrary unit length), we see that (4.137) has again the form of an $\hbar$ expansion and can therefore be identified with the result of the ETF model. Below we shall see that the form of Eq. (4.137) is rather general and holds for cavities of any shape in the so-called Weyl expansion.

At this point, some remarks about the separation (3.41) of the level density into a smooth and an oscillating part are in place. In the present chapter, we are investigating the smooth part $\tilde{g}(E)$ which we identify with the level density obtained in the ETF model. In Sect. 2.7, on the other hand, we have shown how for integrable systems a semiclassical approximation to the level density can be found from the Poisson summation formula (2.157), and that this approximation contains both a smooth and an oscillating part. The smooth part arises from the points M=0 of the topological lattice and the edge correction $f(0)/2$ in (2.157), whereas the oscillating part comes from all the other points M which can be associated to the rational tori in phase space, i.e., to the periodic orbits of the classical system. It therefore appears quite obvious – at least at first sight – to use exactly this fact for the definition of the oscillating part. But then the question arises how the higher-order corrections in the Euler-MacLaurin formula which in general are not zero – polygonal billiards and cavities with piecewise plane boundaries are an exception here – fit into this picture.

In trying to answer this question, it is instructive to realize that the Euler-Mac-Laurin formula (4.133) can, in fact, be derived from the Poisson formula.[4] Integrating twice by parts the last term on the right-hand side of Eq. (4.136), we get:

$$\int_0^\infty f(x)\cos(2\pi Mx)\,dx = -\frac{1}{(2\pi M)^2}\left[f'(0) + \int_0^\infty f''(x)\cos(2\pi Mx)\,dx\right]. \quad (4.138)$$

This process can be iterated K times to find the following result:

$$\begin{aligned}
\sum_{n=0}^\infty f(n) &= \int_0^\infty f(x)\,dx + \frac{1}{2}f(0) + 2\sum_{M=1}^\infty \left\{-\frac{f'(0)}{(2\pi M)^2} + \frac{f^{(3)}(0)}{(2\pi M)^4} - \frac{f^{(5)}(0)}{(2\pi M)^6} + - \cdots\right.\\
&\quad + \left.\frac{(-1)^K}{(2\pi M)^{2K}}\left[f^{(2K-1)}(0) + \int_0^\infty f^{(2K)}(x)\cos(2\pi Mx)\,dx\right]\right\}. \quad (4.139)
\end{aligned}$$

Each sum on the right-hand side of the above can be written in terms of the Riemann zeta function, defined by

$$\zeta(s) = \sum_{k=1}^\infty \frac{1}{k^s}. \quad (4.140)$$

Using furthermore its special values for even-integer arguments, $\zeta(2k)$ (see Ref. [7], p. 807), we find the result

$$\begin{aligned}
\sum_{n=0}^\infty f(n) &= \int_0^\infty f(x)\,dx + \frac{1}{2}f(0) - \sum_{k=1}^K \frac{B_{2k}}{(2k)!}f^{(2k-1)}(0)\\
&\quad + 2\sum_{M=1}^\infty \frac{(-1)^K}{(2\pi M)^{2K}}\int_0^\infty f^{(2K)}(x)\cos(2\pi Mx)\,dx. \quad (4.141)
\end{aligned}$$

We have now reproduced all but the last term of the Euler-MacLaurin summation formula (4.133) for the special case $a = 0$, $b = \infty$, $h = 1$. The last term might have to be transformed further before it takes the form appearing in (4.133), but we are presently not so much interested in its detailed form. What we have just seen, is that any finite

---

[4] We are grateful to M. V. N. Murthy who pointed this out to us.

number K of higher correction terms in the Euler-MacLaurin summation formula can be obtained from the Poisson formula by partial integration of the oscillatory terms. Hereby one creates a series of smooth-looking terms proportional to $f^{(2k-1)}(0)$, and the only oscillatory term that remains is the last one. It is clear that the same procedure could also be used for a function in an arbitrary interval (a, b), with a result equal to the general formula (4.133) but containing oscillatory integrals at the order h^2K.

Not all of the derivative correction terms in (4.133) give average contributions to the level density, however. Many of them will give higher derivatives of delta functions that must contribute to the oscillatory part, although they cannot be used easily in this form. Some of them, in turn, will give finite and smooth contributions to quantities derived from integrals over the level density, such as the particle number N or the total energy E_{sp}. In conclusion, the Euler-MacLaurin formula reveals itself as a tool for extracting the average part of a given sum in the form of an $\hbar$ expansion like in the ETF model, thereby pushing its oscillatory part to increasing orders in $\hbar$. (For a mathematically more rigorous analysis of this situation, see Ref. [86].)

We discuss this point at some length here in order to emphasize that there is no unique place to truncate the Euler-MacLaurin series, and with it the Wigner-Kirkwood series, that gives well-defined average values for all quantities of interest. As already pointed out in Sect. 4.3, the number of terms to be included depends on the quantity one is investigating. For the particle number N, one more term is usually needed than for the level density, namely a constant times $\delta(E)$ (see, e.g., Sect. 3.2.6), whereas for the sum of single-particle energies E_{sp}, a term proportional to $\delta'(E)$ gives a smooth and finite contribution (see, e.g., Problem 4.2 for the harmonic oscillator in 3 dimensions).

The pragmatic convention in the ETF model is, for each individual quantity of interest, to include as many terms of the asymptotic series as possible, as long as they are smooth and finite. The sum of these smooth terms is then usually called the "ETF value" of that quantity. In Section 4.7 below, we shall discuss the Strutinsky averaging method for determining averaged level densities and other quantities derived therefrom. This method is purely numerical and does not know of any $\hbar$ expansion. That it leads to average energies which are identical with those obtained in the ETF model using the above convention – at least within the numerical inaccuracies of either method – gives a strong confirmation of that convention. In Chapters 5 and 6, we are going to develop the periodic orbit theory which allows us to find (approximately) the oscillating part of the level density also for non-integrable systems, in which no *a priori* quantization scheme is known and thus the Poisson summation technique cannot be applied. Nevertheless, we tend to identify the smooth part of the level density of these systems with its ETF value, too, since this can always be derived using either the Wigner-Kirkwood or the Weyl expansion to be discussed next.

4.6.2 The Weyl expansion

A very early analysis of the density of eigenmodes of a cavity with reflecting walls goes back to Weyl [87] in 1911, even before the advent of quantum mechanics. He surmised it to have the form

$$\tilde{g}(E) = \frac{1}{4\pi^2}\left(\frac{2m}{\hbar^2}\right)^{3/2}\sqrt{E}\,\mathcal{V} \mp \frac{1}{16\pi}\left(\frac{2m}{\hbar^2}\right)\mathcal{S}, \qquad (4.142)$$

where $\mathcal{V}$ is the volume and $\mathcal{S}$ the surface area of the cavity. The volume term is easily found from the Thomas-Fermi expression for the level density (4.4), noting that $V(r) \equiv 0$ inside the cavity. In the surface term, the upper sign is for the use of Dirichlet boundary conditions ($\psi = 0$) and the lower for Neumann boundary conditions ($\partial\psi/\partial n = 0$) when solving the Laplace equation in order to determine the eigenmode spectrum. The proof of Eq. (4.142) and the form of the next term was given only in the 1950's; for an account of this, see Balian and Bloch [88, 89] who generalized the result for more general classes of boundary conditions. The generalized Weyl formula for a cavity with an arbitrary smooth convex boundary (i.e., without corners or edges) with Dirichlet conditions reads:

$$\tilde{g}(E) = \frac{1}{4\pi^2}\left(\frac{2m}{\hbar^2}\right)^{3/2}\sqrt{E}\,\mathcal{V} - \frac{1}{16\pi}\left(\frac{2m}{\hbar^2}\right)\mathcal{S} + \frac{1}{12\pi^2}\left(\frac{2m}{\hbar^2}\right)^{1/2}\frac{1}{\sqrt{E}}\,\mathcal{C}\,. \tag{4.143}$$

Hereby $\mathcal{C}$ is the mean curvature of the cavity, which is defined as the surface integral over the algebraic mean of the two principal curvatures κ_1, κ_2 (which are the inverse of the principal curvature radii R_i, i.e., $\kappa_i = 1/R_i$)

$$\mathcal{C} = \oint d\sigma \frac{1}{2}(\kappa_1 + \kappa_2), \qquad \oint d\sigma = \mathcal{S}\,. \tag{4.144}$$

The wave number density of states, defined in Eq. (3.9), becomes

$$\tilde{g}(k) = \frac{k^2}{2\pi^2}\mathcal{V} - \frac{k}{8\pi}\mathcal{S} + \frac{1}{6\pi^2}\mathcal{C}\,. \tag{4.145}$$

For a spherical box with radius R, one has $\mathcal{V} = 4\pi R^3/3$, $\mathcal{S} = 4\pi R^2$ and $\mathcal{C} = 4\pi R$, leading to the formula (6.40) used in Sect. 6.1.6. Waechter [90] has derived the next term in the expansion (4.143), which is a constant times the delta function, $c_2\delta(E)$, with c_2 given by

$$c_2 = \frac{1}{512\pi}\oint d\sigma\,(\kappa_1 - \kappa_2)^2\,. \tag{4.146}$$

This is not a smoothly varying term, but after integration over E it contributes the constant c_2 to the number N of states (or to the particle number in a fermion system).

We refer the reader to a comprehensive book by Baltes and Hilf [91] on the general problem of determining the density of states in finite systems, and in particular on many details and examples of the Weyl expansion. In the following, we compile a few results which are used elsewhere in this book.

For cavities with edges, the first two terms of (4.143) may be used, but the third term is in general different and cannot be interpreted in terms of a mean curvature, since at least one curvature radius is zero along the edges. The case of the cubic box with side length L has been considered by Hill and Wheeler [92]. They obtained the result

$$\tilde{g}(E) = \frac{1}{E_0}\left[\frac{\pi}{4}\sqrt{\frac{E}{E_0}} - \frac{3\pi}{8} + \frac{3}{8}\sqrt{\frac{E_0}{E}}\right]\,, \tag{4.147}$$

which is identical with (4.137), using $\mathcal{V} = L^3$, $\mathcal{S} = 6L^2$, and $\mathcal{L} = 3L$. Waechter [90] has given some surmises of the corrections due to edges and corners, deduced from the two-dimensional results quoted next.

In two-dimensional billiards, a similar expansion can be found. An early result [93] for the smooth part of the level density for billiards with simply connected smooth boundaries is

$$\tilde{g}(E) = \frac{1}{4\pi} \left(\frac{2m}{\hbar^2}\right) \mathcal{A} \mp \frac{1}{8\pi} \left(\frac{2m}{\hbar^2}\right)^{1/2} \frac{1}{\sqrt{E}} \mathcal{L}, \qquad (4.148)$$

or in k space

$$\tilde{g}(k) = \frac{k}{2\pi} \mathcal{A} \mp \frac{1}{4\pi} \mathcal{L}, \qquad (4.149)$$

where the sign of the second term depends on the boundary condition as in Eq. (4.142) above. In these expressions, $\mathcal{A}$ is the area of the billiard and $\mathcal{L}$ the length of its circumference. Again, the leading terms are easily found in the Thomas-Fermi model. For a disk billiard with radius R, one has $\mathcal{A} = \pi R^2$ and $\mathcal{L} = 2\pi R$, leading to Eq. (6.17) used in Sect. 6.1.3. Higher-order terms for the disk billiard with Dirichlet boundary conditions may be found in Ref. [86], and with Neumann boundary conditions in Ref. [94].

Stewartson and Waechter [95] have generalized Eq. (4.148) for billiards with piecewise smooth boundaries to

$$\tilde{g}(E) = \frac{1}{4\pi} \left(\frac{2m}{\hbar^2}\right) \mathcal{A} - \frac{1}{8\pi} \left(\frac{2m}{\hbar^2}\right)^{1/2} \frac{1}{\sqrt{E}} \mathcal{L} + \left[K + \sum_i c(\alpha_i)\right] \delta(E), \qquad (4.150)$$

whereby Dirichlet boundary conditions were used. We have again kept the delta function term which does not strictly contribute to the smooth part of the level density, but to its first integral. In (4.150), K is the average curvature integrated along the boundary

$$K = \frac{1}{12\pi} \oint \kappa(l) \, dl, \qquad (4.151)$$

and the corrections due to the corners with angles α_i are given by the formula [96]

$$c(\alpha_i) = \frac{\pi^2 - \alpha_i^2}{24\pi\alpha_i}. \qquad (4.152)$$

Note that $K = 0$ for polygonal billiards with piecewise straight-lined boundaries. It will be easy for the reader to check that these corner corrections are consistent with the specific results derived in Sect. 2.7.1 for the rectangular billiard and in Sect. 3.2.7 for the equilateral billiard in the case of Dirichlet boundary conditions. We have not found any general formula in the literature, corresponding to (4.152), for the corner corrections with Neumann boundary conditions.

The energy dependent forms of the smooth level densities above reveal that they are $\hbar$ expansions of the ETF type: the leading TF term is of order $\hbar^{-D}$, where D is the spatial dimension of the system, and each successive term is one order higher in $\hbar$. The Weyl expansion can therefore be taken as alternative form of the ETF series.

We have seen above that the Weyl expansion coefficients depend on the geometrical shape of the enclosure. This raises a more general question about the relationship between the exact eigenspectrum of a cavity (or a billiard), and its shape. One may ask, in particular, if the eigenspectrum uniquely fixes the shape of a billiard, given the boundary conditions. This question was posed by Kac [97] in relation to the vibrations of a drum. For a string with fixed ends at a given tension, for example, the normal modes uniquely specify the length. Similarly, can one unambiguously tell the shape of a drum, vibrating with its membrane firmly glued to the boundary, from its frequency spectrum? In mathematical parlance, the question is rephrased as 'whether two isospectral plane domains *must* actually be isometric'? The answer to this question has been found to be *negative*, since Gordon *et al.* [98] have given examples constructed out of seven right triangles which are isospectral, but not isometric. Sridhar and Kudrolli [99] have studied a pair of such shapes experimentally with microwave cavities. An elegant proof of the isospectrality of the pair, together with numerical solutions of the Helmholtz equation, $(\nabla^2 + k^2)\psi_k(\mathbf{r}) = 0$ with Dirichlet boundary conditions, is also given by Wu *et al.* [100]. A non-technical account of the topic is given by Smilansky [101]. We shall not discuss this aspect of the problem any more, but concentrate next on the Weyl expansion for radiation, which also obeys the Helmholtz equation.

4.6.3   Black-body radiation in a small cavity

An interesting application of the methods and results discussed above is to find the corrections to Planck's black-body radiation formula in a cubic or rectangular conducting cavity with side length L. From the electromagnetic wave picture, radiation at equilibrium may be viewed as electric and magnetic standing wave modes in the cavity. When the wavelength of the mode is larger than the spatial extent of the cavity, such standing waves cannot be supported, and the counting of the modes should be done accordingly. In the photon picture, this is just the Weyl expansion in the quantum density of states that takes account of finite size effects. We follow here a derivation by Ross [102] for the semiclassical expansion of the quantum partition function using the Euler-MacLaurin formula (4.133). The electromagnetic field in a cavity may be analyzed in terms of the transverse electric (TE) and the transverse magnetic (TM) modes [103]. In the quantum picture, one may retain the same description, and the eigenenergies of a photon are given by

$$E_{lmn} = \frac{\pi \hbar c}{L} (l^2 + m^2 + n^2)^{1/2}, \qquad (4.153)$$

where the integral quantum numbers $l, m, n \geq 0$, but subject to the following restrictions:

a) For $l, m, n > 0$, both the TE and the TM modes are present, so that each E_{lmn} is doubly degenerate.

b) If any one of $l, m,$ or n equals zero, the TE mode is identically zero, so there is only a single state corresponding to each E_{lmn} in such a case.

c) If more than one of the indices l, m, n are zero, then there exists no mode.

Hence the canonical partition function of a photon is given by

$$Z(\beta) = 2 \sum_{l=1}^{\infty} \sum_{m=1}^{\infty} \sum_{n=1}^{\infty} \exp[-\beta \kappa \sqrt{l^2 + m^2 + n^2}] + 3 \sum_{l=1}^{\infty} \sum_{m=1}^{\infty} \exp[-\beta \kappa \sqrt{l^2 + m^2}], \quad (4.154)$$

where $\kappa = \pi \hbar c / L$. We now apply the Euler-MacLaurin formula (4.133) step by step to the multiple sums. Replacing first the n-sum in the triple sum above by an integral, and retaining only the leading correction term, we get

$$Z_{sc}(\beta) \simeq 2 \sum_{l=1}^{\infty} \sum_{m=1}^{\infty} \left(\int_0^{\infty} \exp[-\beta\kappa\sqrt{l^2 + m^2 + x^2}]\, dx - \frac{1}{2} \exp[-\beta\kappa\sqrt{l^2 + m^2}] \right)$$
$$+ 3 \sum_{l=1}^{\infty} \sum_{m=1}^{\infty} \exp[-\beta\kappa\sqrt{l^2 + m^2}]. \tag{4.155}$$

Similarly, in the next step, we convert the m-sum to a y-integral, and finally the l-sum to a z-integral. The result is

$$Z_{sc}(\beta) \simeq 2 \int_0^{\infty} dx \int_0^{\infty} dy \int_0^{\infty} dz \exp[-\beta\kappa\sqrt{x^2 + y^2 + z^2}] - \frac{3}{2} \int_0^{\infty} dx \, \exp[-\beta\kappa x] + \frac{1}{2}. \tag{4.156}$$

The triple integral may be easily done by going to spherical polar coordinates:

$$\int_0^{\infty} dx \int_0^{\infty} dy \int_0^{\infty} dz \exp[-\beta\kappa\sqrt{x^2 + y^2 + z^2}] = \frac{\pi}{2} \int_0^{\infty} \exp[-\beta\kappa r]\, r^2 dr = \frac{\pi}{(\beta\kappa)^3}.$$

The semiclassical partition function is thus given by

$$Z_{sc}(\beta) \simeq \frac{2}{\pi^2} \frac{L^3}{(\hbar\beta c)^3} - \frac{3}{2\pi} \frac{L}{(\hbar\beta c)} + \frac{1}{2}. \tag{4.157}$$

The Laplace inversion with respect to E immediately yields the smooth density of states $\tilde{g}(E)$:

$$\tilde{g}(E) = \frac{1}{\pi^2} \frac{L^3}{(\hbar c)^3} E^2 - \frac{3}{2\pi} \frac{L}{(\hbar c)} + \frac{1}{2} \delta(E). \tag{4.158}$$

As in the harmonic oscillator case (see Problem 4.2), the surface correction term is absent. This is reasonable, since photons in a cavity are equivalent to a set of noninteracting harmonic oscillators, with the frequency of the oscillator determined by the wave number of the photon. From this, we can immediately obtain the correction terms to Planck's expression for the spectral density and other thermodynamic quantities at an equilibrium temperature T, expressed in units of the Boltzmann constant. In these units, $T = 1$ corresponds to $\simeq 0.86 \times 10^{-4}$ eV. Using Bose statistics, the energy U of the photons in the cavity at an equilibrium temperature T is given by (see Eq. (4.92) for comparison with the Fermi case)

$$U(T) = \int_0^{\infty} \frac{g(E)E}{(\exp[E/T] - 1)}\, dE = \int_0^{\infty} u(\omega, T)\, d\omega. \tag{4.159}$$

where $g(E) = \sum_i \delta(E - E_i)$ as usual. The quantity $u(\omega, T)$ is the spectral density, usually expressed for unit volume, $L^3 = 1$, where we have replaced E by $\hbar\omega$. Replacing the quantum level density by the semiclassical $\tilde{g}(E)$, and recalling that $E = \hbar\omega$, we obtain after a little algebra

$$u(\omega, T) = \frac{\hbar\omega}{[\exp(\hbar\omega/T) - 1]} \left[\frac{1}{\pi^2} \left(\frac{L}{c}\right)^3 \omega^2 - \frac{3}{2\pi} \left(\frac{L}{c}\right) + \frac{\delta(\omega)}{2} \right]. \tag{4.160}$$

The first term is universal, irrespective of the shape of the cavity, and gives Planck's spectral density. The correction is due to the shape-dependent curvature term, proportional to the perimeter of the enclosure. On integrating $u(\omega, T)$ as in Eq. (4.159), we obtain an extension of Stefan's law for the cubic cavity:

$$U(T) = \frac{\pi^2}{15} \left(\frac{L}{\hbar c}\right)^3 T^4 - \frac{\pi}{4} \left(\frac{L}{\hbar c}\right) T^2 + \frac{T}{2} + \dots . \qquad (4.161)$$

Case and Chiu [104] first derived Eq. (4.158) using a different method. The general formula for the semiclassical density of states in a cavity with no sharp edges was obtained by Balian and Bloch (see Ref. [89]; there was an error in the original paper that was corrected in an erratum). With this correction, the density of states for a perfectly conducting spherical cavity is given by

$$\tilde{g}(E) = \frac{d}{2\pi^2} \frac{\mathcal{V}}{(\hbar c)^3} E^2 - \frac{4d}{3\pi} \frac{R}{\hbar c} + \dots . \qquad (4.162)$$

In the above, d is the degeneracy factor ($d = 2$ for photons) and $\mathcal{V}$ is the volume of the cavity. In quantum chromodynamics (QCD), massless gluons are the vector particles that are exchanged between quarks, and these come in 8 distinct varieties. Unlike the photons, gluons interact with each other. If these interactions are neglected, then such an approach may be applied to QCD also with $d = 8 \times 2$ [105]. A similar formula has also been derived for massless quarks obeying the Dirac equation [106]. In this case,

$$\tilde{g}(E) = \frac{d}{2\pi^2} \frac{\mathcal{V}}{(\hbar c)^3} E^2 - \frac{d}{3\pi} \frac{R}{\hbar c} + \dots . \qquad (4.163)$$

Again, $d = 2$ due to spin-degeneracy. Note that these results are of the same form as the general Weyl expansion (4.143) for the level density, but with different energy dependences in the single terms. Interestingly, also, the surface term (proportional to R^2) only appears when the mass of the particle is nonzero [107].

4.7   The Strutinsky method

Although we have outlined in the above sections how the average single-particle properties of a fermion system may systematically and quite uniquely be determined in the extended Thomas-Fermi model, practical applications of this method can sometimes be rather cumbersome. In particular, in a self-consistent mean-field treatment of a many-fermion system the average potential is only found numerically. The Wigner-Kirkwood expansion requires the knowledge of the lowest derivatives of the potential which might not be numerically very stable. In particular, when the mean field is nonlocal, the ETF expansion has a large number of terms and becomes virtually unpracticable. In such cases one is interested in finding the average single-particle properties in a different, numerical way which requires only the knowledge of the spectrum $\{E_n\}$.

Such a numerical method has been introduced to nuclear physics over 30 years ago by Strutinsky [4, 5], and since then proven to be a powerful device for separating the smooth and oscillating parts of the level density, according to Eq. (3.41), for any given single-particle spectrum $\{E_n\}$. It has become very successful for computing the oscillating part of the total energy of a many-fermion system, particularly in connection with the so-called shell-correction method [4, 5, 58].

Strutinsky's averaging method consists of a convolution of the energy spectrum $\{E_n\}$ over a finite interval γ. It is done in such a way that the result does not depend on the averaging interval γ, provided it is chosen large enough to smooth out all the oscillating terms. We shall in the following first present this method in a rather general formulation [85, 108], then briefly discuss the shell-correction method, and finally study the relation between Strutinsky's energy averaging and the extended Thomas-Fermi model developed above. Some applications of the shell-correction method to nuclear binding and deformations energies will be presented in the first section of Chapter 8, where we contrast them with semiclassical calculations using the periodic orbit theory.

4.7.1 The energy averaging method

We start out from any given energy spectrum $\{E_n\}$ and convolute the sum of delta functions $g(E) = \sum_n \delta(E - E_n)$ with an averaging function $f(x)$ over an energy range γ:

$$g_\gamma(E) = \frac{1}{\gamma} \int_{-\infty}^{+\infty} f\left(\frac{E - E'}{\gamma}\right) g(E')\, dE' = \frac{1}{\gamma} \sum_n f\left(\frac{E - E_n}{\gamma}\right). \tag{4.164}$$

Here and in the following, the summation over n always includes the degeneracy of the n-th quantum state. The function $f(x)$ is assumed to be analytic, normalized to unity, and – for the sake of simplicity – symmetric in x with a maximum at $x = 0$. The most obvious and convenient choice [4] is a Gaussian, but other functions may be used as well. We now choose the energy averaging range γ large enough to smooth out all oscillations, i.e., to suppress all of the oscillating part $\delta g(E)$. Recall that we have learned in Sect. 3 from several exact models that the oscillating part $\delta g(E)$ of the level density may be Fourier decomposed into the form of a trace formula (3.42). (The latter will be derived more generally in the following sections 5 and 6.) The shortest period of the energy oscillations is governed by the classical periodic orbits with the smallest actions; it usually leads to characteristic energy gaps between the bunches of energy levels, the so-called energy "shells". Let us call this shell gap $\hbar\Omega$, in obvious reminiscence of the harmonic-oscillator energy level spacing. In order to wipe out all oscillations, we have to choose

$$\gamma \gtrsim \hbar\Omega. \tag{4.165}$$

After this averaging, the quantity $g_\gamma(E)$ in (4.164) will be a smooth function of energy and essentially be determined by the convolution of the smooth part $\tilde{g}(E)$ alone:

$$g_\gamma(E) \simeq \tilde{g}_\gamma(E) = \frac{1}{\gamma} \int_{-\infty}^{+\infty} f\left(\frac{E - E'}{\gamma}\right) \tilde{g}(E')\, dE' = \hat{\Gamma}[\tilde{g}(E)]. \tag{4.166}$$

On the right-hand side of this equation, we have introduced the linear operator $\hat{\Gamma}$ as a short-hand notation for the integral operator that provides the convolution.

Although $\tilde{g}_\gamma(E)$ is a smooth function of energy, it depends in general on the precise value of the parameter γ – for this reason we denote it by the index γ – whereas our aim is to obtain the true smooth part $\tilde{g}(E)$ in a unique way. In other words: the convolution (4.166) should leave the average density $\tilde{g}(E)$ unchanged, but it generally does not. To achieve this aim, Strutinsky modified the convolution by adding a correction which he termed "curvature correction". The name is justified since for any *linear* function $\tilde{g}(E)$ (thus with zero curvature), the result of the convolution in (4.166) will be independent of γ and identical with $\tilde{g}(E)$, by virtue of the assumed symmetry of the averaging function $f(x)$. We shall present this correction here in terms of a moment expansion of the averaging function $f(x)$.

Let us rewrite the folding operator $\hat{\Gamma}$ in Eq. (4.166) as an infinite series of derivatives:

$$\hat{\Gamma} = \sum_{\mu=0}^{\infty} \frac{c_{2\mu}}{(2\mu)!} \gamma^{2\mu} \frac{d^{2\mu}}{dE^{2\mu}} . \qquad (4.167)$$

Hereby, c_m are the moments of the function $f(x)$

$$c_m = \int_{-\infty}^{+\infty} x^m f(x)\, dx . \qquad (4.168)$$

Since we have assumed $f(x)$ to be symmetric, only the even moments appear in Eq. (4.167). From the form (4.167), it is evident that applying $\hat{\Gamma}$ to any polynomial $P(E)$ of degree $\leq 2M + 1$ will create another polynomial of the same degree with coefficients that depend in general on γ, with the exception of the coefficient of the highest power E^{2M+1} which will be the same for both polynomials (since $c_0 = 1$). We next define the inverse operator of $\hat{\Gamma}$ by

$$\hat{\Gamma}^{-1} = \sum_{\mu=0}^{\infty} \frac{a_{2\mu}}{(2\mu)!} \gamma^{2\mu} \frac{d^{2\mu}}{dE^{2\mu}} ; \qquad (4.169)$$

the coefficients $a_{2\mu}$ can easily be found recursively from the $c_{2\mu}$ by imposing the identity $\hat{\Gamma}^{-1} \times \hat{\Gamma} \equiv 1$. Obviously, acting with $\hat{\Gamma}^{-1}$ on $\tilde{g}_\gamma(E)$ in (4.166) will yield the original function $\tilde{g}(E)$ back. If this is assumed to be a polynomial of degree $2M + 1$, then only the first $M + 1$ terms of both $\hat{\Gamma}$ and $\hat{\Gamma}^{-1}$ are operative.

Therefore, we now define the "Strutinsky averaging operator" $\hat{\Sigma}_M$ as

$$\hat{\Sigma}_M = \sum_{\mu=0}^{M} \frac{a_{2\mu}}{(2\mu)!} \gamma^{2\mu} \frac{d^{2\mu}}{dE^{2\mu}} \times \hat{\Gamma} = \sum_{\mu=0}^{M} \frac{a_{2\mu}}{(2\mu)!} \gamma^{2\mu} \frac{d^{2\mu}}{dE^{2\mu}} \frac{1}{\gamma} \int_{-\infty}^{+\infty} f\left(\frac{E - E'}{\gamma}\right) dE', \qquad (4.170)$$

which by construction leaves any polynomial of degree $n \leq 2M + 1$ unchanged, independently of the value of γ, as long as γ is larger than zero:

$$\hat{\Sigma}\left[P_{2M+1}(E)\right] \equiv P_{2M+1}(E) . \qquad (\gamma > 0) \qquad (4.171)$$

The *curvature-corrected Strutinsky-averaged level density* for any given single-particle level spectrum $\{E_n\}$ thus becomes

$$\begin{aligned}
\tilde{g}_M(E) &= \hat{\Sigma}_M[g(E)] = \frac{1}{\gamma} \sum_n \sum_{\mu=0}^{M} \frac{a_{2\mu}}{(2\mu)!} \gamma^{2\mu} \frac{d^{2\mu}}{dE^{2\mu}} f\left(\frac{E - E_n}{\gamma}\right) \\
&= \frac{1}{\gamma} \sum_n f_M\left(\frac{E - E_n}{\gamma}\right),
\end{aligned} \qquad (4.172)$$

where $f_M(x)$ now represents the curvature-corrected averaging function. One usually refers to $2M$ as the order of the curvature correction. Mathematically, $\tilde{g}_M(E)$ turns into the exact level density in the limit $M \to \infty$ (although this may not work numerically). Therefore, one is interested in keeping M as low as possible, in order not to start reproducing its oscillating part.

Let us repeat at this point the two main features of the quantity $\tilde{g}_M(E)$:

(*i*) If the averaging width γ is chosen larger than the main-shell distance $\hbar\Omega$ in the spectrum $\{E_n\}$, which corresponds to the shortest period of the classical orbits contributing to the oscillating part of the level density $\delta g(E)$, then this oscillating part is averaged out and $\tilde{g}_M(E)$ is a smooth function of energy.

(*ii*) If the average part $\tilde{g}(E)$, which in Sect. 4.3 was obtained from the ETF model, is a polynomial of degree $2M + 1$ or lower in the energy E, it is identically reproduced by $\tilde{g}_M(E)$ for any $\gamma > 0$.

Only in exceptional cases, the average level density $\tilde{g}(E)$ is exactly given as a polynomial in the energy E, such as for the harmonic-oscillator potentials in arbitrary dimensions (see Sect. 3.2.4). When $\tilde{g}(E)$ is not a simple polynomial in E, it cannot be identically reproduced by $\tilde{g}_M(E)$. In this case, the Strutinsky averaging amounts to a *local polynomial approximation* of order $2M + 1$ to the true $\tilde{g}(E)$ near the energy E. Then, it is important that this approximation be optimized, both with respect to the precise value of γ and the correction order $2M$. As has been shown in [85], this can be achieved by making the results of the averaging procedure stationary with respect to these two parameters.

In many applications, the focus is not on the average level density but on the average part of the sum E_{sp} of occupied single-particle levels, which is given (including a spin degeneracy factor 2 for fermions) by

$$\tilde{E}_{sp} = 2 \int_0^{\tilde{\mu}} E \, \tilde{g}(E) \, dE \,, \tag{4.173}$$

where $\tilde{\mu}$ is the average Fermi energy determined by the conservation of the particle number N:

$$N = 2 \int_0^{\tilde{\mu}} \tilde{g}(E) \, dE \,. \tag{4.174}$$

The stationary conditions for $\tilde{E}_{sp}$, often referred to as the "plateau conditions", are then

$$\left. \frac{\partial \tilde{E}_{sp}}{\partial \gamma} \right|_{\gamma_0, M_0} = 0 \,, \qquad \left. \frac{\Delta \tilde{E}_{sp}}{\Delta M} \right|_{\gamma_0, M_0} = 0 \,. \tag{4.175}$$

In most practical cases, solutions to the above stationary conditions are found in the ranges $\gamma_0 \sim (1.2 - 1.6) \hbar\Omega$ and $M_0 \sim 2 - 3$, although larger values are sometimes necessary (see Fig. 4.7 below for an example). From Eq. (4.173) it is immediately recognized that if $\tilde{g}(E)$ is a polynomial of degree $2M$ in energy, then $\tilde{E}_{sp}$ will be a polynomial of degree $2M + 2$ in the Fermi energy $\tilde{\mu}$. Therefore, a curvature correction higher by one unit in M is required for the energy $\tilde{E}_{sp}$ to become stationary than for the average level density itself.

The introduction of averaged occupation numbers [58, 85] leads to a very useful representation of the Strutinsky averaging. Since Eq. (4.172) can be written in the form

$$\tilde{g}_M(E) = \sum_n \tilde{g}_n(E), \qquad (4.176)$$

it is straightforward to define averaged occupation numbers $\tilde{\nu}_n$ by

$$\tilde{\nu}_n = \int_{-\infty}^{\tilde{\mu}} \tilde{g}_n(E)\, dE, \qquad N = 2\sum_n \tilde{\nu}_n. \qquad (4.177)$$

It can now be shown by some algebraic manipulations [85] that the energy $\tilde{E}_{sp}$ fulfills the following differential equation, independently of the form of the averaging function $f(x)$ and of the correction order $2M$:

$$\tilde{E}_{sp} = 2\sum_n E_n \tilde{\nu}_n + \gamma \frac{\partial \tilde{E}_{sp}}{\partial \gamma}\bigg|_N. \qquad (4.178)$$

If the plateau conditions (4.175) can be fulfilled, the average total energy is thus just given by the sum $2\sum_n E_n \tilde{\nu}_n$. Then, one may also define a Strutinsky averaged density matrix by

$$\tilde{\rho}(\mathbf{r}, \mathbf{r}') = \sum_n \psi_n^*(\mathbf{r}')\, \psi_n(\mathbf{r})\, \tilde{\nu}_n, \qquad (4.179)$$

and averaged values of any expectation value defined through integrals over $\tilde{\rho}(\mathbf{r}, \mathbf{r}')$.

The equation (4.178) suggests an interesting analogy [108, 109] that gives us an understanding of the physics of the Strutinsky energy averaging. Recall the well-known differential equation of the free Helmholtz energy of a grand canonical ensemble of fermions in a heat bath at temperature T:

$$F = 2\sum_n E_n \nu_n(T) + T \frac{\partial F}{\partial T}\bigg|_{\nu_n}, \qquad (4.180)$$

where $\nu_n(t)$ are the Fermi-Dirac occupation numbers

$$\nu_n(T) = \left[1 + \exp\left(\frac{E_n - \mu}{k_B T}\right)\right]^{-1}; \qquad (4.181)$$

k_B is the Boltzmann constant and μ the chemical potential. We can now make the following one-to-one correspondence between the quantities appearing in the Strutinsky energy averaging method and those of the grand-canonical system:

$$\begin{aligned}
\tilde{E}_{sp} &\longleftrightarrow F, \\
\gamma &\longleftrightarrow kT, \\
\frac{\partial \tilde{E}_{sp}}{\partial \gamma}\bigg|_N &\longleftrightarrow \frac{\partial F}{\partial T}\bigg|_{\nu_n} = -S.
\end{aligned} \qquad (4.182)$$

In the last line above, S is the entropy of the system. The correspondence (4.182) shows us that the Strutinsky averaging is very similar to that brought about by the finite temperature in a grand-canonical ensemble – however, with the one important difference that it leaves the system unexcited since the "effective entropy" $-\partial \tilde{E}_{sp}/\partial \gamma$

can be kept to zero by virtue of the plateau conditions (4.175). This works, of course, only with $M > 0$. We may therefore call it a "cold" averaging, in contrast to that done by a finite temperature where the system is excited and the heat energy $-TS$ is substantially different from zero.

The close relation between Strutinsky and temperature averaging has been exploited by a number of authors [110, 111, 112] to formulate alternative methods for determining the average single-particle energy $\tilde{E}_{sp}$. We should also like to note that the Strutinsky averaging can be introduced into a self-consistent calculation of average properties of an interacting many body system in the Hartree-Fock or Kohn-Sham approximation (see Appendix A), making use of the averaged occupation numbers $\tilde{\nu}_n$. This has been done for nuclei in Ref. [59], where also some formal aspects of such a self-consistent averaging are discussed in detail. We shall come back to this in Sect. 4.7.3 below.

All the above relations and properties are independent of the precise form of the averaging function $f(x)$, as long as it fulfills the specifications mentioned in the beginning of this section. The standard and numerically most efficient choice is that of a Gaussian. The averaging operator $\hat{\Gamma}$ can then be written in a compact way in terms of the dimensionless differential operator

$$\hat{d} = \gamma \frac{d}{dE}, \tag{4.183}$$

and the curvature-correction coefficients can be given analytically. They are:

$$f(x) = \frac{1}{\sqrt{\pi}} e^{-x^2}, \quad \hat{\Gamma} = \exp\left(\frac{\gamma}{2}\hat{d}\right)^2, \quad a_{2\mu} = \frac{(-1)^\mu}{2^{2\mu}\mu!}. \tag{4.184}$$

With this, the curvature-corrected averaging function involves the Hermite polynomials $H_{2\mu}(x)$, whose sum may also be written [113, 114] as an associated Laguerre polynomial $L_M^{1/2}(x^2)$:

$$f_M(x) = \frac{1}{\sqrt{\pi}} \sum_{\mu=0}^{M} \frac{(-1)^\mu}{2^{2\mu}\mu!} H_{2M}(x)e^{-x^2} = \frac{1}{\sqrt{\pi}} L_M^{1/2}(x^2)e^{-x^2}. \tag{4.185}$$

Alternatively, we may use the following function and its associated averaging operator and curvature corrections:

$$f(x) = \frac{1}{2\text{Cosh}^2(x)}, \quad \hat{\Gamma} = \frac{\frac{\pi}{2}\hat{d}}{\sin\left(\frac{\pi}{2}\hat{d}\right)}, \quad a_{2\mu} = \left(\frac{\pi}{2}\right)^{2\mu} \frac{(-1)^\mu}{(2\mu+1)!}. \tag{4.186}$$

Note that the first integral of the function $f(x)$ given in (4.186) is a Fermi function, and thus the occupation numbers obtained by it with $M = 0$ are identical to the Fermi occupation numbers (4.181). Further examples of averaging functions and their curvature-correction coefficients may be found in Ref. [85].

As an illustration, we show in Fig. 4.6 the energy $\tilde{E}_{sp}$ for a system of $N = 70$ particles in a spherical harmonic-oscillator potential, calculated with the Gaussian averaging function (4.184) and plotted as a function of γ for different values of $2M$. Since for this potential, the ETF energy is a polynomial of fourth order in the Fermi energy $\tilde{\mu}$ – see Eq. (4.201) in Problem 4.3 – a fourth-order curvature correction ($2M{=}4$)

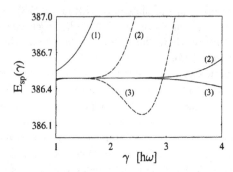

Figure 4.6: Strutinsky-averaged energy $\tilde{E}_{sp}$ of $N = 70$ particles in a
spherical harmonic-oscillator potential versus averaging width γ. (Both
quantities in units of $\hbar\omega$.) *Solid lines:* using the lowest 23 levels, *dashed
lines:* using only 10 levels of the single-particle spectrum $\{E_n\}$. Numbers in parentheses indicate the degree M of the curvature correction.
Gaussian averaging (4.184) was used. (After [85]).

is required. This is confirmed in the figure, where we see that a well-developed plateau
exists for $2M \geq 4$, indeed. Note that nothing is gained here, however, by going beyond
the value $2M = 4$. The length of the plateau is limited by the energy cut-off in the
spectrum used in the numerical calculation. The plateau value is constant up to 7
digits in the range $1.3 \lesssim \gamma \lesssim 2.2$, using 23 levels in the spectrum. Exactly the same
plateau value is also found with Eq. (4.186) or other averaging functions [85]. Within
the same numerical accuracy, the plateau value agrees with the ETF value (4.201) of
the sum of occupied single-particle energies. We shall discuss the relation between
Strutinsky averaging and the ETF model explicitly in Sect. 4.7.3 below.

Alternative energy averaging methods have been studied by Ivanyuk and Strutinsky
[115], who also devised methods of averaging the total single-particle energy E_{sp} over
the particle number N, and compared the relation between the two kinds of averaging.
In the shell-correction method, one replaces the average part of E_{sp} by an empirical
liquid-drop model expansion of the total binding energy of the system under study
(see the next section). The question then arises, to which extent Strutinsky's energy
averaging of the single-particle spectrum is consistent with the number averaging or,
more precisely speaking, the asymptotic expansion for large particle numbers N which
is inherent in the liquid drop model and which corresponds to the Weyl expansion
discussed in Sect. 4.6. This question has been addressed in the book by Bohr and
Mottelson [116], who proposed the use of the large-N expansion to determine the
average total energies in the shell-correction method. The first term in this expansion
is identical to the Thomas-Fermi approximation and corresponds to the volume term
of the Weyl expansion. For the surface term, Siemens and Sobiczewski [117] gave a
general expression in terms of phase shifts in a given potential.

The relation between Strutinsky averaging and the large-N expansion was carefully
examined numerically for Woods-Saxon potentials in Ref. [118]. It is generally found
that the results agree within the numerical accuracy of determining the average total
energy in either method. As we have seen in Sect. 4.3, the ETF model leads naturally

to liquid-drop type expansions of the total average single-particle energy, if the Fermi energy is eliminated from the semiclassical $\hbar$-expansions of energy and particle number. Therefore, the equivalence of energy averaging and large-N expansion is guaranteed, at least formally, as soon as the equivalence of the Strutinsky averaging and the ETF model has been established, to which we will turn in the following section.

In billiard systems, where particles are bound by a reflecting wall and the quantum spectrum $\{E_n\}$ is determined by the free Schrödinger equation with Dirichlet or von Neumann boundary conditions, the average level density $\tilde{g}(E)$ is a function of the wave number $k = \sqrt{2mE}/\hbar$, as we have seen in Sect. 4.6. The average total single-particle energy is then a simple polynomial in k. (Terms with negative powers in k may appear at some high order in the $\hbar$ expansion but are numerically unimportant at not too low energies.) For these systems, the Strutinsky averaging should then preferably be performed over the variable k instead of the energy E, in order to find ideal plateaux. Most conveniently, one may use the k-number density of states defined in Eq. (3.9) and Strutinsky average it over k in a way analogous to (4.172):

$$\tilde{g}_M(k) = \frac{1}{\gamma} \sum_n \sum_{\mu=0}^{M} \frac{a_{2\mu}}{(2\mu)!} \gamma^{2\mu} \frac{d^{2\mu}}{dk^{2\mu}} f\left(\frac{k - \sqrt{2mE_n}/\hbar}{\gamma}\right). \tag{4.187}$$

Finally, we briefly address a problem that arises in potentials with a finite depth which are typical of all self-bound systems such as nuclei or metal clusters, but also of the electrons in the atom. Since a sufficient number of energy levels lying above the Fermi energy must be used in the Strutinsky energy averaging in order to guarantee the existence of a plateau in the averaged quantities, continuum states may have to be included. Formally, the extrapolation of the bound states in the continuum region is given by the resonances, i.e., the scattering states whose phase shifts take the value π at definite energies $E_n > 0$ (cf. Chapter 7). Ross and Bhaduri [113] have included the resonances of a spherical Woods-Saxon potential into the Strutinsky-averaged energy and thereby obtained well-defined plateaux. A pragmatic, but rather efficient way to handle the continuum problem, which has been used by most authors employing finite-depth potentials, is to diagonalize the potentials in a harmonic-oscillator basis. Hereby the true continuum is automatically discretised. For reasonably, but not too large basis sizes it is found that the lowest-lying discrete eigenvalues with $E > 0$ actually provide a fair estimate of the positions of the resonances (see, e.g., Ref. [119]).

4.7.2 The shell-correction method

The energy averaging method described above has most successfully been used in nuclear physics in connection with the shell-correction approach, for which it was originally designed by Strutinsky [4]. This approach can be applied to any finite many-fermion system. Historically, Strutinsky invented the shell-correction method at a time where microscopical Hartree-Fock calculations, and the effective nuclear interactions needed thereby, had not yet reached the sophistication required to provide a realistic description of nuclear binding or deformation energies. The method therefore allowed for the first time for an *approximately* self-consistent way to calculate these properties. Its justification within the framework of the self-consistent mean-field approximation was pointed out by Strutinsky [5]. This formal aspect, now usually referred to as the "Strutinsky energy theorem", is presented in Appendix A.3. Here we will explain the main idea of the shell-correction method as it is practically used.

The total energy $E_{tot}(N)$ of an interacting system of N fermions is written as the sum of two parts:

$$E_{tot}(N) = \tilde{E}(N) + \delta E(N). \qquad (4.188)$$

$\tilde{E}(N)$ represents the smooth part of the energy, and the *shell-correction* $\delta E(N)$ its fluctuating part. The smooth part may be determined self-consistently within the variational ETF model, as discussed in Sect. 4.4.2. Alternatively, $\tilde{E}(N)$ may be taken as a phenomenological mass formula of the type developed by Weizsäcker [28] and by Bethe and Bacher [29] using the liquid-drop model (LDM). This is of the general form given in Eq. (4.69). The LDM parameters a_v, a_s, etc., in (4.69) are hereby determined phenomenologically by fits to experimental binding energies. (In nuclei one has, of course, two kinds of fermions, the neutrons and protons; correspondingly, the energy is a function of two particle numbers, as discussed in Sect. 8.1.1.) Both the ETF and the LDM models cannot account for shell effects which are due to the quantized motion of the particles in their mean fields. That is why the shell-correction term has to be added to the smooth energy in Eq. (4.188).

The shell-correction energy $\delta E(N)$ is evaluated from the eigenvalue spectrum $\{E_n\}$ of the *averaged* mean field in which the fermions move. Again, this field my be determined self-consistently by the ETF variational method, or taken to be a phenomenological shell-model potential whose parameters are adjusted to experimental single-particle data. (In nuclear physics one has to use separate potentials for neutrons and protons; the shell-correction is then the corresponding sum of two terms.) In terms of the spectrum $\{E_n\}$, the shell-correction energy is defined by

$$\delta E(N) = E_{sp}(N) - \tilde{E}_{sp}(N) = 2 \sum_{E_n < \mu} E_n - 2 \int_0^{\tilde{\mu}} E\, \tilde{g}(E)\, dE, \qquad (4.189)$$

i.e., as the difference between the sum (3.39) of the lowest N occupied energy levels and its average part (4.173). The factors 2 in the second line above take explicitly the spin degeneracy into account and N is assumed to be an even number. If the plateau conditions (4.175) are fulfilled, the second term in (4.178) is zero and $\delta E(N)$ can be expressed in terms of the occupations numbers (4.177):

$$\delta E(N) = 2 \sum_n E_n \delta\nu_n, \qquad \delta\nu_n = \Theta(\mu - E_n) - \tilde{\nu}_n. \qquad (4.190)$$

In the above, μ and $\tilde{\mu}$ are the Fermi energies determined by

$$N = 2 \sum_n \nu_n = 2 \int_0^\mu g(E)\, dE = 2 \sum_n \tilde{\nu}_n = 2 \int_0^{\tilde{\mu}} \tilde{g}(E)\, dE. \qquad (4.191)$$

The combination of the two equations (4.188) and (4.189) makes it clear that in (4.188), the average part of the single-particle energy sum E_{sp} is replaced by the average total energy $\tilde{E}(N)$. The shell-correction method is therefore sometimes also called the "Strutinsky renormalization". It corrects the wrong average part of the energy E_{sp}

obtained in the independent-particle model by the correct average binding energy of the interacting system. In fact, it is well-known (see Appendix A) that even if evaluated in terms of self-consistent Hartree-Fock or Kohn-Sham levels E_n, the quantity E_{sp} does not describe correctly the total binding energy of an interacting fermion system. It follows, however, from the variational principle used in the self-consistent mean-field approach, that the *oscillating part* of the total binding energy is correctly described (to leading order in the oscillating part of the density matrix) by the shell-correction energy (4.189). This is the content of the Strutinsky energy theorem discussed in more detail in the Appendix A.3.

For later reference, we give an alternative approximate form for the energy shell-correction, expressed directly in terms of the oscillating part $\delta g(E)$ of the level density and the Fermi energy μ. Writing $\mu = \tilde{\mu} + \delta\mu$, expanding the equations (4.173,4.174) around μ, and using Eqs. (3.38,3.39), one obtains in a straight-forward manner the following approximate expression

$$\delta E(N) \simeq \delta E(\mu) = \int_{-\infty}^{\mu} (E - \mu)\, \delta g(E)\, dE + \mathcal{O}(\delta\mu)^2\,. \tag{4.192}$$

This form allows to relate the oscillations in the total energy of a system with those in the level density (see also Ref. [58]) and will be used in Chapter 8.

In nuclear and metal cluster physics, the mean fields of the fermions are often deformed, as a consequence of the Jahn-Teller effect [120]. Therefore, both the liquid drop model and the shell-model potentials have to be extended to describe the deformed shapes. The above energy expressions thus all depend not only on the particle number(s) but also on some suitably defined deformation parameters (see Sect. 8.1 for a more detailed discussion and examples). The deformed liquid drop model was successfully used by Bohr and Wheeler to explain qualitatively the nuclear fission process [121], and was further developed in extensive numerical studies by Cohen and Swiatecki [122]. These will be further discussed in Chapter 8.

The shell-correction approach, in connection with more elaborate versions of the LDM model (see, e.g., Refs. [30, 123]) and various realistic deformed nuclear shell-model potentials, has been very successful in reproducing experimentally measured nuclear ground-state masses and deformations, as well as static fission barriers. The interested reader may consult Refs. [58, 62, 108, 119, 124, 125] for detailed accounts. As an example of the important role played by the shell effects and of their description by the shell-correction method, we explore in Sect. 8.1.2 the complex deformation energy surface of a typical actinide nucleus. In recent years, the shell-correction method has also found useful applications in the field of metal clusters [126, 127, 128, 129].

4.7.3 Relation between ETF and Strutinsky averaging

The close relation between the Strutinsky averaging and the ETF model has been pointed out early [130, 131]; Strutinsky actually had the statistical nature of the averaging method in his mind from the very beginning [4]. Bhaduri and Ross [6] demonstrated the equivalence numerically for several model potentials, including harmonic oscillators and the rectangular box. Jennings [114] pointed out that the Strutinsky averaging, when applied to the quantum-mechanical level density $g(E)$, yields approximately the same result as when it is applied to the ETF level density:

$$\hat{\Sigma}_M[g(E)] \approx \hat{\Sigma}_M[g_{ETF}(E)]. \tag{4.193}$$

This was shown using the two-sided Laplace transform of the curvature-corrected Gaussian function (4.185), which is of the form

$$\mathcal{L}_\beta[f_M(E/\gamma)] = \sum_{\mu=0}^{M} \frac{(-1)^\mu}{2^{2\mu}\mu!}(\gamma\beta)^{2\mu}\,e^{\frac{1}{4}\gamma^2\beta^2} = P_{2M}(\gamma\beta)\,e^{\frac{1}{4}\gamma^2\beta^2}, \tag{4.194}$$

where $P_{2M}(x)$ is a polynomial of order $2M$ in x. Due to the folding theorem of Laplace transforms, the Laplace transform of the Strutinsky-averaged level density therefore becomes a simple product:

$$\mathcal{L}_\beta[\tilde{g}_M(E)] = \tilde{Z}_M(\beta) = Z(\beta)P_{2M}(\gamma\beta)\,e^{\frac{1}{4}\gamma^2\beta^2}, \tag{4.195}$$

where $Z(\beta)$ is the quantum-mechanical canonical partition function (3.15). Note that $\tilde{Z}_M(\beta)$ is strongly peaked at $\beta = 0$, due to the extra Gaussian factor, and falls off exponentially along the imaginary β axis, provided γ is large enough according to the criterion (4.165). Therefore, when taking the inverse Laplace transform of (4.195) in order to recover $\tilde{g}_M(E)$, the only significant contributions will come from the region around $\beta = 0$. (The precise form of the curvature corrections and thus of the polynomial $P_{2M}(x)$ hereby is immaterial.) Now, in Sect. 4.3 we have seen, that the ETF expansion of the level density can be obtained by a Laurent expansion of $Z(\beta)$ in powers of β (4.39), which is valid for small β. Thus, without making too much of an error, the exact partition function $Z(\beta)$ may be replaced by its semiclassical expansion $Z_{sc}(\beta)$. After Laplace inversion, this gives then directly the equivalence (4.193).

Jennings [114] went on to show, following Gross [131], that for not too large values of γ, the Strutinsky averaging leaves the ETF level density approximately unchanged

$$\hat{\Sigma}_M[g_{ETF}(E)] \approx g_{ETF}(E), \tag{4.196}$$

and that approximately the same averaged total energies can be derived from either of them. This step is, of course, a direct consequence of the way in which the curvature corrections have been constructed; see the Sect. 4.7.1 above. The upper bound of γ for (4.196) to be true is thus just the upper end of the plateaux in the Strutinsky-averaged level density (cf. Fig. 4.6), and the same criteria apply for the averaged total single-particle energies.

In systems for which $g_{ETF}(E)$ is a polynomial in E, the equivalence (4.196) becomes an identity, as a consequence of Eq. (4.171). For harmonic-oscillator potentials one can, in fact, show directly that the Strutinsky-averaged level density $\tilde{g}_M(E)$ is identical to $g_{ETF}(E)$, and that the total energies derived from them are also identical. The analytical proof for this has been given by Brack [85] for the general class of averaging functions discussed in Sect. 4.7.1; it makes use of the Euler-MacLaurin summation formula (4.133), applied directly to the sum over the single-particle spectrum on the right-hand side of Eq. (4.172). The same identity holds for billiard systems, where the converging part of the average ETF energy is a polynomial in the wave number k

(see Sect. 4.6.2), provided the Strutinsky smoothing is done in the variable k and the spectrum $k_n = \sqrt{2mE_n}/\hbar$ is used, as shown in Eq. (4.187).

Jennings *et al.* [24] have tested the Strutinsky-ETF equivalence for realistic nuclear potentials of Woods-Saxon type, hereby also including a spin-orbit coupling term (cf. Problem 4.3). In applications of the shell-correction method to nuclear binding and deformation energies, one is generally able to determine the energy to within about 1 MeV on an absolute scale. It was found [24] that the Wigner-Kirkwood expansion of the total single-particle energy in realistic nuclear potentials converges within $\pm \sim 1$ MeV, when terms up to 4-th order in the $\hbar$-expansion are included and when the particle number N is not too low (say, $N \gtrsim 40$). And within the same limit of $\pm \sim 1$ MeV, the plateau values of the Strutinsky-averaged energies generally agree with the ETF values, the largest uncertainties in the plateau values stemming from the continuum problem mentioned at the end of Sect. 4.7.1.

An example is shown in Fig. 4.7, where an unusually high order ($2M \sim 16$) of the curvature correction had to be used to find a converged plateau value according to Eq. (4.175). Nevertheless, the ETF value for this case falls within the error bars around the plateau value that are shown in the figure. (Note that this figure shows not the average energy $\tilde{E}_{sp}$, but its shell-correction part $\delta E = E_{sp} - \tilde{E}_{sp}$, there denoted by δU.)

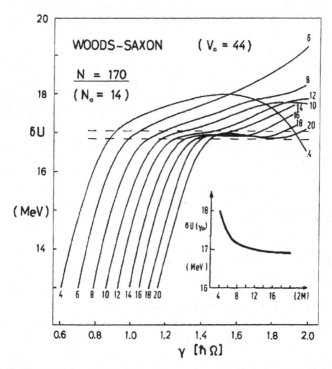

Figure 4.7: Shell-correction energy versus averaging width γ for $N = 170$ neutrons in a spherical Woods-Saxon potential for various values of the curvature-correction order $2M$ (shown at the ends of the curves). The horizontal dashed lines give the error bars for the optimal plateau value. The insert shows the plateau values versus $2M$. (From [118].)

We have seen now that the equivalence of the Strutinsky averaging with the ETF model is well established. This opens the possibility of an interesting combination of the ETF model with Strutinsky's shell-correction method. As mentioned in Sect. 4.7.2, most shell-correction calculations in nuclear physics have been performed with phenomenological shell-model potentials and LDM energies. Ideally, however, one should use a consistent set of mean-field potentials and total energies, both derived from the two-body interaction governing the system, in order to obtain the best approximation to the fully self-consistent microscopical binding energy. This is discussed in Appendix A.3, where we also show that it has been numerically tested with microscopically Strutinsky-averaged Hartree-Fock potentials and energies. Because of the ETF-Strutinsky equivalence, a practical way of achieving this kind of optimal consistency is the use of the variational ETF method described in Sect. 4.4.2. The idea is thus to replace the shell-model potential by the self-consistent potential $V_{ETF}[\rho]$, which is defined by Eq. (4.81) and calculated from the variational ETF densities $\rho(\mathbf{r})$. The shell-correction δE is then evaluated from the eigenvalue spectrum E_n of $V_{ETF}[\rho]$. The LDM energy is replaced by the variational ETF energy (4.79), so that the total energy is written in the form

$$E_{tot} = E_{ETF}[\rho] + \delta E[V_{ETF}], \qquad (4.197)$$

where the notation $\delta E[V_{ETF}]$ indicates that δE is uniquely determined through the eigenvalues of the ETF potential. In this energy expression, now, everything has been derived from the same energy functional.

In this combination, the variational ETF plus shell-correction method is, in fact, a successful and economic substitute for the purely microscopic self-consistent mean-field methods described in the Appendix A, which are numerically much more time consuming. The energy (4.197) has been found to reproduce the shell effects in the total Hartree-Fock energy of nuclei very accurately [26]. An example of this is shown in Fig. 4.2, where the crosses show the energies obtained by the expression (4.197). We recall, however, that the total energy (4.197) suffers from the overbinding effect discussed in Sect. 4.4.2.

4.8   Problems

(4.1) Husimi transform

The Wigner distribution function $\rho_W(x,p)$, given by Eq. (4.18) may oscillate rapidly in the phase space and take on negative values. As such, it is not too instructive to regard it as a probability distribution. One exception, however, is the Husimi distribution function [20, 21] which is non-negative, and may behave smoothly for a suitably chosen smoothing parameter. It is obtained by convoluting $\rho_W(x,p)$ with a minimum uncertainty Gaussian wave packet:

$$\rho_H(x,p) = \frac{1}{\pi\hbar} \int_{-\infty}^{\infty} dx' \int_{-\infty}^{\infty} dp' \, \exp[-m\kappa(x-x')^2/\hbar - (p-p')^2/\hbar m\kappa] \, \rho_W(x',p'), \quad (4.198)$$

where $\hbar\kappa$ has the dimensions of energy, and governs the smoothing. The Husimi distribution function given above has gained some popularity in the quantum study of nonlinear systems that exhibit classical chaos. Show from the above definition that $\rho_H(x, p)$ is non-negative. (You should first perform the p'-integration analytically). For more details, consult the review by Lee [21].

(4.2) Average level density, particle number and energy in the three-dimensional spherical harmonic-oscillator potential at T=0
(a) Determine the ETF level density for the spherical harmonic-oscillator potential, $V(r) = (m\omega^2/2)r^2$, by direct integration of Eq. (4.61) and show that it has the form:

$$g_{ETF}(E) = \frac{1}{2(\hbar\omega)^3}\left[E^2 - \frac{1}{4}(\hbar\omega)^2\right]. \qquad (4.199)$$

By integration according to Eq. (4.67), show the particle number to be given by (including a spin degeneracy factor 2)

$$N_{ETF}(\mu) = \frac{1}{3}\left(\frac{\mu}{\hbar\omega}\right)^3 - \frac{1}{4}\frac{\mu}{\hbar\omega}. \qquad (4.200)$$

(b) For the energy, you must be more careful. Doing the integral according to Eq. (4.66) over the result (4.199), you will only get two terms. However, there is a finite fourth-order contribution to E_{ETF} in this case. If you are patient, you may find explicit general expressions for the fourth-order contribution to the level density and integrate it. However, there is a smarter way to get this term, and also the above results. To this purpose, start from the $\hbar$-expansion of the partition function of the spherical harmonic oscillator, given in Eq. (4.2) of Sect. 4.1. Then, using the right-hand sides of Eqs. (3.38), (3.39) and Laplace inverting term by term with the rules given in Appendix B, re-derive the above results and show that the finite terms contributing to the total energy are (with spin)

$$E_{ETF}(\mu) = \hbar\omega\left[\frac{1}{4}\left(\frac{\mu}{\hbar\omega}\right)^4 - \frac{1}{8}\left(\frac{\mu}{\hbar\omega}\right)^2 - \frac{17}{960}\right]. \qquad (4.201)$$

The explicit contribution of the fourth-order term to the level density is $\frac{17}{1920}\hbar\omega\delta'(E)$ and contains the first derivative of the delta function $\delta(E)$. It therefore does not contribute in a finite way to either $g_{ETF}(E)$ or N, but it gives, after integration by parts, the last term in (4.201).
(c) Evaluate the total energy numerically for $N=70$ and $\hbar\omega=7.896$ MeV, and show that this yields the value F_{ETF} (and its individual contributions) for the free energy at $T=0$ given in Table 4.1.
(d) Eliminate the Fermi energy from Eqs. (4.201,4.200) and show that this yields the expansion

$$E_{ETF}(N) = \frac{1}{4}\hbar\omega\,(3N)^{4/3}\left[1 + \frac{1}{2}\,(3N)^{-2/3} - \frac{47}{240}\,(3N)^{-4/3} + \ldots\right]. \qquad (4.202)$$

Compare its convergence to that of (4.201).
Note: The first semiclassical correction terms in the above results are of relative order $\hbar^2$, or $N^{-2/3}$, with respect to the leading TF terms. There are thus no surface corrections such as they arise in all other three-dimensional systems. This is a peculiarity of harmonic-oscillator potentials.

(4.3) Average number of states in the wedge billiard (After [132])
Consider the wedge billiard given by the Hamiltonian

$$H = \frac{1}{2m}(p_x^2 + p_y^2) + \alpha y , \quad x \geq 0 , \quad y \geq x \cot \phi . \tag{4.203}$$

The slope parameter α has units of E/L. Derive the extended Thomas-Fermi expression for $N(E)$, the number of states between 0 and E:

$$N_{ETF}(E) = \left(\frac{2m}{\hbar^2 \alpha^2}\right) \frac{\tan\phi}{24\pi} E^3 - \left(\frac{2m}{\hbar^2 \alpha^2}\right)^{1/2} \frac{(1 + \sec\phi)}{6\pi} E^{3/2} + \frac{1}{6} . \tag{4.204}$$

(See Sect. 5.5 for the oscillating part of $N(E)$ and Sect. 7.2 for further application of this model.)

(4.4) Finite-temperature Thomas-Fermi calculation in two dimensions
Consider Eq. (4.90) for particles moving in a plane. Show that

$$\frac{\mu}{T} = \frac{1}{T}\left[V(\mathbf{r}) + \frac{\pi\hbar^2}{m}\rho_{TF}(\mathbf{r})\right] + \ln\{1 - \exp(-\pi\hbar^2\rho_{TF}(\mathbf{r})/mT)\} . \tag{4.205}$$

This formula has an interesting application when $V(\mathbf{r})$ is density-dependent [133].

(4.5) Surface tension of a semi-infinite Fermi liquid
Let us study the following simple energy density functional [31, 134] of a Fermi liquid:

$$E[\rho] = \int \mathcal{E}[\rho]d^3r = \int \left[\mathcal{E}_\infty(\rho) + b\frac{(\nabla\rho)^2}{\rho}\right]d^3r , \tag{4.206}$$

where b is a constant and $\mathcal{E}_\infty(\rho)$ describes the bulk energy density of the infinite system. If the latter is saturating, its energy density may be simply parameterized by

$$\mathcal{E}_\infty(\rho) = \rho\left[a_v + \frac{K}{18\rho_0^2}(\rho - \rho_0)^2\right] . \tag{4.207}$$

The significance of the three parameters in (4.207) is the following. The equilibrium bulk density of the saturated system is ρ_0, and its energy per particle, given by $\mathcal{E}_\infty(\rho_0)/\rho_0$, is the so-called bulk energy a_v. For a finite liquid drop, this is the volume energy appearing in the expansion (4.69) of the total energy. For small deviations away from the equilibrium density ρ_0, $\mathcal{E}_\infty(\rho)$ has a quadratic minimum, and the curvature at this minimum is the bulk incompressibility K, in nuclear physics traditionally defined by $K = 9\rho_0^2 d^2(\mathcal{E}_\infty/\rho_0)/d\rho_0^2$. The functional (4.207) contains, in the spirit of the local density approximation (cf. Appendix A.2), the main parts of the kinetic and potential energy of the bulk system. The gradient-correction term in (4.206) is again phenomenological; part of it may be thought of as coming from the kinetic energy in the form of a Weizsäcker correction, part of it simulates [14] the finite range of the attractive forces that hold the particles together. Due to the presence of this gradient term, the functional (4.206) cannot only describe the infinite bulk system, but also finite liquids with varying densities. – Assume now half the space to be filled with the liquid and the location of its infinite plane surface to be given by the (x, y) plane. This is often called the "semi-infinite liquid". Since the integral in (4.206) over the whole space diverges for this system, one must subtract from it the bulk part $\int \rho a_v d^3 r$, so that the integrand goes to zero both for $z \to -\infty$ and for $z \to +\infty$. The one-dimensional integral over z then defines the surface tension σ which is defined as the energy per unit area of the surface:

$$\sigma = \int_{-\infty}^{\infty} \{\mathcal{E}[\rho(z)] - a_v \, \rho(z)\} \, dz \,. \tag{4.208}$$

Use the density variational method to determine the density profile function $\rho(z)$ and to calculate the surface tension σ.

(a) Make the surface tension (4.208) stationary with respect to variations of $\rho(z)$, assuming the boundary conditions $\rho(-\infty) = \rho_0$ and $\rho(+\infty) = 0$. Writing the density in terms of a dimensionless function $y(z)$ as

$$\rho(z) = \rho_0 \, y(z) \,, \tag{4.209}$$

show that the Euler variational equation becomes

$$\frac{K}{18}(3y^2 - 4y + 1) + b \left[\left(\frac{y'}{y} \right)^2 - \frac{2y''}{y} \right] = 0 \,. \tag{4.210}$$

Here the primes on y denote derivatives with respect to z.

(b) Investigate the asymptotic behavior of $\rho(z)$ for large distances away from the surface and show, in particular, that

$$y(z) \sim e^{-z/\alpha} \quad \text{for} \quad z \to \infty, \quad \text{with} \quad \alpha = \sqrt{\frac{18\,b}{K}} \,. \tag{4.211}$$

(c) Find the exact solution of the differential equation (4.210). Since this is a nonlinear, second-order DE, you may need some help. As a first hint, write the derivative of $y(z)$ as $y'(z) = p$ and consider it as a function of y: $p = p(y)$. Since Eq. (4.210) does not contain the variable z explicitly, you can now rewrite it as a differential equation for the function $p(y)$. Doing this, you will find that you can integrate it once over y, with the result

$$p(y) = y'(z) = \frac{1}{\alpha} y(y - 1) \,, \tag{4.212}$$

where α is given in (4.211). The integration constant has been chosen such that $p(y)$ goes to zero at both boundaries. Next, write $dy/dz = 1/(dz/dy)$ and solve Eq. (4.212) for the inverse function $z(y)$. This is done easily by direct integration of the inverse of the r.h.s. of (4.212). Show that after inverting back to $y(z)$, the exact solution of Eq. (4.210) is given by the simple Fermi function

$$y(z) = \frac{1}{[1 + e^{z/\alpha}]} \,, \tag{4.213}$$

which obeys (4.211).

(d) Using the solution (4.213), show that the surface tension σ in (4.208) becomes

$$\sigma = \frac{\rho_0 b}{\alpha} = \rho_0 \sqrt{\frac{Kb}{18}} \,. \tag{4.214}$$

[Another hint: Rather than inserting (4.213) into (4.208) and integrating over z, which is rather tedious, make use of the above substitutions and reduce the integral (4.208) to one over ρ.]

Note: The earliest attempts to calculate the nuclear surface energy in the framework of the (extended) Thomas-Fermi model go back to Swiatecki [135, 136], Skyrme [14], and Berg and Wilets [49]. Swiatecki [136] did not use the density variational method but solved the Schrödinger equation exactly, after linearizing the potential at the classical turning point (see Sect. 2.4), leading to a modified TF equation which could be solved analytically. Skyrme considered a gradient correction of the form $c(\nabla\rho)^2$ instead of the Weizsäcker type term in (4.206), for which the density profile was found to be of the form $\rho(z) \propto \mathrm{Tgh}^2[(z - z_0)/\alpha]$ which goes to zero at a finite distance z_0 rather than at infinity. Berg and Wilets [49] used functionals of the form (4.206). A nice review of all these early investigations of the nuclear surface energy has been given by Wilets [31]. More recent applications of the simple model presented here led to the inclusion of the coupling of surface and bulk vibrations in the semiclassical description of the nuclear breathing-mode excitations [134, 137], and to the determination of the curvature-energy coefficient a_c in the liquid drop expansion Eq. (4.69) of the nuclear binding energy [138, 139].

(4.6) Level density, particle number, and free energy for the three-dimensional spherical harmonic-oscillator potential at T>0

Start from the general relation in Eq. (4.99) and the $\hbar$-expansion of the $T=0$ partition function for the spherical harmonic oscillator in Eq. (4.2). Then obtain the level density, the particle number and the free energy, using inverse Laplace transforms term by term, by the relations (3.37) and (4.101), respectively. This means that you will have to Laplace invert the characteristic temperature factor on the r.h.s. of Eq. (4.98), multiplied by different negative and positive powers of β. Note that the two-sided inverse Laplace transformation formula is the same as that for one-sided transforms, given in Eq. (3.37) (see, e.g., [67]).

(a) As a first exercise, use Eq. (4.98) and the relation (B.2) in Appendix B to show that

$$\mathcal{L}_E^{-1}\left[\frac{\pi T}{\sin(\pi\beta T)}\right] = \frac{1}{4T}\int_{-\infty}^{E}\frac{1}{\mathrm{Cosh}^2\left(\frac{E}{2T}\right)}\,dE = \frac{1}{\left[1+\exp\left(-\frac{E}{T}\right)\right]} = f_F(E/T), \quad (4.215)$$

where $f_F(x)$ is the Fermi function.

(b) Next, having $\mathcal{L}_E^{-1}[1/\beta^2] = E\,\Theta(E)$ and the result (4.215), use the folding theorem for two-sided Laplace transforms [67] (see also Appendix B) to show that

$$\mathcal{L}_E^{-1}\left[\frac{1}{\beta^2}\frac{\pi T}{\sin(\pi\beta T)}\right] = \int_0^{+\infty}\frac{E'}{\left[1+\exp\left(\frac{E'-E}{T}\right)\right]}\,dE' = T^2 F_1\left(\frac{E}{T}\right), \quad (4.216)$$

where $F_1(\eta)$ is the Fermi integral (4.110) for $\xi=1$.

(c) Proceed now in a similar way for the other terms in the $\hbar$ series, to show that the ETF level density is given by

$$g_{ETF}(E) = \frac{1}{\hbar\omega}\left\{\left(\frac{T}{\hbar\omega}\right)^2\tilde{F}_1\left(\frac{E}{T}\right) - \frac{1}{8}\tilde{F}_{-1}\left(\frac{E}{T}\right) + \frac{17}{1920}\left(\frac{\hbar\omega}{T}\right)^2\tilde{F}_{-3}\left(\frac{E}{T}\right) + \dots\right\}.$$
$$(4.217)$$

Hereby $\tilde{F}_\xi(\eta)$ is a modified Fermi integral obtained from its definition (4.110) by division through a Gamma function:

$$\tilde{F}_\xi(\eta) = \frac{1}{\Gamma(\xi+1)} F_\xi(\eta). \tag{4.218}$$

It can be shown [70] that the functions (4.218) are analytic for all real values of ξ and real for real η (whereas the Fermi integrals $F_\xi(\eta)$ diverge for negative integer ξ; this divergence is cancelled in (4.218) by the corresponding divergence of the Gamma function). In fact, for $\xi = 0, -1, -2, \ldots$ they are elementary functions, and in particular,

$$\begin{aligned}
\tilde{F}_0(\eta) &= F_0(\eta) = \ln(1+e^\eta), \\
\tilde{F}_{-1}(\eta) &= (1+e^{-\eta})^{-1} = f_F(\eta),
\end{aligned} \tag{4.219}$$

etc. The two functions obey the differential recurrence relations

$$F_{\xi-1}(\eta) = \frac{1}{\xi}\frac{d}{d\eta}F_\xi(\eta), \quad -\xi \notin (0,1,2,3,\ldots) \tag{4.220}$$

which may also be used to *define* the $F_\xi(\eta)$ for non-integer values $\xi < -1$ where the integral (4.110) does not exist, and

$$\tilde{F}_{\xi-1}(\eta) = \frac{d}{d\eta}\tilde{F}_\xi(\eta). \tag{4.221}$$

(d) Applying the same techniques, show that the particle number is given by

$$N_{ETF}(\mu) = \left(\frac{T}{\hbar\omega}\right)^3 F_2\left(\frac{\mu}{T}\right) - \frac{T}{4\hbar\omega} F_0\left(\frac{\mu}{T}\right) + \frac{17}{960}\frac{\hbar\omega}{T}\tilde{F}_{-2}\left(\frac{\mu}{T}\right) + \ldots, \tag{4.222}$$

and the free Helmholtz energy by

$$F_{ETF}(\mu) = \mu N_{ETF}(\mu) - \frac{T^4}{3(\hbar\omega)^3} F_3\left(\frac{\mu}{T}\right) + \frac{T^2}{4\hbar\omega} F_1\left(\frac{\mu}{T}\right) - \frac{17}{960}\hbar\omega \tilde{F}_{-1}\left(\frac{\mu}{T}\right) + \ldots \tag{4.223}$$

(e) Using the asymptotic expansion for large arguments [25, 69]

$$F_\xi(\eta) \sim \frac{\eta^{\xi+1}}{(\xi+1)}\left[1 + \sum_{k=1}^\infty (2^{2k}-2)|B_{2k}|\binom{\mu+1}{2k}\eta^{-2k}\right], \quad \eta \gg 1 \tag{4.224}$$

where B_{2k} are the Bernoulli numbers (4.134) and to the right of them above are binomial coefficients, show that the results (4.217), (4.222) and (4.223) go over to those given in Eqs. (4.199), (4.200), and (4.201) above in the limit $T \to 0$.

(f) Using the following expansion [69] of the Fermi integrals, valid for negative arguments η:

$$F_\xi(\eta) = \Gamma(\xi+1) \sum_{k=1}^\infty (-1)^{k-1}\frac{1}{k^{\xi+1}}e^{k\eta}, \quad \eta < 0 \tag{4.225}$$

show that the level density (4.217) takes the form given in Eq. (3.76) of Sect. 3.2.5.

(4.7) Thermodynamical properties of fermions and bosons
(a) Prove the relation for the free energy F for fermions given in Eq. (4.96). This is quite straightforward, but requires some algebra using the equations (4.94) and (4.95).
(b) Consider now a grand-canonical system of N ideal bosons in a mean-field potential, with single-particle energies E_n and eigenfunctions $\psi_n(\mathbf{r})$, $n = 0, 1, 2, ..$, where $E_0 > 0$ is the zero-point energy. At a finite temperature T, the particle number N and the Helmholtz free energy F are given by the relations (with $k_B = 1$):

$$N = \sum_n \nu_n^B, \qquad \nu_n^B = \frac{1}{\exp\left(\frac{E_n - \mu}{T}\right) - 1}, \qquad F = \sum_n E_n \nu_n^B - TS, \qquad (4.226)$$

whereby the entropy is expressed in terms of the Bose occupation numbers ν_n^B by

$$S = \sum_n [(1 + \nu_n^B) \log(1 + \nu_n^B)] - \nu_n^B \log \nu_n^B]. \qquad (4.227)$$

Show, analogously to the fermionic case presented in Sect. 4.4.3, that a hot bosonic level density $g_T^B(E)$ can be defined by

$$g_T^B(E) = \sum_n \frac{1}{4T \mathrm{Sinh}^2[(E - E_n)/2T]}, \qquad (4.228)$$

so that the following relations hold:

$$N = \int_{-\infty}^{\mu} g_T^B(E)\, dE, \qquad F = \int_{-\infty}^{\mu} E\, g_T^B(E)\, dE. \qquad (4.229)$$

For proving the last equation for F, a useful intermediate step is to show that

$$S = \sum_n \left\{ \nu_n^B x_n - \log[1 - \exp(-x_n)] \right\}, \qquad x_n = (E_n - \mu)/T. \qquad (4.230)$$

Note: In Ref. [140], the Laplace transform of $g_T^B(E)$ in (4.228) was used to derive the Wigner-Kirkwood expansion of the grand potential for bosons, employing the same techniques as those presented in Sect. 4.4.3 for fermions. The results were applied to obtain quantum corrections to the thermodynamical properties of weakly interacting trapped bosons.
(c) The spatial density of the bosons is given by

$$\rho(\mathbf{r}) = \sum_n |\psi_n(\mathbf{r})|^2 \nu_n^B. \qquad (4.231)$$

By setting $\tilde{E}_n = (E_n - E_0)$ and $\tilde{\mu} = (\mu - E_0)$, show that

$$\rho(\mathbf{r}) = \sum_{l=1}^{\infty} e^{\tilde{\mu}\beta l}\, \tilde{C}(\mathbf{r}, \beta l), \qquad (4.232)$$

where we have defined

$$\tilde{C}(\mathbf{r}, \beta l) = \sum_n |\psi_n(\mathbf{r})|^2 \exp(-\beta l \tilde{E}_n). \qquad (4.233)$$

(4.8) Finite-N correction to the critical temperature in BEC

The critical temperature $T_c^{(0)}$ of an ideal bose gas in a three-dimensional spherical HO was given by Eq. (4.129). This is in the limit when $N \to \infty$, $\omega \to 0$, such that $N^{1/3}\omega = $ constant. In experiments, the number of bosonic atoms in the trap are large but finite (say 10^6). The fraction of particles in the ground state N_0/N, as a function of temperature T for large but finite N, goes to 0 rapidly but continuously (cf. Fig. 4.5). Strictly speaking, there is no phase transition for finite N, but we may still define a critical temperature T_c by extrapolating the rapid fall-off, and ignoring the tail. The difference $\delta T_c = (T_c - T_c^{(0)})$ in a HO trap may be estimated semiclassically as follows.
(a) Using the notation of the previous problem, show that

$$N = \sum_{l=1}^{\infty} e^{\tilde{\mu}\beta l}\, \tilde{Z}(\beta l)\,, \tag{4.234}$$

where the single-particle partition function $\tilde{Z}$ has been calculated with the lowest energy shifted to 0, i.e., with the spectrum $\{\tilde{E}_n\}$, and $\tilde{\mu} = (\mu - E_0)$. For semiclassical applications to bosons, $\tilde{Z}(\beta)$ is more useful. For a three-dimensional spherical HO, show that $\tilde{Z}(\beta)$ is given by

$$\tilde{Z}(\beta) = 1 + \frac{e^{-3\hbar\omega\beta/2}}{[2\operatorname{Sinh}(\hbar\omega\beta/2)]^3} + \frac{3\,e^{-\hbar\omega\beta}}{[2\operatorname{Sinh}(\hbar\omega\beta/2)]^2} + \frac{3\,e^{-\hbar\omega\beta/2}}{[2\operatorname{Sinh}(\hbar\omega\beta/2)]}\,. \tag{4.235}$$

The first term isolates the contribution of the ground state.
(b) Note that for large N, $T_c^{(0)} \gg \hbar\omega$, so that it makes sense to use the semiclassical expansion for the partition function. For a spherical HO, for $\hbar\omega\beta \gg 1$, show by direct expansion of the exact partition function that

$$\tilde{Z}_{ETF}(\beta) = \frac{1}{(\hbar\omega\beta)^3} + \frac{3}{2}\frac{1}{(\hbar\omega\beta)^2} + \frac{1}{\hbar\omega\beta} + \cdots \tag{4.236}$$

Compare this with Z_{ETF} by expanding Eq. (3.67) of the text for $D = 3$. Note that the corrections to the TF density of states are quite different in the two cases.
(c) Show that the ground state population for a temperature $T \le T_c$ is given by (with the Boltzmann constant $k_B = 1$)

$$N_0 = N - \left(\frac{T}{\hbar\omega}\right)^3 \sum_{l=1}^{\infty} \frac{e^{\beta l\tilde{\mu}}}{l^3} - \frac{3}{2}\left(\frac{T}{\hbar\omega}\right)^2 \sum_{l=1}^{\infty} \frac{e^{\beta l\tilde{\mu}}}{l^2} - \cdots \tag{4.237}$$

Note that one obtains a cubic equation for T_c by demanding that $N_0 = 0$ at this temperature, with $\tilde{\mu} = 0$:

$$N - T_c^3\,\zeta(3) - \frac{3}{2}T_c^2\,\zeta(2) = 0\,, \tag{4.238}$$

where the zeta function $\zeta(n) = \sum_{l=1}^{\infty} l^{-n}$ is used. Define $\delta T_c = (T_c - T_c^{(0)})$, and show using the above equations that

$$\frac{\delta T_c}{T_c^{(0)}} \simeq -\frac{\zeta(2)}{2[\zeta(3)]^{2/3}}\, N^{-1/3}\,. \tag{4.239}$$

Bibliography

[1] E. Wigner, Phys. Rev. **40**, 749 (1932).

[2] J. G. Kirkwood, Phys. Rev. **44**, 31 (1933).

[3] B. K. Jennings, Ann. Phys. (N. Y.) **84**, 1 (1974).

[4] V. M. Strutinsky, Yad. Fiz. **3**, 614 (1966) [Sov. J. Nucl. Phys. **3**, 449 (1966)]; Ark. Fys. **36**, 629 (1966); Nucl. Phys. **A 95**, 420 (1967).

[5] V. M. Strutinsky, Nucl. Phys. **A 122**, 1 (1968).

[6] R. K. Bhaduri and C. K. Ross, Phys. Rev. Lett. **27**, 606 (1971).

[7] M. Abramowitz and I. A. Stegun: *Handbook of Mathematical Functions* (Dover Publications, New York, 9th printing, 1970).

[8] L. H. Thomas, Proc. Camb. Phil. Soc. **23**, 195 (1926).

[9] E. Fermi, Z. Phys. **48**, 73 (1928).

[10] N. H. March: *Self-consistent Fields in Atoms* (Pergamon Press, Oxford, 1975).

[11] H. Yukawa, Proc. Phys. Math. Soc. of Japan **17**, 48 (1935).

[12] D. M. Brink: *Nuclear Forces* (Pergamon Press, Oxford, 1965).

[13] R. Serber, Phys. Rev. **C 14**, 718 (1976).

[14] T. H. R. Skyrme, Phil. Mag. **1**, 1043 (1956).

[15] W. Kohn and L. J. Sham, Phys. Rev. **A 137**, 1697 (1965); **140**, 1133 (1965).

[16] P. A. M. Dirac: *The Principles of Quantum Mechanics* (The Clarendon Press, Oxford, 1958, Fourth ed.), pp. 130-135.

[17] B. Grammaticos and A. Voros, Ann. Phys. (N. Y.) **123**, 359 (1979); **129**, 153 (1980).

[18] P. M. Morse: *Thermal Physics* (W. A. Benjamin, New York, 1969, Second ed.), p. 204.

[19] P. Ring and P. Schuck: *The Nuclear Many Body Problem* (Springer Verlag, New York, 1980).

[20] K. Husimi, Prog. Phys. Math. Soc. Jap. **22**, 264 (1940).

[21] Hai-Woong Lee, Phys. Rep. **259**, 147 (1995).

[22] N. L. Balazs and B. K. Jennings, Phys. Rep. **104**, 347 (1984).

[23] B. K. Jennings, Ph.D. thesis, McMaster University (1976, unpublished).

[24] B. K. Jennings, R. K. Bhaduri and M. Brack, Phys. Rev. Lett. **34**, 228 (1975); Nucl. Phys. **A 253**, 29 (1975).

[25] R. K. Pathria: *Statistical Mechanics* (Pergamon Press, New York, 1972), p. 131.

[26] M. Brack, C. Guet, and H.-B. Håkansson, Phys. Rep. **123**, 275 (1985).

[27] M. Brack and B. K. Jennings, Nucl. Phys. **A 258**, 264 (1976).

[28] C. F. v. Weizsäcker, Z. Phys. **96**, 431 (1935).

[29] H. A. Bethe and R. F. Bacher, Rev. Mod. Phys. **8**, 82 (1936).

[30] W. D. Myers and W. J. Swiatecki, Ann. Phys. (N. Y.) **55**, 395 (1969).

[31] L. Wilets, Rev. Mod. Phys. **30**, 542 (1958).

[32] P. Hohenberg and W. Kohn, Phys. Rev. **136**, B864 (1964).

[33] R. O. Jones and O. Gunnarsson, Rev. Mod. Phys. **61**, 689 (1989).

[34] C. H. Hodges, Can. J. Phys. **51**, 1428 (1973).

[35] M. Brack, B. K. Jennings, and Y. H. Chu, Phys. Lett. **65B**, 1 (1976).

[36] C. Guet and M. Brack, Z. Phys. **A 297**, 247 (1980).

[37] D. R. Murphy and W. P. Wang, J. Chem. Phys. **72**, 429 (1980).

[38] R. M. Dreizler and E. K. U. Gross: *Density Functional Theory* (Springer-Verlag, Berlin, 1990).

[39] M. Centelles, X. Viñas, M. Barranco, and P. Schuck, Ann. Phys. (N. Y.) **221**, 165 (1993).

[40] J. Caro, E. Ruiz Arriola, and L. L. Salcedo, J. Phys. G 22, 981 (1996).

[41] B. K. Jennings, Phys. Lett. **74 B**, 13 (1978).

[42] J. C. Slater, Phys. Rev. **81**, 385 (1951).

[43] J. W. Negele and D. Vautherin, Phys. Rev. **C 5**, 1472 (1972).

[44] R. K. Bhaduri, Phys. Rev. Lett. **39**, 329 (1977).

[45] M. Durand, M. Brack, and P. Schuck, Z. Phys. **A 286**, 381 (1978); *ibid.* **A 296**, 87 (1980).

[46] J. Bartel, M. Durand, and M. Brack, Z. Phys. **A 315**, 341 (1984).

[47] N. H. March, Phys. Lett. **64 A**, 185 (1977); and earlier references quoted therein.

[48] R. K. Bhaduri, M. Brack, H. Gräf, and P. Schuck, J. Physique – Lettres **41**, L-347 (1980).

[49] R. A. Berg and L. Wilets, Phys. Rev. **101**, 201 (1956).

[50] E. Engel and J. P. Perdew, Phys. Rev. **B 43**, 1331 (1991).

[51] M. Centelles, M. Pi, X. Viñas, F. Garcias, and M. Barranco, Nucl. Phys. **A 510**, 397 (1990).

[52] M. Brack, Phys. Rev. **B 39**, 3533 (1989).

[53] M. Seidl and M. Brack, Ann. Phys. (N. Y.) **245**, 275 (1996).

[54] C. Fiolhais and J. P. Perdew, Phys. Rev. **45**, 6207 (1992).

[55] W. Ekardt, Phys. Rev. **B 29**, 1558 (1984).

[56] O. Genzken and M. Brack, Phys. Rev. Lett. **67**, 3286 (1991).

[57] J. Bartel, P. Quentin, M. Brack, C. Guet, and H.-B. Håkansson, Nucl. Phys. **A 386**, 79 (1982).

[58] M. Brack, J. Damgård, A. S. Jensen, H. C. Pauli, V. M. Strutinsky, and C. Y. Wong, Rev. Mod. Phys. **44**, 320 (1972).

[59] M. Brack and P. Quentin, Nucl. Phys. **A 361**, 35 (1981).

[60] Y. Aboussir, J. M. Pearson, A. K. Dutta, and F. Tondeur, Nucl. Phys. **A 549**, 155 (1992), and earlier references quoted therein.

[61] Y. Aboussir, J. M. Pearson, A. K. Dutta, and F. Tondeur, At. Data Nucl. Data Tables **61**, 127 (1995).

[62] P. Möller, J. R. Nix, W. D. Myers, and W. J. Swiatecki, At. Data Nucl. Data Tables **59**, 185 (1995).

[63] M. Seidl, K.-H. Meiwes-Broer, and M. Brack, J. Chem. Phys. **95**, 1295 (1991).

[64] M. Brack, Rev. Mod. Phys. **65**, 677 (1993).

[65] A. Bohr and B. R. Mottelson: *Nuclear Structure, Vol. I* (W. A. Benjamin, New York, 1969), p. 151.

[66] J. Bartel, M. Brack, and M. Durand, Nucl. Phys. **A 445**, 263 (1985).

[67] B. Van der Pohl and H. Bremmer: *Operational Calculus* (Cambridge University Press, Cambridge, 1955).

[68] L. D. Landau and E. M. Lifshitz: *Statistical Physics* (Pergamon Press Ltd., London and Paris, 1958).

[69] E. C. Stoner, Phil. Mag. **28**, 257 (1939).

[70] F. J. Fernández Velicia, Phys. Rev. **A 30**, 1194 (1984).

[71] M. Onsi, A. M. Chaara, and J. M. Pearson, Z. Phys. **A 348**, 255 (1995).

[72] F. Perrot, Phys. Rev. **A 20**, 586 (1979).

[73] M. Brack, Phys. Rev. Lett. **53**, 119 (1984); Erratum: Phys. Rev. Lett. **54**, 851 (1985).

[74] C. Guet, E. Strumberger, and M. Brack, Phys. Lett. **205 B**, 427 (1988).

[75] M. Onsi, H. Przysiezniak, and J. M. Pearson, Phys. Rev. **C 55**, 3139 (1997).

[76] M. H. Anderson, J. R. Ensher, M. R. Matthews, C. E. Weiman, and E. A. Cornell, Science **269**, 198 (1995).

[77] K. B. Davis, M.-O. Mews, M. R. Andrews, N. J. van Druten, D. S. Durfee, D. M. Kurn, and W. Ketterle, Phys. Rev. Lett. **75**, 3969 (1995).

[78] C. C. Bradley, C. A. Sackett, J. J. Tollett, and R. G. Hulet, Phys. Rev. Lett. **75**, 1687 (1995);
C. C. Bradley, C. A. Sackett, and R. G. Hulet, Phys. Rev. Lett. **78**, 985 (1997).

[79] S. L. Cornish, N. R. Claussen, J. L. Roberts, E. A. Cornell, and C. E. Weiman, Phys. Rev. Lett. **85**, 1795 (2000);
J. L. Roberts *et al.*, Phys. Rev. Lett. **86**, 4211 (2001).

[80] K. Burnett, Contemp. Phys. **37**, 1 (1996).

[81] F. Dalfovo, S. Giorgini, L. P. Pitaevskii, and S. Stringari, Rev. Mod. Phys. **71**, 463 (1999).

[82] C. J. Pethick and H. Smith: *Bose-Einstein Condensation in Dilute Gases* (Cambridge University Press, Cambridge, U. K., 2002).

[83] K. Huang: *Statistical Mechanics* (John Wiley, New York, 1965), p. 275.

[84] L. P. Pitaevskii, Zh. Eksp. Teor. Fiz. **40**, 646 (1961) [Sov. Phys. JETP **13**, 451 (1961)]; E. P. Gross, Nuovo Cimento **20**, 454 (1961).

[85] M. Brack, Ph.D. thesis, Basel University (1972, unpublished); see also M. Brack and H.-C. Pauli, Nucl. Phys. **A 207**, 401 (1973).

[86] M. V. Berry and C. J. Howls, Proc. Math. Phys. Sci. **447 (A)**, 527 (1994); and related papers quoted therein.

[87] H. Weyl, Nachr. Akad. Wiss. Göttingen (1911), 110.

[88] R. Balian and C. Bloch, Ann. Phys. (N. Y.) **60**, 401 (1970); **63**, 592 (1971).

[89] R. Balian and C. Bloch, Ann. Phys. (N. Y.) **64**, 271 (1971); *Erratum:* **84**, 559 (1974).

[90] R. T. Waechter, Proc. Camb. Phil. Soc. **72**, 439 (1972).

[91] H. P. Baltes and E. R. Hilf: *Spectra of finite systems* (B-I Wissenschaftsverlag, Mannheim, 1976).

[92] D. L. Hill and J. A. Wheeler, Phys. Rev. **89**, 1102 (1953).

[93] F. H. Brownell, J. Math. Mech. **6**, 119 (1957).

[94] U. Smilansky and I. Ussishkin, J. Phys. **A 29**, 2587 (1996).

[95] K. Stewartson and R. T. Waechter, Proc. Camb. Phil. Soc. **69**, 353 (1971).

[96] D. B. Ray, quoted in J. P. McKean and I. M. Singer, J. Diff. Geom. **1**, 43 (1967).

[97] M. Kac, Am. Math. Monthly, **73** Pt. II, 1 (1966).

[98] C. Gordon, D. Webb, and S. Wolpert, Invent. Math. **110**, 1 (1992).

[99] S. Sridhar and A. Kudrolli, Phys. Rev. Lett. **72**, 2175 (1994).

[100] Hua Wu, D. W. L. Sprung, and J. Martorell, Phys. Rev. **E 51**, 703 (1995).

[101] U. Smilansky, Contemp. Phys. **34**, 297 (1993).

[102] C. K. Ross, Ph.D. thesis, McMaster University (1973, unpublished).

[103] R. S. Elliott: *Electromagnetics* (McGraw-Hill, New York, 1966), p. 300.

[104] K. M. Case and S. C. Chiu, Phys. Rev. **A 1**, 1170 (1970).

[105] J. I. Kapusta, Phys. Rev. **D 23**, 2444 (1981); B. K. Jennings and R. K. Bhaduri, Phys. Rev. **D 26**, 1750 (1981).

[106] J. Baacke and Y. Igarashi, Phys. Rev. **D 27**, 460 (1983); R. K. Bhaduri, Jishnu Dey and M. K. Srivastava, Phys. Rev. **D 31**, 1756 (1985); H. T. Elze and W. Greiner, Phys. Lett. **B 179**, 386 (1986).

[107] M. S. Berger and R. L. Jaffe, Phys. Rev. **D 40**, 2128 (1987); **44**, 566(E) (1991); I. Mardor and B. Svetitsky, Phys. Rev. **D 44**, 878 (1991).

[108] M. Brack, in: *Nuclear Structure Models*, eds. R. Bengtsson, J. Draayer, and W. Nazarewicz (World Scientific, Singapore, 1992), p. 165.

[109] M. Brack, in: *Physics and Chemistry of Fission 1979* (IAEA Vienna, 1980) Vol. I, p. 227.

[110] V. S. Ramamurthy, S. S. Kapoor, and S. K. Kataria, Phys. Rev. Lett. **25**, 386 (1990); Phys. Rev. **C 5**, 1124 (1972).

[111] R. Bengtsson, Nucl. Phys. **A 198**, 591 (1972).

[112] R. K. Bhaduri and S. Das Gupta, Phys. Lett. **47 B**, 129 (1973).

[113] C. K. Ross and R. K. Bhaduri, Nucl. Phys. **A 188**, 566 (1972).

[114] B. K. Jennings, Nucl. Phys. **A 207**, 538 (1973).

[115] F. A. Ivanyuk and V. M. Strutinsky, Z. Phys. **A 290**, 107 (1979), and earlier references quoted therein.

[116] A. Bohr and B. Mottelson: *Nuclear Structure, Vol. II* (Benjamin, New York, 1975).

[117] P. J. Siemens and A. Sobiczewski, Phys. Lett. **41 B**, 16 (1972).

[118] A. Sobiczewski, A. Gyurkovich, and M. Brack, Nucl. Phys. **A 289**, 346 (1977).

[119] M. Bolsterli, E. O. Fiset, J. R. Nix, and J. L. Norton, Phys. Rev. **C 5**, 1050 (1972); J. R. Nix, Ann. Rev. Nucl. Sci. **22**, 65 (1972).

[120] H. A. Jahn and E. Teller, Proc. Roy. Soc. (London) **A 161**, 220 (1937).

[121] N. Bohr and J. A. Wheeler, Phys. Rev. **56**, 426 (1939).

[122] S. Cohen and W. J. Swiatecki, Ann. Phys. (N. Y.) **22**, 406 (1963).

[123] W. D. Myers and W. J. Swiatecki, Nucl. Phys. **81**, 1 (1966).

[124] S. G. Nilsson, C. F. Tsang, A. Sobiczewski, Z. Szymański, S. Wycech, C. Gustafsson, I.-L. Lamm, P. Möller, and B. Nilsson, Nucl. Phys. **A 131**, 1 (1969).

[125] I. Ragnarsson, S. G. Nilsson, and R. K. Sheline, Phys. Rep. **45**, 1 (1978).

[126] S. M. Reimann, M. Brack, K. Hansen, Z. Phys. **D 28**, 235 (1993); S. M. Reimann, S. Frauendorf and M. Brack, Z. Phys. **D 34**, 125 (1995).

[127] C. Yannouleas and U. Landman, Phys. Rev. **B 51**, 1920 (1995).

[128] S. Frauendorf and V. V. Pashkevich, Ann. Phys. (Leipzig) **5**, 34 (1996).

[129] A. Vieira and C. Fiolhais, Z. Phys. **D 37**, 269 (1996).

[130] A. S. Tyapin, Sov. J. Nucl. Phys. **11**, 53 (1970); *ibid.* **14**, 50 (1972) [Yad. Fiz. **11**, 98 (1970) and **14**, 88 (1972)].

[131] D. H. E. Gross, Phys. Lett. **42 B**, 41 (1972).

[132] T. Szeredi and D. A. Goodings, Phys. Rev. **E 48**, 3518 (1993); *ibid.*, 3529 (1993).

[133] R. K. Bhaduri, M. V. N. Murthy, and M. K. Srivastava, Phys. Rev. Lett. **76**, 165 (1996).

[134] M. Brack and W. Stocker, Nucl. Phys. **A 388**, 230 (1982).

[135] W. J. Swiatecki, Proc. Phys. Soc. **A 64**, 226 (1951).

[136] W. J. Swiatecki, Proc. Phys. Soc. **A 68**, 285 (1955).

[137] M. Brack and W. Stocker, Nucl. Phys. **A 406**, 413 (1983).

[138] B. Grammaticos, Z. Phys. **A 312**, 99 (1983).

[139] M. Farine, Z. Phys. **A 320**, 337 (1985).

[140] Subhasis Sinha, Phys. Rev. **A 58**, 3159 (1998).

Chapter 5

Gutzwiller's trace formula for isolated orbits

We have pointed out in Chapter 1 that the existence of classical periodic orbits leads to shells in the density of states, and thus to shell effects in the total binding and separation energies and other observables of finite fermion systems. These shell effects are strongest in systems with a high symmetry, and thus with high degeneracies. But they can exist also, at least to some extent, in non-integrable systems – even ones that exhibit classical chaos – due to an intricate relation between the existence of periodic orbits and the quantization of energy levels that is independent of the integrability of a system. It emerges from the so-called periodic orbit theory that the shell effects are dominated by the shortest classical periodic orbits of the system. More precisely speaking: the periodic orbits with the smallest actions and the highest classical degeneracies are responsible for the gross-shell features of a spectrum such as the main periodicity of the bunching of energy levels into shells. If one wants to know the finer details of a spectrum, longer orbits with larger actions can become important, too. In fact, what this theory really does is to give a Fourier decomposition of the oscillating density of single-particle states in terms of classical periodic orbits, their actions, periods, and stabilities. Such representations of the level density are usually called "trace formulae", following Gutzwiller [1]. The quantitative determination of the amplitudes is, however, all but trivial. Furthermore, since this Fourier decomposition is the result of a semiclassical expansion that is *asymptotic* in nature, the summation of the trace is hampered with serious problems of convergence and makes the full quantization of non-integrable systems generally very difficult.

The periodic orbit theory was developed by Gutzwiller in a series of papers that culminated with his presentation in 1971 of a trace formula [1] that applies, however, only to systems with isolated orbits and thus is mostly of use for chaotic systems. Balian and Bloch presented in 1972 a similar formula which they developed for cavities with reflecting walls of arbitrary shape in two and more dimensions [2]. Their formula applies also to integrable systems, such as the spherical cavity studied in detail in Ref. [2]. Berry and Tabor derived in 1976 a trace formula for integrable systems with arbitrary smooth (or infinitely steep) potentials [3], taking the EBK quantization as a starting point, and thereafter rederived the same result from Gutzwiller's semiclassical Green's function [4]. We have already discussed their method in Sect. 2.7.

One of the ultimate goals of the periodic orbit theory is for many of its users the full semiclassical quantization of non-integrable systems. If the periodic orbit sum in the trace formula converges at all, it will have its poles at the position of the (approximate) quantum levels. This goal is, however, rarely achieved in realistic physical systems. Nevertheless, it has led to remarkable successes in the understanding of the quantum spectra of some classically chaotic systems [5, 6] (see also Sects. 1.4 and 5.6.1). Recent developments using partial resummations of the trace formula, leading to a dynamical zeta function, and employing so-called cycle expansions have brought further considerable success in the (partial) quantization of chaotic systems; some of them will be discussed in Chapter 7. To stay on the level of trace formulae, these have been further developed, by extending Gutzwiller's approach to include degenerate orbits [7, 8, 9], and can thus be formulated also for systems with continuous symmetries – including the fully integrable systems. These extensions of the Gutzwiller theory will be discussed in Chapter 6.

In this book, we have taken the view that one of the main strengths of the periodic orbit theory is that it enables one to obtain gross-shell effects in the level density of a quantum system semiclassically, i.e., without solving the Schrödinger equation, independently of its integrability or the possibility of a full quantization, and to interpret shell effects in terms of the interplay of the most important short periodic orbits. This will be a recurrent theme, and several nice manifestations of classical periodic orbits in various physical quantum systems will be discussed in Chapter 8. Some examples have also been presented in Chapter 1.

Although the method is, in principle, applicable to any interacting many-body system, we shall develop and discuss it, apart from illustrative and educational one-body problems given later in this chapter, mainly for physical systems that are treated in the mean-field approach leading to the motion of independent particles in a common average potential, which we furthermore assume to be local. For readers uninitiated in many-body theory, a short and elementary presentation of the mean-field approximation to many-fermion systems is given in Appendix A.

In this chapter, we derive Gutzwiller's trace formula for isolated orbits. We give its derivation in several steps going into quite some detail, because it is a central result of the periodic orbit theory, and the understanding of the main arguments and approximations used in the derivation give keys for further developments. The central result will be stated in Sect. 5.3. After some short discussions of the stability matrix and the Maslov index which appear in the trace formula (and whose more technical aspects are dealt with in the Appendices C and D), and of general convergence problems and ways to circumvent them, we mention in Sect. 5.6 several prominent applications, mostly to chaotic systems where the trace formula has been successfully applied. Finally, we present three examples which are more of an educational nature and easy to understand. In the last example, the Hénon-Heiles potential, we will face the basic problem of the transition from integrability to non-integrability which leads to a divergence. This problem will be dealt with in the second part of Chapter 6.

5.1 The semiclassical Green's function

We recall that the quantum-mechanical single-particle propagator $K(\mathbf{r}, \mathbf{r}'; t)$ defined in Eq. (3.23) of Sect. 3.2.1 fulfills the folding property, given already in Eq. (3.27):

$$K(\mathbf{r}, \mathbf{r}'; t) = \int K(\mathbf{r}, \mathbf{r}''; t - t'') K(\mathbf{r}'', \mathbf{r}'; t'') d^3 r''. \tag{5.1}$$

We also recall here that t above is the *time difference* needed for the particle to propagate from $\mathbf{r}'$ to $\mathbf{r}$. Eq. (5.1) means that a particle propagating from $\mathbf{r}'$ at time $t' = 0$ to $\mathbf{r}$ at time t may be viewed as passing through all possible intermediate points $\mathbf{r}''$ at times t'', over which one has to integrate when dividing the trajectory into two parts. This process can be iterated, by dividing a given trajectory into many small sectors, which eventually leads to Feynman's path integral formulation of quantum mechanics (see Sect. 1.1). If the potential $V(\mathbf{r})$ is sufficiently smooth, it may be neglected locally in a small time and space interval and the particle can be thought of propagating freely. This is the main idea of the semiclassical approximation to the propagator and the quantities derived from it. We therefore first have to know the propagator of a free particle.

In Eq. (3.22) of Sect. 3.2.1 we have given the classical (Thomas-Fermi) approximation to the Bloch function. If the potential V is set to zero, this is the *exact* result for a free particle in three dimensions. With the replacement $\beta \to i(t - t')/\hbar$, we obtain from it the propagator for a free particle in three dimensions. It can be generalized in a straightforward way to D dimensions and takes the form

$$K_0(\mathbf{r}, \mathbf{r}'; t) = \left(\frac{m}{2\pi i \hbar t}\right)^{D/2} \exp\left[\frac{i}{\hbar}\frac{m}{2t}(\mathbf{r} - \mathbf{r}')^2\right]. \tag{5.2}$$

It is easy to check that (5.2) fulfills the Schrödinger equation (3.26) for the free Hamiltonian $\hat{H}_0 = -\hbar^2 \nabla^2/2m$. Note that K_0 also fulfills Eq. (3.27).

We now observe the following: the exponent of K_0 in (5.2) is exactly equal to $i/\hbar$ times Hamilton's principal function $R_0(\mathbf{r}, \mathbf{r}', t)$ (1.21) for a free particle. Furthermore, if we take the determinant of the (negative) second variations of R_0, we obtain

$$\det\left(-\frac{\partial^2 R_0}{\partial r_i \partial r_j'}\right) = \left(\frac{m}{t}\right)^D. \tag{5.3}$$

We can therefore express the exact free propagator in D dimensions in terms of R_0 as

$$K_0(\mathbf{r}, \mathbf{r}'; t) = \left(\frac{1}{2\pi i \hbar}\right)^{D/2} \sqrt{\det\left(-\frac{\partial^2 R_0}{\partial \mathbf{r} \partial \mathbf{r}'}\right)} \exp\left[\frac{i}{\hbar}R_0(\mathbf{r}, \mathbf{r}', t)\right]. \tag{5.4}$$

Van Vleck [10] noticed the relation (5.4) already in 1928. In order to take into account the effect of the potential on the particle's motion, he proposed the following (semi-)classical approximation to K, which consists simply in replacing R_0 by the full principal function $R(\mathbf{r}, \mathbf{r}'; t)$ defined in Eq. (2.5)

$$K_V(\mathbf{r}, \mathbf{r}'; t) = (2\pi i \hbar)^{-D/2} \sqrt{\det C} \, \exp\left[\frac{i}{\hbar}R(\mathbf{r}, \mathbf{r}'; t)\right]. \tag{5.5}$$

We have here used the symbol C for the matrix of the negative second variations of R:

$$C_{ij}(\mathbf{r}, \mathbf{r}'; t) = -\frac{\partial^2 R}{\partial r_i \partial r'_j}. \qquad (i, j = 1, 2, \ldots, D) \qquad (5.6)$$

Gutzwiller [11] re-derived Van Vleck's approximation (5.5) from Feynman's path integral. He noticed that an extra phase $-\phi$ must be added to the exponent of K_V:

$$K_V(\mathbf{r}, \mathbf{r}'; t) \to (2\pi i \hbar)^{-D/2} \sqrt{\det C} \exp\left[\frac{i}{\hbar} R(\mathbf{r}, \mathbf{r}'; t) - i\phi\right]. \qquad (5.7)$$

The necessity for this extra phase arises when checking the folding relation (3.27) which must be fulfilled for any approximation to the propagator to make sense. In doing so, one uses the *stationary phase method*. Since we will use this method repeatedly below, we give here just a few qualitative arguments and state the result of Gutzwiller's analysis. In the space integral (3.27) over the exponential $\exp[(i/\hbar)R]$, the integrand usually oscillates violently since $\hbar$ is small compared to R. (This becomes particularly so when $\hbar$ formally is taken to zero in deriving the classical limit.) As a consequence, the contributions to the integral mostly cancel. The only paths along which the integrand varies slowly are the classical trajectories which make the integral of R stationary according to Hamilton's variational principle (2.6). Therefore, the integrations transverse to the classical paths may, to a good approximation, be performed using the method of steepest descent. This leads to Fresnel type integrals, as we shall demonstrate below in deriving the semiclassical Green's function. At all points where the determinant $\det C$ in (5.5) becomes zero or infinite – the so-called *conjugate points* discussed below – some phase changes occur in the Fresnel integrals which add to the phase ϕ. Finally, since the equations of motion may have several solutions corresponding to distinct classical paths, one has to add the contributions to the integral (3.27) from *all allowed classical paths*, hereby making use of the superposition principle.

The resulting approximation[1] of Gutzwiller is

$$K_{scl}(\mathbf{r}, \mathbf{r}'; t) = \sum_{class.paths} (2\pi i \hbar)^{-D/2} \sqrt{|\det C|} \exp\left[\frac{i}{\hbar} R(\mathbf{r}, \mathbf{r}'; t) - i\kappa\pi/2\right]. \qquad (5.8)$$

The phase index κ is obtained by counting the number of conjugate points along each trajectory from $\mathbf{r}'$ to $\mathbf{r}$ at the fixed transit time t. It is important hereby that the root of the *absolute value* of $\det C$ is used in (5.8). This is a consequence of a theorem established by M. Morse [12]: *The second variation of R, considered as a quadratic form in the displacements $\delta\mathbf{r}(t)$ of all the possible paths around a given trajectory from $\mathbf{r}'$ to $\mathbf{r}$ in the time t, has as many negative eigenvalues as there are conjugate points along the trajectory.*

In order to illustrate the geometrical meaning of the conjugate points, let us consider what happens if we change the initial momentum $\mathbf{p}'$ of a given trajectory $\mathbf{r}(t)$ by a small amount $\delta\mathbf{p}'$. The new trajectory then will spread out from the initial point $\mathbf{r}'$ like a

---

[1] This expression holds, in general, only when all classical paths going through a fixed point $\mathbf{r}$ are *isolated*. For problems which may arise when this is not the case, we refer to Chapter 6.

fan. At the time t, the end point $\mathbf{r}$ will have changed by the amount $\delta\mathbf{r}$ which is related to $\delta\mathbf{p}'$ by

$$\delta r_i = \sum_j \frac{\partial r_i}{\partial p'_j} \delta p'_j = \sum_j \left(\frac{\partial p'_j}{\partial r_i}\right)^{-1} \delta p'_j = \sum_j \left(C^{-1}\right)_{ij} \delta p'_j, \qquad (5.9)$$

where the matrix C_{ij} is given in Eq. (5.6) and use of Eq. (1.26) has been made. If at some time t_c the entire fan of new trajectories crosses in one point, we call this a conjugate time and the point where they cross a *conjugate point*. At such a point, the determinant of C^{-1} becomes zero, i.e., at least one of the eigenvalues of C changes its sign.

From the above one can answer the question that interested students of a first-grade course on classical mechanics may ask: When applying Hamilton's variation principle (2.6), is R a minimum or a maximum? Or, phrasing it in standard terms: is the "principle of least action" really one of *least* action or could it also be one of *largest* action? The answer is that, indeed, the action R is a minimum only as long as the particle does not run into a conjugate point along its trajectory. Then, the second derivative of R changes sign (unless it happens to be a double conjugate point), and the minimum is turned into a maximum.[2]

The next step is the conversion from a fixed transit time t to a fixed energy E. This is achieved by taking the Fourier transform (3.28) of the approximate propagator K_{scl} in Eq. (5.8) and leads to a semiclassical approximation of the Green's function (3.29). The (half-sided) Fourier transform involves a time integral

$$\int_0^\infty \sqrt{|\det C|} \exp\left\{\frac{i}{\hbar}\left[R(\mathbf{r},\mathbf{r}';t) + Et\right]\right\} dt, \qquad (5.10)$$

which is again performed by the method of stationary phases. We will explain this method at this place, as it is the most simple one-dimensional example. The exponent of the integrand is rapidly oscillating since $\hbar$ is small compared to R at most times; therefore most of the contributions cancel. The largest contribution comes from the time interval around the *stationary point* t_0 of the exponent, which is given by $-\partial R/\partial t|_{t_0} = E$. We thus expand up to second order in $(t - t_0)$:

$$R(\mathbf{r},\mathbf{r}';t) + Et = R(\mathbf{r},\mathbf{r}';t_0) + Et_0 + \frac{1}{2}(t - t_0)^2 \left.\frac{\partial^2 R}{\partial t^2}\right|_{t_0} + \dots \qquad (5.11)$$

Inserting this into (5.10) and assuming that $\det C$ is a slowly varying function of t, so that we take it outside the integral at the value of t_0, we find

$$\sqrt{|\det C(t_0)|}\, e^{iS/\hbar} \int_{-t_0}^\infty e^{i\alpha y^2}\, dy\,, \qquad \alpha = \frac{1}{2\hbar}\left.\frac{\partial^2 R}{\partial t^2}\right|_{t_0}. \qquad (5.12)$$

For large enough (positive) t_0, we may replace the integral above by a Fresnel integral of the type given in Eq. (2.169) of Sect. 2.7. Note that an extra phase $\pi/2$ occurs for

---

[2]We point out that in the *Maupertuis principle of least action*, based on the early works of de Maupertuis [13] and Euler [14] in 1744, the action $S(\mathbf{r},\mathbf{r}';E)$ (5.13) is varied: $\delta S = 0$. For detailed discussions of the subtle differences between Maupertuis' and Hamilton's variational principles, we refer to Gutzwiller [5, 15] and a very recent article [16].

each sign change of α; this is important for obtaining the overall phase in the result below. In (5.12), we have used Eq. (2.8) to replace $R(\mathbf{r}, \mathbf{r}'; t_0) + Et_0$ by the action $S(\mathbf{r}, \mathbf{r}'; E)$:

$$S(\mathbf{r}, \mathbf{r}'; E) = \int_{\mathbf{r}'}^{\mathbf{r}} \mathbf{p}(\mathbf{r}'', E) \cdot d\mathbf{r}'' = \sum_i \int_{r_i'}^{r_i} p_i(\mathbf{r}'', E)\, dr_i''. \qquad (5.13)$$

Using Eq. (1.25), one finds

$$\frac{\partial^2 R}{\partial t^2} = -\frac{\partial E}{\partial t} = -\left(\frac{\partial^2 S}{\partial E^2}\right)^{-1}. \qquad (5.14)$$

This leads finally to Gutzwiller's semiclassical approximation to the Green's function:

$$G_{scl}(\mathbf{r}, \mathbf{r}'; E) = \sum_{class.traj.} 2\pi(2\pi i\hbar)^{-(D+1)/2}\sqrt{|\mathcal{D}|}\, \exp\left[\frac{i}{\hbar} S(\mathbf{r}, \mathbf{r}', E) - i\mu\pi/2\right], \qquad (5.15)$$

where

$$\mathcal{D}(\mathbf{r}, \mathbf{r}'; E) = (-1)^D \frac{\partial^2 S}{\partial E^2} \det C = \begin{vmatrix} \dfrac{\partial^2 S}{\partial \mathbf{r}'\partial \mathbf{r}} & \dfrac{\partial^2 S}{\partial \mathbf{r}'\partial E} \\[2mm] \dfrac{\partial^2 S}{\partial E \partial \mathbf{r}} & \dfrac{\partial^2 S}{\partial E^2} \end{vmatrix} \qquad (5.16)$$

is the determinant of a $(D+1)\times(D+1)$ matrix. Like for the propagator (5.8), this approximation for the Green's function is only valid if all trajectories passing through a fixed point $\mathbf{r}$ are isolated. If this is not the case, alternate approximations discussed in Sects. 6.1.1 and 6.1.4 must be used.

In order to demonstrate the equality of the middle and right-hand sides in Eq. (5.16), we first note that the determinant of C in Eq. (5.6) can be rewritten, using Eq. (1.26), as a Jacobian determinant:

$$\det C = \det \left(\frac{\partial \mathbf{p}'}{\partial \mathbf{r}}\right)_{t,\mathbf{r}'}, \qquad (5.17)$$

where the subscripts indicate the variables to be kept constant. Similarly, the final expression (5.16) for $\mathcal{D}$ may, using Eqs. (1.24, 1.25), be written as a Jacobian determinant:

$$\mathcal{D} = (-1)^D \begin{vmatrix} \dfrac{\partial \mathbf{p}'}{\partial \mathbf{r}} & \dfrac{\partial \mathbf{p}'}{\partial E} \\[2mm] \dfrac{\partial t}{\partial \mathbf{r}} & \dfrac{\partial t}{\partial E} \end{vmatrix} = (-1)^D \det \left(\frac{\partial(\mathbf{p}', t)}{\partial(\mathbf{r}, E)}\right)_{\mathbf{r}'}. \qquad (5.18)$$

Straightforward matrix manipulation and the use of standard properties of determinants now allows to transform (5.18) into

$$\mathcal{D} = (-1)^D \det \left(\frac{\partial(\mathbf{p}', t)}{\partial(\mathbf{r}, t)} \frac{\partial(\mathbf{r}, t)}{\partial(\mathbf{r}, E)}\right)_{\mathbf{r}'} = (-1)^D \det \left(\frac{\partial \mathbf{p}'}{\partial \mathbf{r}}\right)_{t,\mathbf{r}'} \left(\frac{\partial t}{\partial E}\right)_{\mathbf{r}',\mathbf{r}}, \qquad (5.19)$$

which, using (5.17) and the inverse of the right-hand side in Eq. (5.14), leads back to the expression in the middle of Eq. (5.16).

To obtain the overall phase index[3] μ in (5.15), one has to count the number of sign changes of the eigenvalues of the matrix of second derivatives of S in Eq. (5.16) along each classical trajectory *at fixed energy* E. [Note that to get the index κ in Eq. (5.8), one had to do the counting at fixed time t. In many cases such as billiards and cavities, μ is actually identical with κ; see Ref. [11] for a detailed discussion of this point.] Again, this way of counting the phases must be done consistently together with the use of the *absolute value* of $\mathcal{D}$ in Eq. (5.15).

In Fig. 5.1 we show as an illustration the conjugate points along a periodic orbit of a particle in the two-dimensional Hénon-Heiles potential discussed in Sect. 5.6.4 below. The periodic orbit of circular type is shown by the solid line. Starting from the point $\mathbf{r}_0$ (marked by a cross) in the positive x direction with a zero slope, the particle returns to the same point with the same momentum and retraces this orbit periodically. Starting with angles ± 1 degree to the horizontal gives the dashed and dotted orbits, relatively, which are no longer periodic. The three orbits intersect at the three conjugate points labeled $\mathbf{r}_1$, $\mathbf{r}_2$, and $\mathbf{r}_3$. (The fact that they do not exactly intersect in the same point $\mathbf{r}_3$ in the figure is due to the instability of this orbit; had we taken smaller variations of the starting direction, the points of intersection would tend to coincide exactly.) The phase index μ appearing in the semiclassical Green's function (5.15) along the unperturbed periodic orbit is therefore $\mu = 3$.

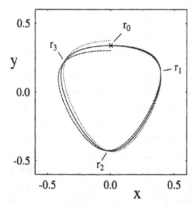

Figure 5.1: Conjugate points $\mathbf{r}_i$ (i=1,2,3) along the periodic orbit C in the Hénon-Heiles potential (cf. Sect. 5.6.4) at fixed energy. The starting point is $\mathbf{r}_0$. Note that the conjugate points approximately coincide here with the turning points of the radial motion. This gives a phase index μ=3 in the Green's function (5.15).

The physical interpretation of the phases occurring at the conjugate points is the following (we quote from Gutzwiller [17]): *In each conjugate point, the trajectory of the particle is tangent to a focal surface or focal line, which represents a boundary for those classical trajectories which start with the energy E at a given point* $\mathbf{r}'$. *Quantum-*

---

[3]Note that the index μ here is different from that appearing in the WKB and EBK quantization of integrable systems (see Chapter 2).

mechanically, however, the particle is allowed to tunnel beyond this boundary. The net effect of this penetration into the classically inaccessible region is a loss of phase. It is as if the particle had been able to penetrate by an eighth of a wavelength before being reflected back into the classically allowed region by an infinitely hard well.

The number μ of conjugate points along a closed orbit at fixed energy E plays an important role in the determination of the Maslov index σ appearing in the trace formula (3.42) which we shall derive below. Note that for the motion of free particles there are no conjugate points, since their trajectories cannot cross, and C becomes m/t times the unit matrix due to Eq. (1.21). Also, in systems with a high symmetry, such as harmonic-oscillator or Coulomb potentials, the periodic orbits come in degenerate families; in such cases Gutzwiller's trace formula does not apply and different methods must be used (see Chapter 6).

5.2   Taking the trace of $G_{scl}(\mathbf{r}, \mathbf{r}'; E)$

The final step for obtaining the level density is to take the trace of $G_{scl}(\mathbf{r}, \mathbf{r}'; E)$, according to Eqs. (3.34,3.36). In performing the integrations over $\mathbf{r}' = \mathbf{r}$, we use again the method of stationary phases. To get the stationary points $\mathbf{r}_0$ of the exponent in (5.15), we use Eq. (1.24) to write

$$\left[\frac{\partial S}{\partial \mathbf{r}}\right]_{\mathbf{r}_0} = \left[\frac{\partial S(\mathbf{r}, \mathbf{r}', E)}{\partial \mathbf{r}'} + \frac{\partial S(\mathbf{r}, \mathbf{r}', E)}{\partial \mathbf{r}}\right]_{\mathbf{r}'=\mathbf{r}=\mathbf{r}_0} = \mathbf{p} - \mathbf{p}' = 0. \qquad (5.20)$$

Thus, when taking the trace of G_{scl} by the stationary phase method, one has both $\mathbf{r}' = \mathbf{r}$ and $\mathbf{p}' = \mathbf{p}$. Therefore both coordinates and momenta of the particle at the initial and final points are equal and thus, only the *periodic closed orbits* from the sum over classical trajectories will essentially contribute to Eq. (5.15). The contributions from all other trajectories tend to cancel by destructive interference.

Mathematically speaking, one also obtains contributions from paths that reach from $\mathbf{r}'$ to $\mathbf{r}$ in zero time, often referred to in the literature as "zero length orbits". Although there are no classical orbits of length zero, these contributions can be shown [2, 18] to yield the smooth part of the level density which is identical to that obtained in the Thomas-Fermi approximation (see also Sect. 2.7). We shall omit these contributions here and concentrate on the oscillating part of the level density. We refer to Chapter 4 for the Thomas-Fermi model and its extensions.

For each periodic orbit, the D-dimensional trace integration according to Eq. (3.36) is now done in a "local" coordinate system

$$\mathbf{r} = (q, r_{\perp 1}, r_{\perp 2}, \ldots, r_{\perp (D-1)}) = (q, \mathbf{r}_\perp) \qquad (5.21)$$

whose first variable q is chosen along the orbit, and the vector $\mathbf{r}_\perp$ of the remaining D–1 variables is transverse to it and equals zero on the orbit. It is furthermore defined such that the Jacobian for this coordinate transformation is unity, so that $d^D r = dq\, d^{D-1} r_\perp$. In this special coordinate system, the determinant $\mathcal{D}$ in (5.16) takes a particularly simple form. In order to show this, we first note that the upper left sub-determinant in the right-hand side of (5.16) is identically zero:

$$\left|\frac{\partial^2 S}{\partial \mathbf{r}' \partial \mathbf{r}}\right| = 0. \tag{5.22}$$

This is easily seen by writing down the energy conservation in the form of the Hamilton-Jacobi equation (2.55)

$$H(\mathbf{r}, \mathbf{p}) = H\left(\mathbf{r}, \frac{\partial S}{\partial \mathbf{r}}\right) = E, \tag{5.23}$$

and taking its partial derivative with respect to r_i', leading to

$$\frac{\partial}{\partial r_i'} H(\mathbf{r}, \mathbf{p}) = \sum_j \frac{\partial H}{\partial p_j} \frac{\partial^2 S}{\partial r_i' \partial r_j} = \sum_j \dot{r}_j \frac{\partial^2 S}{\partial r_i' \partial r_j} = 0, \tag{5.24}$$

where we have used the Hamilton equations of motion (2.12) to introduce the velocity vector $\dot{\mathbf{r}}$. Since this vector is not identically zero everywhere, equation (5.24) tells us that the matrix of mixed second derivatives of S must be singular, and hence its determinant is zero as given in (5.22). In the local coordinate system (5.21) of a given orbit, now, the velocity vector by definition has the components $(\dot{q}, 0, 0, ...)$. It follows then from Eq. (5.24) that the first column and the first row of the determinant in (5.22), i.e., all those elements in which one of the derivatives is taken with respect to q or to q', also vanish. In evaluating the determinant $\mathcal{D}$ (5.16) and making use of Eq. (5.22), the only nonzero contribution then gives

$$\mathcal{D} = -\frac{\partial^2 S}{\partial E \partial q} \frac{\partial^2 S}{\partial E \partial q'} \left|\frac{\partial^2 S}{\partial \mathbf{r}'_\perp \partial \mathbf{r}_\perp}\right| = (-1)^D \frac{1}{\dot{q}\dot{q}'} \det\left(\frac{\partial \mathbf{p}'_\perp}{\partial \mathbf{r}_\perp}\right). \tag{5.25}$$

To get the expression on the right-hand side, we have again used some of the identities in Eqs. (1.24,1.25), and denoted the momentum in the direction transverse to the orbit by $\mathbf{p}_\perp$. This expression for $\mathcal{D}$ separates in the coordinates q, q' and the transverse ones, which facilitates the trace integral.

As we have just seen, the stationary condition (5.20) allows us to select only the periodic orbits at a given energy E. Thus, for each of these orbits, the local coordinate system $(q, \mathbf{r}_\perp)$ defined in (5.21) can be identified with one particular sequence of stationary points $\mathbf{r}_0(t)$. Therefore, the integration *along* the orbit, i.e., over q must be performed exactly since the saddle-point approximation along this direction would fail. For the trace integral over the remaining $D-1$ transverse coordinates $\mathbf{r}_\perp$, we now expand the action S in the exponent of (5.15) around the stationary path $\mathbf{r}_0$ up to second order in $\mathbf{r}_\perp$ and $\mathbf{r}'_\perp$:

$$S(\mathbf{r}, \mathbf{r}', E) = S_{po}(E) + \frac{1}{2} \sum_{i,j=1}^{D-1} W_{ij}(q) r_{\perp i} r_{\perp j}, \tag{5.26}$$

where $S_{po}(E)$ is the action integral along the periodic orbit (which may include multiple retracing, i.e., repeated runs through the same primitive periodic orbit)

$$S_{po}(E) = \oint_{po} \mathbf{p} \cdot d\mathbf{r} = \oint p_q \, dq, \tag{5.27}$$

and W_{ij} is the $(D{-}1)\times(D{-}1)$ matrix of mixed second derivatives of $S(\mathbf{r}, \mathbf{r}', E)$ transverse to the periodic orbit, evaluated on the orbit itself. In vector notation, it reads

$$W(q) = \left(\frac{\partial^2 S}{\partial \mathbf{r}_\perp \partial \mathbf{r}_\perp} + \frac{\partial^2 S}{\partial \mathbf{r}_\perp \partial \mathbf{r}'_\perp} + \frac{\partial^2 S}{\partial \mathbf{r}'_\perp \partial \mathbf{r}_\perp} + \frac{\partial^2 S}{\partial \mathbf{r}'_\perp \partial \mathbf{r}'_\perp} \right)_{\mathbf{r}_\perp = \mathbf{r}'_\perp = 0}. \qquad (5.28)$$

Using the saddle point approximation for each of the transverse directions yields a Fresnel integral for each of them (see Sect. 2.7); hereby we may assume the term $|\mathcal{D}|^{1/2}$ to vary smoothly so that it can be taken outside the $\mathbf{r}_\perp$ integral and evaluated along the orbit. Combining all Fresnel integrals leads to the inverse root of $\det|W(q)|$ and an extra phase $-i\nu\pi/2$ in the exponent, where ν is the number of negative eigenvalues of the matrix W (5.28), including the phase contribution from the pre-factor $i^{-(D+1)/2}$ in the Green's function (5.15). Thus, for each periodic orbit we obtain a one-dimensional integral over q contributing to the trace of the Green's function

$$\delta \operatorname{tr} G_{scl}(E) \;\propto\; e^{i[S_{po}/\hbar - (\mu+\nu)\pi/2]} \int |\mathcal{D}(q)|^{1/2} \, |\det W(q)|^{-1/2} \, dq \qquad (5.29)$$

where $\mathcal{D}(q)$ is given by Eq. (5.25), but evaluated along the orbit, i.e., at $\mathbf{r}_\perp = \mathbf{r}'_\perp = 0$, with $q = q'$. The overall phase factor can be taken outside the integral, since the action S_{po} is constant, and the same can also be shown for the sum of the phase indices $\mu + \nu$, as we will discuss in the next section.

Employing the canonical relations (1.24), we next can express the determinant of $W(q)$ in terms of a Jacobian:

$$\begin{aligned} \det W(q) &= \det \left(\frac{\partial \mathbf{p}_\perp}{\partial \mathbf{r}_\perp} - \frac{\partial \mathbf{p}'_\perp}{\partial \mathbf{r}_\perp} + \frac{\partial \mathbf{p}_\perp}{\partial \mathbf{r}'_\perp} - \frac{\partial \mathbf{p}'_\perp}{\partial \mathbf{r}'_\perp} \right)_{\mathbf{r}_\perp = \mathbf{r}'_\perp = 0} \\ &= -\det \left(\frac{\partial(\mathbf{p}_\perp - \mathbf{p}'_\perp, \mathbf{r}_\perp - \mathbf{r}'_\perp)}{\partial(\mathbf{r}_\perp, \mathbf{r}'_\perp)} \right)_{\mathbf{r}_\perp = \mathbf{r}'_\perp = 0}. \end{aligned} \qquad (5.30)$$

(The correctness of this equality is seen by straight-forward evaluation of the Jacobian determinant.) The modulus of $\mathcal{D}(q)$, using Eq. (5.25), can similarly be written as

$$|\mathcal{D}(q)| = \frac{1}{\dot{q}^2} \left| \frac{\partial \mathbf{p}'_\perp}{\partial \mathbf{r}_\perp} \right|_0 = \frac{1}{\dot{q}^2} \left| \frac{\partial(\mathbf{p}'_\perp, \mathbf{r}'_\perp)}{\partial(\mathbf{r}_\perp, \mathbf{r}'_\perp)} \right|_0 \qquad (5.31)$$

where the index 0 stands for $\mathbf{r}_\perp = \mathbf{r}'_\perp = 0$ from now on. Now, when taking the ratio of $|\mathcal{D}(q)|$ and $|\det W(q)|$, the denominators of the two above Jacobians cancel. Changing the order of arguments in the remaining Jacobian, the complete denominator under the integral (5.29) therefore becomes

$$|\mathcal{D}(q)|^{-1/2} |\det W(q)|^{1/2} = |\dot{q}| \left| \frac{\partial(\mathbf{r}_\perp - \mathbf{r}'_\perp, \mathbf{p}_\perp - \mathbf{p}'_\perp)}{\partial(\mathbf{r}'_\perp, \mathbf{p}_\perp)} \right|_0^{1/2}. \qquad (5.32)$$

Apart from the factor $\dot{q}$, this quantity can be shown to be an invariant property of the periodic orbit, and thus independent of q. To see this most elegantly,[4] we

---

[4] We are grateful to S. Creagh who showed us the way to this shortcut.

introduce a $(2D{-}2)$-component phase space vector $\boldsymbol{\xi}_\perp$, analogous to (2.117), consisting of the transverse components of the radial and the momentum vector:

$$\boldsymbol{\xi}_\perp = (\mathbf{r}_\perp, \mathbf{p}_\perp). \tag{5.33}$$

The determinant in (5.32) then becomes simply $|\partial(\boldsymbol{\xi}_\perp - \boldsymbol{\xi}'_\perp)/\partial\boldsymbol{\xi}'_\perp|_0$. Remember that in (5.26) we have made a Taylor expansion of the action for *small values* of the transverse coordinates $\mathbf{r}, \mathbf{r}'$ near the periodic orbit, and that $\mathbf{r}'$ and $\mathbf{r}$ refer to the starting and the end points of the periodic orbit, respectively, which are taken to be identical under the trace integral. The determinant $|\partial(\boldsymbol{\xi}_\perp - \boldsymbol{\xi}'_\perp)/\partial\boldsymbol{\xi}'_\perp|_0$ can therefore directly be related to the so-called *stability matrix* $\widetilde{M}$ of the orbit, which has the dimension $2D{-}2$ and will be discussed in more detail in Sect. 5.4 below. The relation is simply [cf. Eqs. (5.40,5.45)]

$$\left|\frac{\partial(\mathbf{r}_\perp - \mathbf{r}'_\perp, \mathbf{p}_\perp - \mathbf{p}'_\perp)}{\partial(\mathbf{r}'_\perp, \mathbf{p}'_\perp)}\right|_0 = \left|\frac{\partial(\boldsymbol{\xi}_\perp - \boldsymbol{\xi}'_\perp)}{\partial\boldsymbol{\xi}'_\perp}\right|_0 = |\det(\widetilde{M}_{po} - I)|, \tag{5.34}$$

where I is the $(2D{-}2)\times(2D{-}2)$ unit matrix. Since $\widetilde{M}$ is an invariant property of the classical periodic orbit, the above determinant can be taken outside the q integral (5.29). The only factor left under the integral (5.29) is then the inverse velocity $\dot{q}$ along the orbit, so that the integral simply becomes

$$\int \frac{1}{\dot{q}}\, dq = \int_{ppo} dt = T_{ppo}. \tag{5.35}$$

Here T_{ppo} is the *period* of the *primitive periodic orbit*, that is the time it takes the particle to run through it once. Note that in the sum over periodic orbits to be done in the Green's function (5.15), repeated runs through the same primitive orbit have to be included, too. However, taking the trace integral in the local coordinate system of the periodic orbit as $\int dq\, d^{D-1}r_\perp$, the coordinate q may run only once along the orbit.

Before summing up all the contributions into the famous trace formula of Gutzwiller, we should discuss a fundamental problem occurring in the above development. In order for (5.30) to be nonzero on a continuous set of points, the orbits must be *isolated* in both coordinate space and momentum space. This means that at fixed energy E, there may not exist any neighboring orbits with identical classical actions that can be obtained by a continuous deformation, displacement, or rotation of the orbit. This happens, actually, for most integrable systems where there exist entire families of degenerate orbits connected through one or more continuous parameters. Examples are given in the presence of continuous spatial symmetries, but also of special dynamical symmetries, e.g. in harmonic-oscillator potentials or the Coulomb potential. For such systems, the stationary-phase integrations may not be done in all transverse directions to the orbit, since the classical action then is identically constant and thus the second derivatives appearing in Eq. (5.30) are identically zero in a continuous domain of the space. An exception is given for anisotropic harmonic-oscillator potentials with incommensurable frequencies; they have only isolated orbits and can thus be treated in the way that follows (cf. the example given in Sect. 5.6.2 below.)

This problem arises actually not only at the stage of taking the trace of the semiclassical Green's function. In some cases with high symmetry, already the semiclassical

approximation to the propagator in Eq. (5.8) may become singular along the most degenerate orbits. Remember that in order to arrive at the expression (5.8) with the
correct phases, Gutzwiller started from the folding relation (3.27) of the propagator
and performed the spatial integrals using the stationary phase approximation. For
highly degenerate orbits, this approximation fails and must, at least in some of the
spatial directions, be performed exactly instead [2, 3, 7]. This point will be clarified
explicitly in Chapter 6, where we discuss extensions of Gutzwiller's approach and their
relations to alternative formulations of the periodic orbit theory.

We must therefore exclude systems with degenerate orbits from the present development of Gutzwiller's theory and assume for the following that all periodic orbits of
the system are *isolated*. The determinant in (5.30) then stays finite except in isolated
points at which some of its eigenvalues change sign; these isolated points are of zero
measure in the trace integral and thus do no harm. Still, divergences can also occur
in connection with bifurcations of stable periodic orbits; this point will be discussed at
the end of the following section.

5.3   The trace formula for isolated orbits

After summing over all periodic orbits in (5.15) and including all constants and phases
in the definition (3.36) of the level density, we arrive at Gutzwiller's *trace formula* [1]
for the oscillating part of the level density of a system which has only *isolated periodic
orbits*:

$$\delta g_{scl}(E) = \frac{1}{\hbar\pi} \sum_{po} \frac{T_{ppo}}{\sqrt{|\det(\widetilde{M}_{po} - I)|}} \cos\left(\frac{1}{\hbar}S_{po} - \sigma_{po}\frac{\pi}{2}\right). \tag{5.36}$$

Here *po* denotes a periodic orbit, $\widetilde{M}_{po}$ is its stability matrix as discussed further below,
S_{po} is its action as defined in (5.27), $T_{ppo} = \partial S_{ppo}/\partial E$ is the period of the corresponding
primitive periodic orbit, and σ_{po} is the so-called *Maslov index* which is just the sum of
the two phase indices occurring in (5.29):

$$\sigma_{po} = \mu_{po} + \nu_{po}. \tag{5.37}$$

We recall that μ_{po} is the phase index appearing in the semiclassical Green's function
G_{scl} in Eq. (5.15) and counts the number of conjugate points along the periodic orbit
at constant energy. The index ν_{po} arises when taking the trace of G_{scl} by saddle-
point integrations transverse to the periodic orbits and includes the phases coming
from the dimension-dependent pre-factor in (5.15). Creagh, Robbins, and Littlejohn
[19] have shown that it is only the sum (5.37) that is constant over each orbit and
is a canonical and topological invariant; μ_{po} alone (and thus ν_{po}) may depend on the
starting point along the orbit, or on the choice of coordinates. We discuss explicit
methods for the calculation of Maslov indices in Appendix D. The stability properties
of classical periodic orbits are discussed in the following section, together with the
main features of their stability matrix $\widetilde{M}$. The precise mathematical definition of the
so-called *monodromy matrix* M, whose non-trivial part is contained in $\widetilde{M}$, and of its
general properties are given in Appendix C.

The great formal achievement of the trace formula (5.36) is that it expresses the oscillating part of the quantum level density in terms of properties of *classical* orbits. The relation is semiclassical in the sense that the quantity $\hbar$ appears on the right-hand side, and that the approximation of stationary phase, which is closely related to the classical limit $S \gg \hbar$, has been used repeatedly in its derivation. The formula is thus asymptotic in nature, and serious problems occur in general when one tries to sum over all periodic orbits; this problem will be dealt with in Sect. 5.5 below. It is quite remarkable that the purely *quantum-mechanical* function $\delta g(E)$ on the left-hand side of (5.36), which contains the spectrum of the quantum levels of a given potential, is expressed on the right-hand side through quantities such as action, period, or stability, of a *classical particle* moving in the same potential. The Maslov indices σ_{po}, of course, contain certain elements of wave optics, but they are also derived from the classical trajectories of the particles alone. Another remarkable aspect is that the quantum spectrum of a system, within the limits of applicability of the trace formula, is determined only by the *periodic* orbits, although in chaotic systems, these cover only a minor part of the total available phase space (in the presence of "hard chaos" even a subspace of measure zero), whereas the rest of the phase space is dominated by all the nonperiodic orbits.

λ	$\sqrt{|\det(\widetilde{M}-I)|}$	$\sqrt{|\det(\widetilde{M}^n-I)|}$	$\mathrm{tr}\widetilde{M}$	nature of orbit	stability
$-e^{\pm\chi}$	$2\,\mathrm{Cosh}\chi/2$	$|e^{n\frac{\chi}{2}}-(-1)^n e^{-n\frac{\chi}{2}}|$	$-\infty < \mathrm{tr}\widetilde{M} < -2$	inv. hyperbolic	unstable
-1	2	$1-(-1)^n$	$\mathrm{tr}\widetilde{M} = -2$	inv. parabolic	$\sim$ stable
$e^{\pm i\chi}$	$2\,|\sin(\chi/2)|$	$2\,|\sin(n\chi/2)|$	$-2 < \mathrm{tr}\widetilde{M} < 2$	elliptic	stable
1	0	0	$\mathrm{tr}\widetilde{M} = 2$	dir. parabolic	$\sim$ stable
$e^{\pm\chi}$	$2\,|\mathrm{Sinh}(\chi/2)|$	$2\,|\mathrm{Sinh}(n\chi/2)|$	$2 < \mathrm{tr}\widetilde{M} < \infty$	dir. hyperbolic	unstable

Table 5.1: Eigenvalues λ (with real χ) of the stability matrix $\widetilde{M}$ for two-dimensional periodic primitive orbits of various nature. The second column gives the denominator of the trace formula (5.36) for a primitive orbit, the third column that of the n times repeated primitive orbit, and the fourth column gives the trace of $\widetilde{M}$.

For two-dimensional systems, the denominator in (5.36) can easily be calculated in terms of the eigenvalues of $\widetilde{M}$. It is given in Table 5.1, along with the trace of $\widetilde{M}$ and the stability situation for orbits of various nature. Note that the stability matrix $\widetilde{M}$ of a primitive orbit multiplies itself n times after n repetitions [see Eqs. (C.6,C.7)].

For the contributions of isolated *unstable* orbits with direct hyperbolic character, which are most common in non-integrable two-dimensional systems, the trace formula becomes

$$\delta g_{scl}^{(unst)}(E) = \frac{1}{\hbar\pi} \sum_u T_u \sum_{n=1}^{\infty} \frac{1}{2|\mathrm{Sinh}(n\chi_u/2)|} \cos\left[n\left(\frac{1}{\hbar}S_u - \sigma_u\frac{\pi}{2} \right) \right], \qquad (5.38)$$

where the index u here refers to the primitive unstable periodic orbits and the sum over the repetitions n is written out explicitly. We have made use here of the fact that for unstable orbits, the Maslov index repeats itself in successive repetitions. This had been assumed by Gutzwiller and was proven rigorously much later by Creagh *et al.* [19] for two-dimensional systems (see Appendix D). Note that for inverse hyperbolically unstable orbits, the denominator in (5.38) must be replaced by the expression given in the top line of the third column in Tab. 5.1.

For isolated *stable* orbits, Gutzwiller has shown by a direct bookkeeping of all phases that the Maslov index can be omitted if the absolute sign in the denominator of (5.36) is left out and the cos function is replaced by the sin function [see Eq. (36) of his 1971 paper [1]]. Miller [20] corrected this result by an explicit contribution to the phase coming from possible turning points which were ignored by Gutzwiller. The corrected trace formula for the contributions of all isolated *stable* orbits in two dimensions is

$$\delta g_{scl}^{(stab)}(E) = \frac{1}{\hbar\pi} \sum_s T_s \sum_{n=1}^{\infty} \frac{1}{2\sin(n\chi_s/2)} \sin\left[n\left(\frac{1}{\hbar}S_s - \lambda_s\frac{\pi}{2}\right)\right], \qquad (5.39)$$

where λ_s is the number of turning points of the primitive orbit in which the momentum of the particle changes sign.

We notice that in the parabolic cases (denoted by "$\sim$ stable" in Tab. 5.1), the denominator in the trace formula can become zero and thus the semiclassical approximation breaks down. This is what we pointed out at the end of the previous subsection and which demonstrates directly that the above trace formulae only apply to systems with *isolated* periodic orbits. Parabolic orbits that are members of degenerate families of periodic orbits are characteristic of systems with continuous symmetries and, in particular, of integrable systems. Their treatment will be discussed in Sect. 6.1.

But also for isolated orbits, the denominators in the trace formulae (5.36) and (5.39) become zero, whenever $\text{tr}\widetilde{M}^n$ becomes equal to $+2$. This may happen at specific energies, or at specific values of any other smooth system parameter (such as a deformation or the strength of an external field), and is characteristic of systems with mixed classical dynamics. What happens at such a specific point is that the nth repetition of the isolated orbit under consideration changes its stability and thereby undergoes a bifurcation, usually giving birth to one or more new isolated orbits which do not exist below this point. (We assume for the moment, to be specific, that we vary a system parameter in increasing direction.) A pair of stable and unstable orbits may also be created "out of the blue", i.e., both orbits do not exist on one side of the bifurcation point. At the bifurcation point, parent and new-born orbits are all degenerate. The number and stabilities of the orbits above and below the critical point depend on the type of bifurcation. For a general discussion and classification of orbit bifurcations, we refer to the book by Ozorio de Almeida [21]. In integrable systems, families of degenerate orbits may also be born at a bifurcation; an example for this appears in Problem 5.3, where we discuss the bifurcation of an isolated orbit in the elliptic billiard.

In a finite interval around a bifurcation point, the Gutzwiller trace formula for isolated orbits cannot be used, because it diverges at the bifurcation point. Technically, what happens is that the saddle-point approximation to some of the trace integrations is no longer justified. Balian and Bloch investigated this situation without, however, mentioning the process of bifurcation explicitly (see the section 6.F 'Accidental Degeneracies' of Ref. [2]). They showed how the divergence can be avoided by going to higher orders in the saddle-point expansion, i.e., taking third or higher derivatives of the exponent under the integral into account. They also used complex stationary paths to handle certain situations (see section 12 in Ref. [2]). Similarly, Berry and Tabor [3] proposed a "uniform approximation" which leads to finite amplitudes in the trace formula. Their approximation, which later was further developed by Richens [22], involves contributions from complex tori in phase space, which may lead to complex orbits, the so-called "ghost orbits" (cf. also Ref. [23]). This uniform approximation has recently been applied to some of the orbit bifurcations in the elliptic billiard by Sieber [24]. Ozorio de Almeida and Hannay [25] have systematically developed uniform approximations for the various types of bifurcations, based upon their so-called "normal forms" [26] (see section 4.4 in Ref. [21] for a detailed presentation). – It would exceed the scope of this book to present and discuss the topic of bifurcations in more detail. We will return to it briefly at the end of Sect. 6.3 where we discuss some more recent uniform approximations. Their main virtue is that they smoothly interpolate the semiclassical amplitudes through the critical points and reach the original Gutzwiller amplitudes in the asymptotic region away from the critical points.

Bifurcations of periodic orbits provide the main obstacle of quantizing systems with mixed classical dynamics, i.e., systems in which stable and unstable isolated orbits coexist and the phase space has both regular and chaotic regions. The semiclassical description of such mixed systems is still a subject of much current research.

Before closing this central section on the Gutzwiller trace formula, we mention that corrections with higher orders of $\hbar$ in the amplitudes of (5.36) have been systematically derived by Gaspard and Alonso [27]. For an extensive presentation of this subject, we refer to a recent review article by Gaspard *et al.* [28]. – A relativistic trace formula has recently been derived by Bolte and Keppeler [29], who also discussed earlier approaches to include explicitly the spin degrees of freedom into the periodic orbit theory [30, 31]. Some applications and limitations of these approaches for treating spin-orbit interactions are discussed in [32], and more general approaches making use of spin coherent states have been proposed recently in [33, 34].

5.4 Stability of periodic orbits

The stability of a periodic orbit is closely connected to the second variations of R (2.5). Assume that we have found a closed orbit along a path $\mathbf{r}(t)$ which fulfills the conditions (2.4). Let us now change the initial coordinates by $\delta\mathbf{r}'$ and the momenta by $\delta\mathbf{p}'$, hereby keeping the energy E fixed. After one period T, the final coordinates and momenta will have changed by the amounts $\delta\mathbf{r}$ and $\delta\mathbf{p}$. They can be related to each other by the *monodromy matrix* M discussed in more detail in Appendix C:

$$\begin{pmatrix} \delta\mathbf{r} \\ \delta\mathbf{p} \end{pmatrix} = M \begin{pmatrix} \delta\mathbf{r}' \\ \delta\mathbf{p}' \end{pmatrix} \qquad \text{or} \qquad \delta\boldsymbol{\xi} = M\,\delta\boldsymbol{\xi}', \tag{5.40}$$

where $\boldsymbol{\xi}$ is the $2D$-dimensional phase space vector $\boldsymbol{\xi} = (\mathbf{r}, \mathbf{p})$. In other words, the matrix M describes the *linearized* motion of small perturbations around a periodic orbit in phase space.

Note that M is a real $2D \times 2D$ matrix which in general is not symmetric, but *symplectic*. Its eigenvalues λ_i are found by solving the characteristic equation

$$F(\lambda) = |M - \lambda I| = 0, \tag{5.41}$$

where I is the $2D \times 2D$ unit matrix. In the derivation of Gutzwiller's trace formula (5.36), one needs the values of $F(1)$. The monodromy matrix M has the following important properties:

1) M is symplectic. As a consequence, its eigenvalues λ_i occur in pairs whose product is unity. Consequently, the determinant of M is also equal to unity:

$$\det M = F(0) = 1. \tag{5.42}$$

2) For conservative Hamiltonian systems, two of the eigenvalues are equal to unity as a consequence of the conservation of energy. These two eigenvalues correspond to displacements along the orbit itself and to small changes of energy leading to a neighboring orbit. The other eigenvalues belong to displacements *transverse* to the orbit; if one of these eigenvalues is real and its absolute value is larger than unity, the orbit is unstable.

For a two-dimensional system, M has four eigenvalues. Since two of them are unity and the other two are inverse to each other, there is only one number λ to be found which can be either real or complex. Exploiting the invariance of the trace, λ is simply given by $\mathrm{tr}\, M$, without the need of diagonalization, through the quadratic equation

$$\mathrm{tr}\, M = 2 + \lambda + 1/\lambda. \tag{5.43}$$

If λ is complex or ± 1 the orbit is stable, otherwise it is unstable. One distinguishes the following cases:
(i) *elliptic orbit* for $\lambda = \exp(\pm i\chi)$ with χ real;
(ii) *direct parabolic* for $\lambda = +1$ or *inverse parabolic* for $\lambda = -1$;
(iii) *direct hyperbolic* for $\lambda = \exp(\pm\chi)$ or *inverse hyperbolic* for $\lambda = -\exp(\pm\chi)$, both with real positive χ.

For $\lambda = \pm 1$ the orbits are also called *neutrally stable* or *marginally stable*. The unstable orbits (hyperbolic cases) are governed by one eigenvalue of M which grows exponentially; the exponent χ is called the *Lyapounov exponent* (cf. Eq. (C.14) in App. C).

In three or higher dimensional systems, another type of orbit can occur, the so-called *loxodromic* case. It corresponds to eigenvalues of $\widetilde{M}$ of the form $\lambda = \exp(\pm u \pm iv)$ with independent real u and v (both $\neq 0$). For $D = 3$ one has

$$F(1) = 16 \left[\mathrm{Sinh}^2(u/2)\cos^2(v/2) + \mathrm{Cosh}^2(u/2)\sin^2(v/2) \right]^2. \tag{5.44}$$

The periodic orbit here shows a rather complex behavior; usually such orbits have no simple symmetry properties such as lying in particular symmetry planes.

The stability matrix $\widetilde{M}$ appearing in Gutzwiller's trace formula (5.36) is simply the $2D-2$ dimensional submatrix of the monodromy matrix M which does not contain the trivial eigenvalues $\lambda = 1$:

$$M \sim \begin{pmatrix} \widetilde{M} & \cdots \\ 0 & \begin{pmatrix} 1 & \cdots \\ 0 & 1 \end{pmatrix} \end{pmatrix}. \tag{5.45}$$

In the Appendix C, we will present a general method of determining M for Hamiltonian systems with smooth potentials and, in some more detail, the stability matrix $\widetilde{M}$ for two-dimensional billiards.

Having obtained the monodromy matrix M, one can exploit the equation (5.40), which describes the linearized propagation of small deviations from a periodic orbit, to perform a search of periodic orbits in a multidimensional Newton-Raphson iteration which, essentially, is based upon the (matrix) inversion of Eq. (5.40) (for details, see Ref. [35] or the end of section 3.3 in Ref. [36]). – A quite different approach to the search of periodic orbits has recently been proposed by Schmelcher and Diakonos [37], who transform a given dynamical system in such a way that its (stable or unstable) periodic orbits are made dissipatively stable and can be detected selectively according to their desired stability or geometrical properties.

5.5 Convergence of the periodic orbit sum

The trace formula (5.36) is, as the expansions discussed in the extended Thomas-Fermi model in Chapter 4, an *asymptotic* series due to the assumed limit $S \gg \hbar$ (or $R \gg \hbar$) used at several stages in its derivation. In principle, one might hope that $\delta g_{scl}(E)$ exhibits poles at the quantum energies E_n of the system, or at least at some approximations thereof. This would then provide a means of semiclassical quantization of non-integrable systems, to which it mainly pertains in the version discussed so far. To achieve this goal, however, one would have to sum over *all* periodic orbits, also those with very large actions S_{po}. Keeping in mind that Eq. (5.36) is a Fourier decomposition of the oscillating level density, it is immediately obvious that if one asks for a high resolution in energy space, one has to include high Fourier components, i.e., contributions with large actions. Unfortunately, this goal can only be achieved in a few exceptional cases. Usually, the periodic orbit sum does not converge and the series has to be cut – not only for practical reasons. In chaotic systems, one also faces the problem that the number of periodic orbits increases exponentially with energy. In Chapter 7, we shall take up these problems and present some newer resummation techniques used to improve the convergence of the periodic orbit sum.

As we have stated in the introduction to this chapter, many of the applications of the periodic orbit theory given in this book have another emphasis than that of full quantization (except when this works "for free" in some integrable systems). Inverting the statement just made above, it is also obvious that if one is asking only for a low resolution of the shell structure, only those orbits with the smallest actions and, simultaneously, the largest amplitudes in the Fourier decomposition of $\delta g_{scl}(E)$ are important. Many observable shell effects in real physical systems do not depend on the exact spectrum, or the exact level density, but on a "coarse-grained" level density which may be obtained by averaging over the finer details of the spectrum. We shall become more specific about the exact definition of the coarse-grained level density below, and just state here that it is this quantity that contains the gross-shell effects responsible for the stability of finite fermion systems and for their deformation properties. Several examples of this will be given in Chapter 8.

Already early in the development of the POT, the idea of averaging the single-particle level density has been used. Balian and Bloch [2] introduced a small but finite imaginary part $i\gamma$ to the energy, which is equivalent to smoothing the level density with a Lorentzian of width γ, in order to emphasize its gross-shell behavior. The same technique was also employed by Berry and Tabor [3] and many others. Sieber and Steiner [38] explicitly introduced a convolution of the level density with various averaging functions, including Gaussians, in order to enforce the convergence of the periodic orbit sum, and used the term "sum rules" for the convoluted level densities.

In our investigations of gross-shell structure, we shall be using a Gaussian averaging, thus replacing the exact sum of delta functions (3.4) by

$$g_\gamma(E) = \frac{1}{\gamma\sqrt{\pi}} \sum_i \exp\left\{-[(E - E_i)/\gamma]^2\right\}. \tag{5.46}$$

The quantity $g_\gamma(E)$ will henceforth be called the *coarse-grained level density*. This averaging is, of course, totally different in nature from the Strutinsky averaging discussed in Chapter 4 (see Sect. 4.7), although mathematically closely related. The difference lies in the value of the smoothing width γ and in the so-called curvature corrections used in the Strutinsky method. The purpose of that method was to extract the average part of the level density that does not oscillate as a function of energy, but describes its smooth behavior in the spirit of the extended Thomas-Fermi model. In order to obtain this, γ has to be chosen to be slightly larger than the characteristic main-shell splitting $\hbar\Omega$ of a spectrum, which in the POT may be related to the inverse period T_{ppo}^{min} of the leading shortest periodic orbit. Here, we want to keep the main-shell oscillations of the level density and to average only over the finer details of the spectrum. This means that γ should be chosen in the energy range

$$\Delta E_n \lesssim \gamma \ll \hbar\Omega \simeq 2\pi\hbar/T_{ppo}^{min}, \tag{5.47}$$

where ΔE_n is of the order of the average level spacing within the main shells.

The Gaussian convolution leading to (5.46) can easily be performed on the semiclassical level density in the saddle-point approximation. To this purpose let us write the trace formula in a more compact form (we omit henceforth the index scl)

$$\delta g(E) = \sum_{po} \delta g_{po}(E) \quad \text{with} \quad \delta g_{po}(E) = \Im m\, \mathcal{A}_{po}(E)\, e^{i\left[\frac{1}{\hbar}S_{po}(E)+\phi\right]}, \tag{5.48}$$

where the extra phase ϕ is immaterial here. The convolution of each of the terms in (5.48) gives an integral of the form

$$\frac{1}{\gamma\sqrt{\pi}} \int_{-\infty}^{\infty} e^{-[(E'-E)/\gamma]^2}\, \mathcal{A}_{po}(E')\, e^{i\left[\frac{1}{\hbar}S_{po}(E')+\phi\right]}\, dE'. \tag{5.49}$$

Expanding the action in the exponent up to first order in $(E'-E)$ and using $dS_{po}(E)/dE = T_{po}$, completing the square in the exponent, and taking the amplitude $\mathcal{A}_{po}(E)$ outside the E' integral since it varies very smoothly compared to the rapid oscillations coming from the phase in the exponent, we can do the Gauss integral analytically and obtain finally one simple Gaussian factor, so that the convoluted semiclassical level density becomes

$$\delta g(E) = \sum_{po} e^{-(\gamma T_{po}/2\hbar)^2}\, \delta g_{po}(E). \tag{5.50}$$

Thus, each periodic orbit contribution is damped exponentially, whereby the orbits with longer periods are damped more strongly than the short ones. Note that the period T_{po} here includes the repetitions, so that its value is $T_{po} = nT_{ppo}$ for an n times repeated primitive orbit. In many applications, we will see, in fact, that the first repetitions (i.e., the lowest harmonics) are sufficient to give a well-converging coarse-grained level density.

In connection with quantum billiards or cavities, it is often more natural to consider the k-space level density $g(k)$ defined in Eq. (3.9), where k is the wave number $k = \sqrt{2mE}/\hbar$ of the particle. The single terms in the trace formula are then of the form

$$\delta g_{po}(k) = \Im m \, \mathcal{A}_{po}(k) \, e^{i[kL_{po}+\phi]}, \qquad (5.51)$$

where $L_{po} = S_{po}/\hbar k$ is the length of the periodic orbit. Then, the Gaussian averaging is preferably done over the variable k and can be done analytically without further approximation. One then finds for the convoluted trace formula the expression

$$\delta g(k) = \sum_{po} e^{-(\gamma L_{po}/2)^2} \delta g_{po}(k). \qquad (5.52)$$

A similar type of convergence factor comes about if one considers a system at *finite temperature*. In Sect. 4.4.3 we have dealt with a grand canonical ensemble of fermions in a heat bath at a finite temperature T. (In order to avoid confusion between the temperature T with the period T_{po} of a classical orbit, we re-introduce the Boltzmann constant k_B in the following.) We have seen there that a very useful quantity is the finite-temperature level density $g_T(E)$, obtained from the zero-temperature level density $g(E)$ by the convolution with an inverse Cosh function as shown in Eq. (4.97). From the function $g_T(E)$, all thermodynamical single-particle properties such as average particle number, entropy, and free Helmholtz energy, can be derived according to Eq. (4.96). This temperature convolution can easily be performed also for the semi-classical level density. For the single periodic orbit contributions in (5.48), one obtains integrals

$$\int_{-\infty}^{\infty} \frac{1}{4k_B T \mathrm{Cosh}^2[(E'-E)/2k_B T]} \mathcal{A}_{po}(E') \, e^{i\left[\frac{1}{\hbar}S_{po}(E')+\phi\right]} dE'. \qquad (5.53)$$

Proceeding exactly in the same way as above with the integral in (5.49), i.e., doing the saddle-point approximation, one finds the following trace formula for the finite-temperature level density

$$\delta g_T(E) = \sum_{po} \frac{\tau_{po}}{\mathrm{Sinh}\tau_{po}} \delta g_{po}(E) \quad \text{with} \quad \tau_{po} = \frac{\pi k_B T T_{po}}{\hbar}, \qquad (5.54)$$

where $g_{po}(E)$ is the contribution pertinent to the $T{=}0$ level density (5.48) which is gained from Eq. (5.54) in the limit $T{=}\tau_{po}{=}0$. Again, the extra factor in (5.54) leads to a smoothing which is more efficient for orbits with longer periods and eventually suppresses all oscillations at large temperature. We recall that we have given analytical examples of $\delta g_T(E)$ for harmonic-oscillator potentials at finite temperature in Sect. 3.2.5. Since for these potentials, the action is $S(E) = 2\pi nE/\omega$ for all n times repeated orbits, the saddle-point integration is exact and the quantity τ_{po} becomes exactly equal to $2\pi^2 nk_B T/\hbar\omega$, as given in Eq. (3.75). The result given in Eq. (5.54) has also been derived by Kolomietz, Magner, and Strutinsky in a semiclassical study of shell effects in excited nuclei [39]; it is presently being used also in mesoscopic physics [40].

Better convergence than for the level density, even without temperature or extra smoothing, is obtained for integrated quantities, such as the number of particles N or the sum of occupied single-particle energies given in Eqs. (3.38), (3.39), for a many-fermion system in the mean-field approximation. The particle number as a function of energy, $N(E)$, itself is also of interest simply because it gives the number of single-particle states up to the energy E.

Integrating (5.48) over the energy, using the saddle-point approximation as above, one obtains readily

$$\delta N_{scl}(E) = -\Re e \sum_{po} \left(\frac{\hbar}{T_{po}} \right) \mathcal{A}_{po}(E)\, e^{i\left[\frac{1}{\hbar} S_{po}(E) + \phi\right]}. \tag{5.55}$$

We see that compared to the level density, the amplitude here is smaller by a factor $\hbar/T_{po}$ which accelerates the convergence. For systems with only isolated orbits, we may also start directly from the trace formula (5.36) and use the relation $T_{ppo}dE = dS_{ppo} = dS_{po}/n$ (with n being the repetition number) to do the integration directly over the variable S_{po} for each periodic orbit. Ignoring the energy dependence of the denominator in (5.36) (which is again a saddle-point approximation), this leads to

$$\delta N_{scl} = \frac{1}{\pi} \sum_{po} \frac{1}{n|\det(\widetilde{M} - I)|^{1/2}} \sin\left(\frac{1}{\hbar} S_{po} - \sigma_{po}\frac{\pi}{2} \right). \tag{5.56}$$

When δN_{scl} is added to its smooth extended Thomas-Fermi part N_{ETF} (see Chapter 4), we obtain the total number of quantum states between 0 and E semiclassically. Recalling the diagram of the quantum stair-case in Fig. 1.5, it may be reasonable to locate the quantized energies E_n from the equation:

$$N(E_n) = N_{ETF}(E_n) + \delta N_{scl}(E_n) = n + \frac{1}{2}, \quad n = 0, 1, 2, \ldots . \tag{5.57}$$

The contribution $+1/2$ on the right-hand side above comes from the fact that integrating a delta function up to its zero gives only one-half of unity. This "stair-case quantization" method was suggested by Voros [41] and also discussed by Aurich and Steiner [42]. For the procedure to work, one has to assume that there are no degeneracies in the spectrum, i.e., that $E_n \neq E_{n'}$ for $n \neq n'$. In systems with any symmetry (continuous or discrete), this means that one has to work in subspaces in which the corresponding quantum numbers are fixed. This leads to so-called "symmetry-reduced trace formulae" which we shall not discuss here; the interested reader is referred to Refs. [43, 44, 45, 46].

As an example of the practical application of this procedure, we quote a result of Szeredi and Goodings [47], who have made a careful study of different quantization rules using the wedge billiard. The wedge billiard describes the motion of a particle confined in a two-dimensional wedge, with the boundary lines $x = 0$ and $y = x \cot\phi$. The Hamiltonian of the wedge billiard is given by (see Problem 4.3 with $\alpha=1$)

$$H = \frac{1}{2m}(p_x^2 + p_y^2) + y, \qquad x \geq 0,\ y \geq x \cot\phi . \tag{5.58}$$

Note that the motion in the wedge is *not* force-free, since the gravitational force acts on the particle vertically downwards. The particle makes elastic collisions with the walls of the wedge and exhibits hard chaos for $\phi > 45°$. For the quantum mechanical problem, the wave function goes to zero at the boundary, i.e., Dirichlet boundary conditions have been imposed.

The result of the quantization condition (5.57) for a wedge billiard of 60° inclination is compared with the quantum result in Fig. 5.2. The quantum calculation was done by solving for the first 18 eigenvalues numerically. In the trace formula (5.56), 1621 primitive orbits were included. The smooth part of the level density, g_{ETF}, may be derived analytically (see Problem 4.3), and the leading TF term goes like E^2. Thus the mean number of states between $E = 0$ and E, obtained by integrating $g_{TF}(E)$, goes like $g_{TF}E/3$, which is taken to be the abscissa in the diagram. This ensures that the mean spacing in the stair-case function $N(E)$ is unity. From the diagram, we see that the oscillating part $\delta N_{scl}(E)$ chisels out the steps fairly accurately from the Thomas-Fermi background. According to the authors, however, the method is less successful for the 49° wedge (the wedge is integrable at 45°). Also, sometimes Eq (5.57) possessed multiple solutions, when one had to choose the one nearest to $N_{ETF}(E_n^{ETF}) = n + 1/2$.

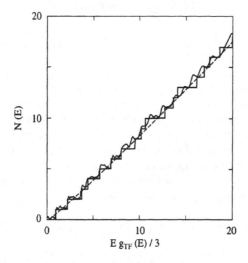

Figure 5.2: The exact quantum stair-case function $N(E)$ for the wedge billiard with $\phi = 60°$ is compared with the function obtained from the quantization rule (5.57). It is shown by the oscillating solid curve superimposed on the quantum stair-case. The dashed line is N_{ETF}. See text for the choice of the abscissa. (After [47].)

Finally, we like to give the semiclassical approximation to the shell-correction energy of a finite fermion system. The oscillating part δE of the sum of occupied single-particle levels has been defined and discussed in Sect. 4.7; a suitable expression for it is given in Eq. (4.192) which we repeat here for convenience:

$$\delta E \simeq \int_{-\infty}^{\mu} (E - \mu)\, \delta g(E)\, dE \tag{5.59}$$

In order to obtain it from the periodic orbit theory, let us insert the trace formula (5.48) into Eq. (5.59) and expand the exponent around μ up to first order, we obtain for each orbit an integral

$$\Im m\, e^{i\left[\frac{1}{\hbar}S_{po}(\mu)+\phi\right]} \int_{-\infty}^{\mu} \mathcal{A}_{po}(E)(E-\mu)\, e^{-\frac{i}{\hbar}(E-\mu)T_{po}}\, dE\,, \tag{5.60}$$

where $T_{po} = dS_{po}/dE$ is evaluated at the Fermi energy μ. The phase above is stationary at $E = \mu$, so that the main contribution comes from this point. Thus taking an amplitude factor $\mathcal{A}_\Gamma(\mu)$ outside the integral and substituting $E' = (E - \mu)$, we are left with the integral

$$\int_{-\infty}^{0} E' e^{\frac{i}{\hbar}T_{po}E'}\, dE' \tag{5.61}$$

which, putting a real regularization factor $e^{\epsilon E'}$ ($\epsilon > 0$) under the integral and letting ϵ go to zero after integration, gives $\hbar^2/T_{po}^2$. Thus, we find for the energy shell-correction the semiclassical expression [7]

$$\delta E \simeq \sum_{po} \left(\frac{\hbar}{T_{po}}\right)^2 \delta g_{po}(\mu)\,. \tag{5.62}$$

We see that the inverse-squared period under the summation accelerates the convergence appreciably, and both the longer orbits and the higher repetitions of all orbits are naturally suppressed. An application of this formula shall be discussed in Sect. 8.1.1 in the context with nuclear deformations. We should like to recall that Eq. (5.59) is only correct up to first order in $(\mu - \tilde{\mu})$, where μ is the Fermi energy of the exact sum of single-particles and $\tilde{\mu}$ that of its average part obtained in the extended Thomas-Fermi model (see Chapter 4). In recent studies of shell-correction energies δE in cavities with various multipole deformations [48], it was found that this approximation holds well in regions near the minima and maxima of δE, but becomes rather bad in the regions in between where δE is small in absolute value. In that case one has to calculate the exact single-particle energy sum semiclassically by

$$E_{scl} = \int_{-\infty}^{\mu} E\left[\delta g_{scl}(E) + g_{ETF}(E)\right] dE\,, \tag{5.63}$$

which necessitates the iteration of the Fermi energy μ from the particle number condition

$$N_{ETF}(\mu) + \delta N_{scl}(\mu) = N\,. \tag{5.64}$$

This can be numerically difficult, as illustrated by Fig. 5.2.

Note that all the formulae given in this section [with the exception of Eq. (5.56)] pertain also to systems with continuous symmetries. For these, the degeneracies of the periodic orbits have to be dealt with separately, as will be discussed in Chapter 6. But we shall see there that the resulting trace formulae always have the form given by Eq. (5.48).

5.6 Examples

5.6.1 Applications to chaotic systems

Gutzwiller's trace formula has scored remarkable successes for chaotic systems and thereby stimulated a revival of the old dreams about semiclassical quantization of non-integrable systems. Full quantization remains an exception, though, reserved to some simple model potentials. But the partial quantization of certain quantum states, preferentially with high quantum numbers, in classically chaotic systems, has become a reality. Some examples have already been given in Chapter 1, others will be mentioned in Chapter 7 after introducing the dynamical zeta function. Let us here just flash through a few selected landmarks: the anisotropic Kepler problem [49]; the motion of a particle on a surface with constant negative curvature [5, 50]; atoms in strong magnetic fields [35] (see also Sect. 1.4); the hyperbolic billiard [51]; the wedge billiard [52] (see also Sect. 7.4); the helium atom and related three-body systems [53] (see also Sect. 1.5.2). For a special "Chaos Focus Issue" on periodic orbit theory, including applications and extensions of Gutzwiller's trace formula, see also Ref. [54].

For the educational aspect of this book, we have selected below two simple but instructive examples of *integrable* systems with isolated orbits, where the quantization with Gutzwiller's trace formula becomes exact: harmonic oscillators in two dimensions. Although mathematically quite trivial, the successful analytical application of the trace formula to these systems does not appear to be common knowledge.

Our main focus is, however, not on full quantization, but rather on the use of the periodic orbit theory for the determination of gross-shell structures in the mean field properties of complex many-body systems. These may be integrable or non-integrable, chaotic or ordered, or – which occurs, in fact, most frequently in real systems – exhibit mixed classical dynamics. Some key examples for this will be given in Chapter 8; for further recent reviews, see also Refs. [36, 55]. As a prelude to this, we discuss in the third example below a beat structure in the coarse-grained level density of the Hénon-Heiles system that is well reproduced using only the three shortest periodic orbits in the Gutzwiller trace formula.

5.6.2 The irrational anisotropic harmonic oscillator

As already pointed out, the Gutzwiller trace formula, as derived and discussed above, only applies to isolated orbits. These are typical for non-integrable systems, and therefore this theory has most often been studied so far in connection with chaos. However, there are integrable systems with only isolated periodic orbits: harmonic oscillator potentials with incommensurable (i.e., irrational) frequency ratios. We shall here discuss the two-dimensional case (see also Ref. [56])

$$H = \frac{1}{2m}(p_x^2 + p_y^2) + \frac{m}{2}(\omega_x^2 x^2 + \omega_y^2 y^2), \tag{5.65}$$

assuming the ratio $\omega_x : \omega_y$ to be an irrational number. The only periodic orbits with a finite period are the one-dimensional oscillations along the x and y axes; they are isolated and stable. In Appendix C.2.1 we have derived the eigenvalues (C.23) of the stability matrix for these orbits; the stability angles are found to be $\chi_x = 2\pi\omega_y/\omega_x$ and $\chi_y = 2\pi\omega_x/\omega_y$, respectively. Inserting them into Eq. (5.39) and using $\lambda_s = 2$ for the number of turning points, we get directly the oscillating part of the level density:

$$\delta g\left(E\right) = \frac{1}{\hbar\omega_x} \sum_{k=1}^{\infty} \frac{(-1)^k}{\sin\left(k\pi\frac{\omega_y}{\omega_x}\right)} \sin\left(k\frac{2\pi E}{\hbar\omega_x}\right) + \frac{1}{\hbar\omega_y} \sum_{k=1}^{\infty} \frac{(-1)^k}{\sin\left(k\pi\frac{\omega_x}{\omega_y}\right)} \sin\left(k\frac{2\pi E}{\hbar\omega_y}\right). \quad (5.66)$$

The same result is also obtained from Eq. (5.36) using the Maslov indices derived in Appendix D.1.3.

The result (5.66) is identical with the *exact* oscillating part of the quantum-mechanical level density of the incommensurable harmonic oscillator given in Eq. (3.113). We therefore see that in this example, the Gutzwiller theory gives the exact level density.

An interesting limit is that where the frequency ratio $\omega_x : \omega_y$ becomes rational. In this case, the periodic orbits become non-isolated and form families of degenerate orbits. It is easily seen that in Eq. (5.66), there are always values of the cycle number k for which some amplitudes diverge when $\omega_x : \omega_y$ is a ratio of integers. A careful investigation [56] shows, however, that these divergent amplitudes always occur pairwise with opposite signs in the two sums of Eq. (5.66), so that they cancel and the exact oscillating part of the quantum-mechanical level density given in Eq. (3.116) is obtained.

This cancellation of the diverging amplitudes is an exception which occurs only for the harmonic oscillator potential. It is connected with the fact that the $SU(2)$ symmetry of the two-dimensional isotropic harmonic oscillator is *preserved* when it is deformed with a rational frequency ratio $\omega_x : \omega_y = n : m$ (see, e.g., Ref. [57]). The periodic orbits, which are Lissajous figures with $2n$ zeros in y and $2m$ zeros in x direction, still form a degenerate family in this case. Only for irrational frequency ratios, the $SU(2)$ symmetry is destroyed; then the only periodic orbits left are the isolated linear oscillations along the principal axes. – Typical cases where the diverging Gutzwiller amplitudes do not cancel are given for the Hénon-Heiles potential below and in Sect. 6.2.1.

The same is found for harmonic-oscillator potentials in more than two dimensions [56]. The three-dimensional case is treated in Problem 5.1 where, again, Gutzwiller's trace formula is found to be exact.

5.6.3   The inverted harmonic oscillator

We shall next study the motion of a particle moving on a saddle surface defined by an inverted harmonic oscillator in the y direction and a confining harmonic oscillator in the x direction [58]. The Hamiltonian is

$$H = \frac{1}{2m}(p_x^2 + p_y^2) + \frac{m}{2}(\omega_x^2 x^2 - \omega_y^2 y^2). \quad (5.67)$$

In this potential, there is classically only one periodic orbit, namely the one-dimensional oscillation along the x axis. Since this is a libration on a ridge, the orbit is highly unstable and isolated. The results of the previous example can easily be applied to the

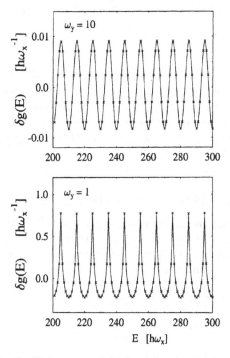

Figure 5.3: Oscillating part of the level density for the inverted harmonic oscillator potential (5.67) for $\hbar\omega_x=10$ and the two values $\hbar\omega_y=10$ (above) and $\hbar\omega_y=1$ (below). *Solid lines*: semiclassical result using the trace formula (5.68). *Crosses*: quantum-mechanical results. (From [58].)

present case if one realizes that the motion in the y direction is described by a harmonic oscillation with an imaginary frequency, $i\omega_y$. In the stability matrix (C.21) derived in Appendix C.2.1, we thus have to replace the cos and sine functions of the y motion by the corresponding hyperbolic functions Cosh and Sinh, respectively. The eigenvalues for the isolated x orbit then become $\exp(\pm 2\pi\omega_y/\omega_x)$. The Maslov index σ_x for the primitive orbit can be calculated as a winding number, as specified in Appendix D.1.1, and becomes $\sigma_x = 2$. The trace formula for the oscillating part of the level density then is found to be

$$\delta g\,(E) = \frac{1}{\hbar\omega_x} \sum_{n=1}^{\infty} \frac{(-1)^n}{\mathrm{Sinh}\left(n\pi\frac{\omega_y}{\omega_x}\right)} \cos\left(n\frac{2\pi E}{\hbar\omega_x}\right). \tag{5.68}$$

Quantum mechanically, this system has a continuum of eigenstates. We write its exact density of states for $E > 0$ in a convoluted form

$$g\,(E) = \int_{-\infty}^{E} g_x(E - E')\,g_y(E')\,dE', \tag{5.69}$$

where

$$g_x(E) = \sum_{n_x=0}^{\infty} \delta[E - \hbar\omega_x(n_x + 1/2)] \tag{5.70}$$

is the level density of the one-dimensional harmonic oscillator in the x direction, and

$$g_y(E) = -\frac{1}{\pi} \Re e \, \Psi(1/2 + iE/\hbar\omega_y) \tag{5.71}$$

is the quantum level density of an inverted harmonic oscillator; $\Psi(z) = d/dz \ln \Gamma(z)$ is the digamma function. The expression (5.71) has been carefully studied by Barton [59]. For comparison of the quantum result with the semiclassical trace formula (5.68), the smooth Thomas-Fermi part has to be subtracted. The numerical comparison was done in Ref. [58]. In Fig. 5.3 we show the results of this comparison. The agreement is perfect within the accuracy of the figure.

5.6.4   The Hénon-Heiles potential

As an example of a non-integrable system with isolated orbits and mixed classical dynamics, we shall study a two-dimensional potential which was proposed [60] in 1964 by Hénon and Heiles (HH) as a simple model for the galactic mean field felt by a star. Its form in cartesian coordinates is

$$V(x,y) = \frac{m}{2} \omega^2 (x^2 + y^2) + \alpha \left(x^2 y - \frac{1}{3} y^3\right) \tag{5.72}$$

or, in polar coordinates (r, ϕ),

$$V(r,\phi) = \frac{m}{2} \omega^2 r^2 + \alpha \frac{1}{3} r^3 \sin(3\phi). \tag{5.73}$$

It has the discrete symmetries of reflection at the three axes with the polar angles $\phi = \pi/2$ and $\phi = \pm \pi/6$, and of rotations about the angles $\pm 2\pi/3$. The potential is harmonic for $\alpha = 0$ and has a minimum at $r = 0$. There are three saddle points lying on the symmetry axes at a saddle point energy of $E^* = (m\omega^2)^3/6\alpha^2$. For $E > E^*$, the particle can escape if it has the right direction of its momentum.

The HH potential is a typical example of a system which is classically chaotic for high energies (close to and above E^*), and quasi-regular at small energies. We shall first discuss the classical motion in this potential, which consists only of isolated orbits, and then compare its quantum-mechanical level density to that predicted by the Gutzwiller trace formula.

There is only one constant of motion in the HH potential: the energy E. The classical trajectories thus depend on the two parameters E and α. We can, however, introduce *scaled variables* u, v:

$$u = \alpha x, \qquad v = \alpha y, \tag{5.74}$$

such that the classical motion only depends on one combination of these parameters, namely a scaled energy e

$$e = 6\alpha^2 E = E/E^* = 3(\dot{u}^2 + \dot{v}^2) + 3(u^2 + v^2) + 6v u^2 - 2v^3, \tag{5.75}$$

which is simply the energy in units of the saddle point energy E^*. In (5.75) and below we use simplified units such that $m = \omega = \hbar = 1$.

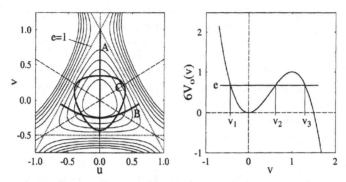

Figure 5.4: The Hénon-Heiles potential. *Left:* Equipotential contour lines in scaled energy units e (5.75) in the plane of scaled variables u, v (5.74). The dashed lines are the symmetry axes. The three shortest periodic orbits A, B, and C (evaluated at the energy $e = 1$) are shown by the heavy solid lines. *Right:* Cut of the scaled potential along $u = 0$. (After [61].)

On the left-hand side of Fig. 5.4, we show the equipotential surface of the HH potential as a function of u and v. The equipotentials for $e = 1$ (shown by dotted lines) are straight lines given by $v = -1/2$ and $v = 1 \pm \sqrt{3}\,u$, intersecting at the saddle points in an equilateral triangle. Along the three symmetry axes with $\phi = \pi/2$ (v-axis) and $\phi = \pm\pi/6$ (dashed lines), the potential is a cubic parabola, as shown on the right-hand side of Fig. 5.4.

The Newton equations of motion in the scaled coordinates u, v become

$$\begin{aligned}
\ddot{u} &= -u - 2uv, \\
\ddot{v} &= -v + (v^2 - u^2).
\end{aligned} \qquad (5.76)$$

These equations depend only on the scaled energy e. It is thus sufficient to solve them only once for a given value of e; the scaling relation (5.75) can then be used later to get solutions for any pair of the variables E and α.

Hénon and Heiles [60] have already observed that the classical motion in this potential is quasi-regular up to energies $e \sim 0.5$ and then becomes increasingly chaotic; when one reaches the saddle energy ($e = 1$), more than 95% of the phase space is covered ergodically. The shortest periodic orbits in the HH potential have been classified and investigated in detail by Churchill *et al.* [62] and more recently by Davies *et al.* [63]. Up to a scaled energy of $e \simeq 0.9$, there exist only three kinds of periodic orbits whose primitive periods are of the order of $T_0 = 2\pi/\omega$ ($= 2\pi$ in the present units), which is the fundamental period of the underlying harmonic oscillator potential obtained for $\alpha = 0$. These three periodic orbits are shown in Fig. 5.4 (left) by the heavy lines.

The orbit A is a libration along one of the symmetry axes, corresponding to the motion in the one-dimensional potential shown on the right-hand side of Fig. 5.4. [It is easy to see that, e.g., $u(t) = 0$ is a solution of the equations of motion (5.76).] This orbit exists only for $e < 1$; its period T_A diverges for $e = 1$, since the particle there needs an infinite time to reach the saddle point with zero momentum. All the other orbits can only be found numerically by solving the full two-dimensional equations of motion (5.76). The orbit B, which is also a libration, is orthogonal to the mode A at low energies. Orbit C is a rotation. The orbits A and B occur in three orientations, corresponding to the symmetries of the HH potential. Orbit C maps unto itself under rotations of the potential about $2\pi/3$ and $4\pi/3$, but the particle can run through it in two opposite time orientations, whereas time reversal maps the orbits A and B onto themselves. Thus, the overall discrete degeneracies are 3 for orbits A and B, and 2 for orbit C.

Orbit C is stable up to $e = 0.89$ where it undergoes a period-doubling bifurcation and becomes unstable. A stable double-loop orbit D is born at this energy with exactly the double period: $T_D(e = 0.89) = 2T_C(e = 0.89)$. It remains stable up to an energy $e \simeq 1.24$. – Orbit B is always unstable. – Orbit A is stable up to above $e \simeq 0.97$ and then undergoes an infinite cascade of increasingly dense-lying bifurcations [63], alternating between being stable and unstable up to $e = 1$ where it disappears. This bifurcation cascade has a fractal structure; the orbits born at the bifurcations exhibit self-similarity with explicit analytical scaling constants [64] and can be well approximated by periodic Lamé functions [65].

In order to illustrate the mixed classical dynamics of a particle in the Hénon-Heiles potential, which is reminiscent of the Walker-Ford model discussed in Sect. 2.8, we give in Fig. 5.5 two examples of more complex orbits: a quasi-regular orbit and a chaotic orbit. In the upper left panel, we show a quasi-regular orbit, obtained at $e = 1$, which

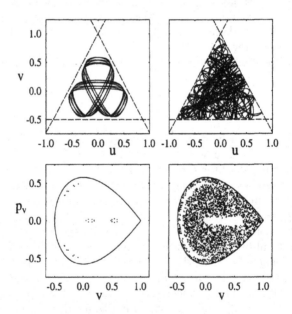

Figure 5.5: Examples of a quasi-regular orbit (left panels) and a chaotic orbit (right panels) in coordinate space (u, v) (upper panels) and in the Poincaré surface of section (v, p_v) (lower panels). Note the irregular filling of most of the phase space by the chaotic orbit in the lower right panel. The empty islands correspond to the quasi-regular motion seen on the left; elliptic fixed points at their centers correspond to the stable double-loop orbit D. (From [66].)

approximately closes after 28 revolutions around the origin. Its trace on the Poincaré surface of section given by the (p_v, v) plane $(u=0)$ is shown in the lower left panel of Fig. 5.5. (The maximum available phase space in this plane is given by the drop-like solid boundary.) It is seen to lie on sections of invariant tori similar to those discussed in Sect. 2.8.2. At the centers of the sections of these tori, there are four elliptic fixed points corresponding to the stable orbit D which has the same topological structure as the orbit seen in the figure above, lying at the center of the "band" covered by it.

In the right panels of Fig. 5.5, we show the trajectory of one chaotic orbit at $e = 1$ in coordinate space (u, v) (above) and in the Poincaré surface of section (p_v, v) (below). Here the available phase space is filled almost ergodically, but we recognize two empty islands, belonging to orbit D and quasi-periodic orbits like the orbit shown on the left.

In spite of its already long and famous history, the potential of Hénon and Heiles has only recently been investigated within the context of periodic orbit theory in Ref. [45], and in Refs. [61, 67, 68, 69] where the focus was laid on the *coarse-grained* level density in the sense discussed in Sect. 5.5 above. The quantum-mechanical energy eigenvalues E_i were obtained by diagonalization of the HH Hamiltonian in the basis of the unperturbed harmonic oscillator, i.e., of the potential (5.73) with $\alpha = 0$. The oscillating part of the quantum level density was then defined by

$$\delta g(E) = g_\gamma(E) - \tilde{g}(E), \qquad (5.77)$$

where $\tilde{g}(E)$ is the numerically Strutinsky-averaged level density, obtained as defined in Sect. 4.7, Eq. (4.172), and $g_\gamma(E)$ is slightly Gaussian-averaged, see Eq. (5.46).

The value of γ is chosen such as to suppress the higher harmonics; in the present case of the HH spectrum with a main spacing $\hbar\omega = 1$ of the unperturbed harmonic-oscillator spectrum, we use $\gamma_{sh} = 0.25$. The resulting $\delta g(E)$ for the HH spectrum is shown in the uppermost part of Fig. 5.6 on the next page for the value $\alpha = 0.04$, for which the saddle point energy is $E^* = 104.6\,\hbar\omega$. It shows a very pronounced beating pattern.

This quantum beat can be well explained in terms of POT, as is shown in the lower two parts of Fig. 5.6. The central part shows the result of the Gutzwiller trace formula, Eq. (5.36), including the Gaussian averaging factors $\exp[-(\gamma T_{po}/2\hbar)^2]$ which are consistent with the energy averaging used for the quantum part (5.46) (cf. the discussion in Sect. 5.5). Only the lowest harmonics $(n = 1)$ of the three shortest periodic orbits A, B, and C discussed above were included. Its actions, stabilities and Maslov indices $(\sigma_A=5, \sigma_B=4, \text{ and } \sigma_C=3$ below the energies where the first bifurcations occur) were determined numerically in Ref. [67] (see also the Problem 6.5).

The agreement of the Gutzwiller results with the quantum-mechanical ones is quantitative in the region $40 \lesssim E \lesssim 70$, as can be seen more clearly in the lowest part of the figure where the two results are superposed on an enlarged scale. For $E \gtrsim 70$, the agreement becomes worse for several reasons. First, other periodic orbits than A, B, and C become important when one approaches the saddle-point energy. Second, the bifurcations of the orbit A that occur in this region have not been treated correctly. They lead to divergences which were avoided in Ref. [67] by a simple smooth interpolation of the amplitude of orbit A in the region above the bifurcation points (cf. Sect. 6.2.4, where a similar procedure is used). A better treatment of the first few of these

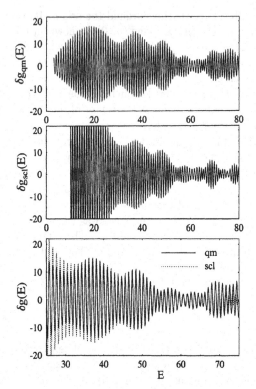

Figure 5.6:   Oscillating part of the level density for the Hénon-Heiles potential
with $\alpha = 0.04$, Gaussian averaged over an energy range $\gamma = 0.25\,\hbar\omega$. *Upper part:*
quantum-mechanical result. *Central part:* result obtained with Gutzwiller's trace
formula for isolated orbits, including the lowest harmonics ($k = 1$) of the three shortest periodic orbits A, B, and C. *Lower part:* superposition of quantum-mechanical
and semiclassical results. The energy E is in units of $\hbar\omega$. (From [67].)

bifurcations is possible with the recently developed uniform approximations discussed
in Sect. 6.3.3. (Note that the period-doubling bifurcation of orbit C poses no problem,
as long as only its primitive $n = 1$ is included. At the bifurcation point it goes from
stable to unstable with a value tr $\widetilde{M} = -2$ where the denominator of the Gutzwiller
amplitude stays finite with the value 2, see the second line in Table 5.1. Its second
repetition $n = 2$ would lead to a diverging amplitude, because at the bifurcation
point the twice repeated orbit C is degenerate with the newborn orbit D.) Finally,
the quantum-mechanical probability of tunneling through the barriers, which leads to
imaginary parts of the energy spectrum, has been neglected.

For energies $E \lesssim 40$, the amplitudes start diverging, since one approaches in the
limit $E \to 0$ the degenerate orbits of the harmonic oscillator with its SU(2) symmetry.
This divergence is generic and cannot be overcome within the Gutzwiller theory for
isolated orbits. We shall come back to the general problem of symmetry breaking (or
restoring) and present its solution for Hénon-Heiles type potentials in a perturbative
treatment [68] in Sect. 6.2.4, and in a uniform approximation [69] in Sect. 6.3.2.

5.7   Problems

(5.1) Anisotropic three-dimensional harmonic-oscillator potential (After [56])
Consider the Hamiltonian of a three-dimensional harmonic oscillator

$$H = \frac{m}{2}(p_x^2 + p_y^2 + p_z^2) + \frac{m}{2}(\omega_x^2 x^2 + \omega_y^2 y^2 + \omega_z^2 z^2). \tag{5.78}$$

(a) First assume that the oscillator frequencies ω_i are *incommensurable*, i.e., all their ratios are irrational. The only classical periodic orbits (with a finite period) are the three linear oscillations along the coordinate axes. They are isolated and stable, and thus the trace formula (5.39) can be used. Hence show, with the tools used and discussed in Sect. 5.6.2, that the oscillating part of the level density is given by:

$$
\begin{aligned}
\delta g\,(E) \;=\;& \frac{1}{2\hbar\omega_x} \sum_{k=1}^{\infty} \frac{(-1)^{k+1}}{\sin(k\pi\omega_y/\omega_x)\sin(k\pi\omega_z/\omega_x)} \cos\!\left(k\frac{2\pi E}{\hbar\omega_x}\right) \\
+\;& \frac{1}{2\hbar\omega_y} \sum_{k=1}^{\infty} \frac{(-1)^{k+1}}{\sin(k\pi\omega_x/\omega_y)\sin(k\pi\omega_z/\omega_y)} \cos\!\left(k\frac{2\pi E}{\hbar\omega_y}\right) \\
+\;& \frac{1}{2\hbar\omega_z} \sum_{k=1}^{\infty} \frac{(-1)^{k+1}}{\sin(k\pi\omega_x/\omega_z)\sin(k\pi\omega_y/\omega_z)} \cos\!\left(k\frac{2\pi E}{\hbar\omega_z}\right).
\end{aligned}
\tag{5.79}
$$

Note, by comparing with Eq. (3.124) in Problem 3.1, that this is the exact quantum-mechanical result.

(b) Now assume that the potential has *axial symmetry* by writing $\omega_x = \omega_y = \omega_\perp$, and furthermore that the frequency ratio $\omega_\perp : \omega_z$ is *rational* by writing $\omega_\perp = p\omega_0$ and $\omega_z = n\omega_0$, with n and p being relatively prime. In this case, Eq. (5.79) cannot be used since for some values of k, the amplitudes will diverge. Therefore, take $\omega_x : \omega_y = 1 + \delta$ and $\omega_x : \omega_z = p/n + \epsilon$, with δ and ϵ being some inifintesimally small (irrational) numbers. Now, show that in the limit $\delta \to 0$ and $\epsilon \to 0$, the above result turns into:

$$
\begin{aligned}
\delta g\,(E) \;=\;& g_{ETF}(E) \times 2 \sum_{k=1}^{\infty} (-1)^{kn} \cos\!\left(k\frac{2\pi E}{\hbar\omega_0}\right) \\
+\;& \frac{E}{(\hbar\omega_\perp)^2} \sum_{k\neq lp}^{\infty} \frac{1}{\sin(k\pi n/p)} \sin\!\left(k\frac{2\pi E}{\hbar\omega_\perp}\right) \\
+\;& \frac{\hbar\omega_z}{2(\hbar\omega_\perp)^2} \sum_{k\neq lp}^{\infty} \frac{\cos(k\pi n/p)}{\sin^2(k\pi n/p)} \cos\!\left(k\frac{2\pi E}{\hbar\omega_\perp}\right) \\
+\;& \frac{1}{2\hbar\omega_z} \sum_{k\neq ln}^{\infty} \frac{(-1)^{k+1}}{\sin^2(k\pi p/n)} \cos\!\left(k\frac{2\pi E}{\hbar\omega_z}\right), \qquad (l \in \mathbf{N})
\end{aligned}
\tag{5.80}
$$

where $g_{ETF}(E)$ is the smooth part which, for arbitrary frequencies, is given by (cf. Sect. 4.3)

$$g_{ETF}(E) = \frac{1}{2\,\hbar\omega_x\hbar\omega_y\hbar\omega_z}\left\{ E^2 - \frac{1}{12}\left[(\hbar\omega_x)^2 + (\hbar\omega_y)^2 + (\hbar\omega_z)^2\right]\right\}. \tag{5.81}$$

The derivation of Eq. (5.80) is not trivial and requires a careful Laurent expansion in δ and ϵ. The leading terms in this expansion all diverge, but they cancel mutually and the above finite expression is left (cf. Ref. [56]) if you don't make any mistakes.

(c) Show, finally, that for $\omega_\perp = \omega_z = \omega$ (i.e., $n = p = 1$), we obtain from this the exact quantum-mechanical result for the *spherical* three-dimensional harmonic oscillator, contained in the oscillating part of Eq. (3.69).

(d) If you have a PC, do the following numerical experiment. Give the potential in (5.78) a tiny triaxial deformation by choosing the frequency ratios to be, say, $\omega_x : \omega_y : \omega_z = 1 : 1.0001 : 1.0000001$. These ratios are practically irrational, so that the above trace formula (5.79) for isolated orbits applies if the summation over the harmonics k is restricted to a number k_{max} that is not too large. Compute Eq. (5.79), add to it the smooth part (5.81), and renormalize the sum by a factor $1/(2k_{max} + 1)$. Plot $g(E)$ for increasing values of k_{max}. By the time you have reached $k_{max} \sim 500$, you will find a sum of delta functions with the correct degeneracies $d_n = (n + 1)(n + 2)/2$ of the spherical harmonic oscillator quantum levels. (You might need to program this in double precision for the above cancellations to take place.) The result is actually the same as if you had programmed the much simpler formula (3.69) for the spherical harmonic oscillator (see Fig. 3.2) – but the amazing fact is here that you have reproduced the exact quantum spectrum from a trace formula that makes use of only three isolated periodic orbits!

(5.2) Isolated orbit in the equilateral triangle billiard

Use the trace formula (5.36) of Gutzwiller to calculate the contribution of the isolated orbit $(M = 1)$ with length $3L/2$ in the middle of the equilateral triangle billiard with side length L, shown on the left side of Fig. 6.1 in Sect. 6.1.2, to the level density of this system. Assume both Dirichlet and Neumann boundary conditions for the definition of the quantum spectrum (cf. Problem 3.6).

(a) Show, with the formalism provided in App. C.5, that the stability matrix for the primitive orbit is

$$\widetilde{M}^{(1)} = \begin{pmatrix} -1 & -2L \\ 0 & -1 \end{pmatrix}, \tag{5.82}$$

so that both its eigenvalues are equal to -1. Hence the trace is $\mathrm{tr}\widetilde{M}^{(1)} = -2$ and the orbit is marginally stable. For its odd repetitions the denominator in Eq. (5.36) equals 2 (see Table 5.1), so that we get finite amplitudes. For even repetitions, the amplitudes diverge. However, the even repetitions of this orbit are degenerate with a family of orbits with length $3L$, denoted by (11), and its repetitions – see the right side of Fig. 6.1 – that cannot be treated by the Gutzwiller trace formula and will be dealt with in Sect. 6.1.2. The even repetitions must therefore be left out.

(b) By the methods discussed in Appendix D, show that the Maslov index is $\sigma = \mu + \nu = 6 + 0 = 6$ for the Dirichlet case and $\sigma = 0$ for the Neumann case, and that this holds for all repeated periods of the orbit. This is most easily found if you visualize the orbit in the way used in Sect. 6.1.2 as a straight line passing through three adjacent triangles of the tesselated plane (cf. Fig. 6.3).

(c) Combining all these results, prove that the contribution to the level density is

$$\delta g^{(1)}(E) = \mp \frac{1}{4\pi} \left(\frac{2m}{\hbar^2} \right)^{1/2} \frac{3L}{\sqrt{E}} \sum_{M=1,3,5,\ldots}^{\infty} \cos\left(\frac{1}{\hbar} S_M \right) = \mp \frac{1}{\sqrt{EE_\Delta}} \sum_{M=1,3,5,\ldots}^{\infty} \cos\left(\frac{1}{\hbar} S_M \right), \tag{5.83}$$

where m is the mass of the particle in the billiard, $E_\Delta = (16/9)(\hbar^2 \pi^2)/(2mL^2)$, and $S_M = pL_M = \sqrt{2mE}\, 3ML/2$ is the action of the M times repeated orbit. The upper sign in Eq. (5.83) holds for Dirichlet and the lower for Neumann boundary conditions. The result (5.83) is identically contained in the oscillating part of the second terms in the exact trace formulae (3.102) and (3.106) derived in Sect. 3.2.7.

(5.3) Isolated orbits in the elliptic billiard

Derive the contributions of the two isolated linear orbits along the principal axes of the elliptical billiard to its trace formula (5.36) for the level density. The geometry of the ellipse with eccentricity ϵ is discussed in Appendix C.5.1.

(a) Using Eq. (C.52), show that the the curvature radii at the end points of the short and the long diameter are $R_< = A^2/B$ and $R_> = B^2/A$, respectively. Their lengths are $\rho_< = 2B$ and $\rho_> = 2A$. From Eq. (C.47), show that the longer diameter orbit is unstable and the shorter one is stable. Using Eq. (C.46), show that the eigenvalues of their monodromy matrices can be written as

$$\lambda_> = e^{\pm \chi_>}, \qquad \chi_> = 4\,\text{arctanh}\,\epsilon, \tag{5.84}$$

and

$$\lambda_< = e^{\pm i \chi_<}, \qquad \chi_< = 4\,\text{arcsin}\,\epsilon. \tag{5.85}$$

Thus, the long diameter orbit is isolated and unstable, whereas the short one is isolated and stable except at specific points to be discussed below. Using Table 5.1, show that the denominators $D_{po} = |\det(\widetilde{M}_{po}^n - I)|^{1/2}$ of the Gutzwiller amplitudes for these orbits in the trace formula (5.36) are given by

$$\begin{aligned} D_> &= 2\,\text{Sinh}(2n\,\text{arctanh}\,\epsilon), \\ D_< &= 2\,|\sin(2n\,\text{arcsin}\,\epsilon)|, \end{aligned} \tag{5.86}$$

where n is the repetition number of the primitive orbits. *Note:* In the limit $\epsilon \to 0$ the Gutzwiller amplitudes diverge, which is due to the fact that one here reaches the integrable limit of the circular disk billiard whose periodic orbits are not isolated (cf. Sect. 6.1.3). We shall come back to this problem in Sect. 6.2.2.

(b) Note that all nth repetitions of the short diameter orbit with $n \geq 2$ go through $n-1$ bifurcations, each of them at a specific value ϵ_{nm} of the eccentricity. At these bifurcation points, new families of degenerate orbits are born that only exist for $\epsilon \geq \epsilon_{nm}$. (See Ref. [70] for a detailed classification of the periodic orbits in an elliptic billiard and their bifurcations.) These families consist of marginally stable librating orbits that pendulate between the focal points and have $2n$ reflection points at the boundary; at the bifurcation points they collapse into the n-fold repeated short diameter orbits. Knowing that these new orbits are marginally stable, show from the above result (5.86) that the bifurcation points ϵ_{nm} are given by

$$\epsilon_{nm} = \sin\left(\frac{\pi m}{2n}\right), \qquad m = 1, 2, \ldots, (n-1). \qquad (n \geq 2) \tag{5.87}$$

The bifurcation of the second repetition happens thus at $\epsilon_2 = 1/\sqrt{2}$, corresponding to an axis ratio $q = A/B = \sqrt{2}$. The new family of orbits that appears at this bifurcation point consists of librating oscillations with four reflection points. Two of its members, the "bird" and "butterfly" orbits, are investigated in Problem C.2 and shown in Fig. C.3.

(c) Show, using the prescriptions given in Appendix D, that the Maslov indices for the two isolated primitive diameter orbits are

$$\sigma_< = 5, \qquad \sigma_> = 6. \qquad (5.88)$$

Note: All orbits in the elliptic billiard that turn around the two focal points also occur in marginally stable families and undergo no bifurcations (cf. Ref. [70]). They can therefore, like all the librating orbit families appearing at the bifurcation points of the short diameter orbit, not be described by the trace formula (5.36) for isolated orbits.

Bibliography

[1] M. C. Gutzwiller, J. Math. Phys. **12**, 343 (1971).

[2] R. Balian and C. Bloch, Ann. Phys. (N. Y.) **69**, 76 (1972).

[3] M. V. Berry and M. Tabor, Proc. R. Soc. Lond. **A 349**, 101 (1976).

[4] M. V. Berry and M. Tabor, J. Phys. **A 10**, 371 (1977).

[5] M. C. Gutzwiller: *Chaos in Classical and Quantum Mechanics* (Springer Verlag, New York, 1990).

[6] M. Gutzwiller: *Quantum Chaos*, Scientific American, January 1992, p. 78.

[7] V. M. Strutinsky, Nukleonika (Poland) **20**, 679 (1975); V. M. Strutinsky and A. G. Magner, Sov. J. Part. Nucl. **7**, 138 (1976) [Elem. Part. & Nucl. (Atomizdat, Moscow) **7**, 356 (1976)].

[8] S. C. Creagh and R. G. Littlejohn, Phys. Rev. **A 44**, 836 (1991).

[9] S. C. Creagh and R. G. Littlejohn, J. Phys. **A 25**, 1643 (1992).

[10] J. H. Van Vleck, Proc. Natl. Acad. Sci. USA **14**, 178 (1928).

[11] M. C. Gutzwiller, J. Math. Phys. **8**, 1979 (1967).

[12] M. Morse: *The Calculus of Variations in the Large*, Am. Math. Soc. Colloquium Publ. **18** (New York, 1934).

[13] P. L. N. de Maupertuis: *Accord de différentes lois de la nature qui avaient jusqu'ici paru incompatibles* (Mém. As. Sc., Paris, 1744), p. 267.

[14] L. Euler: *Methodus Inveniendi Lineas Curvas Maximi Minimive Proprietate Gaudentes: Additamentum II* (Bousquet, Lausanne and Geneva, 1744). Modern edition by C. Carathéodry: *Leonhardi Euleri Opera Omnia: Series I, vol. 24* (Orell & Fuessli, Zürich, 1952).

[15] M. Gutzwiller, Physics Scripta **T 9**, 184 (1985).

[16] C. G. Gray, G. Karl, and V. A. Novikov, Ann. Phys. (N. Y.) **251**, 1 (1997).

[17] M. C. Gutzwiller, J. Math. Phys. **10**, 1004 (1969).

[18] M. V. Berry and K. E. Mount, Rep. Prog. Phys. **35**, 315 (1972).

[19] S. C. Creagh, J. M. Robbins and R. G. Littlejohn, Phys. Rev. **A 42**, 1907 (1990).

[20] W. H. Miller, J. Chem. Phys. **63**, 996 (1975).

[21] A. M. Ozorio de Almeida: *Hamiltonian Systems: Chaos and Quantization* (Cambridge University Press, Cambridge, 1988).

[22] P. J. Richens, J. Phys. **A 15**, 2101 (1982).

[23] M. Kuś, F. Haake, and D. Delande, Phys. Rev. Lett. **71**, 2167 (1993)

[24] M. Sieber, J. Phys. **A 30**, 4563 (1997).

[25] A. M. Ozorio de Almeida and J. H. Hannay, J. Phys. **A 20**, 5873 (1987).

[26] K. R. Meyer, Trans. Am. Math. Soc. **149**, 95 (1970).

[27] P. Gaspard and D. Alonso, Phys. Rev. **A 47**, R3468 (1993).

[28] P. Gaspard, D. Alonso, and I. Burghardt, Adv. Chem. Phys., Vol. XC (1995), p. 105.

[29] J. Bolte and S. Keppeler, Phys. Rev. Lett. **81**, 1987 (1998); Ann. Phys. (N. Y.) **274**, 125 (1999).

[30] R. G. Littlejohn and W. G. Flynn, Phys. Rev. **A 44**, 5239 (1991); *ibid.* **A 45**, 7697 (1992).

[31] H. Frisk and T. Guhr, Ann. Phys. (N. Y.) **221**, 229 (1993).

[32] Ch. Amann and M. Brack, J. Phys. **A 35**, 6009 (2002).

[33] A. Alscher and H. Grabert, Eur. Phys. J. **D 13**, 127 (2001).

[34] M. Pletyukhov, Ch. Amann, M. Mehta, and M. Brack, Phys. Rev. Lett. **89**, 116601 (2002).

[35] H. Friedrich and D. Wintgen, Phys. Rep. **183**, 39 (1989).

[36] M. Brack, in: *Atomic Clusters and Nanoparticles*, Les Houches Summer School (NATO Advanced Study Institute), Session LXXIII, eds. C. Guet *et al.* (Springer-Verlag, Berlin, 2001), pp. 161-219.

[37] P. Schmelcher and F. K. Diakonos, Phys. Rev. Lett. **78**, 4733 (1997); Phys. Rev. **E 57**, 2739 (1998).

[38] M. Sieber and F. Steiner, Phys. Rev. Lett. **67**, 1941 (1991); M. Sieber, Chaos **2**, 35 (1992).

[39] V. M. Kolomietz, A. G. Magner, and V. M. Strutinsky, Yad. Fiz. **29**, 1478 (1979); A. G. Magner, V. M. Kolomietz, and V. M. Strutinsky, Izvestiya Akad. Nauk SSSR, Ser. Fiz. **43**, 142 (1979).

[40] K. Richter, D. Ullmo, and R. A. Jalabert, Phys. Rep. **276**, 1 (1996).

[41] A. Voros, Ph.D. thesis, University of Paris (1976, unpublished).

[42] R. Aurich and F. Steiner, Phys. Rev. **A 45**, 583 (1992).

[43] B. Lauritzen, Phys. Rev. **A 43**, 603 (1991).

[44] S. C. Creagh, J. Phys. **A 26**, 95 (1993).

[45] B. Lauritzen and N. D. Whelan, Ann. Phys. (N. Y.) **244**, 112 (1995).

[46] N. D. Whelan, J. Phys. **A 30**, 533 (1997).

[47] T. Szeredi and D. A. Goodings, Phys. Rev. **E 48**, 3518, 3529 (1993).

[48] P. Meier, M. Brack, and S. C. Creagh, Z. Phys. **D 41**, 281 (1997).

[49] M. C. Gutzwiller, Physica **5D**, 183 (1982).

[50] A. Selberg, J. Indian Math. Soc. **20**, 47 (1956).

[51] M. Sieber and F. Steiner, Physica **44D**, 248 (1990).

[52] T. Szeredi and D. A. Goodings, Phys. Rev. **E 48**, 3513 and 3529 (1990).

[53] J. M. Rost and G. Tanner, in: *Classical, Semiclassical and Quantum Dynamics in Atoms*, eds. H. Friedrich and B. Eckhardt (Lecture Notes in Physics 485, Springer, Berlin, 1997), p. 274; see also G. Tanner, K. Richter, and J. M. Rost, Rev. Mod. Phys. **72**, 497 (2000).

[54] *Chaos Focus Issue on Periodic Orbit Theory*, ed. P. Cvitanović: Chaos **2**, pp. 1-158 (1992).

[55] M. Brack, Adv. Solid State Phys. **41**, 459 (2001).

[56] M. Brack and S. R. Jain, Phys. Rev. **A 51**, 3462 (1995).

[57] P. Kramer, M. Moshinsky, and T. H. Seligman, in: *Group Theory and its applications*, ed. by E. M. Loebl (Academic Press, New York, 1975), Vol. III, p. 249; see, in particular, Sect. VII.A.

[58] R. K. Bhaduri, Avinash Khare, S. M. Reimann, and E. L. Tomusiak, Ann. Phys. (N. Y.) **254**, 25 (1997).

[59] G. Barton, Ann. Phys. (N. Y.) **166**, 322 (1986).

[60] M. Hénon and C. Heiles, Astr. J. **69**, 73 (1964).

[61] M. Brack, R. K. Bhaduri, J. Law, and M. V. N. Murthy, Phys. Rev. Lett. **70**, 568 (1993).

[62] R. C. Churchill, G. Pecelli, and D. L. Rod, in: *Stochastic Behavior in Classical and Quantum Hamiltonian Systems*, ed. by G. Casati and J. Ford (Springer-Verlag, N.Y., 1979) p. 76.

[63] K. T. R. Davies, T. E. Huston, and M. Baranger, Chaos **2**, 215 (1992).

[64] M. Brack, in: *Festschrift in honour of the 75th birthday of Martin Gutzwiller*, eds. A. Inomata *et al.*; Foundations of Physics **31**, 209 (2001).

[65] M. Brack, M. Mehta, K. Tanaka, J. Phys. **A 34**, 8199 (2001).

[66] M. Brack, in: *Many-body Physics*, ed. by C. Fiolhais, M. Fiolhais, C. Sousa, and J. N. Urbano (World Scientific, Singapore, 1994), p. 233.

[67] M. Brack, R. K. Bhaduri, J. Law, M. V. N. Murthy, and Ch. Maier, Chaos **5**, 317 (1995); Erratum: Chaos **5**, 707 (1995).

[68] M. Brack, S. C. Creagh, and J. Law, Phys. Rev. **A 57**, 788 (1998).

[69] M. Brack, P. Meier, and K. Tanaka, J. Phys. **A 32**, 331 (1999).

[70] S. Okai, H. Nishioka, and M. Ohta, Mem. Konan Univ., Sci. Ser. **37**, 29 (1990).

Chapter 6

Extensions of the Gutzwiller theory

The trace formula of Gutzwiller discussed in Chapter 5 can only be applied to systems in which all classical periodic orbits are isolated. Often, however, many or all of the periodic orbits occur in degenerate families. This happens when a system has *continuous symmetries:* the nature of the classical orbits then remains invariant under the group of transformations that create the corresponding symmetries (cf. Figs. 6.7 and 6.16). Consequently, the orbits are not isolated; typically they are only marginally stable. Then, as we have discussed in Sect. 5.2 when taking the trace over the Green's function in Eq. (5.15), the stationary phase integration perpendicular to the periodic orbits leads to divergences. In fact, what happens is that the action $S(\mathbf{r}, \mathbf{r}, E)$ of a closed orbit going through some point $\mathbf{r}$ is identically constant over the whole continuous domain of the $\mathbf{r}$-space covered by the orbit when it is subject to any of these symmetry operations. Hence, its second derivatives appearing in Eq. (5.30) become identically zero over a continuous domain, and the stationary phase integral diverges. This situation is typical for most integrable systems. An exception is given by anisotropic harmonic oscillators with incommensurable frequencies (see Sect. 5.6.2 and Problem 5.1). In some cases, already the semiclassical expression of the propagator (5.8), which is the starting point of Gutzwiller's derivations, turns out to be singular along the periodic orbits with the highest degree of degeneracy. As we shall discuss in the following, the degeneracy of a periodic orbit family, which is proportional to the phase space volume occupied by all members of the family, plays an important role.

Balian and Bloch used a multi-reflection expansion of the energy-dependent Green's function to investigate first the smooth part [1] and then the oscillating part [2] of the density of eigenstates in a cavity with ideally reflecting walls. For the oscillating part they also derived a trace formula involving classical periodic orbits. Their method applies to both nonintegrable and integrable systems, and thus also to degenerate orbits. In order to avoid the above mentioned divergences, they replaced the stationary phase integrations by exact integrations in those spatial directions along which the actions of the classical paths are identically constant. (Although their method is quite different from that of Gutzwiller, the use of stationary phase integrations for deriving approximate results in terms of classical paths is common to both approaches.)

Strutinsky and coworkers [3] generalized the Gutzwiller theory for smooth potentials with arbitrary symmetries and correspondingly degenerate periodic orbits. They removed the divergences by going back to the folding integral of the time-dependent propagator (3.27) and performing exactly as many integrations in the folding integral and in the trace integral over the Green's function, as there are independent parameters describing the degeneracy of a periodic orbit family going through a given point r. Creagh and Littlejohn [4, 5] later pursued the same idea using general phase space variables, performing the trace integral exactly over the measure of the group that characterizes the continuous symmetry of the system. For integrable systems, their procedure is analogous to that of Berry and Tabor [6] who derived a trace formula for integrable systems employing the action-angle variables, which led to the same result as their earlier approach [7] which we have discussed in Sect. 2.7.

In the present chapter, we discuss some of these extensions of Gutzwiller's theory to systems with degenerate orbits. We do not enter into much technical detail here but give only the essential ideas. In Sect. 6.1, we present several trace formulae for systems with continuous symmetries and illustrate them by some specific examples. Most of these examples happen to be integrable; they were mainly chosen because they allow for a comparison with results found by other methods or with the exact quantum-mechanical solutions.

Integrable systems are, of course, quite trivial from the point of view of the EBK quantization discussed in Sect. 2.5. We mean this here in the sense that the torus quantization leads directly to a quantum spectrum and no trace formula is needed for the level density. Still, it is quite non-trivial to investigate to which extent EBK quantization and periodic orbit theory lead to similar or even identical results, since they a priori are two quite independent approaches – the first was developed from the very beginning of quantum theory (see Chapter 1), whereas the second only took its shape in the early 1970s. A first clue to the relation between the two approaches was given by Berry and Tabor [7]. Starting from the EBK quantization, they arrived at a trace formula similar in form to that of Gutzwiller, but different in detail and applicable only to integrable systems since only for these, EBK quantization is possible a priori (see Sect. 2.7). The relations between the method of Berry and Tabor and the extensions of Gutzwiller's approach discussed in this chapter will become manifest in several of the chosen examples.

Even when the treatment of degenerate periodic orbits has been made possible, there still persists a generic problem with all these trace formulae. When there exists a continuous parameter in the potential (or in the definition of the boundary of a cavity) that breaks its highest symmetry – let us call it a deformation parameter – then the corresponding transition cannot be handled continuously in most cases. Although the highest-symmetric and the deformed case can each be treated separately by one of the appropriate trace formulae, the zero-deformation limit of the deformed case usually gives diverging amplitudes of some – or all – of the periodic orbit contributions. We have already encountered this situation in the Hénon-Heiles potential discussed in Sect. 5.6.4: there the limit $\alpha=0$, and with it the limit of low energy, leads back to the higher $SU(2)$ symmetry of the harmonic oscillator, and all of the Gutzwiller amplitudes diverge in this limit. This symmetry-breaking transition is associated with the destruction of rational tori in phase space, as discussed in Sect. 2.8.1, and constitutes a rather serious problem for the semiclassical approaches. An intuitively appealing way to handle this transition by classical perturbation theory has been proposed by Creagh [8]; it is discussed and illustrated in Sect. 6.2.

6.1 Trace formulae for degenerate orbits

The essential idea of Strutinsky and Magner [3] to overcome the problem of degenerate orbits was to remove the degeneracy by fixing a specific point $\mathbf{r}$ on the orbit, and then to perform some integrations exactly instead of using the saddle-point approximation. In some cases, such as the billiard systems discussed in the next two sections, this can be done directly starting from Gutzwiller's semiclassical Green's function given in Eq. (5.15). More precisely speaking: when the orbit family has a one-dimensional degeneracy – i.e., when its degeneracy can be described by a single parameter – then only isolated members of the family pass through a fixed point $\mathbf{r}$, and the Green's function for isolated orbits (5.15) can be directly used. Taking the trace over the Green's function, one now must integrate exactly over the area covered by the orbit family, because the action is identically constant over this area and the saddle-point approximation in the direction perpendicular to the orbit, as used for isolated orbits by Gutzwiller, would fail.

For higher-degenerate orbits, however, which occur, e.g., in systems with SU(2), SO(3) or higher symmetries, already the semiclassical propagator $K_{scl}(\mathbf{r}, \mathbf{r}'; t)$ given in Eq. (5.8) and with it the Green's function (5.15) diverges, and a more powerful technique has to be developed. To this purpose, Strutinsky and Magner [3] went back to the folding integral (3.27) for the propagator. In the considered cases of high symmetry, the divergence of K_{scl} can be removed by fixing the intermediate point $\mathbf{r}''$ and performing the folding integrals over some components of $\mathbf{r}''$ exactly instead of by saddle-point approximation, hereby exploiting the continuous symmetry of the problem. This leads, after the Fourier integration in Eq. (3.28), to a symmetry-adapted form of the semiclassical Green's function whose trace integral is furthermore taken exactly over those coordinates which coincide with symmetry parameters of the periodic orbit families.

The precise number and choice of exact integrations depends on the overall degeneracy of a given orbit family and the particular nature and symmetry of the given problem. The cases treated by Strutinsky and coworkers [3, 9, 10] include the spherical cavity (confirming the earlier results of Balian and Bloch [2]), the spheroidal cavity (approximately), and spherical harmonic oscillators in two and three dimensions. A more general discussion of symmetries and the related degeneracies of periodic orbits is given in Sect. 6.1.4 below.

A question will now arise in the alert reader's mind: if the stationary phase approximation is not applied to all integrations perpendicular to the orbits, how does the condition of periodicity then arise, which allows one to eliminate all non-periodic closed orbits from the trace formula? We recall that it is exactly the stationary condition (5.20) for the action that yields the condition for periodicity of the orbits contributing to the trace. This condition is, indeed, not automatically fulfilled if more than one integration of the trace integral is done exactly. (Remember that the integration *along* the orbit is always taken exactly.) Mathematically, non-periodic orbits then also contribute to the trace. But for nonperiodic orbits the mismatch of momenta still leads to a mismatch of phases in the integrand, so that their contributions essentially cancel through rapid phase oscillations when the trace integral is performed. Therefore, nonperiodic orbits usually give small contributions which are of the same order as other terms that are neglected in the periodic orbit theory. There are cases, however, where such contributions are not negligible [11, 12, 13] (see, e.g., the elliptic billiard with a small eccentricity discussed in Sect. 6.2.2).

Therefore, in the extensions of Gutzwiller's theory discussed in this chapter, one always makes the *assumption* that only periodic orbits are important. In this sense, the stationary phase assumption is always made, even if it is not used technically in some of the integrals.

6.1.1 Two-dimensional systems, singly degenerate orbits

In the following two examples, we shall illustrate these ideas for systems with a one-dimensional degeneracy of the periodic orbits that can be overcome by a simple exact trace integration. We choose integrable billiard systems for which the exact quantum-mechanical spectra are known, so that we can test the quality of the semiclassical approximation. Before discussing them explicitly, let us first rewrite the trace formula for two-dimensional billiards, from which the specific examples can easily be derived. We start from the semiclassical Green's function (5.15) for $D = 2$

$$G_{scl}(\mathbf{r},\mathbf{r}';E) = \frac{2\pi}{(2\pi\hbar)^{3/2}} \sum_{class.traj.} \sqrt{|\mathcal{D}|} \exp\left[\frac{i}{\hbar}S(\mathbf{r},\mathbf{r}',E) - i\mu\frac{\pi}{2} - i\frac{3\pi}{4}\right], \qquad (6.1)$$

and express the determinant $\mathcal{D}$ in the local coordinate system (5.21) which here is two-dimensional: $\mathbf{r} = (q, r_\perp)$, where q is measured along the orbit and $r_\perp$ perpendicular to it. The expression for $\mathcal{D}$ is given in Eq. (5.25). For a billiard system, all orbits are straight lines between successive reflections, so that the total energy E is just the kinetic energy of the free motion along the orbit: $E = p_q^2/2m = p_q'^2/2m = p^2/2m$. The modulus of $\mathcal{D}$ thus simplifies to

$$|\mathcal{D}(q, r_\perp; E)| = \left(\frac{m}{p}\right)^2 \left|\frac{\partial p_\perp'}{\partial r_\perp}\right|. \qquad (6.2)$$

Including only periodic orbits, whose actions and Maslov indices do not depend on the starting point $\mathbf{r}$, the trace formula becomes

$$\delta g(E) = \frac{1}{(2\pi\hbar)^{3/2}} \frac{2m}{p} \sum_{po} \cos\left[\frac{1}{\hbar}S_{po}(E) - \sigma_{po}\frac{\pi}{2} - \frac{\pi}{4}\right] \int |\mathcal{J}_{po}|^{-1/2} \, dq \, dr_\perp, \qquad (6.3)$$

where we have expressed the integrand through the Jacobian

$$\mathcal{J}_{po} = \left(\frac{\partial r_\perp}{\partial p_\perp'}\right)_{po}, \qquad (6.4)$$

which determines the variation of the final perpendicular coordinate due to a change of the initial perpendicular momentum, and thus is a measure of the stability of the periodic orbit. The Jacobian (6.4) becomes zero at caustic points; for a specific orbit these are, however, only isolated points which have a zero measure in the integral over x and y so that the trace in Eq. (6.3) is well defined. The phase changes occurring at the caustic and reflection points are contained, as in the trace formula (5.36) for isolated orbits, in the overall Maslov index σ_{po} which here is the same as the index μ appearing in the semiclassical Green's function (6.1). Note that, since no saddle-point integration has been made in taking the trace, the contribution ν in the general form of the Maslov index (5.37) is zero here.

6.1.2   Example 1: The equilateral triangular billiard

Let us first consider a particle with mass m in a two-dimensional billiard whose boundary forms an equilateral triangle with side length L. The particle is reflected at the boundary according to the specular law: incoming angle = outgoing angle. Between two reflections, the motion is that of a free particle, so that it moves along straight lines with constant momentum $p = \sqrt{2mE}$, where E is the total (kinetic) energy and $p^2 = p_x^2 + p_y^2$. Although this system is not separable because of its non-cartesian boundary, it is actually integrable (cf., e.g., Ref. [14]). In Sect. 3.2.7, we have written down the wavefunctions and the energy spectrum of this system and derived an exact quantum-mechanical trace formula, which is given in Eq. (3.100). Presently we shall show that the leading terms of this trace formula can, in their asymptotic expansion for large actions, be obtained from the extended Gutzwiller theory.

The three simplest types of periodic orbits in the triangular billiard are shown in Fig. 6.1. On the left, we see a triangular orbit of length $3L/2$ which is isolated and marginally stable. Its contribution can be derived from the usual Gutzwiller trace formula for isolated orbits; the explicit evaluation is posed as Problem 5.2 in Chapter 5. In the middle of Fig. 6.1, we see five different orbits with the length $L_{10} = \sqrt{3}L$ that corresponds to twice the height of the triangle. All these orbits belong to a continuous family of degenerate orbits with the same length and thus the same action $S_{10} = pL_{10}$. We shall define the nomenclature for a general orbit family $(M_1 M_2)$ in a moment. On the right, we show three members of a degenerate family of orbits with the length $L_{11} = 3L$ corresponding to the triangle's circumference. These two families, as well as all other families whose orbit lengths we will determine below, can be characterized by one degeneracy parameter which we choose to be the position of one of the reflection points along the boundary. Moving this point over one half of a side will yield all members of the family exactly once, not counting their discrete degeneracies.

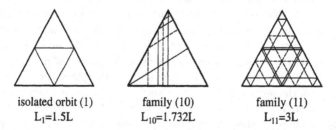

isolated orbit (1)   family (10)   family (11)
$L_1=1.5L$ $L_{10}=1.732L$ $L_{11}=3L$

Figure 6.1: The three shortest (families of) orbits in the equilateral triangle with unit side length. *Left:* isolated orbit with length $L_1=3L/2$; *center:* five members of the orbit family (10) with length $L_{10}=\sqrt{3}L$; *right:* three members of the orbit family (11) with length $L_{11}=3L$.

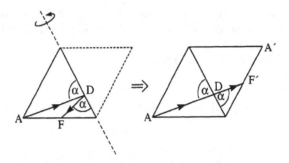

Figure 6.2: Diagram for illustrating the reflection rule. Reflection at the point D and arriving at point F of the original triangle is equivalent to passing on a straight line through point D into the reflected triangle and arriving at its point F'.

The forms of the other periodic orbits in this system are rather complicated and not easy to guess. For example, the next longer orbit family (12) has already 10 reflections on the boundary. In order to find the lengths and other characteristics of all periodic orbits of this system, we use a technique that is rather powerful and simplifies the task enormously. It is sometimes called the 'reflection rule' and is illustrated in Fig. 6.2. Take the trajectory which starts at the point A, is reflected in the point D and from there reaches point F. Now reflect the triangular boundary at the side containing the point D, which is equivalent to rotating it by 180 degrees about this side, leading to the dotted triangle shown on the left-hand side of Fig. 6.2. Denoting the images of points A and F by A' and F', respectively, we note that the points A, D and F' lie on a straight line (see the right-hand side of the figure). Thus, being reflected at the point D inside the original boundary is equivalent for a particle to passing straight through this point into the reflected triangle. The next portion of the trajectory after reflection in the point F can now be found by reflecting the new triangle at the side containing the point F' and letting the particle continue on a straight line through point F'. This process can be repeated as long as the trajectory lasts. If one is hitting an image of the starting point A after a finite number of steps, the trajectory is that of a periodic orbit, and the length of the primitive orbit is the distance between the starting point A and its image encountered for the first time along the mapped-out straight orbit.

Similarly, by reflecting the equilateral triangle at all its sides and repeating this *ad infinitum*, we can tesselate the entire plane, covering it exactly once. (It can be shown rather generally that any polygon which tesselates the plane under the reflection rule is integrable [14].) In Fig. 6.3, we show one sixth of the plane corresponding to the first 60 degrees of the polar angle. To the lower left, we have the original triangle with its corners numbered clockwise by the letters A, B and C. The corners of the reflected triangles are numbered correspondingly with the same letters. Note that each group of six triangles joining at one corner C forms a regular hexagon; each triangle within such a hexagon has a different orientation of its corners A, B and C.

The periodic orbits occurring in degenerate families now correspond to straight lines which connect two equivalent points of any pair of triangles which have identical orientations and thus are positioned at equivalent places within the hexagons. We have labeled them in Fig. 6.3 by the numbers (mn) which correspond to a 'coordinate system' that allows to uniquely identify them on the plane. The numbers m count the successive corners C along horizontal lines and the numbers n those along lines with an inclination of 30 degrees. Note that the numbers (mn) are different from those used in Fig. 6.1; their relation will be be given in a moment.

Since all orbit families contain a member that passes through any of the corners, we find *all* periodic orbit families by connecting the corner C (00) of the original triangle systematically with *all* reflected corners C on the plane, as is shown by the arrows in Fig. 6.3. All members of one degenerate orbit family are obtained by a parallel shift of the corresponding arrow such that its starting point C is moved along any side of the triangle over the distance $L/2$. Reverting the direction of an arrow corresponds to a time reversal operation. Reflecting an arrow at the n or the m axis leads to the same orbit family, but with a different orientation within the triangle, corresponding to the discrete symmetry of the boundary concerning rotations about 120 and 240 degrees around its center. We come back to these discrete symmetries further below.

The length of an orbit of the family denoted by (mn) is easily found by applying the law of Pythagoras. In order to reach from (00) to (mn), we have to walk $m + n/2$ times the distance $3L$ (i.e., the circumference of the triangle) horizontally to the right and n times the distance $\sqrt{3}L/2$ (i.e., the height of the triangle) in the perpendicular direction, i.e., vertically up. This gives the following length L_{mn} of the orbit (mn):

$$L_{mn} = L\sqrt{\left(3m + \frac{3}{2}n\right)^2 + \left(\frac{\sqrt{3}}{2}n\right)^2} = L\sqrt{3\left(3m^2 + 3mn + n^2\right)}. \tag{6.5}$$

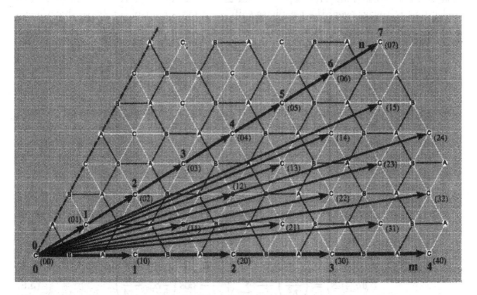

Figure 6.3: The plane tesselated with equilateral triangles. Each straight line with an arrow pointing from the lower left corner C (00) to a corner C (mn) corresponds to a family of degenerate orbits with given length L_{mn} and action $S_{mn} = pL_{mn}$.

For later comparison with the quantum result discussed in Sect. 3.2.7, we make the following transformation:

$$M_1 = m, \quad M_2 = m+n, \qquad M_1 = 0, 1, 2, \ldots, \quad M_2 = M_1, M_1+1, M_1+2, \ldots. \quad (6.6)$$

Inserting this into Eq. (6.5), we get for the orbit lengths:

$$L_{M_1 M_2} = L \sqrt{3 \left(M_1^2 + M_2^2 + M_1 M_2 \right)}. \qquad (6.7)$$

We now can identify the two families with shortest lengths shown in Fig. 6.1. The family $(M_1 M_2) = (10)$ and its repetitions is found on any line that forms a 30 degree angle with the horizontal in Fig. 6.3, i.e., on any line parallel to the n axis containing the points $(mn) = (0n)$. The family $(M_1 M_2) = (11)$ and its repetitions is lying on any horizontal line parallel to the m axis which contains the points $(mn) = (m0)$. It is also easy to verify that the number of reflections of a given orbit is $r = 6m+4n = 2M_1+4M_2$, whereby reflections in a corner must be counted twice for the $(m0)$ and three times for all other orbits, as seen from Fig. 6.3 by slightly shifting the orbits away from a corner.

Having determined the lengths and therefore the actions $S_{M_1 M_2} = pL_{M_1 M_2}$ of the two-dimensional periodic orbit families, we now have to find their Maslov indices and the amplitudes of the trace formula (6.3). The Maslov indices are easy to calculate. From Fig. 6.3 it is immediately clear that there are no self-crossings of the orbits after slight changes of their initial direction, since they always remain straight lines. Thus, the only contribution to σ_{po} comes from the reflections at the boundary. Since each reflection gives a phase shift of $-\pi$ for the Dirichlet and zero for the Neumann boundary conditions, corresponding to $\Delta\sigma_{po} = 2$ and $\Delta\sigma_{po} = 0$, respectively, and the number of reflections is always even, the net phase shift is always a multiple of 2π, so that we can put $\sigma_{po} = 0$ for all orbit families. The Jacobian $\mathcal{J}_{po}$ in Eq. (6.4) is also found easily from Fig. 6.3, bearing in mind that a periodic orbit corresponds to a finite straight line in this figure. Thus, the local coordinate system to be used is q along the orbit (with momentum p), and $r_\perp$ and $p_\perp$ in the transverse direction. Writing an infinitesimal change of the initial transverse momentum as $dp'_\perp = p\,d\alpha$, where $d\alpha$ is the change of the polar angle of the orbit, and the change of the final transverse coordinate as $dr_\perp = L_{po}d\alpha$, where L_{po} is the length of the orbit, we get immediately

$$\mathcal{J}_{po} = \left(\frac{\partial r_\perp}{\partial p'_\perp} \right)_{po} = \frac{L_{po}}{p}. \qquad (6.8)$$

This Jacobian is indepedent of the coordinates, which holds for all polygonal billiards. It can thus be taken outside the spatial integral in (6.3), so that the latter simply yields the area $\int dq\, dr_\perp = \mathcal{A}_\Delta = \sqrt{3}L^2/4$ of the billiard which is fully covered by all orbit families. Putting all ingredients together, we obtain the following result

$$\delta g_{scl}^{(2)}(E) = \left(\frac{2m}{\hbar^2} \right) \frac{\mathcal{A}_\Delta}{4\pi} \sum_{po} \sqrt{\frac{2\hbar}{\pi S_{po}}} \, \cos\left(\frac{1}{\hbar}S_{po} - \frac{\pi}{4} \right). \qquad (6.9)$$

This agrees with the leading oscillating term of the exact quantum-mechanical level density (3.102) derived in Sect. 3.2.7, after an asymptotic expansion of the Bessel function for large arguments $x \gg 1$, giving $J_0(x) \sim (2/\pi x)^{1/2} \cos(x - \pi/4)$, and explicit insertion of the quantities $\mathcal{A}_\Delta$ above and E_Δ as defined in Eq. (3.94). We see thus that the periodic orbit theory, in the present form for simply degenerate orbits, gives a result that becomes exact for sufficiently large $S_{po}/\hbar$. To estimate the quality of the asymptotic expansion of $J_0(x)$ in this context, we take the contribution of the shortest orbit family (10) with length $L_{10} = \sqrt{3}L$ at the lowest energy that is of interest, i.e., at the energy of the lowest quantum state $E = E_{12} = 3E_\Delta$, see Eq. (3.94) in Sect. 3.2.7. The argument of the Bessel function is then found to be $x = S_{10}/\hbar = 4\pi$ for which the asymptotic expansion is already good to within about two percent; for larger energies and for contributions of longer orbits it will work even better.

It remains to check that the double sum over M_1 and M_2 in the first term of Eq. (3.102) indeed counts all periodic orbit families, denoted briefly by '*po*' in (6.9) above, including their discrete degeneracies. For that we must make sure that we correctly take into account the discrete symmetries of the problem, i.e., the reflection symmetry about the three symmetry axes, the rotational symmetry about 120 and 240 degrees, and the time reversal symmetry – without double counting, however, when combining these operations. This is again done most simply by inspecting Fig. 6.3. In the first sixth of the entire plane shown there (not counting the line at 60 degrees), all orbits for which one of m and n is zero occur only once, whereas those with both nonzero m and n occur twice. Since we have to include *all* periodic orbits passing through a given point **r** under the trace integral, this means that we have to rotate the direction of the orbit arrow around the full 360 degrees of the (x, y) plane. In doing so, we get 6 times the orbit families $(M0)$ and (MM) with $M > 0$, and 12 times the families $(M_1 M_2)$ with $0 < M_1 < M_2$. These are, indeed, the factors appearing in front of the corresponding sums given explicitly on the first line of Eq. (3.100) for the exact quantum-mechanical result. Note that the latter also contains the Thomas-Fermi part of the level density which arises (with weight 1) from the point $(mn) = (M_1 M_2) = (00)$ (the so-called 'zero-length orbit') at the center of the plane. We recall that the cartesian mesh of points $(M_1 M_2)$ is exactly the topological mesh used by Berry and Tabor [6, 7] in their derivations of a semiclassical trace formula for integrable systems (see Sect. 2.7). When deriving the exact trace formula for the triangle in Sect. 3.2.7, we have used the same starting point as Berry and Tabor and employed the Poisson summation formula to transform the exact quantum level density, without making any approximation, from a sum over delta functions (3.95) into Eq. (3.102). Note that this result, which is a mathematical identity, contains as the *leading term* the sum over the mesh $(M_1 M_2)$, but also a term with a single sum which is of higher order in $\hbar$. The method of Berry and Tabor reproduces the leading term (see Problem 6.3) in the same approximation Eq. (6.9) as we have obtained it here from the extended Gutzwiller theory, but it does not yield the $\hbar$-correction (unless the edge terms in the Poisson formula are included, which is often not done when using this method). As we have already indicated, the $\hbar$-correction term includes the contributions from the isolated orbit (1) shown at the left of Fig. 6.1, but also other contributions which we will interpret semiclassically below.

At this point it is instructive, and actually revealing a surprise, to discuss the relative importance of the isolated orbit and, more generally, of the $\hbar$-corrections contained in the exact quantum result (3.102). Remember that Eq. (6.9) only represents the leading term of the oscillating part of the level density, coming from the degenerate orbit families. The isolated orbit (M) and its repetitions with lengths $L_M = 3ML/2$ (with odd M) give the contribution (5.83) evaluated in Problem 5.2 using the standard trace formula of Gutzwiller. We repeat it here for convenience, inserting the value of E_Δ in (3.94) in order to exhibit the power of $\hbar$ in its amplitude:

$$\delta g_{scl}^{(1)}(E) = \mp \frac{1}{4\pi} \left(\frac{2m}{\hbar^2} \right)^{1/2} \frac{3L}{\sqrt{E}} \sum_{M=1,3,5,\ldots}^{\infty} \cos \left(\frac{1}{\hbar} p L_M \right), \qquad (6.10)$$

where the upper and lower sign holds for Dirichlet and Neumann boundary conditions, respectively. Let us now take the ratio $\mathcal{R}$ of the amplitude of the correction term (6.10) and that of the leading term in (6.9). Apart from numerical constants, it is

$$\mathcal{R} = \sqrt{\hbar L_{M_1 M_2}} / \sqrt{p} L. \qquad (6.11)$$

The isolated orbit thus gives a contribution of higher order in $\hbar$ than the degenerate orbit families and goes to zero for large energies if the length of the orbit is fixed. One might want to ignore it in view of the fact that other terms of higher order in $\hbar$ have already been neglected at various places in the derivation of the semiclassical result. In general, such $\hbar$ corrections are indeed small, and that is the reason why they are usually neglected. We shall see, however, that in the present case the isolated orbit contribution is not negligible; the exact quantum-mechanical trace formula (3.102) derived in Sect. 3.2.7 being very helpful for a quantitative analysis. Indeed, the result (3.102) does exactly contain the contribution (6.10) from the isolated orbit, but the sum over M there also includes the even values.

Before investigating numerically the importance of these higher $\hbar$ terms, we want to give a qualitative semiclassical interpretation to the even-M contributions in the second term of the quantum result (3.102). First we note that the length $L_M = 3ML/2$ for even M is identical to the length L_{MM} of the $M/2$-th repetition of the orbit family (11) shown to the right of Fig. 6.1. Now, in evaluating the leading-order term (6.9) of the trace formula, one is implicitly integrating over all members of each family (see also Sect. 6.1.4 below, where this is done explicitly, and Problem 6.3). The orbit family (11) has two 'edge terms': one is the equilateral triangle connecting the midpoints of the boundary, which looks like the isolated orbit (1) but is run through twice during one period (a nearby orbit is shown in the figure), and the other is an orbit that 'grazes' the boundary, being reflected twice in the corners (this must be taken as a limit of the outermost orbit shown in the figure). Both these orbits are contained in the leading-order term (6.9). However, the orbit that grazes the boundary should not be admitted from a quantum-mechanical point of view when Dirichlet boundary conditions are imposed, which force the wavefunctions and thus the probability amplitudes of all states to be zero along the boundary. Even if we picture the classical orbit as a standing wave, it should not be allowed along the walls of the billiard at a distance smaller than its de Broglie wavelength. This contribution therefore ought to be cancelled at some higher order in $\hbar$, as a quantum correction to the lowest-order semiclassical periodic orbit term (cf. the discussion of grazing and diffraction corrections in Sect. 7.5).

This cancellation is, indeed, found to take place in a careful asymptotic analysis [15] of the integral over the degenerate orbit families. Whereas the degenerate periodic orbits themselves arise from the stationary points inside the integration interval, the end-point corrections to the integral – evaluated by a formula which we give in Eq. (6.86) of Sect. 6.2.4 – yield contributions of order $\hbar^{1/2}$ (relative to the leading-order terms) that correspond to the grazing orbits along the boundary and are found to be identical to the terms with even M in the second part of (3.102), with the desired negative overall sign. – If one starts from the Neumann boundary conditions which force the probability amplitudes to have a maximum along the boundary, one may want to give the grazing orbits more weight, since in the leading-order result (6.9) they have only been counted with the same average weight as the rest of the orbit family, and thus to *add* rather than to subtract the corresponding correction terms. Semiclassically these corrections arise, indeed, with the opposite sign – like the contributions of the isolated central orbit (M) with odd M which receive no phase shifts (i.e., Maslov indices zero) from the wall reflections with Neumann boundary conditions.

It is instructive now to assess the relative importance of the different contributions to the level density by numerical calculations. We first investigate the gross-shell behavior of the semiclassical trace formula (6.10) for $\delta g(E)$ by Gaussian averaging it over the wave number $k = p/\hbar$ with a width γ. As discussed in Sect. 5.5, this leads to an extra Gaussian factor $\exp[-(\gamma L_{po}/2)^2]$ under the periodic orbit sum. Using an averaging range $\gamma = 1/L$, only the shortest orbits should contribute significantly. This is confirmed in the top part of Fig. 6.4 on the next page, where we have plotted the coarse-grained $\delta g(E)$ as a function of $\sqrt{E/E_\Delta}$ which is proportional to the wave number. The dotted line results when all orbits are included. Within the accuracy of the figure, this cannot be distinguished from the oscillating part of the exact quantum-mechanical level density obtained from Eq. (3.102) with the same Gaussian averaging. The solid line in the top part gives the result obtained when only the three shortest primitive orbits, shown in Fig. 6.1, are included. Except at the lowest energies, three orbits give thus already an excellent approximation to the coarse-grained level density, the influence of the longer orbits being sufficiently suppressed by the Gaussian averaging factor. The beating pattern of the oscillations in $\delta g(E)$ is therefore seen to come about through the interference of these three shortest orbits. (This situation is quite similar to that in the Hénon-Heiles potential discussed in Sect. 5.6.4.)

In the center part of Fig. 6.4, we give the sum of the contributions of the two shortest orbits (10) and (11), and in the bottom part that of the isolated orbit (1) alone; both are shown over a larger energy range than their sum in the top part. We see that the contribution of the isolated orbit, although being of higher order in $\hbar$ and going to zero for large energies due to the factor $1/\sqrt{p}$ given in Eq. (6.11), is essential in bringing about the correct beating pattern of the quantum result, as seen in the top part of the figure.

The $\hbar$-correction term of the exact trace formula becomes even more important when one asks for a higher resolution of the shell structure or, which in the present case poses no problem, when one sums over all orbits in the trace formula in order to obtain the exact quantum spectrum. To this purpose, let us study the behavior of the amplitude ratio $\mathcal{R}$ (6.11) in the limit of infinite orbit length. Denoting the larger of M_1 and M_2 simply by M, $L_{M_1 M_2}$ will go to infinity like ML. For large energies, the EBK quantization is exact and we may identify M with the larger of the quantum numbers, so that the quantized energies (3.94) go like $E \sim M^2 E_\Delta \sim M^2/L^2$. Therefore, both numerator and denominator of $\mathcal{R}$ (6.11) diverge in the same way for large energies and

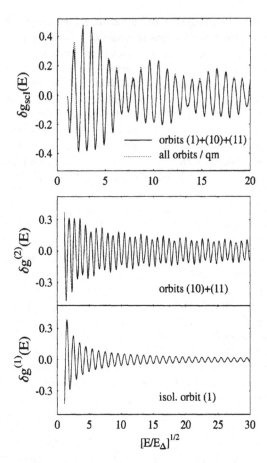

Figure 6.4:   Oscillating part of the level density in the equilateral triangular billiard, obtained using the semiclassical trace formula (6.10) and plotted versus $\sqrt{E/E_\Delta}$ (Gaussian averaging width: $\gamma = 1/L$). *Top:* using all orbits (dotted line; the result cannot be distinguished from the full quantum-mechanical result 'qm' using Eq. (3.102)), and using only the three shortest primitive orbits shown in Fig. 6.1 (solid line). *Center:* separate contribution of the two orbit families (10) and (11). *Bottom:* separate contribution of the primitive isolated orbit (1). The energy unit E_Δ is defined in Eq. (3.94).

long orbits, so that we must expect it to stay finite when both limits are taken simultaneously. (In fact, one finds that $R \sim 1$ when all numerical constants are inserted.) For the exact quantization, the $\hbar$-correction term is thus equally important as the leading-order term.

This is illustrated in Fig. 6.5, where we show the total level density $g(k)$, including its average part, as a function of E/E_Δ (Dirichlet boundary conditions assumed). It is here Gaussian averaged using a very small width of $\gamma = 0.02/L$ and normalized such that each quantum level has the height of its expected degeneracy (i.e., like the k-space level density in Eq. (3.9), so that each single delta function appears with the height 1).

In the lower part of the figure, the exact trace formula (3.102) has been used. It yields, as expected, the correct quantum spectrum with the correct degeneracies according to Eq. (3.94). In the upper part, the leading-order term given in Eq. (6.9) and the TF part (i.e., the first term) of the average level density (3.107) have been used, which would be the standard approximation of the POT. Besides yielding the exact quantum spectrum, this gives an extra series of spurious quantum energies with degeneracies 1, positioned at the energies $M^2 E_\Delta$ with $M = 1, 2, 3...$ and marked by the small arrows. These spurious states now are exactly cancelled by adding the $\hbar$-correction term in Eq. (3.102). This is not surprising, if one notices that the latter is identical with the level density of a one-dimensional box, see Eq. (3.64), with the length $3L/4$ that yields exactly the spurious spectrum marked by the arrows.

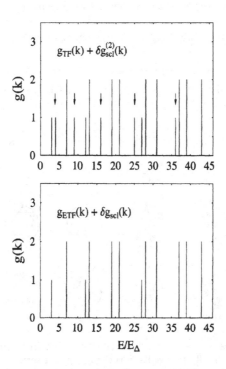

Figure 6.5: Level density of the equilateral triangle billiard for Dirichlet boundary conditions, Gaussian-averaged with a width $\gamma = 0.02/L$. *Above:* using the leading-order semiclassical trace formula (6.9) and adding the average TF part. The arrows mark those levels which are spurious. *Below:* using the exact quantum-mechanical trace formula (3.102).

6.1.3   Example 2: The two-dimensional disk billiard

In this example, we come back to the circular billiard with radius R which we have already discussed in Chapter 2. In Sect. 2.6.3, we have seen that the EBK quantization does not yield its exact quantum spectrum (2.155). In the following we sketch the derivation of the same result from the extended Gutzwiller theory given in Sect. 6.1.1 above.

The classical periodic orbits for this system, some of which are shown in Fig. 6.6, are the same as those in a spherical cavity with reflecting walls discussed by Balian and Bloch [2] (see Sect. 6.1.6). They may be uniquely labeled by two integers (v, w), where v is the number of vertices (or sides) and w the winding number around the center. They must fulfill $v \geq 2w$. For $w=1$ one has ordinary regular polygons, and for $w > 1$ star-like polygons that intersect themselves, except if v and w are coprime. In this case, the labels $(v, w) = (na, nb)$ describe the nth repetition of a primitive orbit with a vertices (sides) and winding number b. The lengths L_{vw} of the orbits are given by

$$L_{vw} = 2\, v R \sin \varphi_{vw}\,, \qquad \varphi_{vw} = \pi\, w/v\,. \tag{6.12}$$

The angle φ_{vw} is one-half of the polar angle covered by one segment of the polygon (v, w).

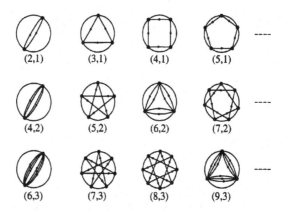

Figure 6.6:   Classical periodic orbits, labeled (v, w), in a circular billiard or a spherical cavity with reflecting walls. v is the number of vertices (or sides) and w is the winding number around the center. All orbits are piecewise straight-lined; the curved portions are just drawn to emphasize the repeated passages along the same path. (After [2].)

Due to the $U(1)$ symmetry of the boundary, all periodic orbits occur in degenerate families whose degeneracy can be characterized by the polar angle ϕ by which one can rotate their members into each other. This is illustrated in Fig. 6.7 for the case of the triangular orbit $(3, 1)$. All these orbits are marginally stable, the eigenvalues of their stability matrix being equal to unity. Their amplitudes in the Gutzwiller trace formula (5.36) therefore diverge. But as already mentioned above, finite amplitudes are obtained by performing the trace integral over the Green's function exactly over the area covered by each of the orbit families, leading to the trace formula (6.3). In the present case, this is just the ring area between the boundary and the inner caustic of the orbit family (shown by the dashed line in Fig. 6.7). The Jacobian $\mathcal{J}_{vw}$ (6.4) is,

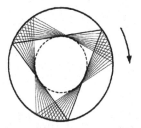

Figure 6.7: Degeneracy of the periodic orbits in the circular billiard, illustrated for the triangle, $(v, w) = (3, 1)$. The actions and stabilities of the orbits are invariant under arbitrary rotations around the center. The family $(3,1)$ of orbits covers the ring area between the outer boundary and the dashed caustic circle. (From [16].)

however, not independent of the point $\mathbf{r}$ in which it is evaluated – differently from the triangle billiard discussed above. Its somewhat lengthy geometrical evaluation has been done following Ref. [3]. Expressed in polar coordinates (r, ϕ) it becomes

$$\mathcal{J}_{vw} = \left(\frac{\partial r_\perp}{\partial p'_\perp}\right)_{vw} = \frac{2v(r^2 - R^2 \cos^2 \varphi_{vw})}{pR \sin \varphi_{vw}}, \tag{6.13}$$

which is independent of the polar angle ϕ due to the axial symmetry. Its integration over the indicated ring area is elementary. The Maslov index can be computed with the rules given in Appendix D and is evaluated in Problem 6.1. Inserting the results into Eq. (6.3), one finds [16] the following trace formula for the oscillating part of the level density

$$\delta g_{scl}(E) = \frac{1}{E_0} \sqrt{\frac{\hbar}{\pi pR}} \sum_{w=1}^{\infty} \sum_{v=2w}^{\infty} f_{vw} \frac{\sin^{3/2} \varphi_{vw}}{\sqrt{v}} \sin \Phi_{vw}, \tag{6.14}$$

where the overall phase Φ_{vw} and the factor f_{vw} are given by

$$\Phi_{vw} = pL_{vw}/\hbar - 3v\frac{\pi}{2} + \frac{3\pi}{4}, \qquad f_{vw} = \left\{ \begin{array}{ll} 1 & \text{for } v = 2w \\ 2 & \text{for } v > 2w \end{array} \right\}, \tag{6.15}$$

and the natural energy unit E_0 here is

$$E_0 = \frac{\hbar^2}{2mR^2}. \tag{6.16}$$

In (6.14), E is the energy and $p = \sqrt{2mE}$ the momentum of the particle, and pL_{vw} is the classical action S_{vw} of the orbit (v, w). The factor f_{vw} counts the fact that for all orbit families except that of the diameter ($v=2w$), one always finds two different specific orbits going through one fixed point **r**. We note that the result (6.14) is identical to that derived in 1972 by Bogachek and Gogadze [17] by a method equivalent to that of Berry and Tabor (see the discussion at the end of Sect. 2.7). It has also been derived by Tatievski *et al.* [18] employing the Green's function reflection expansion of Balian and Bloch [2].

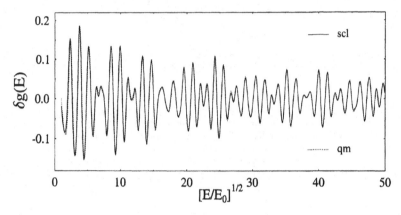

Figure 6.8: Oscillating part of the level density in the disk billiard, averaged over k with a width $\gamma = 0.4/R$. The solid line shows the semiclassical and the dashed line the quantum-mechanical result. (After [16].)

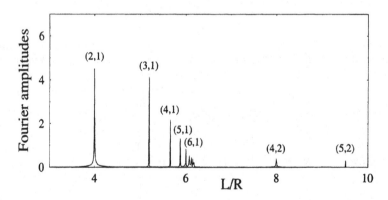

Figure 6.9: Fourier spectrum of the level density $\delta g (E)$ of the disk billiard obtained with $\gamma = 0.4/R$, plotted versus relative orbit length L/R. The labels (v, w) of the leading orbits are indicated.

In Fig. 6.8 we show the oscillating part of the level density obtained with the trace formula (6.14) after Gaussian averaging over k with a width $\gamma = 0.4/R$ (cf. Sect. 5.5). It is compared to the correspondingly averaged quantum-mechanical result which is shown by the dashed line. The difference can hardly be recognized. The rather complicated oscillatory pattern of the level density comes about by the superposition of the three shortest orbits, $(v, 1)$, with $v = 2$ (diameter), 3 (triangle) and 4 (square).

Fig. 6.9 contains the Fourier transform of the averaged quantum-mechanical $\delta g\,(E)$, showing us the length spectrum of these leading orbits. Some small peaks indicate minor contributions of longer orbits, such as the second repetition (4,2) of the diameter and the (5,2) star orbit (cf. Fig. 6.6). The decrease of amplitudes is due to the factor $1/\sqrt{v}$ in (6.14) and the Gaussian smoothing factor $\exp(-\gamma^2 L_{vw}^2/4)$.

It is interesting now to study the convergence of the periodic orbit sum in Eq. (6.14). As discussed in Sect. 5.5, the convergence of the trace formula is in most cases a difficult mathematical problem. The result (6.14) for the disk billiard contains a series that is not absolutely convergent, and we know of no exact study of its convergence nor of any exact way of doing the summation. Numerically, its convergence has been studied in Ref. [16] by letting the smoothing width go to very small values. If one chooses a Gaussian smoothing with $\gamma = 0.02/R$, which is smaller than any of the quantum-mechanical level spacings up to $kR \sim 11$, one obtains the result shown in Fig. 6.10. Here we have added to $\delta g\,(k)$ in (6.14) the smooth part of the level density which is given by

$$\tilde{g}(k) = \frac{R}{2}\,(kR - 1) \qquad (6.17)$$

(see Sect. 4.6), and normalized the total level density by the factor $\gamma\sqrt{\pi}$ like in Fig. 6.5. What we see in Fig. 6.10 is, indeed, a quantum spectrum with the correct degeneracies, 1 for the levels with $l = 0$ and 2 for the levels with $|l| > 0$. The position of the levels closely agrees with the quantum-mechanical ones, given in Eq. (2.155), but not exactly. In fact, a careful numerical convergence study [16] reveals that the peak positions seen in Fig. 6.10 agree with the EBK spectrum of the disk billiard, as given by Eq. (2.153) in Sect. 2.6.3, up to eight digits. This agreement is not trivial. In principle, the trace formula (6.14) can be derived from the EBK spectrum, as shown by Bogachek and Gogadze [17]. However, in this derivation one uses a saddle-point approximation to an exact integral, which usually entails the neglect of some higher-order terms in $\hbar$. Why and how these neglected terms should cancel is not evident at all.

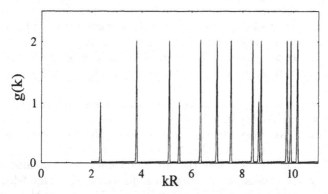

Figure 6.10: Total level density obtained from the semiclassical trace formula with a smoothing width $\gamma = 0.02/R$ and adding the smooth part (6.17). The sum over periodic orbits in Eq. (6.14) is truncated at $v_{max} = w_{max} = 200$. (After [16].)

The disk billiard might appear as a quite academic study system at this place. However, we shall see in Sect. 8.3 that it finds a nice physical application in connection with nanostructures, where it can serve as a realistic lowest-order approximation to the mean field of the electrons in a circular quantum dot.

6.1.4   More general treatment of continuous symmetries

In the above examples of the triangle and disk billiards, we have illustrated the idea of taking an exact trace integral over the semiclassical Green's function. This method of exploiting the symmetry of a given system is not always easy, in particular if the symmetry cannot be directly linked to a simple spatial operation like a rotation or translation. A more abstract mathematical approach to deal with continuous symmetries and to derive trace formulae for such cases has been developed by Creagh and Littlejohn [4, 5]. We will briefly summarize here the essential ideas of their approach and then illustrate it by re-deriving the amplitudes in the trace formulae for the circular and rectangular billiards.

Assume a system with D degrees of freedom to possess a continuous symmetry that can be characterized by f independent parameters, such as the polar angle ϕ for the $U(1)$ symmetry ($f=1$), or the three Euler angles for the spherical $SO(3)$ symmetry ($f=3$). The periodic orbits of this system will then occur in continuous degenerate families rather than as isolated trajectories. If the infinitesimal generators of the group G associated with the symmetry are all linearly independent, then the degree of degeneracy f of these orbit families is equal to the dimension of the group G. The members of each orbit family can thus be transformed into each other by acting with the group elements g on any of them, and the properties of the orbits, in particular their action integrals and Maslov indices, remain invariant under the group operations. In conservative systems where the energy is a constant of motion, the time t plays the analogous role of a group parameter, associated to the generator H of the time translation symmetry. Then, the periodic orbit families cover $(f+1)$-dimensional hypersurfaces in phase space; following Ref. [4] we will denote them henceforth by Γ. The sum over periodic orbits in the trace formula thus turns into a sum over discrete orbit families which may be labeled by Γ, each summand containing an $(f+1)$-dimensional integral over the parameters describing their degeneracies. It is this integral that must be done exactly, since the phase of the integrand is stationary over the whole surface Γ. Only in the integrations transverse to the orbit families, i.e., orthogonal to Γ, may one use the stationary phase approximation.

The most obvious way to do this exact $(f+1)$-dimensional integration – although not necessarily the easiest one in practice – is that proposed by Creagh and Littlejohn [4]: one chooses as integration variables the time t plus the f group parameters themselves. This leads to the $(f+1)$-dimensional measure $dt\, d\mu(g)$, where $d\mu(g)$ is the natural group measure, usually called the "Haar measure". Since we always have to take the trace over the D-dimensional **r**-space in order to recover the level density $g(E)$ from the Green's function $G(\mathbf{r}, \mathbf{r}, E)$, one may face an additional problem: if some of the f group parameters coincide with spatial coordinates, one must avoid a repetition of the corresponding integrations. Therefore, the actual way to do the exact $f+1$ group integrations and the remaining spatial stationary phase integrations (if any are left) will depend on the value of $f+1$. If this value is less or equal to D, there will be no extra problem since the $(f+1)$-dimensional integration can, at least formally, be transformed into a corresponding part of the spatial trace integral. In particular, if $f+1$ is equal to the number D of degrees of freedom, the integral over the measure $dt\,d\mu(g)$ exactly exhausts the spatial trace integral, and no further integration need be done. A two-dimensional system with axial symmetry is an example for this case: here we have one parameter, the polar angle ϕ associated to the $U(1)$ symmetry, so that $f=1$. The two-dimensional integral $\int dt\,d\phi$ over the Green's function is performed exactly, and the only approximation coming into the resulting trace formula stems from the semiclassical approximation to the Green's function itself and from the exclusive use of periodic orbits. This is exactly equivalent to the way in which we have derived the trace formula (6.14) for the disk billiard above by doing the two-dimensional trace integral exactly in coordinate space. It represents just another realization of the exact trace integration in Eq. (6.3), whereby the integral $\int dx\,dy$ is transformed to the integral $\int dt\,d\phi$. We shall show explicitly below that the latter leads to the same result. Before doing so, we will briefly outline what is done if $f+1$ is larger than D and then give the general structure of the trace formulae obtained by Creagh and Littlejohn [4, 5].

When $f+1$ is larger than the number D of degrees of freedom, one cannot avoid the problem of surreptitious integrations if one wants to use the standard spatial trace integration over the Green's function. Creagh and Littlejohn [4] therefore resort to a mixed representation of the Green's function

$$G(\mathbf{p}, \mathbf{r}'; E) = \langle \mathbf{p} \left| \frac{1}{E - \hat{H}} \right| \mathbf{r}' \rangle, \tag{6.18}$$

where $\mathbf{p}$ is the final momentum and $\mathbf{r}'$ the initial coordinate of the particle. The level density is then found by a trace over the entire phase space:

$$g(E) = -\frac{1}{\pi} \Im m \int\int d^D r\, d^D p\, e^{i\mathbf{p}\cdot\mathbf{r}/\hbar} G(\mathbf{p}, \mathbf{r}; E + i\epsilon). \qquad (\epsilon > 0) \tag{6.19}$$

They next use a semiclassical approximation for the mixed Green's function $G(\mathbf{p}, \mathbf{r}'; E)$ in terms of classical orbits which is consistent with that discussed in Section 5.1. Now it becomes possible in all cases to single out $f+1$ of the $2D$ phase-space variables and transform their integral to the integral over $dt\,d\mu(g)$, which is done exactly; the remaining $2D-f-1$ integrals transverse to the orbit families are then left for the stationary phase integration.

We should note that this approach is completely analogous to that of Strutinsky and Magner [3] already discussed at the beginning of Sect. 6.1.1. These authors removed the degeneracy of the orbits at the level of the semiclassical propagator K_{scl} (5.8), by fixing an intermediate point $\mathbf{r}''$ in the folding integral (5.1). In the above approach, the same is achieved by fixing a point $(\mathbf{r}, \mathbf{p})$ in phase space. In either method, the price to pay is a set of D extra integrals. But the advantage is that as many of these integrations can be done exactly as there are continuous symmetries in the system. The same idea was also used by Magner [10], who derived semiclassical trace formulae for harmonic-oscillator potentials by integrating directly over the phase angles α_i in their harmonic solutions $r_i(t) = r_{0i}\cos(\omega_i t + \alpha_i)$. Following up the ansatz in Eq. (6.19), Magner et al. have recently proposed a general phase-space trace formula [13] which in principle allows one to handle all symmetries and applies also to integrable systems.

The general form of the trace formulae given by Creagh and Littlejohn [5] for a system with f-dimensional degeneracy is

$$\delta g_{scl}(E) = -\frac{1}{\pi} \Im m \left\{ \frac{1}{i\hbar} \frac{1}{(2\pi i\hbar)^{f/2}} \sum_{\Gamma} \int_{\Gamma} dt \, d\mu(g) \, |K_{\Gamma}|^{-1/2} e^{i\,[S_{\Gamma}(E)/\hbar - \sigma_{\Gamma}\pi/2]} \right\}. \quad (6.20)$$

In contrast to the trace formula (5.36) for isolated orbits, the discrete sum is here over distinct *families* Γ of periodic orbits which are assumed to have all the same degeneracy f; S_{Γ} and σ_{Γ} are their actions and Maslov indices, respectively. Note that the repetitions of each primitive orbit family must be included explicitly in the sum over Γ. The continuous summation over the degenerate orbits within each family is given by the integrals over dt and $d\mu(g)$. The factor K_{Γ} in the denominator is an invariant of the family, determined by a linearization of the dynamics about one typical orbit of the family. Its detailed form, which we shall not discuss here, is given in Ref. [4]. In the case of isolated orbits with $f=0$, it is just the term $\det(\widetilde{M}^n - I)$ discussed in Sect. 5.3; the integral over the group measure $d\mu(g)$ then is void and one recovers from (6.20) the old Gutzwiller trace formula (5.36). One important point is that K is determined purely classically and therefore does not depend on $\hbar$. The amplitudes in the trace formula therefore go with a factor of $\hbar^{-1/2}$ for each independent symmetry parameter; the overall factor is thus $\hbar^{-(f/2+1)}$. This shows us immediately that the amplitude of the level density oscillations – and with it the strength of the shell effects – grows with increasing degeneracy of the periodic orbits, i.e., with increasing degree of the symmetry. This point had been noted and discussed extensively already by Strutinsky and Magner [3]. It is the reason why in many cases, it is sufficient to include only the highest-degenerate class or orbit families in the studies of gross-shell effects. An example for this is the spherical cavity mentioned in Sect. 6.1.6 below. However, there are also exceptions to this rule; examples are the triangular billiard (see Sect. 6.1.2) and spheroidal cavities with large deformations [11, 19, 20].

Rather than going into further technical details, let us just give here some of the trace formulae which were given by Creagh and Littlejohn. For symmetries created by *Abelian groups*, they obtained the formula [4]

$$\delta g_{scl}(E) = \frac{1}{\hbar\pi} \frac{1}{(2\pi\hbar)^{f/2}} \sum_{\Gamma} \frac{T_{\Gamma} V_{\Gamma}}{|\mathcal{J}_{\Gamma}|^{1/2} |\det(\widetilde{M}_{\Gamma} - I)|^{1/2}} \cos\left(\frac{1}{\hbar} S_{\Gamma} - \sigma_{\Gamma}\frac{\pi}{2} - \frac{f\pi}{4}\right). \quad (6.21)$$

Here, $\widetilde{M}_{\Gamma}$ is the symmetry-reduced surface of section mapping matrix, rather than the stability matrix of an isolated orbit, and T_{Γ} is the period of a primitive orbit of the family. V_{Γ} is the unit cell volume of the phase-subspace Γ, which is proportional to the invariant integral V_G over the group measure

$$V_G = \int d\mu(g) = \int d\Theta_1 d\Theta_2 \dots d\Theta_f, \quad (6.22)$$

and $\mathcal{J}_{\Gamma}$ in the denominator is a Jacobian determinant defined by

$$\mathcal{J}_{\Gamma} = \det\left(\frac{\partial\Theta}{\partial \mathbf{J}}\right). \quad (6.23)$$

Thereby $\mathbf{J} = (J_1, J_2, \ldots, J_f)$ is the vector of the commuting first f constants of motion that are associated to the generators of the symmetry group G, and $\boldsymbol{\Theta} = (\Theta_1, \Theta_2, \ldots, \Theta_f)$ is the f-dimensional vector of group variables. More precisely, Θ_i are coordinates on the symmetry group that correspond to additional symmetry transformations required to close the orbits in full phase space. Thus, $\mathcal{J}_\Gamma$ measures the amount by which periodic orbits of the reduced dynamics, which are close to Γ, fail to be periodic in the full phase space. The matrix $\partial\boldsymbol{\Theta}/\partial\mathbf{J}$ in (6.23) can be shown to be symmetric.

At this point, it might be helpful to specify the above for a very simple symmetry: that of rotational invariance about a given symmetry axis, which leads to a conserved angular momentum L. This 1-dimensional $U(1)$ symmetry is obviously parametrized by the polar angle of rotation ϕ. Thus, we can choose $d\mu = d\phi$ and $f=1$ in the above equations. In principle, the range of integration over ϕ is from 0 to 2π. However, a periodic orbit may have an a_Γ-fold discrete rotational symmetry. In that case, only an angle of $2\pi/a_\Gamma$ should be covered by the integration, so that the unit cell volume is

$$V_\Gamma = \int d\phi = \frac{V_{U(1)}}{a_\Gamma} = \frac{2\pi}{a_\Gamma}. \tag{6.24}$$

The Jacobian determinant $\mathcal{J}_\Gamma$ (6.23) becomes simply

$$\mathcal{J}_\Gamma = \frac{\partial\Theta\,(L_\Gamma)}{\partial L_\Gamma}, \tag{6.25}$$

where L_Γ is the angular momentum of the orbit family, and $\Theta(L_\Gamma)$ is the amount by which periodic orbits of the family Γ fail to close in the angle ϕ when L_Γ is varied. (This will be illustrated in the following figure.)

For a *two-dimensional system with axial symmetry*, the integral over $dt\,d\phi$ exhausts all degrees of freedom, and the term $|\det(\widetilde{M}_\Gamma - I)|^{1/2}$ in the denominator becomes unity. The trace formula then becomes

$$\delta g\,(E) = \frac{1}{\hbar}\frac{2}{(2\pi\hbar)^{1/2}}\sum_\Gamma \frac{T_\Gamma}{a_\Gamma\,|(\partial\Theta/\partial L)_\Gamma|^{1/2}} \cos\left(\frac{1}{\hbar}S_\Gamma - \sigma_\Gamma\frac{\pi}{2} - \frac{\pi}{4}\right). \tag{6.26}$$

The determination of $\mathcal{J}_\Gamma$ (6.25) is illustrated in Fig. 6.11 on the next page for the triangular orbit (3,1) in the disk billiard (cf. Fig. 6.6). The angle of incidence (and reflection) of the unperturbed periodic orbit, shown by the solid line, is denoted by θ_{inc}. Increasing this angle by an amount $\Delta\theta_{inc}$ leads to the dashed orbit whose angular momentum is increased by $\Delta L_\Gamma = pR\cos\theta_{inc}\cdot\Delta\theta_{inc}$. After v bounces, the resulting trajectory falls behind by a total angle $\Delta\Theta = 2v\Delta\theta_{inc}$. In the limit of an infinitesimal $\Delta\theta_{inc}$ we find thus easily for an orbit with v bounces

$$|\mathcal{J}_v| = \frac{2v}{pR\cos\theta_{inc}}. \tag{6.27}$$

Inserting the result (6.27) into the trace formula (6.26) and using $\theta_{inc} = \pi/2 - \pi w/v$ [see Eq. (6.12)], one finds exactly the amplitudes of the trace formula for the disk billiard given in Eq. (6.14). The factor f_{vw} in that formula arises in the present derivation from the fact that when summing over the orbit families (v, w), the two distinct time orientations (clockwise or anticlockwise) of all orbits except the diameter ($v=2w$) have to be counted separately. This discrete degeneracy is related to the time reversal symmetry: whereas time reversal maps the diameter orbit onto itself, it maps the other distinct orbit pairs, running in opposite directions, mutually onto each other.

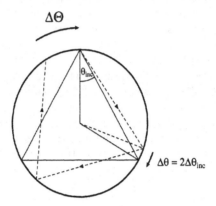

Figure 6.11: Determination of the Jacobian $\mathcal{J}_\Gamma$ (6.25) for the triangle orbit (3,1) in a disk billiard. (From [8].)

For integrable systems, one always finds at least D constants of motion, and we have seen in Sect. 2.5 that the so-called torus quantization is always possible using the so-called action-angle variables $(\mathbf{I}, \boldsymbol{\phi})$. It is advantageous to use these action-angle variables also in the derivation of the trace formula. This had actually been proposed for the first time by Berry and Tabor [6]. In the approach explained above, we have in this case $f+1=D$, and the D first integrals $(H, \mathbf{J})$ can be used to compute the action-angle variables, so that the Hamiltonian $H = H(\mathbf{I})$ is a function only of the actions $\mathbf{I} = (I_1, I_2, \ldots, I_D)$ and independent of the angles $\boldsymbol{\phi} = (\phi_1, \phi_2, \ldots, \phi_D)$. The periodic orbit families Γ are rational D-tori in phase space and may be labeled by the topological lattice vectors $\mathbf{M}_\Gamma$ consisting of D integers $(M_1, M_2, \ldots, M_D)$ given by (cf. Eq. (2.166))

$$\boldsymbol{\omega} T_\Gamma = 2\pi \mathbf{M}_\Gamma, \qquad (6.28)$$

where T_Γ is the fundamental period of the family and $\boldsymbol{\omega} = (\omega_1, \omega_2, \ldots, \omega_D)$ is the vector of the angular frequencies defined by

$$\boldsymbol{\omega} = \frac{\partial H(\mathbf{I})}{\partial \mathbf{I}}. \qquad (6.29)$$

The action integral S_Γ in the phase of the trace formula (6.21) becomes simply

$$S_\Gamma(E) = \oint_\Gamma \mathbf{p} \cdot d\mathbf{r} = \sum_{i=1}^{D} \oint_\Gamma I_i \, d\phi_i = 2\pi \, \mathbf{M}_\Gamma \cdot \mathbf{I}. \qquad (6.30)$$

Like in the 2-dimensional disk example given above, the symmetry-reduced space becomes vacuous so that the factor $\det(\widetilde{M}_\Gamma - I)$ is unity. The Jacobian determinant $\mathcal{J}_\Gamma$

(6.23) is also most conveniently expressed in terms of the action-angle variables, and after some algebraic manipulations, Creagh and Littlejohn [4] find the following trace formula[1] for *integrable systems*

$$\delta g_{scl}(E) = \frac{1}{\pi} \left(\frac{2\pi}{\hbar}\right)^{(D+1)/2} \sum_{\Gamma} \frac{\cos[2\pi M_{\Gamma} \cdot I/\hbar - \sigma_{\Gamma}\pi/2 + (D+1)\,\pi/4]}{T_{\Gamma}^{(D-1)/2} \, |\det \mathcal{K}|^{1/2} \, |\omega_{\Gamma} \cdot \mathcal{K}^{-1}\omega_{\Gamma}|^{1/2}}, \qquad (6.31)$$

where $\mathcal{K}$ is the symmetric $D \times D$ matrix defined by

$$\mathcal{K} = \frac{\partial \omega}{\partial I} = \frac{\partial^2 H}{\partial I\,\partial I}, \qquad (6.32)$$

which must be assumed to be nonsingular. Here ω_{Γ} stands for $2\pi M_{\Gamma}/T_{\Gamma}$. The trace formula (6.31) is equivalent to that derived by Berry and Tabor [6, 7] which we discussed in Sect. 2.7. In fact, Eq. (6.31) may also be derived from Eq. (2.170), taking the τ integral by the saddle-point method and generalizing to D dimensions. We recognize in the above, indeed, the classical actions $M \cdot I$ which appeared in Eq. (2.170), and a determinant of second derivatives of H similar to that given in Eq. (2.171). The Maslov index σ_{Γ} here is related to the indices μ_i of the EBK quantization (2.119) by

$$\sigma_{\Gamma} = M_{\Gamma} \cdot \mu + \beta_{\Gamma}. \qquad (6.33)$$

The first part, $M_{\Gamma} \cdot \mu$, is the same that appears in Eq. (2.170). The second term, β_{Γ}, is equal to the number of positive eigenvalues of the matrix $\mathcal{K}$ (6.32) if $\omega_{\Gamma} \cdot \mathcal{K}^{-1}\omega_{\Gamma}$ is positive; otherwise one has to add one unit to it. β_{Γ} contains the phases coming from the Fresnel integrals (2.169). We recognize also the analogy of (6.33) to the general expression (5.37) for the Maslov index in Gutzwiller's trace formula: the first part, there called μ, counts the number of conjugate points along the periodic orbit. For integrable systems it is the sum of the individual numbers μ_i of conjugate points (including hardwall reflections) along the elementary loops on the D-torus, weighted with the elements M_i of the topological lattice vector. The second part, there called ν, comes from the stationary phase integrals transverse to the orbit; here β_{Γ} contains the sum of extra phases collected by the integrals over the actions I_i. Note that Eq. (6.31) cannot be used for systems with extra dynamical symmetries such as harmonic oscillators and the Coulomb potential, since $\mathcal{K}$ becomes singular for these cases. Also, Eq. (6.31) only yields the contributions of the periodic orbits with the *highest* degeneracy in the system. Contributions of orbits with lower degeneracy, which have amplitudes of higher order in $\hbar$ in the trace formula, must be derived by other means (see Sects. 6.1.2, 6.1.6).

Finally, we briefly touch the case of continuous symmetries created by *non-Abelian* groups G. The most prominent examples are systems with rotational symmetry, $SO(3)$, the harmonic-oscillator potentials in D dimensions which have the $SU(D)$ symmetry, and the Coulomb problem with $O(4)$ symmetry. The f first integrals J then do not always Poisson-commute and the process of symmetry reduction becomes more complicated. In particular, it does not allow for a simple elimination of ignorable coordinates and as a result, the dimension of the symmetry-reduced phase space can be larger than $2D-2f-2$. We shall not discuss here the group theoretical complications but refer to

---

[1]Eq. (4.13) of Ref. [4] contains some misprints (S. C. Creagh, private communication, 1996).

Creagh and Littlejohn [5] who derived a trace formula for systems with general symmetries including non-Abelian cases. It is very similar in its form to Eq. (6.21), but some of its ingredients are more complicated to determine. The Jacobian $\mathcal{J}_\Gamma$ must be evaluated with the help of local coordinates suitable to describe the subgroup of symmetry operations that leave the vector $\mathbf{J}$ of first integrals fixed. Pictorially speaking, it tells us to which extent orbits that are periodic in the symmetry-reduced dynamics fail to be periodic when lifted back to the full phase space. Besides this, an additional continuous degeneracy factor appears which reflects the dimension of the irreducible representations of the symmetry group. In the case of $SO(3)$ it is just the angular momentum J_Γ of the orbit family, and for Abelian groups it equals unity.

When the symmetry is that of the three-dimensional rotational group $SO(3)$, all above terms have a relatively simple geometrical interpretation. The Jacobian $\mathcal{J}_\Gamma$ becomes equal to $(\partial\Theta_\parallel/\partial J)_\Gamma$, where $\Theta_\parallel$ is the angle of rotation about the direction of the total angular momentum necessary to close an orbit; this is analogous to the Jacobian $\mathcal{J}_\Gamma$ in Eq. (6.25) discussed above for the axial case. The group integral $\int d\mu\,(g)$ can be represented in terms of the three Euler angles (α,β,γ)

$$\int_{SO(3)} d\mu\,(g) \times = \int_0^{2\pi} d\gamma \int_0^\pi \sin\beta\,d\beta \int_0^{2\pi} d\alpha \times , \qquad (6.34)$$

so that the unit cell volume of the group is simply $V_{SO(3)} = 8\pi^2$. As in the example of the disk billiard given above, it has to be divided by a_Γ for periodic orbits that have a discrete a_Γ-fold rotational symmetry. The trace formula for systems with *three-dimensional rotational symmetry* is then [5]

$$\delta g\,(E) = \frac{1}{\pi\hbar}\frac{1}{(2\pi\hbar)^{3/2}} \sum_\Gamma \frac{8\pi^2\,J_\Gamma\,T_\Gamma}{a_\Gamma\,|(\partial\Theta_\parallel/\partial J)_\Gamma|^{1/2}\,|\det(\widetilde{M}_\Gamma - I)|^{1/2}} \cos\left(\frac{1}{\hbar}S_\Gamma - \sigma_\Gamma\frac{\pi}{2} - \frac{3\pi}{4}\right) .$$
$$(6.35)$$

The term $|\det(\widetilde{M}_\Gamma - I)|^{1/2}$ in the denominator is nontrivial only in a many-body case, i.e., for several particles moving in a spherical potential. For a single particle in a spherical potential, it must be replaced by unity.

We should like to point out that the trace formulae given here for cases with special symmetries always represent the oscillating part of the *total* level density. They are different in spirit from symmetry-reduced trace formulae that give the density of a subset of energy levels with a fixed quantum number such as, e.g., the z component of the angular momentum $L_z = m\hbar$. Whereas such partial level densities have been derived semiclassically [21, 22, 23, 24, 25] and applied to physical systems (see the example of atoms in strong magnetic fields given in Sect. 1.4), we shall not further discuss them here. They are mainly of interest if one is aiming at the semiclassical determination of energy levels with fixed quantum numbers. For the study of gross-shell effects, it is always the total level density that is of main interest.

6.1.5 Example 3: The 2-dimensional rectangular billiard

We shall now illustrate the use of the trace formula (6.21) for the simple example of the rectangular billiard, which has already been discussed in Sect. 2.7.1 and in Sect. 3.2.6, and take this occasion to give a semiclassical interpretation of the earlier results obtained there. (For the inclusion of a magnetic field in this system, see Sect. 6.2.3 below.) The periodic orbits of a particle in a rectangular two-dimensional billiard with sides a_1 (taken along the x axis) and a_2 (along the y axis) can be classified by two integers (M_1, M_2), one of which has to be nonzero. These integers indicate the number of times the particle covers the lengths $2a_1$ and $2a_2$ in x and y direction, respectively, during one period of the orbit. $M_1 = 0$ gives straight-line orbits in the y direction and $M_2 = 0$ straight-line orbits in the x direction. Each of the orbits has a one-dimensional degeneracy corresponding to the position of the starting point of any of its reflection points along one of the sides. In Fig. 6.12 we show four examples of the (1,1) family of orbits; this family includes the two diagonals. (As discussed in Sect. 7.5, the reflection from corners with angles that are *integer* fractions of π may be simply treated as the continuous limit of the reflection points reaching the corner, but other angles lead to additional diffraction effects. In the case of a rectangle, the particle is returning on the same trajectory.)

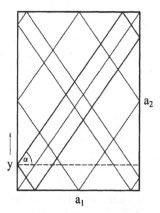

Figure 6.12: Four examples of the degenerate (1,1) orbit family in the rectangular billiard with side lengths a_1 and a_2. The symmetry parameter characterizing a family member is taken to be the y coordinate of the starting point on the left side. The angle α is given by $\tan \alpha = a_2/a_1$.

Due to the separability of the problem in x and y coordinates, the system has as constants of the motion the kinetic energies E_x and E_y in the two directions (cf. Sect. 2.7.1) or, equivalently, the absolute values of the momenta p_x and p_y. Changing the signs of both p_x and p_y corresponds to a time reversal operation; the corresponding symmetry gives a discrete double degeneracy to each orbit which reflects itself in a sign change of both values of M_1 and M_2. In the following, we choose one given time orientation of each orbit, using only non-negative numbers M_1 and M_2, and positive values of both p_x and p_y when starting the motion from a reflection point y on the left wall (positioned at $x = 0$). When M_1 and M_2 are not relatively prime, they denote a repeated orbit. We shall therefore write $M_1 = nm_1$ and $M_2 = nm_2$ and assume in the following that m_1 and m_2 are coprime, so that n is the repetition (or cycle) number of the primitive orbit (m_1, m_2), over which we will have to sum separately at the end.

Keeping the total energy $E = E_x + E_y$ fixed, the one-dimensional degeneracy of the orbit families can thus be characterized by the coordinate y of the starting point, and the corresponding symmetry is the translational symmetry along the y axis. In the language of Section 6.1.4 above, we have $f=1$; the symmetry parameter characterizing a member of an orbit family $\Gamma = (m_1, m_2)$ is y and the constant of the motion created by it is $J = p_y$. Since the translational group is Abelian, the relevant trace formula for the level density of this system is given by Eq. (6.21). The Jacobian $\mathcal{J}_\Gamma$ defined in Eq. (6.23) is thus given by $\mathcal{J}_\Gamma = d\Theta/dJ = dy/dp_y$, where $d\Theta = dy$ is the amount by which an orbit overshoots the starting point in the y direction after one period when its momentum p_y is increased by an amount dp_y at fixed energy E. (Note that we do not have to worry about the overall sign of $\mathcal{J}_\Gamma$, since only its modulus is used in the trace formula.) For a general orbit (m_1, m_2), the angle α of the starting direction with respect to the x axis is given by the condition

$$\tan \alpha = \frac{a_2 m_2}{a_1 m_1}. \tag{6.36}$$

The projection of its total length on the x axis is $L_x = 2 m_1 a_1$. Increasing α by an amount $d\alpha$, the orbit therefore overshoots its original path in y direction after one period by $dy = L_x\, d(\tan \alpha) = (2m_1 a_1 / \cos^2 \alpha)\, d\alpha$. On the other hand, we have $p_y = p \sin \alpha$, where $p = \sqrt{p_y^2 + p_x^2} = \sqrt{2mE}$ and m is the mass of the particle, so that we get $dp_y = dJ = p \cos \alpha\, d\alpha$. Thus, the Jacobian becomes

$$\mathcal{J}_\Gamma = \frac{d\Theta}{dJ} = \frac{dy}{dp_y} = \frac{L_\Gamma^3}{p(2m_1 a_1)^2}, \tag{6.37}$$

where $L_\Gamma = 2\sqrt{a_1^2 m_1^2 + a_2^2 m_2^2}$ is the length of the orbit [cf. Eq. (2.181)] and we have made use of Eq. (6.36) to write $\cos \alpha = (2m_1 a_1)/L_\Gamma$.

Like for any piecewise straight-line motion, the action of the periodic orbit is $S_\Gamma = pL_\Gamma$ and its period is $T_\Gamma = S_\Gamma/(2E)$. The Maslov indices σ_Γ in Eq. (6.21) are multiples of 4 (since each orbit has an even number of reflections, each giving a contribution 2 to σ_Γ) and can thus be omitted. Note that to get the factor V_Γ appearing in Eq. (6.21), we have to integrate over y only from zero to a_2/m_1 because of the extra *discrete* m_1-fold translational symmetry of the orbit in the y direction. Thus, we have $V_\Gamma = a_2/m_1$. Inserting all this into Eq. (6.21) and noting that $\det(\widetilde{M}_\Gamma - I)$ is unity, we obtain for the oscillating part of the level density

$$\delta g(E) = \frac{m a_1 a_2}{\hbar^2 \pi^{3/2}} \sqrt{2} \sum_\Gamma \sum_{n=1}^\infty \sqrt{\frac{\hbar}{n S_\Gamma}} \cos\left(\frac{n}{\hbar} S_\Gamma - \frac{\pi}{4}\right), \tag{6.38}$$

where Γ refers to the primitive periodic orbits with non-negative m_i and n counts their repetitions. Counting all orbits with $m_1, m_2 > 0$ twice, due to their time reversal degeneracy, this result becomes identical with Eq. (2.185) which we have obtained in

Sect. 2.7.1 using Poisson resummation of the EBK quantized spectrum, retaining only the leading-order term in its asymptotic form for large actions.

The situation is the same as for the triangular billiard discussed in Sect. 6.1.2 above. Both, the extended Gutzwiller theory and the method of Berry and Tabor yield the same result, which is the asymptotic form of the leading term in an exact trace formula.[2] In Sect. 2.7.1 we had also derived some extra contributions of higher order in $\hbar$, given in Eq. (2.186), which correspond exactly to minus $1/2$ times the level density of a one-dimensional box with side lengths a_i. These come from the "grazing orbits" along the boundaries that are included in the lowest-order approximation of the POT and are cancelled through semiclassical end-point corrections of the next order in $\hbar^{1/2}$ (cf. the discussion in Sect. 6.1.2).

6.1.6   Example 4: The three-dimensional spherical cavity

As a paradigm of a spherically symmetric system, we briefly discuss the case of a particle bound in an infinite spherical potential with ideally reflecting walls, which later will be used to explain a rather spectacular experimental phenomenon in the physics of metal clusters, the so-called "supershells" (see Sect. 8.2). Balian and Bloch [2] have derived a trace formula for the level density of the spherical cavity, using a multi-reflection expansion of the Green's function that is very similar in spirit, although quite different in the practical way of calculation, to the semiclassical approach developed by Gutzwiller. It also leads to a sum over the periodic orbits of the classical system, which are the same polygons already discussed in Sect. 6.1.3 for the disk billiard. The trace formula reads:

$$
\delta g_{scl}(E) = \frac{1}{E_0}\sqrt{\frac{pR}{\hbar\pi}}\sum_{vw}(-1)^w\sin(2\varphi_{vw})\sqrt{\frac{\sin\varphi_{vw}}{v}}\sin\Phi_{vw}
$$
$$
- \frac{1}{E_0}\frac{1}{2\pi}\sum_{w=1}^{\infty}\frac{1}{w}\sin(4wkR)\,, \tag{6.39}
$$

where the summation and the values of φ_{vw} and Φ_{vw} are identical to those in the disk formula (6.14). Note that with the amplitudes expressed purely in terms of classical quantities and using $E_0 = \hbar^2/2mR^2$, the overall power in $\hbar$ is $-5/2$ for the first term in (6.39) and -2 for the second terms. The first term comes from all polygon orbits with $v > 2w$; they have a three-fold degeneracy ($f=3$) corresponding to the full $SO(3)$ symmetry of the system. The diameter orbit, leading to the second term, has only a two-fold degeneracy ($f=2$), since rotation about its own direction does not lead to any different orbit. The powers of the amplitudes thus agree with the general expression $-(f/2+1)$ given above after Eq. (6.20). Therefore, due to its lower degree of degeneracy, the contribution from the diameter orbit appears as a correction of relative order $\hbar^{1/2}$ compared to those of the other orbits. It here plays a relatively small role for the total level density, differently from the case of the triangular billiard considered in Section 6.1.2. Note that the situation is different here from that of the two-dimensional disk

---

[2]This comes from the fact, pointed out already by Gutzwiller [22], that for free particles, the semiclassical Green's function (5.15) in two dimensions agrees only with the asymptotic form of the exact one (3.32) (cf. Sect. 7.5); in one and three dimensions, it is exact (cf. Problem 6.4).

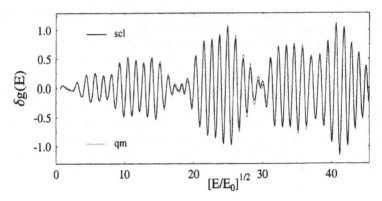

Figure 6.13: Oscillating part of the level density in the spherical cavity, averaged over k with a width $\gamma = 0.4/R$. The solid line shows the semiclassical and the dashed line the quantum-mechanical result. (After [26].)

billiard. Although the periodic orbits are identical in both cases, they are here embedded in a larger space, which leads to higher degeneracies; in the disk billiard they all have the degeneracy $f=1$ and correspondingly, all contributions to Eq. (6.14) are proportional to $\hbar^{-3/2}$. We note that Strutinsky and Magner [3] have also derived the result (6.39) from their extension of the Gutzwiller theory.

In Fig. 6.13 we show the oscillating part of the level density of the spherical cavity, slightly Gaussian averaged as for the disk billiard. Note the pronounced beating pattern, which had already been observed by Balian and Bloch and explained in terms of the interference between the triangular and the square orbits. As we have just discussed, the diameter orbit – although being the shortest one – has a small amplitude of relative order $\hbar^{1/2}$ and therefore the triangular and squared orbits play the leading role for the gross-shell structure of the level density. Fig. 6.13 actually shows a comparison between the quantum-mechanical and the semiclassical results. Although some tiny differences can be observed near the beat minima, the agreement is excellent.

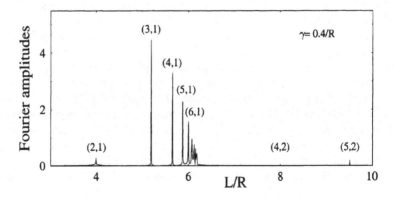

Figure 6.14: Fourier spectrum of the level density for the spherical cavity. As in Fig. 6.9.

That the diameter orbit has a smaller amplitude becomes evident also when one takes a Fourier transform of the quantum-mechanical $\delta g\,(E)$, which results in a length spectrum of the contributing classical orbits. This is shown in Fig. 6.14. We clearly recognize the descending amplitudes of the triangle (3,1), the square (4,1), the pentagon (5,1), etc., which are all stronger than that of the diameter orbit (2,1). (Orbits with a length much larger than $6R$ have been suppressed by the averaging with $\gamma = 0.4/R$.)

Like in the case of the disk billiard, it is possible to sum the trace formula (6.39) numerically by Gaussian averaging it over a very small range $\gamma = 0.02/R$. The result for $g(k)$ is shown in Fig. 6.15. Here we have again added the smooth part of the level density, which is given by the generalized Weyl expansion [see Eq. (4.143) in Sect. 4.6]

$$\tilde{g}(k) = \frac{2R}{3\pi}\left(k^2R^2 - \frac{3\pi}{4}kR + 1\right),\qquad(6.40)$$

and normalized the sum such that the contributions to each single level should add up to the quantum-mechanical degeneracy. As can be seen in the figure, the degeneracies have, indeed, the correct values $2\ell + 1$ for the states with angular momentum ℓ; we recognize in the figure the different hierarchies belonging to the different radial quantum numbers. The positions of the spikes in Fig. 6.15 are again different from the exact quantum-mechanical energies, which are indicated by the circles along the kR axis, and very close to the energies found by the EBK quantization.

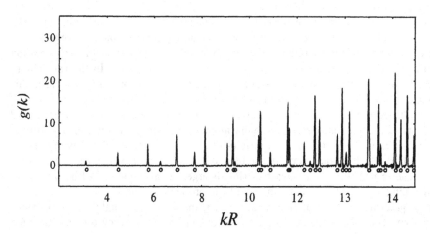

Figure 6.15: Total level density of the spherical billiard, obtained from the semiclassical trace formula (6.39) with a smoothing width $\gamma = 0.02/R$ and adding the smooth part (6.40). The sum over periodic orbits in Eq. (6.39) is truncated at $v_{max} = w_{max} = 200$. Circles underneath the horizontal axis indicate the positions of the exact eigenvalues $k_iR = \sqrt{E_i/E_0}$. (From [26].)

6.2   The problem of symmetry breaking

We shall now discuss a way to treat the phenomenon of symmetry breaking semiclassically. We consider a particle in a potential with a given continuous symmetry which is broken, when some continuous parameter is varied. To be specific, one may think of a deformation parameter or the strength of an external field. Classically, the breaking of symmetry is accompanied by the destruction of resonant tori, which are replaced by chains of resonant islands. This has been discussed and illustrated extensively in Sect. 2.8. Thus the most degenerate periodic orbits disappear, as soon as the perturbation is turned on. Now, from the development in Sect. 6.1.4 it has become evident that each degeneracy class of periodic orbits must be treated separately, since the degree of degeneracy f shows up explicitly in the trace formula (6.21). More specifically, the amplitude of the contribution of an f-fold degenerate family of orbits to the level density is proportional to $\hbar^{-(f/2+1)}$. From this we immediately recognize the problem that arises when the symmetry of the potential is broken: there is a discrete change of the power in $\hbar$ of the corresponding periodic-orbit contributions to the level density. This usually is accompanied by a divergence of some or all amplitudes in the trace formula, when the zero-deformation limit of the perturbed system is taken, i.e., when trying to restore the higher symmetry. Technically, what happens is that some of the saddle-point integrations performed in the perturbed system are no longer allowed.

We have already encountered this situation in two examples in Sect. 5.6. In the anisotropic two-dimensional harmonic oscillator (Sect. 5.6.2), such divergences occur whenever the two oscillator frequencies are commensurate. In this particular case, the divergences cancel exactly for all deformations including the isotropic limit of equal frequencies (see Ref. [27] for details). This cancellation is, however, a fortunate exception which marks the "unbearable easiness" of the harmonic oscillator. In the the the Hénon-Heiles potential (Sect. 5.6.4), one also starts from a two-dimensional isotropic harmonic oscillator, see Eq. (5.72). There, the perturbation breaks all continuous symmetries, so that all periodic orbits become isolated. We therefore have a transition from $f=2$ to $f=0$ for all orbits, as soon as the anharmonicity term proportional to α is turned on. Accordingly, the Gutzwiller amplitudes of all these isolated orbits diverge in the limit $\alpha=0$ or, which is equivalent, in the limit of small energy E where the anharmonic term becomes negligible. The onset of this divergence can be seen in Fig. 5.6. Here the diverging amplitudes do not cancel and the problem is generic.

In a quantum-mechanical treatment, such transitions from higher to lower symmetry (or in the inverse direction) pose no problem. Although they are accompanied by the breaking of degeneracies of some of the eigenenergies E_i, the corresponding level splittings can be very well described by perturbation theory, and the results vary smoothly as functions of the symmetry-breaking parameter. This suggests a treatment of the symmetry-breaking (or symmetry-restoring) transition by perturbative methods also on the semiclassical level. Important first steps in this direction have been taken by Ozorio de Almeida and Hannay [28], who introduced a "uniform approximation" that allows for a smooth, divergence-free transition from the unperturbed to the symmetry-broken case (see also Ref. [29]). Creagh [8] has recently proposed an alternative generalization of the semiclassical perturbation method that allows one to start from arbitrary symmetries. We shall describe and illustrate his method in the remainder of this section.

In the limit of strong perturbations, both the uniform approximation of Ref. [28] and the perturbative approach [8] are not able to recover the correct Gutzwiller amplitudes of the isolated orbits in the symmetry-broken system. This has been achieved recently by Tomsovic *et al.* [30] and by Brack *et al.* [31] in improved "global" versions of the uniform approximation, which we will discuss in Sect. 6.3.

6.2.1 A trace formula for broken symmetry

Let us start from the Hamiltonian of a system that is slightly perturbed by a symmetry-breaking term δH:

$$H = H_0 + \delta H = H_0 + \epsilon H_1 \,. \tag{6.41}$$

Here H_0 describes the unperturbed system and ϵ is a small dimensionless parameter. The idea of Creagh [8] is to start from the trace formula (6.21) appropriate for the Hamiltonian H_0 and to study the perturbative effects of δH only to first order in ϵ. The place where the perturbation plays the most important role is the classical action S_Γ in the phase of the trace formula. It has been shown that S_Γ to first order in ϵ is changed by

$$\Delta S_\Gamma \simeq - \int_{\Gamma_g(t)} \delta H \, dt = -\epsilon \int_{\Gamma_g(t)} H_1 \, dt = \epsilon \, F_\Gamma(g, E) \,, \tag{6.42}$$

where Γ_g is the unperturbed orbit of the family Γ, labeled by the group element g. This expression can only be used for Hamiltonians with smooth potentials $V(\mathbf{r})$. For billiard systems with reflecting walls, the change in action for a periodic orbit with a_Γ reflections is shown to be

$$\Delta S_\Gamma \simeq p \sum_{j=1}^{a_\Gamma} 2 \cos \theta_j \delta r_j = \epsilon \, F_\Gamma(g, E) \,, \tag{6.43}$$

where j counts the single reflections; θ_j and δr_j are the angle of incidence and the perpendicular distance of the perturbed boundary from the unperturbed one, respectively, at the j-th reflection point; and $p = \hbar k$ is the particle momentum. In either case, the action change can be written as ϵ times a function $F_\Gamma(g, E)$ which depends only on the orbit label g and the energy E. (The proofs of the two above equations have been given in the appendices of Ref. [8].)

In the general trace formula (6.20) for systems with continuous symmetries in Sect. 6.1.4, one has to perform an integral over the group measure, $\int d\mu \, (g)$, to take account of the degeneracies of the orbit families. The main step now is to expand the action in the exponent of Eq. (6.20) up to first order in the perturbation, i.e., up to the term ΔS_Γ, and to assume that all the prefactors keep their unperturbed values. This is once again the stationary phase approximation. The integral over the group measure, $\int d\mu \, (g)$, then includes only the exponential containing $i\Delta S_\Gamma/\hbar$, leading to a "modulation factor"

$$\mathcal{M}_\Gamma(\epsilon/\hbar, E) = \langle e^{i\Delta S_\Gamma/\hbar} \rangle_{g \in G} = \langle e^{i\epsilon/\hbar \, F_\Gamma(g,E)} \rangle_{g \in G} = \frac{1}{V_G} \int d\mu(g) e^{i\epsilon/\hbar \, F_\Gamma(g,E)}, \tag{6.44}$$

where V_G is the invariant group volume given in Eq. (6.22). This modulation factor appears as the only new ingredient in the trace formula, which now reads

$$\delta g_{scl}(E) = -\frac{1}{\pi} \Im m \left\{ \frac{1}{i\hbar} \frac{1}{(2\pi i\hbar)^{f/2}} \sum_{\Gamma} \mathcal{M}_{\Gamma}(\epsilon/\hbar, E) \int_{\Gamma} dt \, d\mu(g) \, |K_{\Gamma}|^{-1/2} \, e^{i\,[S_{\Gamma}(E)/\hbar - \sigma_{\Gamma}\pi/2]} \right\}.$$
(6.45)

Note that in general, $\mathcal{M}_{\Gamma}$ (6.44) is complex. By construction, it becomes unity in the limit of zero perturbation: $\mathcal{M}_{\Gamma}(\epsilon/\hbar, E) \to 1$ for $\epsilon \to 0$. In this way, Eq. (6.45) interpolates smoothly between the degenerate case with full symmetry and the "deformed" case where the symmetry is lost.

In general the modulation factor (6.44) can be written as a function of a single argument x:

$$\mathcal{M}_{\Gamma}(\epsilon/\hbar, E) = \mathcal{M}_{\Gamma}(x),$$
(6.46)

whereby x depends not only linearly on the perturbation parameter ϵ and the repetition number n of the unperturbed orbit family Γ, but it depends also on the energy or a characteristic action of the unperturbed system. By construction, we must have $x \lesssim 1$ for the above approximations to be justified. Therefore we cannot do the summation over the repetitions n to arbitrary order; we must rather restrict it to relatively short orbits. This excludes the present perturbative approach from the purpose of a full quantization. However, if one is interested mainly in the gross-shell structure of a system, for which only the shortest orbits are needed, it may well be applied.

For $x \gg 1$, the trace formula (6.45) breaks down, since the non-linear changes in the actions become important and the perturbations of the periodic orbits and the corresponding changes of the amplitudes in the trace formula can no longer be neglected. In contrast to the uniform approximations to be discussed in Sect. 6.3, the correct amplitudes of the relevant trace formula for the symmetry-broken case are not recovered here. It turns out, however, that there usually is an intermediate range of arguments $x \gtrsim 1$ of the modulation factor in which both the trace formula (6.45) and the one pertinent to the symmetry-broken limit give very similar results, so that one can switch forth and back between them. It should be said, though, that it is hard in general to predict the precise boundaries of the region in which both methods agree.

In spite of the restriction $x \lesssim 1$, the asymptotic behavior of the modulation factor for larger arguments $x > 1$ still does give some information about the existence of the isolated orbits that are created by the symmetry breaking perturbation δH. As discussed in Ref. [8], an asymptotic analysis of the group average integral (6.44) exhibits some new isolated orbits at critical points, which may be stationary points or the boundary values, of the group parameters over which one is integrating. Rather than discussing this analysis on a general formal basis, we shall illustrate it in some of the following examples.

6.2.2   Example 1: The two-dimensional elliptic billiard

As a first illustration of the above ideas, we discuss a very simple example of symmetry breaking [8]: the deformation of the two-dimensional circular billiard, whose trace formula has been given in Eq. (6.14), into an elliptic billiard with eccentricity ϵ. This example is interesting from one particular point of view: although the starting symmetry $U(1)$ of the circle is broken when ϵ is different from zero, the elliptic billiard is still integrable. Its symmetry is, however, not obvious. It is linked to the second constant of motion (besides energy), which is the product of the two angular momenta of the particle with respect to the two focal points. In the Appendix C.5.1, we discuss the geometry of the ellipse and give the definitions of the relevant quantities used below. An exhaustive classification of the periodic orbits in the elliptic billiard has been given in Ref. [32], and its trace formula has recently been derived using the Berry-Tabor method in Refs. [12, 13, 33].

The periodic orbits of the circular billiard are labeled by (v, w), as shown in Fig. 6.6. When the circle is deformed into an ellipse, all the polygons (v,w) with $v > 2w$ keep their 1-fold degeneracy, although they are distorted. This is illustrated in Fig. 6.16 for the case of the triangular (3,1) and rhombic (4,1) orbits (the latter corresponding to the squares in the disk billiard). Thus, they still occur in degenerate families.

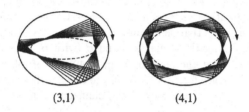

(3,1)   (4,1)

Figure 6.16: Two examples of degenerate periodic orbit families in an elliptic billiard: triangles (left) and rhombi (right). The dashed lines indicate the elliptic caustics of each family, which are confocal with the elliptic boundary. (Courtesy S. M. Reimann.)

The diameter orbit of the unperturbed disk billiard, however, is broken up into two linear orbits along the symmetry axes of the elliptic billiard, whose degeneracy dimension is $f = 0$, i.e., they are isolated. The longer diameter orbit and all its repetitions are unstable for all $\epsilon > 0$. The shorter diameter orbit is stable except at some isolated values ϵ_{nm} of the eccentricity, given by

$$\epsilon_{nm} = \sin\left(\frac{\pi m}{2n}\right), \qquad m = 1, 2, \ldots, (n-1), \qquad (n \geq 2) \qquad (6.47)$$

where n is the repetition number. ϵ_{nm} are the bifurcation points at which new families of librating orbits are "born"; these exist only above the deformations ϵ_{nm} and again have a one-fold degeneracy. At the bifurcation points, the n-th repetitions (with $n \geq 2$) of the short diameter orbit become marginally stable and degenerate with the new librating orbit families. Away from the bifurcation points, the contributions of the diameter orbits to the level density can be obtained from the original Gutzwiller trace formula (5.36), and it is a relatively easy exercise to evaluate these contributions analytically (cf. also Ref. [33]).

We are presently only interested in small excentricities ϵ, since we want to investigate the perturbative approach of Creagh [8]. In order to avoid the divergences connected to the bifurcations, we only consider eccentricities smaller than those of the lowest bifurcation point ϵ_{n1} for each repetition number n, i.e., we assume

$$\epsilon < \epsilon_{n1} = \sin\left(\frac{\pi}{2n}\right). \tag{6.48}$$

(We note that some of the bifurcations in the elliptic billiard have recently been included into the trace formula by Sieber [33] using a uniform approximation, and by Magner et al. [13] using an improved stationary phase method.) To lowest order in ϵ, the contributions of the two diameter orbits to the trace formula become

$$\delta g_{diam}^{<}(E) \;\simeq\; \frac{R}{\hbar \pi p} \sum_{n}^{n_{max}} \frac{1}{n\epsilon} \cos(4nkB - 3n\pi + \pi/2),$$

$$\delta g_{diam}^{>}(E) \;\simeq\; \frac{R}{\hbar \pi p} \sum_{n}^{n_{max}} \frac{1}{n\epsilon} \cos(4nkA - 3n\pi). \tag{6.49}$$

Here p is the momentum of the particle (with unit mass) and $k = p/\hbar$ its wave number. The sums over the repetition numbers n have been restricted to an upper limit n_{max} in order for the following perturbation expansion to be valid. We have used the relations given in Appendix C.5.1 for the semiaxes A and B (with $B \leq A$) and the eccentricity parameter ϵ. We note that the amplitudes in (6.49) are of relative order $\hbar^{1/2}$ with respect to those in the trace formula (6.14) of the circular disk billiard, which reflects the discontinous change in degeneracy of the diameter orbit when switching on the eccentricity. Accordingly, the amplitudes in (6.49) both diverge for $\epsilon \to 0$ where the circular symmetry is restored. Since the oscillations are out of phase by $\pi/2$, the diverging terms cannot cancel (unlike for the harmonic-oscillator case discussed in Sect. 5.6.2). Thus the problem caused by the symmetry breaking is severe: the above results cannot be used for very small eccentricities.

We now want to compute the perturbative trace formula presented in the previous section. For small ϵ, the boundary of the ellipse [see Eq. (C.48)] can be expanded in polar coordinates (r, ϕ) around a circle with radius R

$$r(\phi) = R\left[1 + \frac{1}{4}\epsilon^2 \cos(2\phi) + \mathcal{O}(\epsilon^4)\right], \tag{6.50}$$

and the two semiaxes become

$$A = R\left[1 + \frac{1}{4}\epsilon^2 + \mathcal{O}(\epsilon^4)\right], \qquad B = R\left[1 - \frac{1}{4}\epsilon^2 + \mathcal{O}(\epsilon^4)\right]. \tag{6.51}$$

Next we compute the changes in the actions of the periodic orbits in the unperturbed disk billiard using Eq. (6.43). The angle of incidence θ_j (illustrated by θ_{inc} in Fig. 6.11 for the triangle) is related to the angle φ_{vw} in (6.12) by $\theta_j = \pi/2 - \varphi_{vw} = \pi/2 - \pi w/v$. The radial distances δr_j of the perturbed boundary (6.50) at the reflection points of the orbit (v, w) occur at polar angles $\phi_j = 2j\varphi_{vw} = 2\pi j\, w/v$ with $j = 1, 2, \ldots, v$. At these points, the radial distance δr_j of the perturbed boundary (6.50) becomes

$$\delta r_j = \frac{1}{4}\epsilon^2 R \cos\left(2\theta + 4\pi j\frac{w}{v}\right). \tag{6.52}$$

Here we have included the polar angle θ at which we start counting the reflections of an orbit. This angle defines the position of a specific orbit of the degenerate family

(v, w) within the boundary and takes values from 0 to 2π. It is the parameter we will use to characterize the $U(1)$ symmetry of the unperturbed system and to perform the group integral over $d\mu(g) = d\theta$. Taking now the sum in Eq. (6.43), we obtain from the above

$$\Delta S_{vw} \simeq \epsilon^2 F_{vw}(\theta, E) = \frac{1}{2} \epsilon^2 \, pR \, \sin \varphi_{vw} \sum_{j=1}^{v} \cos \left(2\theta + 4\pi j \frac{w}{v} \right)$$

$$= \begin{cases} \epsilon^2 \, npR \, \cos(2\theta), & v = 2w \\ 0. & v > 2w \end{cases} \quad (6.53)$$

We first note that to lowest order in the perturbation parameter (which here is ϵ^2), only the diameter orbit changes its action. This is consistent with the fact, already pointed out above, that all the polygons ($v > 2w$) stay in degenerate families. Although they change their geometrical shapes, their lengths – and thus their actions – will only be changed by amounts of order ϵ^4. This holds because the area πAB of the ellipse is constant and equal to πR^2 up to terms of order ϵ^4, as follows from Eq. (6.51). The modulation factor $\mathcal{M}_{vw}$ is therefore unity for all the polygons: $\mathcal{M}_{v>2w} = 1$. That of the diameter orbit ($v = 2w$) becomes

$$\mathcal{M}_{diam} = \frac{1}{2\pi} \int_0^{2\pi} e^{i \epsilon^2 npR \cos(2\theta)/\hbar} \, d\theta = J_0(\epsilon^2 nkR), \quad (6.54)$$

where $J_0(x)$ is the cylindrical Bessel function of order zero. If we insert these modulation factors under the summation of the disk trace formula (6.14), we obtain a smooth interpolation formula that gives the level density also away from the circular limit without any divergence. For the diameter contribution we get, in particular,

$$\delta g_{diam}(E) = \frac{2R^2}{\hbar^2} \sqrt{\frac{1}{\pi kR}} \, \Im m \sum_{n=1}^{n_{max}} \frac{1}{\sqrt{2n}} J_0(\epsilon^2 nkR) \, e^{i(4nkR - 3n\pi + 3\pi/4)}. \quad (6.55)$$

Since this expression is valid only as long as the argument x of the Bessel function is small compared to unity, the condition of applicability of the result (6.55) is then $\epsilon \ll \sqrt{1/n_{max}kR}$. Hence, the quality of the approximation is bound to become worse with increasing wave number (or energy). (Note that this is in contrast to the usual tendency of semiclassical approximations.)

In order to test these limitations numerically, we present in the following two figures the level densities $\delta g(k)$ of the elliptic billiard, Gaussian averaged over the wave number k with a width $\gamma = 0.64/R$ in order to emphasize the gross-shell structure. The solid lines in both figures are the quantum-mechanical results [12]. It should be noted that compared to the level densities $\delta g(E)$ given in the formulae of the text, the quantities $\delta g(k)$ shown in the figures carry an extra power k (apart from constants), as given in Eq. (3.10) of Sect. 3.1.1.

In Fig. 6.17 we show the results predicted by the approach of Creagh [8]. The trace formula (6.14) is used, whereby the diameter contribution includes the modulation factor as given in Eq. (6.55). For the smallest eccentricity (upper part), the agreement is seen to be excellent. As the deformation increases (middle and lower parts), some deviations become visible which, indeed, become larger also with increasing wave num-

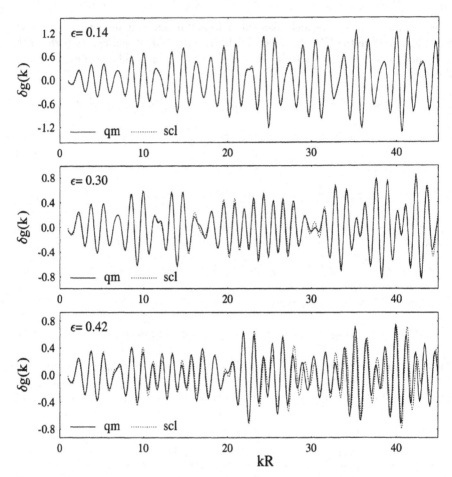

Figure 6.17: Level density $\delta g(k)$ of the elliptic billiard for three different eccentricities ϵ, Gaussian averaged with $\gamma = 0.64/R$. *Solid lines:* quantum-mechanical results. *Dotted lines:* semiclassical results of the perturbative method of Creagh, i.e., using Eq. (6.55) for the diameter orbits (with $n \leq 2$) and the unperturbed spherical disk formula (6.14) for the polygonal orbit families with up to $v=7$ reflections ($w=1$). (After [12].)

ber. Still, the overall agreement is quite good for the gross-shell structure in the regions of ϵ and k shown here.

In order to see now what is gained by the perturbation approach, we compare these results with those of the standard POT. Standard means here to use either the Berry-Tabor method [7] described in Sect. 2.7 or the extended Gutzwiller method described in Sect. 6.1 for the polygonal orbit families, and the Gutzwiller trace formula for the isolated diameter orbits. The Berry-Tabor trace formula for the ellipse billiard has been discussed explicitly in Refs. [12, 13, 33]; in Fig. 6.18 we present some results of Ref. [12]. The same kind of comparison is made here as in Fig. 6.17, but over a larger region of deformations. Clearly, for the smallest deformation (upper part, $\epsilon=0.14$) the

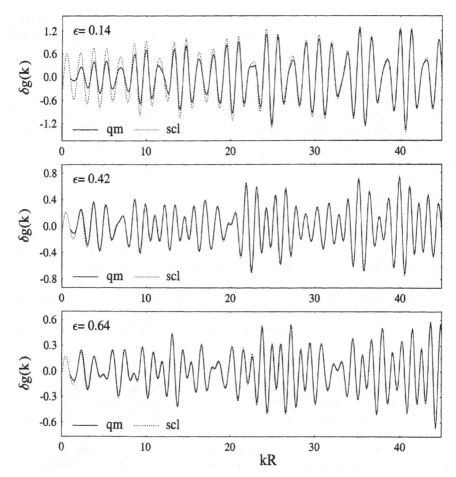

Figure 6.18: The same as in Fig. 6.17, but the semiclassical results (dotted lines) were calculated using the Berry-Tabor amplitudes for the degenerate polygon families up to $v=7$ $(w=1)$, and using the standard Gutzwiller amplitudes for the isolated diameter orbits (with $n \leq 2$). (After [12].)

agreement with the quantum-mechanical result becomes bad at low energies. According to Eq. (6.49), we expect the contribution from the diameters to diverge like $1/\epsilon$. The reason why this is not apparent at larger k numbers is that the polygon orbits have amplitudes larger by a factor $\sqrt{k/\hbar}$ than those of the diameters. Therefore they dominate at larger energies, where one would have to go to even smaller deformations ϵ in order to see the divergence. Inversely, at $k=0$ the polygonal amplitudes in $\delta g(k)$ go to zero like $\sqrt{k}$ so that there, only the diverging diameter contributions survive.

Clearly, the perturbation results given in Fig. 6.17 are better than the standard POT results in Fig. 6.18 near $k=0$, where the diameter orbits dominate, at all deformations. On the other hand, sufficiently large kR and ϵ, the standard POT results are in almost perfect agreement with quantum mechanics. There is thus, at each deformation, a range in energy (or k number) for which both methods agree reasonably well, so that in all cases one has a valid semiclassical result. As already said above, this could not work for a high resolution of the spectrum, but for the gross-shell structure as investigated here, the restriction to a maximum number n_{max} of repetitions does not pose any serious problem.

It is now instructive, nevertheless, to look at the asymptotic behavior of the modulation factor (6.54) for large arguments $x = \epsilon^2 nkR$. A stationary phase analysis of the integral over the angle θ yields directly the asymptotic expansion of the Bessel function $J_0(x)$ for large arguments: $J_0(x) \sim \sqrt{2/(\pi x)} \cos(x - \pi/4)$ for $x \gg 1$. Rewriting the cosine function in terms of exponentials and inserting this into Eq. (6.55), one obtains two contributions:

$$\delta g_{diam}(E) \simeq \left(\frac{R}{\hbar \pi p}\right) \Im m \sum_{n=1}^{n_{max}} \frac{1}{n\epsilon} e^{i(4nkR - 3n\pi + 3\pi/4)} \left[e^{i\epsilon^2 nkR - i\pi/4} + e^{-i\epsilon^2 nkR + i\pi/4} \right]. \quad (6.56)$$

These two contributions can be identified with the two isolated diametrical orbits that arise in the elliptic billiard. Since they correspond to two stationary points of the exponent in (6.54) at $\theta_1 = 0$ and $\theta_2 = \pi/2$, these two angles give their rectangular orientations. Indeed, the leading terms in the exponentials of (6.56) now combine to give the lengths of the two diameters $4nkA$ and $4nkB$, respectively, in their expansions (6.51) up to order ϵ^2. Taking the imaginary part in (6.56) gives therefore exactly the Gutzwiller result (6.49) for the isolated orbits with the correct phases, and with the amplitudes correct to leading order in ϵ.

We see thus that the asymptotic analysis of the modulation factor *predicts* correctly, to the leading order in the perturbation parameter ϵ, both the amplitudes and the phases of the isolated periodic orbits created by the perturbation of the system. Although these amplitudes diverge in the original Gutzwiller approach, their prediction from the perturbative approach will provide a valuable help for finding some of the leading isolated orbits in non-integrable systems whose dynamics may *a priori* be unknown. Note that this can, of course, only work for those isolated orbits that are created by the breaking of a degenerate orbit family in the first order of the perturbing Hamiltonian δH, i.e., if the modulation factor for this family is different from unity. Quantitatively, these asymptotic amplitudes cannot be used at the deformations where they have been obtained. Also, as already pointed out earlier, the perturbative trace formula (6.55) does not lead over to the correct amplitudes of the isolated short diameter in the limit of large deformation ϵ. For this, a uniform approximation [30, 33] has to be used which will be discussed in Sect. 6.3.1.

Similar applications of the perturbative trace formula (6.45) to small multipole deformations of a three-dimensional spherical cavity have been reported in Refs. [26, 34], and their relevance to the physics of metal clusters is discussed in Refs. [35, 36].

An altogether different solution to the symmetry breaking problem in the ellipse billiard has recently been suggested by Magner *et al.* [13], and has also been successfully applied to the three-dimensional spheroidal billiard [20]. Remember that technically, it is the stationary phase approximation made in the derivation of the Gutzwiller amplitudes of the diameter orbits, that causes their divergence in the zero-deformation limit. The idea therefore is to improve their treatment by including in the trace formula explicitly *non-periodic closed orbits* that have two bounces on the boundary and go through a third point $\mathbf{r}_0$ anywhere within the elliptic boundary, where this is allowed according to the laws of spectral reflection. (These non-periodic orbits still have to be solutions of the classical equations of motion.) The Jacobian $\mathcal{J}$ in the trace formula (6.3), which is defined analogously to that in Eq. (6.4) but is no longer allowed to carry the suffix '*po*', is then calculated along these non-periodic orbits, and the intermediate point $\mathbf{r}_0$ is integrated over exactly. If the intermediate points were taken only along the diameter orbits and the integrations transverse to them done in the saddle-point approximation, one would arrive back at the Gutzwiller amplitudes which diverge for $\epsilon \to 0$. By integrating exactly, but only over the classically allowed region, and by separating the two types of non-periodic orbits with hyperbolic and elliptic caustics, one arrives at improved amplitudes which stay finite and in the limit $\epsilon \to 0$ go over to the diameter amplitude in the disk trace formula (6.14) (for details, see Ref. [13]). This is certainly a very attractive improvement of the standard POT. It remains to be investigated how difficult this method becomes for less simple systems.

6.2.3   Example 2: Inclusion of weak magnetic fields

As is discussed in Sects. 8.3 and 1.5, the theory of periodic orbits plays an important role in the investigation of electronic transport properties in two-dimensional nanostructures (see also Ref. [38, 39]). In the corresponding experiments, one usually applies a homogeneous magnetic field perpendicular to the plane to which the conductance electrons are confined. For the investigation of the so-called quantum dots (see Sect. 8.3), it is therefore necessary to treat the classical dynamics in the presence of an external magnetic field. In general, the periodic orbits will change rather dramatically under the influence of the Lorentz forces – we recall the case of a free particle which in a homogeneous magnetic field will move on cyclotron orbits; quantum-mechanically this leads to the Landau quantization (cf. Problem 2.7). However, in a bound system whose size is much smaller than the magnetic length $l_B = \sqrt{\hbar c/eB}$, the effects of the magnetic field may be treated perturbatively to lowest order in B. One can then apply the method described in Sect. 6.2.1 above in which the changes of the periodic orbits of the system are ignored and only the first-order change in the action is included through the modulation factor $\mathcal{M}_\Gamma$ defined in Eq. (6.44).

The Hamiltonian of a particle with charge e moving in a potential $V(x, y)$ in the (x, y)-plane under the influence of a homogeneous magnetic field $\mathbf{B} = Be_z$ in the z-direction is given by

$$H = \frac{1}{2m} \left(\mathbf{p} - \frac{e}{c} \mathbf{A} \right)^2 + V(x, y), \tag{6.57}$$

where $\mathbf{A} = (1/2) \mathbf{B} \times \mathbf{r}$ is the vector potential. To first order in B, this gives the Hamiltonian [cf. Eq. (1.129) in Problem 2.7]

$$H = \frac{p^2}{2m} + V(x, y) - \omega_L L_z, \tag{6.58}$$

where ω_L is the Larmor frequency $\omega_L = eB/2mc$. In first-order perturbation theory of quantum mechanics, the last term gives the Zeeman splitting. In the notation of Sect. 6.2.1, the first-order change in the classical action (6.42) is easily calculated, using the fact that the angular momentum L_z is proportional to the area $\mathcal{A}$ per unit time which is surrounded by the orbit Γ:

$$L_z = 2m\frac{d\mathcal{A}_\Gamma}{dt}\,. \tag{6.59}$$

(Applied to the motion of the planets around the sun, this is Kepler's second law!) Integrating Eq. (6.42) over one basic period, we thus obtain

$$\Delta S_\Gamma(g) = \omega_L 2m\mathcal{A}_\Gamma(g) = \frac{e}{c}B\mathcal{A}_\Gamma(g)\,, \tag{6.60}$$

where we have added the argument (g) to indicate that the area may depend on the group parameter(s) characterizing the particular members of the periodic orbit family Γ. Alternatively, we may derive Eq. (6.60) from the very definition of the action $S = \oint_\Gamma \mathbf{p} \cdot d\mathbf{q}$, noting that the canonical momentum of a particle in a vector potential $\mathbf{A}$ is given by $\mathbf{p} = m\mathbf{v} + e\mathbf{A}/c$, where $\mathbf{v}$ is the particle velocity. The first term gives the unperturbed action, and the second term gives the change in action ΔS_Γ which, after applying Stokes' theorem, yields the expression (6.60) and is, of course, just proportional to the magnetic flux through the area $\mathcal{A}_\Gamma$.

We are now ready to include the appropriate modulation factor into the trace formula for a two-dimensional system with degenerate orbit families Γ, characterized by the parameter g of an Abelian group. Starting from Eq. (6.21) with $f=1$, noting that $\det(\widetilde{M}_\Gamma - I)$ is unity, we get

$$\delta g\,(E) = \frac{1}{\pi\hbar}\frac{1}{(2\pi\hbar)^{1/2}} \sum_\Gamma \frac{T_\Gamma V_\Gamma}{|\mathcal{J}_\Gamma|^{1/2}} \,\Re e\left\{\mathcal{M}_\Gamma \exp\left[\frac{i}{\hbar}S_\Gamma - i\sigma_\Gamma\frac{\pi}{2} - i\frac{\pi}{4}\right]\right\}, \tag{6.61}$$

where the modulation factor $\mathcal{M}_\Gamma$ is given by

$$\mathcal{M}_\Gamma = \frac{1}{V_\Gamma}\int \exp\left[i\frac{eB}{\hbar c}\mathcal{A}_{\Gamma(g)}\right]d\mu(g). \tag{6.62}$$

We next assume that the unperturbed system has time reversal symmetry. The periodic orbits then occur in pairs which run in opposite time orientations, whereby the magnetic flux through their areas changes sign. The other ingredients of the trace formula (6.61) will remain unchanged under the time reversal operation. Summing over the pairs of time reversed orbits leads to replacing the modulation factor effectively by

$$\mathcal{M}_\Gamma = \frac{1}{V_\Gamma}\int \cos\left[\frac{eB}{\hbar c}\mathcal{A}_{\Gamma(g)}\right]d\mu(g) \tag{6.63}$$

and including it into the trace formula

$$\delta g\,(E, B) = \frac{1}{\pi\hbar}\frac{1}{(2\pi\hbar)^{1/2}} \sum_\Gamma \frac{T_\Gamma V_\Gamma}{|\mathcal{J}_\Gamma|^{1/2}} \,\mathcal{M}_\Gamma \cos\left(\frac{1}{\hbar}S_\Gamma - \sigma_\Gamma\frac{\pi}{2} - \frac{\pi}{4}\right). \tag{6.64}$$

The time reversal symmetry factor 2 is hereby assumed to be included explicitly in the summation over the distinct families Γ in Eq. (6.64). Those orbits which map onto themselves under time reversal and thus occur only once in the sum over Γ – these are straight-line oscillations or, more generally, librations which have an average angular momentum zero – do not surround any net area over one period and will thus have a modulation factor of unity.

In the special case of a system with *axial symmetry* (but with no extra dynamical symmetry like for the harmonic oscillator), the result becomes particularly simple. Here the flux does not change when rotating the periodic orbits around the symmetry axis. Thus the group averaging over the angle θ is trivial and the modulation factor becomes

$$\mathcal{M}_\Gamma = \cos\Phi_\Gamma, \qquad \Phi_\Gamma = \frac{eB}{\hbar c}\mathcal{A}_\Gamma, \tag{6.65}$$

where Φ_Γ is the magnetic flux through the periodic orbit Γ in units of $\hbar c/e$. Starting from Eq. (6.26), we thus get the trace formula

$$\delta g\,(E,B) = \frac{1}{\hbar}\frac{2}{(2\pi\hbar)^{1/2}}\sum_\Gamma\frac{T_\Gamma}{a_\Gamma\,|(\partial\Theta/\partial L)_\Gamma|^{1/2}}\cos\Phi_\Gamma\,\cos\left(\frac{1}{\hbar}S_\Gamma - \sigma_\Gamma\frac{\pi}{2} - \frac{\pi}{4}\right), \tag{6.66}$$

where a_Γ is the discrete rotational symmetry factor defined in Eq. (6.24).

Note that if the homogeneous magnetic field were replaced by that of an infinitesimally thin solenoid (i.e., a singular magnetic flux line) placed at the origin, the dimensionless flux Φ_Γ would just be the Aharonov-Bohm phase acquired by the particle by turning around the flux line (cf. Problem 3.4).

In the case of the two-dimensional isotropic harmonic oscillator in a magnetic field, the formula (6.26) does not apply because of its dynamical $SU(2)$ symmetry which is broken by the magnetic field, although the rotational $U(1)$ symmetry persists. The trace formula for this system can be given exactly (see Problem 3.3). After expanding it for weak magnetic fields, the modulation factor turns out to be a spherical Bessel function $j_0(\Phi_\Gamma)$ instead of the cosine function in (6.65) (see also Problem 6.2). In the following, we give some simple applications of the above results.

The 2-dimensional rectangular billiard

We now apply a homogeneous magnetic field B perpendicular to the two-dimensional rectangular billiard treated in Sect. 6.1.5. This case has been discussed in Refs. [37, 38, 39]. We first have to determine the area surrounded by a periodic orbit Γ, when its starting point on the left side of the boundary is y, chosen as in Sect. 6.1.5. Note that when the orbit is crossing its own path during one period, some parts of the area cancel. This is illustrated in Fig. 6.19 on the next page, for a member of the orbit family (5,3). In fact, it can be easily seen that for the primitive orbit (m_1, m_2), the surrounded area can be decomposed into $m_1 m_2$ non-overlapping congruent rectangles that touch at their corners. These partial rectangle areas come in pairs with opposite orientations, so that the contributions of each pair to the total flux is zero. Consequently, the area $\mathcal{A}_\Gamma(y)$ is zero for all orbits for which the product $m_1 m_2$ is even (including zero). For the orbits where this product is odd, the remaining partial area is

$$\mathcal{A}_\Gamma(y) = \frac{2a_1}{m_2}y\left(1 - y\frac{m_1}{a_2}\right), \tag{6.67}$$

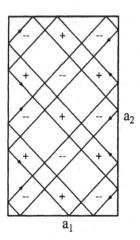

Figure 6.19: A specific orbit of the (5,3) family in the rectangular billiard. The area surrounded once by the orbit during one period falls into 15 congruent rectangles with pairwise opposite orientations (indicated by a '+' or a '−−' sign, respectively); only one of them contributes to the total magnetic flux.

its sign changing when time is reversed. Inserting this into (6.63), we get

$$\mathcal{M}_\Gamma = \frac{m_1}{a_2} \int_0^{(a_2/m_1)} \cos\left[\frac{eB}{\hbar c} \mathcal{A}_\Gamma(y)\right] dy \,. \tag{6.68}$$

With the substitution $t = y m_1/a_2$, we obtain the dimensionless integral

$$\mathcal{M}_\Gamma = \int_0^1 \cos\left[2\Phi_{12}(t^2 - t)\right] dt = 2 \int_0^{1/2} \cos\left(2\Phi_{12}t^2 - \Phi_{12}/2\right) dt \,, \tag{6.69}$$

where Φ_{12} is the dimensionless flux

$$\Phi_{12} = \frac{a_1 a_2}{m_1 m_2} \frac{eB}{\hbar c} \quad (m_1, m_2 \text{ odd}) \,, \qquad \Phi_{12} = 0 \quad (\text{otherwise}) \,. \tag{6.70}$$

Eq. (6.69) can be expressed in terms of the Fresnel integrals (see Ref. [40], p. 300):

$$C(x) = \int_0^x \cos\left(\frac{\pi}{2}t^2\right) dt \,, \qquad S(x) = \int_0^x \sin\left(\frac{\pi}{2}t^2\right) dt \,. \tag{6.71}$$

This allows us finally to write the modulation factor as

$$\mathcal{M}_\Gamma = \sqrt{\frac{\pi}{\Phi_{12}}} \left[\cos(\Phi_{12}/2)\, C\left(\sqrt{\frac{\Phi_{12}}{\pi}}\right) + \sin(\Phi_{12}/2)\, S\left(\sqrt{\frac{\Phi_{12}}{\pi}}\right)\right] \,. \tag{6.72}$$

For small magnetic fields, we may expand the Fresnel integrals:

$$C(x) = x + \mathcal{O}(x^5) \,, \qquad S(x) = \frac{\pi}{6}x^3 + \mathcal{O}(x^7) \,, \tag{6.73}$$

and find

$$\mathcal{M}_\Gamma = \cos(\Phi_{12}/2)\,. \tag{6.74}$$

Thus, to lowest order in the external field B, the modulation factor of a primitive orbit (m_1, m_2) just gives an extra phase which corresponds to a fraction $1/(2m_1m_2)$ of the total dimensionless flux through the rectangle. Summing now separately over the repetitions n of the primitive orbits Γ, we finally obtain the trace formula

$$\delta g(E, B) = \frac{ma_1a_2}{\hbar^2\pi^{3/2}}\sqrt{2}\sum_\Gamma\sum_{n=1}^\infty\sqrt{\frac{\hbar}{nS_\Gamma}}\cos(n\Phi_{12}/2)\cos\left(\frac{n}{\hbar}S_\Gamma - \frac{\pi}{4}\right)\,. \tag{6.75}$$

The sum over Γ in Eq. (6.75) goes only over primitive periodic orbits, i.e., over non-negative pairs (m_1, m_2) of relatively prime integers whereby $m_1 = m_2 = 0$ must be omitted, since this contribution is included in the smooth part of the level density; those orbits with both $m_i > 0$ have to be counted twice due to their two distinct time orientations.

The orbital contribution to the magnetic susceptibility of the system is governed by the factor $\cos(\Phi_\Gamma)$ in Eq. (6.75). Note that there are lots of cancellations, in particular for all orbits with an even product m_1m_2. The largest contribution will come from the $(1,1)$ orbit, the next smaller from the $(1,3)$ orbit whose flux is already a factor 3 smaller and, due to the $1/\sqrt{S_\Gamma}$ dependence of the amplitude, further suppressed. It has been checked by comparison with quantum-mechanical calculations of the susceptibility [38], that the main period of its oscillations as a function of the magnetic field is, indeed, coming from the $(1,1)$ orbit.

A spherical system with magnetic field

Another application shall be briefly mentioned here: that of a three-dimensional spherical billiard with radius R in a homogeneous magnetic field B [41]. It has not found any practical applications, since the metal clusters for which it is a good model, as discussed in Sect. 8.2, are much smaller in size than the magnetic length of the presently available experimental field strengths. We present this case here, because it illustrates another typical case of a modulation factor connected with symmetry breaking.

The spherical billiard has $SO(3)$ symmetry; its periodic orbits thus have an $f=3$ dimensional degeneracy, except for the diameter orbit with $f=2$ (see Sect. 6.1.6). An homogeneous magnetic field B in the z direction breaks this symmetry; as a result, we have the axial $U(1)$ symmetry with energy E and z component of the angular momentum, $L_z = \hbar l$ as constants of the motion. The group parameters g characterizing the unperturbed $SO(3)$ symmetry may be taken as the Euler angles: $g = (\alpha, \beta, \eta)$. We choose them here in such a way that β is the angle between the plane of motion of a given orbit and the the z direction. The other angles α and η then describe rotation about the z axes before and after the β rotation (usually done about the old x axis).

The first-order change in the actions, $\Delta S_\Gamma(g)$ (6.42) of the orbit $\Gamma_{(g)}$, then is the magnetic flux through the periodic orbit like in Eq. (6.60), but projected here onto the x, y plane. Using the properties of the polygonal orbits (v, w) in the sphere or the disk (see Sect. 6.1.3), this can easily be calculated and becomes

$$\Delta S_{\Gamma(g)} = \pm \frac{e}{c} BR^2 \frac{v}{2} \sin(2\pi w/v) \cos\beta, \tag{6.76}$$

where the sign depends on the clockwise or counterclockwise motion of the particle around the z axis. The modulation factor (6.44) becomes

$$\mathcal{M}_\Gamma = \frac{1}{8\pi^2} \int_0^{2\pi} d\alpha \int_0^{2\pi} d\eta \int_0^\pi \sin\beta \exp\left[i\, F_\Gamma \cos\beta\right] d\beta, \tag{6.77}$$

with $F_\Gamma = \kappa\,(v/2)\sin(2\pi w/v)$. Hereby κ is the dimensionless flux $(eBR^2)/(\hbar c)$ which may also be expressed in terms of the magnetic length l_B:

$$\kappa = \frac{eB}{\hbar c} R^2 = \left(\frac{R}{l_B}\right)^2; \qquad l_B = \sqrt{\frac{\hbar c}{eB}}. \tag{6.78}$$

The integrals over α and η in (6.77) are trivial, and that over β reduces to an integral representation of the spherical Bessel function of order zero, $j_0(x) = \sin x/x$, giving finally [41]

$$\mathcal{M}_\Gamma = j_0\left[\kappa\, \frac{v}{2} \sin(2\pi w/v)\right]. \tag{6.79}$$

This modulation factor can now be inserted under the Γ summation in the trace formula of the spherical billiard, Eq. (6.39), to obtain the level density $\delta g\,(E, B)$ to lowest order in the magnetic field strength B. Note that $\mathcal{M}_\Gamma$ is unity for the diameter orbits ($v=2w$), giving the second term in Eq. (6.39), since these enclose no area and hence no magnetic flux. In Ref. [41] the resulting trace formula has been numerically computed and compared to the exact quantum-mechanical result.

In Fig. 6.20 we show a comparison of the coarse-grained level density for three different field strengths expressed in terms of κ and drawn as a function of kR, where k is the Fermi wave number. The agreement for not too strong fields and not too low energies is excellent. The increasing disagreement with decreasing energy can be understood in terms of the new Landau orbits that are created in the magnetic field but cannot be reproduced by the perturbative trace formula. Even for small κ, there is always a lower limit in energy for which cyclotron orbits exist that have a radius smaller than the cavity radius R; this happens simply because the cyclotron radius is proportional to the wave number k and thus to $E^{1/2}$. The above perturbative treatment will always fail below this lower energy limit. For small fields (e.g., $\kappa = 2$ in Fig. 6.20), this limit falls below the position of the lowest quantum state of the unperturbed system and thus has no physical consequence. In stronger fields, however, the lower limit increases. The consequence is most dramatically seen for the case $\kappa = 10$, where the quantum level density already starts building up the lowest Landau orbits. The missing of the Landau orbits in the perturbative trace formula thus explains its complete failure for $kR \lesssim 25$. The small vertical arrows indicate the wave number k for which the cyclotron radius is five times the sphere radius R; above this 'confidence limit', the semiclassical approach is seen to work very well.

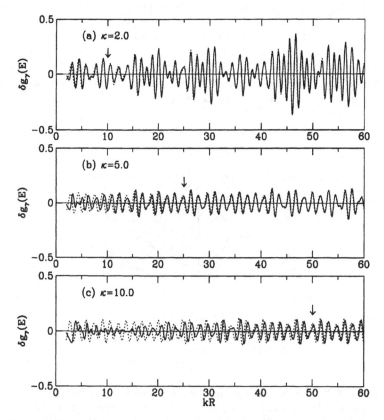

Figure 6.20: Oscillating part $\delta g_\gamma(E)$ of the level density for a spherical billiard in a homogeneous magnetic field B. Both are Gaussian averaged over a range $\gamma = 0.4/R$. *Solid lines:* quantum-mechanical results; *dotted lines:* semiclassical results using the trace formula (6.39) including the modulation factor (6.79). The field strengths given by κ (6.78) are: (a) $\kappa = 2$, (b) $\kappa = 5$, and (c) $\kappa = 10$. (From [41].)

6.2.4 Example 3: The quartic Hénon-Heiles potential

We conclude this chapter with the application of the perturbative trace formula to a system that classically becomes chaotic at high energies, and that can be treated perturbatively with a nontrivial result [42]. We consider a particle in the following anharmonic potential in two dimensions, written in polar coordinates (r, ϕ)

$$V(r, \phi) = \frac{m}{2}\,\omega^2\,r^2 - \alpha\,\frac{1}{4}\,r^4 \cos(4\phi)\,. \tag{6.80}$$

This potential resembles very much the Hénon-Heiles (HH) potential (5.73) which we have studied in Sect. 5.6.4. Potentials with anharmonicities of the general form

$r^n \cos(\mu\phi)$ have been investigated in astrophysics as simple models for elliptical galaxies. The standard HH potential [43] corresponds to $n = \mu = 3$; other values for n and μ were suggested later [44, 45]. The above potential with $n = \mu = 4$ was shown in Ref. [46] to exhibit classical chaos in a similar way as the HH potential. The same type of potential $V(r, \phi) = ar^2 + br^3 \cos(3\phi)$ has recently been proposed [47] for the parametrization of the confinement potential in a two-dimensional quantum dot with triangular boundaries and open contacts (see Sect. 8.3).

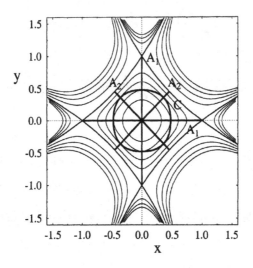

Figure 6.21: Equipotential contours of the quartic HH potential (6.80) for α=1. Also shown (by heavy lines) are the three shortest types of periodic orbits in this potential: A_1, A_2, and C (evaluated at the saddle point energy $E = E^*$). (From [42].)

In Fig. 6.21 we show the equipotential surface of (6.80) in the (x, y)-plane for α=1. Different from the original HH potential, we have here a four-fold discrete rotation symmetry and four saddle points; we therefore call it the "quartic Hénon-Heiles potential". The saddle point energy is $E^* = 1/(4\alpha)$ in units $m = \omega = 1$. We refer to Sect. 5.6.4 for the scaling of the coordinates and the energy; the classical motion depends only on the scaled energy which here is $e = E/E^* = 4E\alpha$. We also show in Fig. 6.21 the shortest periodic orbits (at $e = 1$). They must be evaluated numerically by solving the nonlinear equations of motion. A_1 denotes a straight-line oscillation along a symmetry axis that goes through two saddle points and is very similar to the orbit A of the HH potential. It is stable up to a scaled energy $e \simeq 0.85$ and then undergoes a cascade of increasingly dense lying bifurcations [48], alternating between being stable and unstable, until at $e = 1$ is ceases to exist. A_2 denotes another straight-line oscillation along the symmetry axes at angles $45°$ or $135°$; this orbit is stable up to energies well above $e = 2$. It represents attempts of the particle to go "uphill" along the steepest way and resembles in this respect the orbit B of the HH potential (although the latter is curved and always unstable). The third orbit C is a rotation similar to that in the

HH potential. Here, however, it is always unstable. Each of these three orbits has a discrete degeneracy of two: the straight-line librations because of their two possible orientations with a relative angle of 90°, and the rotation orbit C because of its two time orientations. These are the only short orbits with periods of the order of the unperturbed harmonic-oscillator period $T_0 = 2\pi/\omega$; other more complicated periodic orbits are found only at energies $e \gtrsim 0.85$.

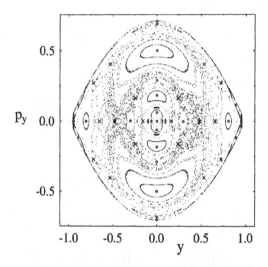

Figure 6.22: Poincaré surface of section of the quartic HH potential, evaluated at the saddle-point energy ($e = 1$). The small circles and crosses indicate elliptic and hyperbolic fix points, respectively, that correspond to isolated periodic orbits. (From [42].)

Like the HH potential, the present system also has chaotic orbits at energies close to and above the saddle point energy E^*. Fig. 6.22 shows a phase-space picture in the form of a Poincaré surface of section, obtained for thirty different classical orbits at $e = 1$. Clearly, we can distinguish regular and irregular domains; the circles in the middle of the regular regions are elliptic fix points corresponding to stable periodic orbits, and the crosses are hyperbolic fix points belonging to unstable periodic orbits. The major part of the phase space is already ergodic at this critical energy, although to a lesser extent as for the HH potential. This system with its mixed dynamics is another nice example of the breaking of rational tori (which here are the 2-tori of the 2-dimensional isotropic harmonic oscillator) discussed in Sect. 2.8. Similarly to the HH case, problems arise with the semiclassical approximation in the limit of small energy (or α), where one reaches the unperturbed harmonic-oscillator potential. In Ref. [42], the quantum-mechanical level spectrum has been calculated for the potential (6.80), and a semiclassical analysis of the level density has been performed. For high enough energies, the original Gutzwiller trace formula for isolated orbits applies and gives excellent results for the averaged level density. The Maslov indices were found, with the methods given in Appendix D.1, to be $\sigma_{A_1}=5$, $\sigma_{A_2}=3$, and $\sigma_C=4$.

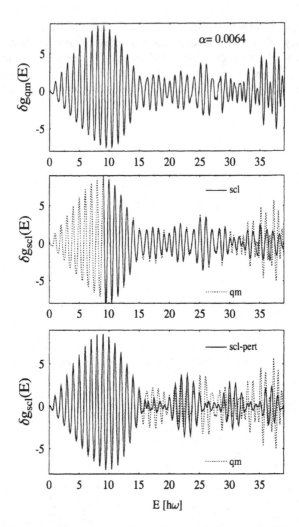

Figure 6.23:   Oscillating part of level density of the quartic HH potential with $\alpha = 0.0064$, Gaussian averaged with $\gamma = 0.25\,\hbar\omega$. *Top:* quantum-mechanical result. *Middle:* semiclassical result of the Gutzwiller trace formula for isolated orbits (solid line) compared to the quantum result (dotted line); the semiclassical result diverges for $E \to 0$ and is not shown below $E = 9\,\hbar\omega$. *Bottom:* semiclassical perturbative result (6.84) (solid line), compared to the quantum result (dotted line). (From [42].)

Some typical results are shown in Fig. 6.23 for an anharmonicity value of $\alpha = 0.0064$, for which the critical energy is at $E = 39\,\hbar\omega$. In the top part of the figure, the quantum-mechanical level density is shown. It exhibits again a beat structure. In the middle of the figure, the result of the Gutzwiller trace formula is given by the solid line and the quantum-mechanical result is overlayed by the dotted line. In the region $10\,\hbar\omega < E < 30\,\hbar\omega$, the agreement is perfect, even up to fine details of the beating pattern. For low energies, the Gutzwiller result diverges (it is not shown in the figure for

$E \lesssim 9\,\hbar\omega$), since one there approaches the higher $SU(2)$ symmetry of the unperturbed harmonic oscillator. The onset of differences for $E > 30\,\hbar\omega$ is partially due to the missing of the more complicated orbits that arise from the bifurcations of the orbit A_1 that start at $e \simeq 0.85$. Like in the calculations for the HH potential reported in Sect. 5.6.4, the singularities occurring at the bifurcation points have been circumvented by a smooth extrapolation of the amplitude of orbit A_1 for $e > 0.65$. (The amplitudes are shown in Fig. 6.24 below.) Another reason for the discrepancies at higher energies is the neglect of the imaginary parts of the quantum levels due to tunneling through the barriers.

So far, the situation is very similar to that of the standard HH potential. We now apply the perturbative approach to the quartic HH potential. For the original HH potential (5.73), a first-order perturbative treatment gives zero results because of the odd parity of the nonlinear term. In the present case (6.80), however, a nonzero result is obtained. The treatment of the $SU(2)$ symmetry is not quite trivial and shall be omitted here. An elegant way to perform the group average is described in Ref. [8] (cf. Problem 6.2). It leads to an integral over the surface of a sphere that can be done by the usual angles (ϕ, θ) of a polar coordinate system. For the anharmonicity term in Eq. (6.80), this gives a modulation factor [42]

$$
\begin{aligned}
\mathcal{M}(x) &= \frac{1}{4\pi} \int_0^{2\pi} d\phi \int_0^{\pi} \sin\theta\, e^{i\,x[\cos^2\phi - \cos^2\theta\,(1+\cos^2\phi)]}\, d\theta \\
&= \frac{1}{2\pi} \int_0^{2\pi} d\phi \int_0^1 e^{i\,x[\cos^2\phi - u^2\,(1+\cos^2\phi)]}\, du \tag{6.81} \\
&= \frac{\pi}{2\sqrt{2}}\, J_{-1/4}\!\left(\frac{x}{2}\right) J_{1/4}\!\left(\frac{x}{2}\right), \tag{6.82}
\end{aligned}
$$

where $J_\mu(x)$ are Bessel functions of fractional order, and their argument x is given by

$$
x = \alpha\, n\, 3\pi E^2/4, \tag{6.83}
$$

with the energy E taken in units of $\hbar\omega$. Here n is the repetition number of the unperturbed orbit. Note that x is proportional not only to the perturbation parameter α, but also to the square of the energy E. Therefore, the limit $x \to 0$ in which the symmetry is restored implies $\alpha \to 0$ or $E \to 0$, or both.

Inserting now this modulation factor into the trace formula (3.68) for the unperturbed harmonic oscillator, we obtain for the oscillating part of the level density

$$
\delta g_{scl}(E) = 2E \sum_{n=1}^{n_m} \mathcal{M}(\alpha\, n\, 3\pi E^2/4) \cos(2\pi n E), \tag{6.84}
$$

where we again have restricted the sum to a finite number n_m of repetitions. A numerical result obtained by this formula is shown in the lowest part of Fig. 6.23 by the solid line, superposed with the quantum-mechanical result (dotted line). Both were averaged over an energy range $\gamma = 0.25\,\hbar\omega$. Clearly, one obtains an excellent agreement in the low-energy limit. Note that the two lowest harmonics ($n_m=2$) had to be used to reproduce correctly the asymmetry of the quantum result below $E \sim 12\,\hbar\omega$. (For larger energies, the contributions from $n=2$ are already negligible.) Above $E \simeq 14\,\hbar\omega$, the perturbative result starts differing from the exact one; it reproduces some of the beat structure but with wrong amplitudes. At $E \simeq 14\,\hbar\omega$ we have $x \simeq 3.0$ for the lowest harmonic. Thus, the perturbative trace formula works actually up to a surprisingly high energy where x is much larger than unity.

It is, of course, desirable to join in some way the two approaches applicable in the two limits: the perturbative trace formula (6.84) valid for low energies (or values of α), and the Gutzwiller trace formula for the isolated orbits, valid at higher energies (or α). This can be achieved by a uniform approximation which analytically interpolates between the two trace formulae [31] and will be discussed in the next section. Before doing so, and as a preparation to it, we investigate again the asymptotic behavior of the modulation factor $\mathcal{M}(x)$ for large arguments x (6.83). For this one has to evaluate all critical points of the exponent under the double integral (6.81). The inner u integral is of the form

$$I(x) = \int_0^1 g(u)\, e^{ix\, h(u)}\, du\,, \tag{6.85}$$

whereby $u = 0$ is a stationary point of the function $h(u)$, but $u = 1$ is not. An asymptotic expansion of the integral (6.85) for $x \to \infty$ has been given by Wong [49]:

$$I(x) \sim g(0)\sqrt{\frac{\pi}{2x\,|h''(0)|}}\, e^{ix\,h(0)}\, e^{i\frac{\pi}{4}\mathrm{sign}[h''(0)]} + \frac{g(1)}{x\,h'(1)}\, e^{ix\,h(1)}\, e^{-i\frac{\pi}{2}} + \dots \tag{6.86}$$

The first term is just the standard stationary phase integral according to Eq. (2.169), multiplied by $1/2$ since the stationary point $u = 0$ is an end point, and is proportional to $x^{-1/2}$. Since x is proportional to $\alpha/\hbar$ (see the general discussion in Sect. 6.2.1), the first term is proportional to $\hbar^{1/2}$. The second term in (6.86) is the leading correction from the non-stationary end point $u = 1$ and is proportional to $\hbar$. (Higher-order corrections are given in [49].)

The ϕ integral in (6.81) has no end points since ϕ is a cyclic variable; its phase has stationary points at $\phi = 0$, $\pi/2$, π, and $3\pi/2$. Applying now the Wong formula (6.86) to the u integral in (6.81), and then performing the ϕ integration – by stationary phase over the first and exactly over the second term (see [31] for the detailed calculation) – one finds the three leading contributions to (6.84) which can be associated to the three isolated orbits A_1, A_2 and C, created by the symmetry-breaking term proportional to α in (6.80). Their asymptotic amplitudes for $n=1$ are found to go like

$$\mathcal{A}_C(E) \sim \frac{E}{2x} = \frac{8}{3\pi e}\,, \qquad \mathcal{A}_{A_1}(E) \sim \mathcal{A}_{A_2}(E) \sim \frac{E}{2\sqrt{2}x} = \frac{8}{3\pi\sqrt{2}\,e}\,. \tag{6.87}$$

Each of them has a two-fold degeneracy as explained above, and the phases obtained from (6.86) and the ϕ integrations yield their correct Maslov indices.

The result (6.87) could also have been derived more easily using the asymptotic expansion of the Bessel functions in (6.82) for large arguments x and expanding the product of cosine functions, but the above asymptotic analysis tells us exactly how the isolated orbits emerge from the critical points of the integrations in (6.81): the orbit A_2 comes from the end point at $u = 1$, which actually corresponds to the two end points $\theta = 0$ and $\theta = \pi$ of the original angle integral – hence the double degeneracy of this orbit. The orbits A_1 and C, each also doubly degenerate, arise from the two stable ($\phi = 0$, π) and the two unstable stationary points ($\phi = \pi/2$, $3\pi/2$), respectively, of the ϕ integration of the first term in (6.86). This information will be needed in Sect. 6.3.2 below. – The prediction (6.87) of the amplitudes is compared in Fig. 6.24 to the numerically evaluated amplitudes from the original Gutzwiller trace formula. The latter are, indeed, converging to the analytical expressions (6.87) for small values of the scaled energy e, exhibiting their $1/e$ divergences.

These results are presently of no practical use, because the asymptotic amplitudes (6.87) are no longer valid in the domain where the periodic orbits are sufficiently isolated to serve the standard Gutzwiller trace formula. We see, however, that the information about the harmonic oscillator SU(2) limit, correctly brought about by the analytical modulation factor (6.82), is encoded in the numerical Gutzwiller amplitudes. This will be exploited in the construction of the uniform approximation in in Sect. 6.3.2.

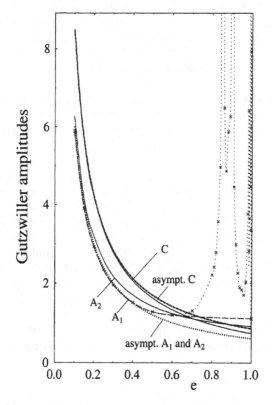

Figure 6.24: Amplitudes in the trace formula of the orbits A_1, A_2 and C in the quartic HH potential. Thin lines are the numerical Gutzwiller amplitudes of the isolated orbits. (A_1 is shown by crosses; the divergences from the bifurcations are numerically regularized by the smooth long-dashed thin line.) Heavy dashed and dotted lines are the asymptotic predictions (6.87) for C (dashed), and for A_1 and A_2 (dotted). (From [42].)

6.3   Uniform approximations

In the context of periodic orbit theory, uniform approximations make use of improved evaluations of the trace integration in cases where the simple saddle-point approximation fails. In the present section, we shall mainly discuss uniform approximations in connection with symmetry breaking, which comes as a natural extension of the previous section where this problem was discussed in the framework of perturbation theory.

We start again from a Hamilton function of the form

$$H = H_0 + \epsilon H_1\,, \tag{6.88}$$

where H_0 describes an integrable system whose periodic orbits are degenerate families, and ϵH_1 is a perturbing Hamiltonian that breaks the symmetry (or several symmetries) of H_0, but now ϵ need not be small. The contribution of one orbit family of the unperturbed system H_0 to its trace formula is of the general form

$$\delta g_0 = \mathcal{A}_0 \cos(\Phi_0) = \Re e\, \mathcal{A}_0\, e^{i\Phi_0}\,, \tag{6.89}$$

where $\mathcal{A}_0$ is a (real) amplitude and $\Phi_0 = S_0/\hbar - \sigma_0\,\pi/2$ an overall phase including the action S_0 and the Maslov index σ_0. We now ask ourselves how the trace formula is affected by the term ϵH_1. In Sect. 6.2 we have found an answer within a perturbative treatment for small ϵ, which consisted in multiplying the contribution (6.89) with a modulation factor.

In Sect. 6.3.1 we will give a simple derivation, based upon the methods discussed earlier in this chapter, of the uniform approximation of Tomsovic *et al.* [30] for two-dimensional systems with U(1) symmetry (cf. also Refs. [31, 33]). Their result is universal in the sense that it holds for any system with singly-degenerate families of periodic orbits that are broken into pairs of stable and unstable isolated orbits. It appears that no universal uniform approximation can be found for the breaking of higher symmetries that create families with f-dimensional denegeracies where $f > 1$. In Sect. 6.3.2 we present some recent uniform approximations for special cases of the breaking of SU(2) symmetry, one of which can also be applied to a particular SO(3) breaking Hamiltonian.

In the final Section 6.3.3, we will briefly discuss some recent uniform approximations for bifurcations without, however, presenting their details which would lead beyond the scope of this book.

6.3.1   U(1) symmetry breaking

We restrict ourselves to the generic case that a 1-dimensionally degenerate periodic orbit family on a 2-torus is broken into a non-degenerate pair of stable and unstable isolated orbits. As discussed in connection with the general treatment of continuous symmetries in Sect. 6.1.4, the unperturbed orbit family can be viewed as possessing the U(1) symmetry that is parametrized by a cyclic variable $\gamma \in [0, 2\pi]$ on which the unperturbed action S_0 does not depend. (If the system H_0 does not possess axial symmetry to start with, a suitable canonical transformation can be found that makes the variable γ cyclic.) The contribution of the orbit family to the trace formula of the perturbed system therefore has the form

$$\delta g = \Re e \left\{ \frac{1}{2\pi} \int_0^{2\pi} \mathcal{A}(\gamma) \, e^{i\Phi(\gamma)} \, d\gamma \right\}, \tag{6.90}$$

where $\Phi(\gamma)$ is the overall phase containing the perturbed action and a Maslov index. In the generic case under consideration, $\Phi(\gamma)$ will in the interval $[0, 2\pi]$ have one minimum and one maximum which correspond, for sufficiently large perturbations, to the pair of stable and unstable isolated periodic orbits.

Our first step is now to map the variable γ onto a new cyclic variable $\phi \in [0, 2\pi]$, as a function of which $\Phi(\phi) = \Phi(\gamma(\phi))$ becomes linear in $\cos\phi$, so that we can write it as $\Phi(\phi) = \bar{\Phi} + (\Delta S/\hbar) \cos\phi$. (This mapping was called "pendulum mapping" in [30].) The quantity ΔS is constant as a function of the variable ϕ, but it depends generally on the energy and on the parameters appearing explicitly in H_1. The transformation $\gamma \to \phi$ defines, in general, a non-linear function $\gamma(\phi)$ that is not explicitly known (but assumed here to exist). Equation (6.90) is then transformed into

$$\delta g = \Re e \left\{ \frac{1}{2\pi} \int_0^{2\pi} \mathcal{A}(\phi) \mathcal{J}(\phi) \, e^{i\left(\bar{\Phi} + \frac{1}{\hbar}\Delta S \cos\phi\right)} \, d\phi \right\}, \tag{6.91}$$

where $\mathcal{A}(\phi) = \mathcal{A}(\gamma(\phi))$ and $\mathcal{J}(\phi) = \partial\gamma/\partial\phi$ is the Jacobian due to the variable mapping. The functions $\mathcal{A}(\phi)$ and $\mathcal{J}(\phi)$ are usually not known explicitly and may be combined into one new amplitude function $\tilde{\mathcal{A}}(\phi) = \mathcal{A}(\phi)\mathcal{J}(\phi)$.

In the perturbative treatment of Sect. 6.2 we have seen that for small ϵ, one may replace $\tilde{\mathcal{A}}(\phi)$ by the unperturbed amplitude $\mathcal{A}_0$ and set $\bar{\Phi} = \Phi_0$. The ϕ integral in (6.91) can then be done analytically, and one obtains

$$\delta g_{pert} = \mathcal{A}_0 \cos(\Phi_0) \, J_0(\Delta S/\hbar), \tag{6.92}$$

where $J_0(x)$ is the cylindrical Bessel function of order zero and plays the role of the modulation factor $\mathcal{M}(x)$. (An example of this is the elliptically deformed circular billiard discussed in Sect. 6.2.2.) When ΔS is zero, we get the trace formula (6.89) in the symmetric limit. For large deviations from the U(1) symmetry, i.e., for $\Delta S \gg \hbar$, we can use the asymptotic expansion of $J_0(x) \sim \sqrt{(2/\pi|x|)} \cos(x - \pi/4)$ to find

$$\delta g \sim \mathcal{A}_0 \sqrt{\hbar/2\pi|\Delta S|} \left\{ \cos(\Phi_0 + \Delta S/\hbar - \pi/4) + \cos(\Phi_0 - \Delta S/\hbar + \pi/4) \right\}. \tag{6.93}$$

This corresponds to two isolated orbits with action shifts $\pm\Delta S$ and some corrections to the Maslov indices. [Note that the two terms above arise also from the saddle-point approximation to the integral in (6.91) at the stationary points $\phi_0 = 0$ and $\phi_0 = \pi$, respectively.] However, the amplitudes in (6.93) are not valid in the asymptotic limit of large perturbations. The purpose of the uniform approximation is to correct this wrong asymptotic limit. For $\Delta S \to 0$, on the other hand, the amplitudes in (6.93) describe exactly the diverging Gutzwiller amplitudes, as shown for an explicit example in Fig. 6.24 of the previous section.

We want to construct an approximation that in the limit $\Delta S \gg \hbar$ leads to the correct Gutzwiller trace formula for the pair of isolated orbits, which is of the form

$$\delta g_{Gutz} = \mathcal{A}_u \cos(S_u/\hbar - \sigma_u \, \pi/2) + \mathcal{A}_s \cos(S_s/\hbar - \sigma_s \, \pi/2), \tag{6.94}$$

where the indices s and u refer to the stable and unstable orbits, respectively. To this purpose, note [28, 30] that the phase in (6.91) formally represents the first two terms in a Fourier expansion of the perturbed function $\Phi(\phi)$. It is therefore consistent to replace also the unknown amplitude function $\tilde{A}(\phi)$ by the two first terms of its Fourier expansion, so that we are led to the following ansatz for the uniform approximation:

$$\delta g_u = \sqrt{2\pi|\Delta S|/\hbar}\; \Re e\left\{\frac{1}{2\pi}\int_0^{2\pi}(\bar{A}+\Delta A\cos\phi)\,e^{i\left(\bar{\Phi}+\frac{1}{\hbar}\Delta S\cos\phi\right)}\,d\phi\right\}. \qquad (6.95)$$

Note that we have multiplied the integral in (6.95) by the inverse of the root factor appearing in (6.93). Altogether we have introduced the four unknown constants $\bar{\Phi}$, ΔS, $\bar{A}$, and ΔA in the ansatz (6.95). In the Gutzwiller trace formula (6.94), on the other hand, we have two amplitudes, two actions and two Maslov indices which should be reproduced by (6.95) in the limit of large perturbations. As we have seen in Sect. 6.2.4 and in Problem 6.5, the Maslov indices obtained in the asymptotic evaluation of the perturbative trace formula are the correct ones (this always happens). Therefore, there remain only four constants in (6.94) to be reproduced, which can be uniquely determined by the above four unknowns.

It is rather straightforward, now, to see that the limit (6.94) is correctly reproduced if we fix the four unknown constants in (6.95) by the following relations:

$$\bar{A} = \frac{1}{2}(\mathcal{A}_u+\mathcal{A}_s)\,, \quad \Delta A = \frac{1}{2}(\mathcal{A}_u-\mathcal{A}_s)\,, \quad \bar{S} = \frac{1}{2}(S_u+S_s)\,, \quad \Delta S = \frac{1}{2}(S_u-S_s)\,, \quad (6.96)$$

$$\bar{\Phi} = \bar{S}/\hbar - \bar{\sigma}\,\pi/2\,, \quad \bar{\sigma} = \frac{1}{2}(\sigma_u+\sigma_s) = \sigma_0\,. \qquad (6.97)$$

For $\epsilon \to 0$, we get $\Delta S \to 0$, $\Delta A \to 0$, and $\bar{S} \to S_0$, and therefore (6.95) leads to the correct symmetric limit (6.89). Hereby the divergence of the Gutzwiller amplitudes, given by the first factor in Eq. (6.93), is cancelled by the factor in front of (6.95). In the limit $|\Delta S| \gg \hbar$, on the other hand, the stationary-phase evaluation of the integral in (6.95) will lead to the amplitudes $\bar{A} \pm \Delta A$ which, using (6.96), are precisely the Gutzwiller amplitudes $\mathcal{A}_u$ and $\mathcal{A}_s$, respectively, and thus give the form (6.94).

The integral in (6.95) can be done analytically, using

$$\frac{1}{2\pi}\int_0^{2\pi}\cos\phi\,e^{ix\cos\phi}\,d\phi = iJ_1(x)\,, \qquad (6.98)$$

and leads to the approximation of Tomsovic, Grinberg, and Ullmo [30] in the compact form given by Sieber [33]:

$$\delta g_u = \sqrt{2\pi|\Delta S|/\hbar}\left\{\bar{A}\,J_0(\Delta S/\hbar)\,\cos(\bar{\Phi}) - \Delta A\,J_1(\Delta S/\hbar)\,\sin(\bar{\Phi})\right\}. \qquad (6.99)$$

The uniform approximation to the Gutzwiller trace formula thus consists in evaluating numerically the actions, Maslov indices, and amplitudes of the two isolated periodic orbits, inserting them into Eqs. (6.96,6.97), and then applying them to Eq. (6.99). Using the asymptotic forms of the Bessel functions, it is straightforward to see that (6.99) leads back to the Gutzwiller formula (6.94) in the limit $\Delta S \gg \hbar$.

Note that the formula (6.99) holds for all generic non-integrable systems in two dimensions that arise from an integrable system with U(1) symmetry through a symmetry-breaking term in the Hamiltonian. Particular examples are two-dimensional billiards obtained by deforming the circular billiard; the nature of the deformation parameter hereby plays no role. The only assumption made is that the original orbit families (i.e., polygons in the case of the circular billiard) are broken into pairs of stable and unstable isolated orbits. The modifications that become necessary, when extra degeneracies due to discrete symmetries are present, are quite trivial (see, e.g., the examples given in Ref. [31] and the discussion in the following section). The generic breaking into a pair of stable and unstable isolated orbits is the most frequent situation. Exceptions occur, e.g., in billiards with octupole or hexadecapole deformations (see Problem C.3 for an example). These would have to be treated with different (and more complicated) uniform approximations that do not seem to be known.

We also note that the deformation away from the integrable case should be small enough so that no bifurcation of the stable isolated orbit has taken place or is about to arise. Near the bifurcation points, other uniform approximations apply, which we shall briefly discuss in Sect. 6.3.3.

6.3.2 An example of SU(2) symmetry breaking

As we have just stated above, the uniform approximation given in Eq. (6.99) for the generic U(1) symmetry breaking is universal. It seems that no universal uniform approximations exist for the breaking of higher symmetries leading to families of orbits with two- or higher-dimensional degeneracies. In Ref. [31], specific uniform approximations have been given recently for two examples of SU(2) symmetry breaking. This is the symmetry of the two-dimensional isotropic harmonic oscillator, which is explicitly broken by the nonlinear parts of the Hénon-Heiles type potentials discussed in this book and treated also in Ref. [31].

We present here the result for the quartic Hénon-Heiles potential (6.80) which we have investigated in connection with the perturbative approach in Sect. 6.2.4. The trace formula for the unperturbed harmonic oscillator is of the form (6.89) with $\mathcal{A}_0 = 2E/(\hbar\omega)^2$, $\Phi_0 = S_0 = 2\pi E/\omega$, and $\sigma_0 = 0$ (we ignore here the summation over the repetitions of the primitive orbit). The modulation factor $\mathcal{M}(x)$ for the perturbed trace formula in first-order perturbation theory is given in Eq. (6.81) and holds for small values of the parameter x given in (6.83). As shown at the end of Sect. 6.2.4, the asymptotic evaluation of the integrals in (6.81) for large x allows us to write $\mathcal{M}(x)$ as a sum of three contributions that can be associated to the three shortest orbits:

$$\mathcal{M}(x) \sim \mathcal{M}_C(x) + \mathcal{M}_{A_1}(x) + \mathcal{M}_{A_2}(x), \qquad (x \gg 1) \qquad (6.100)$$

where

$$\mathcal{M}_C(x) = \frac{1}{2x} \qquad \Rightarrow \qquad \mathcal{A}_C \sim \frac{\mathcal{A}_0}{2x},$$

$$\mathcal{M}_{A_1}(x) = \frac{1}{2\sqrt{2}\,x} \cos(+x - \pi/2) \qquad \Rightarrow \qquad \mathcal{A}_{A_1} \sim \frac{\mathcal{A}_0}{2\sqrt{2}\,x},$$

$$\mathcal{M}_{A_2}(x) = \frac{1}{2\sqrt{2}\,x} \cos(-x + \pi/2) \qquad \Rightarrow \qquad \mathcal{A}_{A_2} \sim \frac{\mathcal{A}_0}{2\sqrt{2}\,x}. \qquad (6.101)$$

The first term corresponds to the loop orbit C, the second and third terms to straight-line libration orbits A_1 and A_2, respectively; each of them has a two-fold discrete degeneracy which is included in the amplitudes. Note that the first-order action shifts predicted in (6.101) are zero for the orbit C and $\pm \hbar x$ for the orbits A_1 and A_2, respectively, and the Maslov indices predicted from the phase shifts are $\sigma_C = 0$ (or 4), $\sigma_{A_1} = 1$ (or 5), and $\sigma_{A_2} = -1$ (or 3), which are exactly the ones quoted in Sect. 6.2.4. We also recall that the asymptotic terms in (6.101) arise from the critical points of the integrals in (6.81) as follows:

$$
\begin{array}{lll}
\text{orbit } A_2: & \text{from } u = 1, & (\text{any } \phi), \\
\text{orbit } A_1: & \text{from } u = 0, & \phi = 0 \text{ and } \pi, \\
\text{orbit } C: & \text{from } u = 0, & \phi = \pi/2 \text{ and } 3\pi/2.
\end{array}
\tag{6.102}
$$

The asymtotic amplitudes in (6.101), which are the same as those given in (6.87), are not correct in the limit of large x. Their values should, instead, be those appearing in the Gutzwiller trace formula for the isolated primitive orbits:

$$
\delta g_{Gutz} = \mathcal{A}_{A_1} \cos \left(\frac{S_{A_1}}{\hbar} - \sigma_{A_1} \frac{\pi}{2} \right) + \mathcal{A}_{A_2} \cos \left(\frac{S_{A_2}}{\hbar} - \sigma_{A_2} \frac{\pi}{2} \right) + \mathcal{A}_C \cos \left(\frac{S_C}{\hbar} - \sigma_C \frac{\pi}{2} \right).
\tag{6.103}
$$

We now follow exactly the same philosophy for the construction of the uniform approximation as in the previous subsection. After a suitable remapping of the variables ϕ, θ – we keep their notations unchanged – the phase of the integrand in (6.81) can be kept in the same form. The new amplitude function $\tilde{\mathcal{A}}(\phi, \theta)$ is expanded up to the same order in $\cos^2 \phi$ and $u^2 = \cos^2 \theta$ as the phase. This leads to the following uniform approximation to the trace formula for the primitive orbits:

$$
\delta g_u = \left(\frac{\Delta S_A}{\pi \hbar} \right) \Re e \left\{ \int_0^{2\pi} d\phi \int_0^1 du \, e^{i \Delta S_A / \hbar [(1 - u^2) \cos^2 \phi - u^2]} \left[\sqrt{2} \, \mathcal{A}_{A_2} u^2 \, e^{i \bar{S}_A / \hbar} \right. \right.
$$
$$
\left. \left. + (1 - u^2) \left(\sqrt{2} \, \mathcal{A}_{A_1} \cos^2 \phi \, e^{i \bar{S}_A / \hbar} + \mathcal{A}_C \sin^2 \phi \, e^{i S_C / \hbar} \right) \right] \right\}, \tag{6.104}
$$

where

$$
\Delta S_A = \hbar x = \frac{1}{2} (S_{A_1} - S_{A_2}), \qquad \bar{S}_A = \frac{1}{2} (S_{A_1} + S_{A_2}).
\tag{6.105}
$$

The constants have already been chosen such that (6.104) leads to the correct Gutzwiller limit (6.103) for large ΔS_A. For $x \to 0$, on the other hand, we get $\Delta S_A \to 0$, and $S_C \to S_0$, $\bar{S}_A \to S_0$, as seen from (6.101), so that (6.104) yields the correct symmetry limit (6.89).

The integrals occurring in (6.104) can all be done analytically [31]. The final form of the uniform level density for the quartic Hénon-Heiles potential, including now also the summation over a finite number n_m of repetitions of the primitive orbits, becomes

$$
\delta g_u = \sum_{n=1}^{n_m} (2nx) \left\{ \sqrt{2} \left[\mathcal{M}_+(nx) \, \bar{A}_A \cos(n \bar{S}_A / \hbar) + \mathcal{M}'(nx) \, \Delta A_A \sin(n \bar{S}_A / \hbar) \right] \right.
$$
$$
\left. + \mathcal{M}_-(nx) \, \mathcal{A}_C \cos(n S_C / \hbar) \right\},
\tag{6.106}
$$

with $x = \Delta S_A / \hbar$ according to (6.105). Hereby we have defined

$$
\bar{A}_A = \frac{1}{2} (\mathcal{A}_{A_1} + \mathcal{A}_{A_2}), \qquad \Delta A_A = \frac{1}{2} (\mathcal{A}_{A_1} - \mathcal{A}_{A_2}),
\tag{6.107}
$$

and the following combinations of Bessel functions have been introduced:

$$\mathcal{M}_{\pm}(x) = \frac{\pi}{4\sqrt{2}}\left[J_{-1/4}\left(\frac{x}{2}\right)J_{1/4}\left(\frac{x}{2}\right) \pm J_{-3/4}\left(\frac{x}{2}\right)J_{3/4}\left(\frac{x}{2}\right)\right],$$

$$\mathcal{M}'(x) = -\frac{\pi}{4\sqrt{2}}\left[J_{-1/4}\left(\frac{x}{2}\right)J_{5/4}\left(\frac{x}{2}\right) + J_{1/4}\left(\frac{x}{2}\right)J_{3/4}\left(\frac{x}{2}\right)\right]. \qquad (6.108)$$

In Fig. 6.25, we show some numerical results of the coarse-grained level density obtained with this approximation for two different values of the nonlinearity parameter α. In the upper part, a large smoothing width of $\gamma = 0.6\,\hbar\omega$ was used, so that only the primitive orbits ($n_m = 1$) were needed. In the lower part, a smaller γ was used and the second repetitions were included ($n_m = 2$). Even with this resolution, an excellent agreement with the quantum-mechanical results is achieved, except close to the critical barrier energy E^*, where bifurcations of the orbit A_1 occur and new orbits are missing.

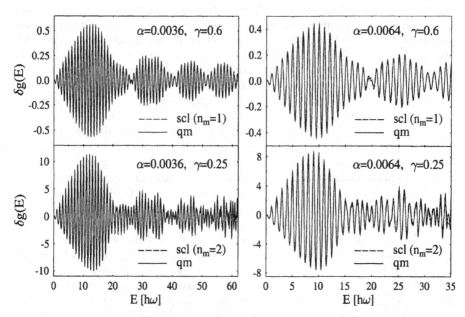

Figure 6.25: Oscillating part $\delta g_\gamma(E)$ of level density of the quartic Hénon-Heiles potential. Solid lines: quantum-mechanical results. Dashed lines: semiclassical results in the uniform approximation (6.106). *Left:* $\alpha = 0.0036$ ($E^* \simeq 69\,\hbar\omega$); *right:* $\alpha = 0.0064$ ($E^* \simeq 39\,\hbar\omega$). *Top:* Gaussian smoothing width $\gamma = 0.6\,\hbar\omega$; *bottom:* $\gamma = 0.25\,\hbar\omega$. (From [31].)

We see from this example that the uniform approximation yields a smooth interpolation between the integrable limit $E \to 0$ and the asymptotic domain where the Gutzwiller trace formula for isolated orbits applies. By construction, Eq. (6.106) goes over to the perturbative trace formula (6.84) at intermediate energies.

A similar uniform approximation was derived in [31] for the standard Hénon-Heiles potential discussed in Sect. 5.6.4. The problem is more complicated for this potential, since the first-order perturbation theory gives a null result for the actions shifts. The

C orbit is obtained at second order, and the isolated orbits A and B emerge only from the fourth-order perturbation. (The modulation factor in the perturbative trace formula for this potential is discussed in Problem 6.5.) We do not give here the result of that uniform approximation, which is expressible in terms of Fresnel integrals. Let us merely mention that similar good numerical results were obtained as in the above figure. The same uniform approximation could also be applied to the breaking of the SO(3) symmetry in a thee-dimensional cavity with axially-symmetric quadrupole deformation. We refer the interested reader to Ref. [31] for the results.

6.3.3 Uniform approximations for bifurcations

At the end of Sect. 5.3, we have already discussed the problems arising in connection with bifurcations of stable isolated orbits, and mentioned some earlier uniform approximations introduced to solve them. These approximations are "local" in the sense that they lead to finite semiclassical amplitudes at and near the bifurcations, but they do not allow one to recover the correct Gutzwiller amplitudes in the asymptotic limit far from the bifurcations. In a sense, they are similar to the perturbative approach discussed in Sect. 6.2.1 in connection with symmetry breaking. More recently, "global" uniform approximations for bifurcations have been developed [50, 51] that do extrapolate smoothly to the correct Gutzwiller amplitudes.

We cannot present the details here, because this would involve an explicit discussion of all generic types of bifurcations. The techniques used in the derivation of these uniform approximations are very similar in spirit to what we have presented above for the symmetry breaking. Rather than starting from the group integral corresponding to the integrable symmetry limit, one starts from a phase-space representation of the trace formula und uses the so-called normal forms (see, e.g., [29]) of the action integral or, more generally speaking, of a suitable generating function corresponding to a given type of bifurcation. In the first step this leads to the local uniform approximations. The global approximations are then found by a mapping of the phase-space variables and an expansion of the unknown amplitude function up to the same order as that appearing in the phase of the normal form. – A characteristic feature is that on one side of a bifurcation, complex "ghost orbits" [52] have to be included which may also bifurcate [53]. Close-lying bifurcations have been treated up to co-dimension two [51], but in general they still present a major unsolved problem in the semiclassical theory.

6.4   Problems

(6.1) Maslov indices for the periodic orbits in the disk billiard

Compute the Maslov indices of the periodic orbits in the two-dimensional disk billiard, treated in Sect. 6.1.3, leading to the phase in Eq. (6.15). Use the fact that $\nu = 0$ and thus $\sigma = \mu$, where μ is the number of conjugate points appearing in the semiclassical Green's function (5.15), which can be used for this two-dimensional problem with simple $U(1)$ symmetry. For any fixed point $\mathbf{r}$ inside the billiard, only isolated orbits pass through this point, so that the rules given in Appendix D can be used.

(6.2) Two-dimensional harmonic oscillator with a weak homogeneous magnetic field *(Difficult!)*

Use the perturbative method described in Sect. 6.2.1 to derive the trace formula for a charged particle moving in a two-dimensional harmonic oscillator potential subjected to a perpendicular external homogeneous magnetic field. The unperturbed trace formula is given in Eq. (3.68). Find the modulation factor that takes the effect of the magnetic field into account in first order of the field strength B.

(a) Use the group average for the $SU(2)$ symmetry of the harmonic oscillator, mentioned in Sect. 6.2.4, to show that the modulation factor (6.63) becomes a Bessel function

$$\mathcal{M}_\Gamma = j_0 \left(\frac{eB}{\hbar c} \mathcal{A}_\Gamma \right), \tag{6.109}$$

where $\mathcal{A}_\Gamma$ is the area surrounded by the periodic orbit Γ, averaged over the family. The integration over the measure of the group $SU(2)$ is not easy. An elegant way to do this was given by Creagh [8]. One can map the 4-dimensional phase space onto the complex 2-dimensional space of the Pauli spinors of a spin-1/2 particle, which have the same $SU(2)$ group structure. Exploiting this analogy, Creagh shows that the group average over the orbits of the unperturbed isotropic harmonic oscillator at fixed energy can be expressed as a solid angle integration over the unit sphere in spinor space. We do not guide you here through the single steps, but ask you to study the section 4.2 of Creagh's article [8]. He has explained the procedure explicitly for the anisotropic two-dimensional harmonic oscillator. But since we know from Sect. 3.2.8 that this is equivalent to the isotropic oscillator in a homogeneous magnetic field, you can really apply his results directly to this problem.

(b) Solving the equations of motion for the two-dimensional harmonic oscillator, show that the area $\mathcal{A}_\Gamma$ covered by the entire family of periodic orbits equals $n\pi E/(m\omega^2)$, so that the argument of the Bessel function in (6.109) may be rewritten as

$$\mathcal{M}_\Gamma = j_0 \left(n \frac{\omega_L}{\omega} \frac{2\pi E}{\hbar\omega} \right). \tag{6.110}$$

Here n is the number of repetitions of the periodic orbits. Note that in Problem 3.3, this modulation factor is derived from an *exact* trace formula for this system, after taking the limit of a small field strength B.

(6.3) Equilateral triangle billiard

Derive the leading oscillating term of the exact trace formula (3.100) for a particle with mass m in an equilateral triangle billiard with side length L, derived in Sect. 3.2.7, but here using the semiclassical methods discussed in Sects. 6.1.4, 6.1.5. Show that it leads to the result given in Eq. (6.9).

(a) First, use the general trace formula (6.21). Like for the rectangular billiard, use as constant of the motion J any component of the particle momentum p and as the continuous group parameter the corresponding coordinate component. Argue that the most convenient choice here is the component perpendicular to the chosen orbit. With this, calculate the Jacobian $\mathcal{J}_\Gamma$ defined in (6.23) and show that it becomes identical to that given in Eq. (6.8) leading, finally, to the result (6.9) given in Sect. 6.1.2.

(b) Now, use the trace formula (6.31). Start from the exact quantum levels given in Eq. (3.94). Then rewrite the classical Hamiltonian in terms of two actions I_1, I_2 and show that Eq. (6.31) yields again the result (6.9). The calculations are elementary and no further hints should be necessary.

Note: The EBK quantization of the equilateral triangular billiard has been shown [14, 54] to yield the exact quantum spectrum (3.94). First, this result is not trivial since the problem is not separable. Second, it is interesting because naturally, one has *three* different actions along the three directions of the boundary. Any two of these actions are independent, and quantizing each pair of them as integer multiples of $\hbar$ leads to the exact spectrum, albeit in different orderings. These orderings are connected by the same symmetry relations as we have discussed in Eqs. (3.91,3.92) for the associated wavefunctions.

(6.4) Three-dimensional rectangular box

Proceed as in Problem (6.3) (b) above to derive a semiclassical trace formula for the three-dimensional rectangular box. Show that the result agrees *exactly* with the leading oscillating term of the quantum result, Eq. (3.84), derived in Sect. 3.2.6. Argue why, differently from the two-dimensional billiards, the result is exact here. (Hint: read the footnote at the end of Sect. 6.1.5.)

(6.5) Asymptotic Gutzwiller amplitudes and Maslov indices in the Hénon-Heiles potential (After [31, 42])

The (semi-)classical perturbation theory for the Hénon-Heiles potential (5.72) gives a zero contribution at the first order in the nonlinear perturbation parameter α. (This follows from the odd power of the anharmonic term and the symmetry of the harmonic oscillator, like in quantum mechanics.) The modulation factor for the perturbative trace formula was given in Refs. [31, 42] to be

$$
\begin{aligned}
\mathcal{M}(x,y) &= \frac{1}{4\pi} \int_0^{2\pi} d\phi \int_0^{\pi} \sin\theta \, e^{i/\hbar \, [\delta S_2(\theta) + \delta S_4(\theta,\phi)]} \, d\theta \\
&= \int_0^1 du \, e^{ix(5-7u^2)/6} \frac{1}{2\pi} \int_0^{2\pi} d\phi \, e^{-iy(1-u^2)^{3/2} \cos(3\phi)} \\
&= \int_0^1 du \, e^{ix(5-7u^2)/6} \, J_0[y(1-u^2)^{3/2}] .
\end{aligned}
\tag{6.111}
$$

Hereby $\delta S_2(\theta) = \hbar x \, (5 - 7 \cos^2\theta)/6$ is the action shift in second-order perturbation, with x proportional to e^2, and $\delta S_2(\theta) = -\hbar y \sin^3\theta \cos(3\phi)$ the fourth-order action shift, with y proportional to e^3, where e is the scaled energy (5.75).

(a) Using the Wong formula (6.86) and the procedures discussed thereafter, show that up to second order in the perturbation theory (i.e., putting $y = 0$), only one doubly-degenerate isolated orbit C is obtained, with an asymptotic amplitude

$$
\mathcal{A}_C \sim c_C \frac{1}{e} ,
\tag{6.112}
$$

and the Maslov index $\sigma_C = -1$ (or 3). Argue why one may say that the orbits A and B remain "degenerate on the circle $\phi \in [0, 2\pi]$" at this order.

(b) Including the fourth-order term $y \neq 0$, show that the isolated orbits A and B arise from the stationary-phase integration over ϕ, with asymptotic amplitudes

$$
\mathcal{A}_A \sim \mathcal{A}_B \sim c_{AB} \frac{1}{e^{3/2}} ,
\tag{6.113}
$$

and the Maslov indices $\sigma_A = 1$ (or 5) and $\sigma_B = 0$ (or 4). Argue why the orbits A and B now have a three-fold discrete degeneracy.

Bibliography

[1] R. Balian and C. Bloch, Ann. Phys. (N. Y.) **60**, 401 (1970); **63**, 592 (1971).

[2] R. Balian and C. Bloch, Ann. Phys. (N. Y.) **69**, 76 (1972).

[3] V. M. Strutinsky, Nukleonika (Poland) **20**, 679 (1975); V. M. Strutinsky and A. G. Magner, Sov. J. Part. Nucl. **7**, 138 (1976) [Elem. Part. & Nucl. (Atomizdat, Moscow) **7**, 356 (1976)].

[4] S. C. Creagh and R. G. Littlejohn, Phys. Rev. **A 44**, 836 (1991).

[5] S. C. Creagh and R. G. Littlejohn, J. Phys. **A 25**, 1643 (1992).

[6] M. V. Berry and M. Tabor, J. Phys. **A 10**, 371 (1977).

[7] M. V. Berry and M. Tabor, Proc. R. Soc. Lond. **A 349**, 101 (1976).

[8] S. C. Creagh, Ann. Phys. (N. Y.) **248**, 60 (1996).

[9] V. M. Strutinsky, A. G. Magner, S. R. Ofengenden, and T. Døssing, Z. Phys. **A 283**, 269 (1977).

[10] A. G. Magner, Sov. J. Nucl. Phys. **28**, 764 (1978); and private communications (1995,1996).

[11] A. G. Magner, S. N. Fedotkin, F. A. Ivanyuk, P. Meier, M. Brack, S. M. Reimann, and H. Koizumi, Ann. Physik (Leipzig) **6**, 555 (1997).

[12] Th. Schachner, Diploma thesis, Regensburg University (1996, unpublished).

[13] A. G. Magner, S. N. Fedotkin, K. Arita, T. Misu, K. Matsuyanagi, Th. Schachner, and M. Brack, Prog. Theor. Phys. (Japan) **102**, 551 (1999).

[14] P. J. Richens and M. V. Berry, Physica **2D**, 495 (1981).

[15] Ch. Amann, unpublished results (1999).

[16] S. M. Reimann, M. Brack, A. G. Magner, J. Blaschke, and M. V. N. Murthy, Phys. Rev. **A 53**, 39 (1996).

[17] E. N. Bogachek and G. A. Gogadze, Sov. Phys. JETP **36**, 973 (1973) [Zh. Eksp. Teor. Fiz. **63**, 1839 (1973)].

[18] B. Tatievski, P. Stampfli, and K. H. Bennemann, Comp. Mat. Sci. **2**, 459 (1994).

[19] H. Frisk, Nucl. Phys. **A 511**, 309 (1990).

[20] A. G. Magner, S. N. Fedotkin, K. Arita, K. Matsuyanagi, and M. Brack, Phys. Rev. **E 63** 065201(R) (2001).

[21] H. Friedrich and D. Wintgen, Phys. Rep. **183**, 39 (1989).

[22] M. C. Gutzwiller, J. Math. Phys. **8**, 1979 (1967); *ibid.* **11**, 1791 (1970).

[23] A. G. Magner, V. M. Kolomietz, and V. M. Strutinsky, Soc. J. Nucl. Phys. **28**, 764 (1978) [Yad. Fiz. **28**, 1487 (1978)].

[24] B. Lauritzen, Phys. Rev. **A 43**, 603 (1991).

[25] S. C. Creagh, J. Phys. **A 26**, 95 (1993).

[26] M. Brack, S. C. Creagh, P. Meier, S. M. Reimann, and M. Seidl, in: *Large Clusters of Atoms and Molecules*, ed. by T. P. Martin (Kluwer, Dordrecht, 1996), p. 1.

[27] M. Brack and S. R. Jain, Phys. Rev. **A 51**, 3462 (1995).

[28] A. M. Ozorio de Almeida and J. H. Hannay, J. Phys. **A 20**, 5873 (1987).

[29] A. M. Ozorio de Almeida: *Hamiltonian Systems: Chaos and Quantization* (Cambridge University Press, Cambridge, 1988).

[30] S. Tomsovic, M. Grinberg, and D. Ullmo, Phys. Rev. Lett. **75**, 4346 (1995); D. Ullmo, M. Grinberg, and S. Tomsovic, Phys. Rev. **E 54**, 136 (1996).

[31] M. Brack, P. Meier, and K. Tanaka, J. Phys. **A 32**, 331 (1999).

[32] S. Okai, H. Nishioka, and M. Ohta, Mem. Konan Univ., Sci. Ser. **37**, 29 (1990).

[33] M. Sieber, J. Phys. **A 30**, 4563 (1997).

[34] P. Meier, M. Brack, and S. C. Creagh, Z. Phys. **D 41**, 281 (1997).

[35] M. Brack, J. Blaschke, S. C. Creagh, A. G. Magner, P. Meier, and S. M. Reimann, Z. Phys. **D 40**, 276 (1997).

[36] V. V. Pashkevich, P. Meier, M. Brack, and A. V. Unzhakova, Phys. Lett. **A 294**, 314 (2002).

[37] F. v. Oppen, Phys. Rev. **B 50**, 17151 (1994).

[38] D. Ullmo, K. Richter, and R. A. Jalabert, Phys. Rev. Lett. **74**, 383 (1995).

[39] K. Richter, D. Ullmo, and R. A. Jalabert, Phys. Rep. **276**, 1 (1996).

[40] M. Abramowitz and I. A. Stegun: *Handbook of Mathematical Functions* (Dover Publications, New York, 9th printing, 1970).

[41] K. Tanaka, S. C. Creagh, and M. Brack, Phys. Rev. **B 53**, 16050 (1996).

[42] M. Brack, S. C. Creagh, and J. Law, Phys. Rev. **A 57**, 788 (1998).

[43] M. Hénon and C. Heiles, Astr. J. **69**, 73 (1964).

[44] M. Schwarzschild, Astrophys. J. **232**, 236 (1979).

[45] J. Goodman and M. Schwarzschild, Astrophys. J. **245**, 1087 (1981).

[46] Nan He, Master thesis, McMaster University (1991, unpublished); R. Pudritz, private communication (1995).

[47] P. Bøggild, A. Kristensen, P. E. Lindelof, S. M. Reimann, and C. B. Sørensen, in: *23rd International Conference on the Physics of Semiconductors (ICPS), Berlin 1996* (World Scientific, Singapore, 1997), p. 1533.

[48] M. Brack, in: *Festschrift in honour of the 75th birthday of Martin Gutzwiller*, eds. A. Inomata *et al.*; Foundations of Physics **31**, 209 (2001).

[49] R. Wong: *Asymptotic Approximation of Integrals* (Academic Press, Inc., San Diego, 1989).

[50] M. Sieber, J. Phys. **A 29**, 4715 (1996); H. Schomerus and M. Sieber, J. Phys. **A 30**, 4537 (1997); M. Sieber and H. Schomerus, J. Phys. **A 31**, 165 (1998).

[51] H. Schomerus, Europhys. Lett. **38**, 423 (1997); J. Phys. **A 31**, 4167 (1998); J. Main and G. Wunner, Phys. Rev. **A 55**, 1753 (1997); J. Main and G. Wunner, Phys. Rev. **E 57**, 7325 (1998).

[52] M. Kuś, F. Haake, and D. Delande, Phys. Rev. Lett. **71**, 2167 (1993)

[53] T. Bartsch, J. Main, and G. Wunner, Ann. Phys. (N. Y.) **277**, 19 (1999).

[54] J. B. Keller and S. I. Rubinow, Annals of Phys. (N. Y.) **9**, 24 (1960).

Chapter 7

Quantization of nonintegrable systems

In Chapter 2 we had seen that there is a well-defined procedure for semiclassical quantization of an integrable system. This runs into trouble for nonintegrable Hamiltonians, where the number of constants of motion is less than the number of spatial dimensions. In Chapter 5 the semiclassical Gutzwiller trace formula was derived for the Green's function and the density of states. Since the poles of the Green's function [or its trace, see Eq. (3.34)], correspond to the eigenvalues of the Hamiltonian, the trace formula itself may be regarded as a method of semiclassical quantization. For hard chaos, however, it is known that the number of periodic orbits grows exponentially with increasing periods of the primitive orbits [see Eq. (7.23) below], and this makes the trace formula divergent unless the longer orbits are damped out by a smoothing function (cf. Sect. 5.5). This causes not only a loss of resolution in the determination of the individual eigenenergies, but makes the whole procedure of using the trace formula unpractical with increasing energy. This is in contrast to the EBK (energy-independent) quantization condition (2.119) which actually becomes more accurate with larger quantum numbers. It is therefore worthwhile to investigate more sophisticated procedures of semiclassical quantization, although the reader should be warned that this problem has not been satisfactorily resolved. One such method is the zeta function approach, where the quantized energies are determined by constructing a function whose zeros correspond to the eigenvalues of the Hamiltonian. The idea that such a function may be constructed comes from the analytical properties of one of the most beautiful and mysterious functions of pure mathematics, the Riemann zeta function of a complex variable. A function (often called the Selberg zeta function) will be defined resembling the Riemann zeta function, whose zeros correspond to the eigenvalues of the dynamical system. We shall see that a semiclassical expression for this may be obtained from our knowledge of the periodic orbit theory. The convergence properties of this function for real energies still pose problems in practical applications. Nevertheless, some applications to dynamical systems will be given and discussed critically. In the first section, we elucidate the properties of the Riemann zeta function in some detail to unfold its connection to the subsequent developments on the dynamical theory. The

311

later sections of this chapter will be devoted to scattering theory, and to the inclusion of diffractive orbits to the trace formula.

7.1   The Riemann zeta function

7.1.1   The zeros of the Riemann zeta function

The Riemann zeta function is of great importance in number theory [1]. First consider the so-called Dirichlet series:

$$\zeta(s) = \sum_{n=1}^{\infty} \frac{1}{n^s}, \qquad (7.1)$$

which converges only for $\Re e\, s > 1$. This function may be analytically continued to the whole complex plane by defining an appropriate contour integral on the complex plane [1] which yields the same value as (7.1) for $\Re e\, s > 1$. This analytically continued function on the complex plane is called the Riemann zeta function, and henceforth will also be denoted by $\zeta(s)$. Before studying its analytical properties, it is important to note that any positive integer n may be expressed in terms of the prime numbers in a unique combination

$$n = \prod_{p} p^{m_p(n)}, \qquad (7.2)$$

where the index p runs over all distinct primes, with each prime having an appropriate integer power m_p that may include zero. For example, $n = 50 = 2^1 3^0 5^2 7^0....$ Because of this fundamental property, Eq. (7.1) may be expressed in a product form:

$$\zeta(s) = \prod_{p} \left(1 - \frac{1}{p^{\sigma+it}}\right)^{-1}, \qquad (7.3)$$

where the infinite product is over all primes p, and $s = \sigma + it$, σ and t being real. The above is called the 'Euler product' form for $\zeta(s)$, and only converges for $\sigma > 1$.

The Riemann zeta function $\zeta(s)$ has remarkable analytical properties, as shown in Fig. 7.1. Its only pole (which is simple) is at $s = 1$ on the real axis. To bring out its intrinsic properties on the complex plane, it is often more convenient to write it as [1]

$$\zeta(s) = \frac{\pi^{s/2}}{\Gamma(s/2+1)} \frac{1}{(s-1)} \, \xi(s). \qquad (7.4)$$

In the above, the prefactor multiplying $\xi(s)$ has the simple pole at $s = 1$, consequently the function $\xi(s)$ is analytic everywhere on the complex plane (an 'entire' function of s). Moreover, the poles of the gamma function in the prefactor reproduce all the zeros of $\zeta(s)$ on the real axis at $\sigma = -2\kappa$, $\kappa = 1, 2, ...$ Thus $\xi(s)$ contains only the complex zeros of $\zeta(s)$. The distribution of these complex zeros is our principal interest. It can be shown that $\xi(s)$ obeys the functional relationship

$$\xi(s) = \xi(1-s). \qquad (7.5)$$

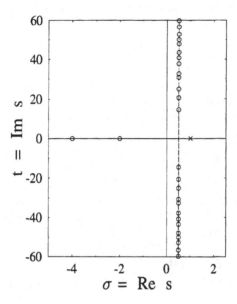

Figure 7.1: Analytical structure of the Riemann zeta function in the complex plane. The only pole is at $t = 0$, $\sigma = 1$ (cross), and an infinite series of zeros (circles) lies on the real axis with negative even-integer values of σ. According to Riemann's conjecture, all the other zeros (circles) lie in complex-conjugate pairs on the (dashed) line with $\sigma = 1/2$.

Riemann had conjectured that there is an infinite number of complex zeros of $\zeta(s)$ on the line $\sigma = 1/2$, and none on other locations. The first part of the statement was proved by Hardy [2] in 1914, but an analytical proof that there are no other complex zeros away from the $\sigma = 1/2$ line is still lacking. It remains one of the most famous unsolved problems of classical mathematics. Computer calculations have numerically demonstrated that all the zeros so far found obey Riemann's conjecture, and we shall assume it to be true. From Eq. (7.5), it follows that the zeros on the $\sigma = 1/2$ line are symmetrically placed about the real axis. These zeros are known to be nondegenerate, i.e., no two of them have the same value of t. The information about the density of the zeros is contained in the phase of the entire function $\xi(s)$. To see this, let us write

$$\xi(s) = \xi(0) \prod_n \left(1 - \frac{s}{s_n}\right), \tag{7.6}$$

where $s = \sigma + it$, and $s_n = 1/2 + it_n$ is the location of the nth zero. Let us denote the density of these zeros on the $\sigma = 1/2$ line by $g(t)$:

$$g(t) = \sum_n \delta(t - t_n), \quad t > 0, \tag{7.7}$$

where t_n is the position of the nth zero on the upper half of the t-line. For a fixed σ, we write $\xi(s)$ in the polar form,

$$\xi(s) = |\xi(s)| \exp[-i\phi_\sigma(t)], \tag{7.8}$$

where $\phi_\sigma(t)$ is the phase angle for a given σ as a function of t. It follows that

$$\frac{d\phi_\sigma}{dt} = -\frac{d}{dt}\left[\Im m \ln \xi(s)\right]. \tag{7.9}$$

Using Eq. (7.6), the right-hand side of the above equation is given by

$$\Im m \ln \xi(s) = \Im m \ln \xi(0) + \sum_n \left[\tan^{-1}\left(\frac{t - t_n}{\sigma - 1/2}\right) + \pi - \tan^{-1}(2t_n)\right]. \tag{7.10}$$

From Eq. (7.9), we then find

$$\frac{d\phi_\sigma}{dt} = -\sum_n \frac{(\sigma - 1/2)}{(\sigma - 1/2)^2 + (t - t_n)^2}. \tag{7.11}$$

Recalling that the Dirac delta function is the limit of a Lorentzian of width γ going to zero:

$$\delta(t - t_n) = \lim_{\gamma \to 0}\left(\frac{\gamma}{\pi}\right)\frac{1}{\gamma^2 + (t - t_n)^2}, \tag{7.12}$$

we see from Eqs. (7.7) and (7.11) that

$$\lim_{\sigma \to 1/2}\frac{d\phi_\sigma}{dt} = -\pi g(t). \tag{7.13}$$

As one moves away to values of $\sigma > 1/2$, the memory of these zeros fades slowly through Lorentzian smoothing [3], given by Eq. (7.11). For later reference, let us denote this Lorentz-smoothed density by $g_\sigma(t)$:

$$g_\sigma(t) = \frac{1}{\pi}\sum_n \frac{(\sigma - 1/2)}{(\sigma - 1/2)^2 + (t - t_n)^2}. \tag{7.14}$$

We are now in a position to obtain an analogous relation for the phase angle of the zeta function $\zeta(s)$ using Eq. (7.4). As before, let us write

$$\zeta(s) = |\zeta(s)| \exp\left[-i\theta_\sigma(t)\right]. \tag{7.15}$$

In the above, θ_σ denotes the phase of the zeta function along the t-line for a fixed value of σ. As in Eq. (7.9), its derivative obeys the relation

$$\frac{d\theta_\sigma}{dt} = -\frac{d}{dt}\left[\Im m \ln \zeta(s)\right]. \tag{7.16}$$

From Eq. (7.4), the logarithm of $\zeta(s)$ involves taking the logarithm of $\xi(s)$ that was obtained earlier, as well as of the gamma function $\Gamma(s/2 + 1)$. For the latter, we use the asymptotic Stirling formula [1]

$$\ln \Gamma(s/2 + 1) = \frac{1}{2}(s + 1)\ln(s/2) - (s/2) + \frac{1}{2}\ln \pi + \frac{B_2}{s} + \mathcal{O}(s^{-3}) + \dots . \tag{7.17}$$

$B_2 = 1/6$ is the Bernoulli number of order two. After some algebra, we finally obtain the relation

$$\frac{1}{\pi}\frac{d}{dt}\left[\Im m \ln \zeta(s)\right] = -\frac{1}{\pi}\frac{d\theta_\sigma}{dt} = \left[g_\sigma(t) - \tilde{g}_\sigma(t)\right] = \delta g_\sigma(t), \tag{7.18}$$

where

$$\tilde{g}_\sigma(t) = \frac{1}{2\pi} \ln \frac{[(\sigma - 1/2)^2 + t^2]^{1/2}}{2\pi}.$$

On the critical line $\sigma = 1/2$, we shall omit the subscript σ. Then $\tilde{g}(t)$ is given by

$$\tilde{g}(t) = \frac{1}{2\pi} \ln \frac{t}{2\pi} + \mathcal{O}(t^{-2}) + \dots. \tag{7.19}$$

The phase $\theta(t)$ has abrupt discontinuities of π at the zeros of $\zeta(1/2 + it)$. The smooth part of $\theta(t)$ is denoted in Problem 7.1 by $\bar{\theta}(t)$, and its negative derivative divided by π yields $\tilde{g}(t)$. To clarify further, note that if $g(t)$ given by Eq. (7.7) was Strutinsky-smoothed by energy averaging (see Sec. 4.7.1), one would obtain $\tilde{g}(t)$. The important Eq. (7.18) then asserts that on the critical line $s = 1/2 + it$, the derivative of the phase contains all the information about the oscillating part of the density of the zeros. Alternately, on this line, we may write

$$\frac{1}{\pi} \frac{d}{dt} \Im m \left[\ln \zeta(1/2 + it) \right] = g(t) - \tilde{g}(t) = \delta g(t) = \sum_n \delta(t - t_n) - \frac{1}{2\pi} \ln \left(\frac{t}{2\pi} \right). \tag{7.20}$$

7.1.2 A trace formula for the zeros

The focus in Chapters 5 and 6 was to find the oscillating part of the density of eigenvalues of a Hamiltonian, that was expressed as a difference of two quantities as in the right-hand side of Eq. (7.20). In the semiclassical periodic orbit theory, this oscillating density of states is related through the trace formula to classical periodic orbits. Can we similarly express $\delta g(t)$ above in the form of a "trace formula"? Following [4], we proceed to derive such a formula. This suggests that the zeros of $\zeta(1/2 + it)$ behave in the same way as the eigenvalues of a dynamical Hamiltonian. Indeed, the intriguing conjecture was made that there exists a dynamical Hamiltonian with an infinite number of bound states, whose eigenvalues are identical to the zeros of the Riemann zeta function on the critical line. Such a Hamiltonian has not yet been discovered. We shall come back to this point again after deriving the trace formula for the zeros.

The results above for $g(t)$ were obtained from Eqs. (7.4) and (7.6), which are exact. Now, however, we shall use Eq. (7.3), which converges only for $\sigma > 1$. Taking the logarithm of both sides of this equation, and recalling that $\ln(1 - x) = -\sum_{k=1}^\infty x^k/k$, we obtain

$$\ln \zeta(s) = \sum_p \sum_k \frac{1}{kp^{ks}} \exp\left(-ikt \ln p\right). \tag{7.21}$$

Taking the imaginary part of both sides of the above equation, and performing differentiation with respect to t, we obtain the analogue of Eq. (7.20) on the line along t for a fixed $\sigma > 1$:

$$\delta g_\sigma = \frac{1}{\pi} \frac{d}{dt} [\Im m \ln \zeta(s)] = -\frac{1}{\pi} \sum_p \sum_k \frac{(\ln p)}{(p)^{k\sigma}} \cos(kt \ln p). \tag{7.22}$$

This is exactly of the same form as the Gutzwiller trace formula Eq. (5.36), with t acting as the energy variable, and $\hbar = 1$. With this interpretation, there is one primitive periodic orbit for every prime number, having a classical action $S_p(t) = t \ln p$, and a period $T_p = \ln p$. Because of the overall negative sign in (7.22), it is as if every Maslov index is 2, corresponding to a Maslov phase of π, although one could also choose these to be $\pi, 3\pi, 5\pi$.. etc. Note that Eq. (7.22) is strictly valid only for $\sigma > 1$, where it is convergent. In this region of the complex plane, there are no zeros in $\zeta(s)$. Nevertheless, the trace formula contains information about the zeros on the critical line, with the shortest orbits (corresponding to the smallest primes) contributing the most. This is apparent from the denominator $(p)^{k\sigma}$ in Eq. (7.22), that damps out the contribution of the larger primes.

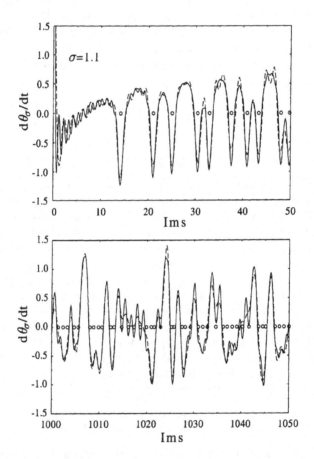

Figure 7.2: The derivative of the phase θ_σ of Riemann zeta function (7.22), plotted as a function of $\Im m\, s = t$ for $\sigma = 1.1$ in the range $0 \le t \le 50$ (upper panel) and $1000 \le t \le 1050$ (lower panel). The dashed line is calculated by using the first ten primes and truncating the sum over the repetitions k at $k_{max} = 10$. The solid line includes the first 100 prime numbers, and $k_{max} = 100$. The exact positions of the Riemann zeros on the $\sigma = 1/2$ axis are shown by open circles. (From [3].) Because of the extra minus sign in Eq. (7.18) before the derivative of the phase, the peaks in the oscillatory part of $\delta g_\sigma(t)$ correspond to valleys in $d\theta_\sigma/dt$.

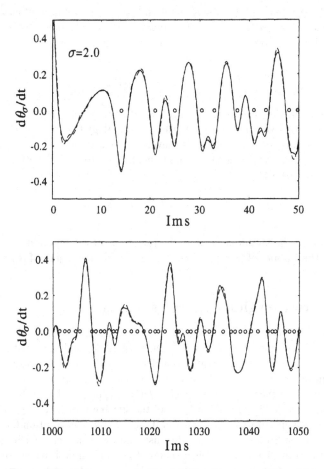

Figure 7.3: The same as in Fig. 7.2, but for $\sigma = 2$. (From [3].)

In Figs. 7.2 and 7.3, the actual locations of the zeros on the critical line are compared with the semiclassical oscillating density of states $\delta g_\sigma(t)$ as obtained through Eq. (7.22). Two features are noteworthy. First, even the lowest ten primes yield the undulations that follow the density of the zeros closely. Secondly, the resolution of the zeros in the trace formula becomes worse as one moves away from the critical line by increasing σ. Since the zeros of the Riemann zeta function on the critical line have a random pattern, the undulations in the curves shown in Figs. 7.2 and 7.3 also follow a random pattern.

We have seen that the number of "periodic orbits" in the trace formula (7.22) is the same as the number of primes, which is infinity. It is known that the number of primes $\mathcal{N}(p < x)$ grows asymptotically as $x/\ln x$. But we have seen that the period T of the orbit is $\ln x$, so $\mathcal{N}$ grows asymptotically as

$$\mathcal{N}(T_p < T) \simeq \frac{\exp T}{T}.$$

This exponential growth of periodic orbits is the same as found in hard chaos where all orbits are unstable. More precisely, in the dynamical case with hard chaos, the number of periodic orbits with period less than T grows asymptotically as:

$$\mathcal{N}(T_p < T) \simeq \frac{\exp[h(E)T]}{hT}, \tag{7.23}$$

where $h(E)$ has the dimensions of $E/\hbar$, and is called the "topological entropy". Comparison of the above two equations strengthens the analogy between the Riemann zeros and dynamical chaos. At this point, it is convenient to introduce the concept of pseudo-orbits for later application. Using the fundamental relation (7.2) between an integer and primes, we note that

$$\mathcal{S}_n = t \ln n = \sum_p m_p(n)\, t \ln p = \sum_p m_p(n)\, \mathcal{S}_p, \tag{7.24}$$

where $\mathcal{S}_n$ is called the pseudo-action for the pseudo-orbit corresponding to the integer n. Note that the number of pseudo-orbits is larger than the number of primitive orbits.

7.1.3   Nearest-neighbor spacings and chaos

It is interesting to examine the correlations between the zeros along the t-axis, i.e., if there is a zero of the zeta function at t_0, then what is the probability of encountering the next zero at $(t_0 + S)$? To elaborate, consider the spacing $S(t)$ between two adjacent zeros in a bin of zeros centered about t. The width of the bin is much smaller than t, but is much larger than the mean spacing $D(t) = 1/\tilde{g}(t)$ of the levels, where $\tilde{g}(t)$ is defined by Eq. (7.19). The global variation of the spacing may be adjusted by defining the dimensionless quantity $y = S/D(t)$, and its probability distribution for the spacing to be between y and $y + dy$ is $P(y)dy$. This distribution of the zeros for large t is found numerically to be the same as the eigenvalues of a Gaussian unitary ensemble (GUE) matrix [5]. A very accurate approximation to the GUE distribution is given by:

$$P(y) \simeq \frac{32}{\pi^2}\, y^2 \exp\left(-\frac{4}{\pi}y^2\right), \qquad \int_0^\infty P(y)\, dy = 1. \tag{7.25}$$

In Fig. 7.4 we show the quality of the fit for the spacings of the Riemann zeros and the exact GUE distribution. This result has added significance because this is the same distribution as the nearest neighbor spacings between the eigenvalues of a Hamiltonian that is classically chaotic and violates time-reversal symmetry.

To appreciate the significance of this result, it should be mentioned that there is a long history of studying the distribution of nearest-neighbor spacings of the eigenvalues of physical systems, and in particular of nuclear levels. An excellent account of this will be found in Bohr and Mottelson's text book [6]. Wigner [7] had conjectured long back that $P(y)$ for the energy level spacings in nuclei for levels of the same angular momentum and parity is given by

$$P(y) \simeq \frac{\pi}{2}\, y \exp\left(-\frac{\pi}{4}y^2\right), \qquad \int_0^\infty P(y)\, dy = 1. \tag{7.26}$$

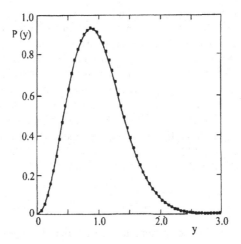

Figure 7.4: GUE distribution (7.25) for Riemann zeros. The sample is from 79 million zeros around the 10^{20}th zero. (From [5].)

For uncorrelated levels, on the other hand, $P(y) = \exp(-y)$. Wigner's distribution given by Eq. (7.26) is a very close approximation to the so-called GOE distribution of the eigenvalues of a Hamiltonian matrix with random matrix elements. The Hamiltonian is taken to be Hermitian, with rotational and time reversal symmetry [8]. A detailed and careful analysis of the experimental data (of the same symmetry class) on the neutron capture resonances was made by Haq et al. [9], and it was shown that the fluctuation properties of these were in agreement with the random-matrix theory. This showed the existence of long-range order in the level spacing correlations, and suggested a universal behavior, independent of the details of nuclear structure. After all, the atomic nucleus is a complicated many-body system, and one could argue that its Hamiltonian behaved like a random matrix. More surprising, however, was McDonald and Kaufman's [10] finding that even the eigenvalues of a very simple system like the stadium billiard seemed to behave similarly. In a now classic paper, Bohigas et al. [11], after carefully checking the consistency of the spectral properties of the stadium billiard with GOE, emphasized the point that such a simple, but classically chaotic system, can give a universal signature in the quantal behavior. This has been a fruitful tool in the study of quantum chaos [12, 13, 14].

7.1.4 The Riemann-Siegel relation

Although the trace formula for the zeros reproduces the rough locations of the zeros, one would like to have a prescription for locating them with precision. This has considerable significance for quantum chaos if we accept the tenet that the Riemann zeros behave similarly to the eigenvalues of a Hamiltonian that is classically chaotic, as evidenced from their distribution. If we devise a method of locating the zeros with precision, it will then be a pointer to finding a rule for quantization of a Hamiltonian that is classically chaotic. This is a central problem in the understanding of the quantum-classical link of nonintegrable systems. As seen above, the trace formula given by Eq. (7.22) was only convergent for $\sigma > 1$, resulting in a loss of resolution of the location of the zeros in the t-axis. One would like to use a convergent formula by some resummation technique. We proceed to do so now, following the treatment of Berry and Keating [15]. We have noted before that the "primitive orbits" that occur in the trace formula (7.22) correspond to the prime numbers. Noting the fundamental relation (7.2) between an integer and primes, it may be more convenient to use "pseudo-orbits", each associated with an integer n, with a "pseudo-action" corresponding to $\ln n$ rather than $\ln p$. To obtain an expression for $\zeta(1/2 + it)$ from which we can find the zeros, we go back to the Dirichlet series (7.1), and express it in terms of $\ln n$:

$$\zeta(1/2 + it) = \sum_{n=1}^{\infty} \frac{1}{n^{1/2+it}} = \sum_{n=1}^{\infty} \frac{1}{\sqrt{n}} \exp\left(-it \ln n\right).$$

Unfortunately, this series is not convergent on the critical line, so we split up the sum over n into two parts. The idea is to convert the second part, which is divergent and contains an infinite number of terms, into a finite sum. The procedure is beautiful and deserves elucidation on its own merit. Splitting the sum, we have

$$\zeta(1/2 + it) = \sum_{n=1}^{n^*} \frac{1}{\sqrt{n}} \exp\left(-it \ln n\right) + \sum_{n^*+1}^{\infty} \frac{1}{\sqrt{n}} \exp\left(-it \ln n\right). \tag{7.27}$$

In the second sum above, use the truncated form of the Poisson formula (2.157)

$$\sum_{n=a}^{n=b} f(n) = \sum_{m=-\infty}^{\infty} \int_{a-1/2}^{b+1/2} f(x) \exp\left(2\pi i m x\right) dx. \tag{7.28}$$

The second sum in (7.27) then becomes

$$\sum_{n^*+1}^{\infty} \frac{1}{\sqrt{n}} \exp\left(-it \ln n\right) = \sum_{m=-\infty}^{\infty} \int_{n^*}^{\infty} \frac{1}{\sqrt{x}} \exp\left[-i\mathcal{S}(x) + 2\pi i m x\right] dx, \tag{7.29}$$

where we have neglected $1/2$ in comparison to n^*, and $\mathcal{S}(n) = \mathcal{S}_n$ is defined by Eq. (7.24). The integral in Eq. (7.29) is now performed by the saddle-point method. The integrand in question is a rapidly varying oscillating function and contributes dominantly only in the vicinity of the stationary point x_0, where the slope of $\mathcal{S}(x)$ with respect to x is zero, i.e., $\mathcal{S}'(x_0) = 0$. This gives $x_0 = t/(2\pi m)$, about which $\mathcal{S}(x)$ is Taylor-expanded to order $(x - x_0)^2$. It is then straightforward to perform the integral using the Fresnel integral formula (2.169) for large t, for which $x_0 \to \infty$.

The important point is that the infinite sum over m is now reduced to a finite sum. Since the saddle-points occur at $x_0 = t/(2\pi m)$, only positive values of m are allowed. Moreover, the saddle exists only up to a maximum $m_{max} = t/(2\pi n^*)$. Thus, now the sum over m in Eq. (7.29) is reduced to $\sum_{m=1}^{[t/(2\pi n^*)]}$, where the square bracket in the upper limit denotes the integer part. After some algebra, Eq. (7.27) reduces to

$$\zeta(1/2 + it) = \sum_{m=1}^{m^*} \frac{1}{\sqrt{m}} \exp\left(-it \ln m\right) + \sum_{m=1}^{[t/2\pi m^*]} \frac{1}{\sqrt{m}} \exp\left(it \ln m\right) \exp\left[-2\pi i \widetilde{N}(t)\right]. \tag{7.30}$$

In the above, $\widetilde{N}(t)$ is given by

$$\widetilde{N}(t) = \frac{t}{2\pi} \ln \frac{t}{2\pi} - \frac{t}{2\pi} + \frac{7}{8}. \tag{7.31}$$

$\widetilde{N}(t)$ is just the integral of $\tilde{g}(t)$ given by Eq (7.19) and may be interpreted as the number of zeros of $\zeta(1/2 + it)$ between 0 and t (see also Problem 7.1). It is remarkable that in Eq. (7.30), the second sum on the right-hand side is like the complex conjugate of the first sum, if an overall factor of $\exp[-i\pi\widetilde{N}(t)]$ is taken out, and the number of terms in both are chosen to be the same. Thus the terms with the longer pseudo-periods are seen to contain the same information as those with the shorter ones. This is the idea of *resurgence* emphasized by Berry and Keating [15]. Following this idea, let us choose m^* such that both sums in Eq. (7.30) have the same number of terms, so that $m^* = [t/(2\pi m^*)]$, i.e., $m^* = [\sqrt{(t/2\pi)}]$. The terms in the two sums may then be matched pair-wise, giving the final form

$$\zeta(1/2 + it) = 2\exp[-i\pi\widetilde{N}(t)] \sum_{m=1}^{[\sqrt{(t/2\pi)}]} \frac{1}{\sqrt{m}} \cos\left[t\ln m - \pi\widetilde{N}(t)\right]. \tag{7.32}$$

This is the main term in the Riemann-Siegel formula. The most remarkable point about this formula is that it has only a *finite* number of terms in the sum on the right-hand side, and is therefore convergent even on the critical line. In the above formula, the term with the highest pseudo-period $\mathcal{T}^* = \ln m^*$ is such that the argument of the cosine term in Eq. (7.32) is stationary, i.e.,

$$\frac{\partial}{\partial t}\left(\mathcal{S}_m(t) - \pi\widetilde{N}(t)\right)_{m=m^*} = 0, \tag{7.33}$$

where $\mathcal{S}_m(t) = t\ln m$, and $\mathcal{T} = \partial S/\partial t$. This criterion yields the maximum pseudo-period in the Riemann-Siegel sum to be

$$\mathcal{T}_{m^*} = \pi\,\tilde{g}(t). \tag{7.34}$$

This has the suggestive form of the Heisenberg uncertainty relation, i.e., the maximum pseudo-period times the average level spacing equaling $h/2$, with Planck's constant taken to be $h = 1$. This criterion of truncation will be used later in the dynamical problem.

There are several correction terms that are added to it to obtain the zeros of $\zeta(1/2+it)$ accurately, but even the leading term above yields an accurate estimate. For example, the lowest zero for $t > 0$ comes at 14.135 (rounded off at the third decimal place), whereas the formula above gives 14.550. Of course, the accuracy improves for larger t. Writing

$$\zeta(1/2 + it) = Z(t)\,\exp[-i\pi\widetilde{N}(t)], \tag{7.35}$$

we see from Eq. (7.32) that $Z(t)$ is given by (see Problem 7.1)

$$Z(t) = 2\sum_{m=1}^{[\sqrt{(t/2\pi)}]} \frac{1}{\sqrt{m}} \cos\left[t\ln m - \pi\widetilde{N}(t)\right]. \tag{7.36}$$

The zeros of $Z(t)$ are the same as of $|\zeta(1/2 + it)|$, but $Z(t)$ changes sign at each zero. Moreover, $Z(t)$, defined on the critical line $\sigma = 1/2$, is real, whereas the zeta function $\zeta(1/2 + it)$ is complex. From Eq. (7.35), the functional relation

$$\zeta(s) = \exp\left[-2i\pi\widetilde{N}(t)\right]\zeta^\star(s), \qquad s = \frac{1}{2} + it \qquad (7.37)$$

follows at once. The Selberg zeta function, which we discuss next, will obey similar relations.

7.2   The quantization condition

7.2.1   The Selberg zeta function

We are guided by the relations derived in the previous section for the Riemann zeta function in constructing a corresponding function for dynamical application. It is often called the Selberg zeta function and denoted by ζ_s, after Selberg, who derived [16] much earlier a trace formula for the motion of a particle on a surface of constant negative curvature. Aurich and Steiner [17] constructed a prototype of the zeta function from this trace formula, for which the Gutzwiller form (5.38) is exact. We shall briefly discuss this case in the next subsection in relation to the construction of pseudo-orbits. In this section, however, we focus on the more general formulation of the Selberg zeta function [21] ζ_s, and its semiclassical expression in terms of the periodic orbits.

For the Riemann zeta function, we saw that the trace formula (7.22) only converged for $\sigma > 1$, but $\zeta(s)$ could be analytically continued to the critical line where the zeros reside. Similarly, the idea is to improve the convergence of the Gutzwiller trace formula by constructing a function whose complex zeros on the critical line correspond to the eigenvalues of the given Hamiltonian. We shall now use the energy variable E instead of t of the previous section, and $\zeta_s(E)$ will denote the Selberg zeta function on the critical line as a function of E. The important relation that we wish $\zeta_s(E)$ to follow is analogous to Eq. (7.20), that relates the phase of the Riemann zeta function to the oscillating part of the density of states. From Eq. (7.22), however, we recognize that there is an extra overall negative sign in the trace formula for the Riemann zeta function, which is absent in the dynamical case. Accordingly, we define $\zeta_s(E)$ to obey the relation

$$\delta g(E) = -\frac{1}{\pi}\frac{d}{dE}\left[\Im m \ln \zeta_s(E)\right]. \qquad (7.38)$$

Here $\delta g(E) = g(E) - \tilde{g}(E)$, where $g(E)$ is the density of states generated by H, and the smoothed $\tilde{g}(E)$ is to be identified with the extended Thomas-Fermi (ETF) density of states. The number of quantum states up to E (assuming that the bottom of the energy scale starts at $E = 0$) is given by $N(E) = \int_0^E g(E')\, dE'$, so it follows from Eq. (7.38) that

$$\Im m \ln \zeta_s(E) = -\pi\, \delta N(E), \qquad (7.39)$$

assuming that $N(0) = 0$. Since $g(E)$, and therefore $N(E)$ may be related to the trace of the Green's function through Eq. (3.36), $\zeta_s(E)$ is often expressed as

$$\zeta_s(E) = B(E) \exp \left\{ \int_0^E [G(E') - \tilde{G}(E')] \, dE' \right\}, \tag{7.40}$$

where $B(E)$ is a real-valued function for real E, that is chosen appropriately for ensuring the convergence of ζ_s [22]. In Eq. (7.40), $G(E)$ is the trace of the Green's function as defined in Eq. (3.34), and $\tilde{G}(E)$ is its smoothed form:

$$G(E) = \lim_{\epsilon \to 0} \sum_n \frac{1}{E - E_n + i\epsilon}, \quad \tilde{G}(E) = \lim_{\epsilon \to 0} \int \frac{\tilde{g}(E')}{E - E' + i\epsilon} \, dE'. \tag{7.41}$$

A knowledge of $G(E)$ requires the entire spectrum of the quantum eigenvalues, which is lacking in general, and therefore a semiclassical approximation is needed. For this purpose, Eq. (7.39) is particularly suitable, since the Gutzwiller trace formula gives $\delta N_{scl}(E)$ in Eq. (5.56). In what follows, for brevity of notation, we shall denote primitive periodic orbits by the subscript γ, instead of the label *ppo* that was used in earlier chapters. Repetitions of an orbit will generally be labeled by n. Combining these two relations, it at once follows that

$$\ln \zeta_s(E) = - \sum_\gamma \sum_{n=1}^\infty \frac{1}{n \left| \det \left(\tilde{M}_\gamma^n - I \right) \right|^{1/2}} \exp \left[i \left(\frac{n}{\hbar} S_\gamma - \sigma_{n\gamma} \frac{\pi}{2} \right) \right]. \tag{7.42}$$

For hard chaos, all the periodic orbits are unstable, and $\sigma_{n\gamma} = n\sigma_\gamma$. Assuming direct hyperbolic instability, the last row of Table 5.1 is applicable, and

$$\ln \zeta_s(E) = - \sum_\gamma \sum_{n=1}^\infty \frac{1}{2n \operatorname{Sinh}(n\chi_\gamma/2)} \exp \left[in \left(\frac{1}{\hbar} S_\gamma - \sigma_\gamma \frac{\pi}{2} \right) \right]. \tag{7.43}$$

Making use of the identity

$$\frac{1}{2 \operatorname{Sinh}(x/2)} = \sum_{n=0}^\infty \exp \left[- \left(n + \frac{1}{2} \right) x \right], \tag{7.44}$$

and the expansion $\ln(1 - x) = - \sum_{n=1}^\infty x^n/n$, we obtain the expression for $\zeta_s(E)$:

$$\zeta_s(E) = \prod_\gamma \prod_{n=0}^\infty \left\{ 1 - \exp \left[i \left(\frac{S_\gamma}{\hbar} - \frac{\pi}{2} \sigma_\gamma \right) - \left(n + \frac{1}{2} \right) \chi_\gamma \right] \right\}. \tag{7.45}$$

Before using this semiclassical form for the quantization of energies, we demonstrate that $\zeta_s(E)$ obeys the same functional relation (7.37) as obeyed by the Riemann zeta function $\zeta(E)$. To this end, we start with the Gutzwiller trace formula for the trace of the Green's function G(E), as defined in Eq. (7.41):

$$G(E) = \tilde{G}(E) + \sum_\gamma \sum_n \frac{T_\gamma}{i\hbar} \frac{\exp \left[in(S_\gamma/\hbar - \sigma_\gamma \pi/2) \right]}{2 \operatorname{Sinh}(n\chi_\gamma/2)}. \tag{7.46}$$

From the definition of $G(E)$, it may be written as $(d/dE) \ln \prod_n (E - E_n)$. The first term on the right-hand side of Eq. (7.46) is

$$\tilde{G}(E) = \lim_{\epsilon \to 0} \int_{-\infty}^{\infty} \frac{\tilde{g}(E')\, dE'}{E - E' + i\epsilon} = P \int \frac{\tilde{g}(E')\, dE'}{E - E'} - i\pi \tilde{g}(E) \tag{7.47}$$

The smoothed density of states $\tilde{g}(E)$ has no poles, so the principal-value integral above does not contribute. Recalling that $\tilde{g}(E) = (d/dE)\widetilde{N}(E)$, we may express $\tilde{G}(E)$ in the form:

$$\tilde{G}(E) = \frac{d}{dE} \left\{ \ln \exp \left[-i\pi \widetilde{N}(E) \right] \right\}.$$

Finally, it is clear by differentiating Eq. (7.43) that the last term in Eq. (7.46) is $(d/dE) \ln \zeta_s(E)$. Therefore Eq. (7.46) reduces to the form

$$\frac{d}{dE} \ln \prod_n (E - E_n) = \frac{d}{dE} \ln \exp \left[-i\pi \widetilde{N}(E) \right] + \frac{d}{dE} \ln \zeta_s(E). \tag{7.48}$$

Integrating with respect to E, and regularising the expression, we get

$$\prod_n \left(1 - \frac{E}{E_n} \right) = \exp \left[-i\pi \widetilde{N}(E) \right] \frac{\zeta_s(E)}{\zeta_s(0)}. \tag{7.49}$$

The function in the left-hand side above is just like $\xi(t)$ defined in connection with the Riemann zeta function in Eq. (7.6), and shows that the zeros of $\zeta_s(E)$ are the eigenvalues of the Hamiltonian. Since for real E the left-hand side of the above equation is real, it follows that so is the right-hand side, and therefore it should equal its complex conjugate. This yields the desired functional relation:

$$\exp \left[-i\pi \widetilde{N}(E) \right] \zeta_s(E) = \exp \left[i\pi \widetilde{N}(E) \right] \zeta_s^*(E). \tag{7.50}$$

To quantize the energies of a given dynamical system, we may now calculate $\zeta_s(E)$ from Eq. (7.45), and look for its zeros. In practice, of course, the sum over the primitive periodic orbits has to be truncated. It should be realized, however, that the semiclassical expression for $\zeta_s(E)$ developed here is only from the leading order term in the Gutzwiller trace formula, and therefore the right-hand side of Eq. (7.49) is not purely real in general. Consequently, the zeros of $\zeta_s(E)$ will have a small complex part even for real E, and it is more convenient to calculate $|\zeta_s(E)|$ and to locate its minima.

To illustrate the effectiveness of the formalism, we take the example of the wedge billiard that was studied extensively by Szeredi and Goodings [23] (see also Sect. 5.5). Fig. 7.5 displays $|\zeta_s(E)|$ for the 49° wedge, calculated from 1048 primitive periodic orbits using formula (7.45), taking n up to 9. As the authors point out, the result is not too satisfactory, with double minima showing up near the fourth energy eigenvalue, and the resolution getting noticeably worse after the eighteenth eigenvalue.

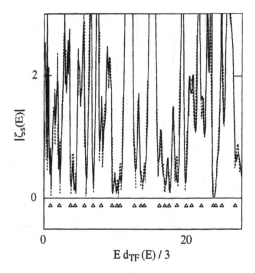

Figure 7.5: The function $|\zeta_s(E)|$ for the 49° wedge billiard, calculated semiclassically from Eq. (7.45), is shown by the solid curve. Its minima are to be compared with the exact energy eigenvalues, shown by the triangles along the normalized energy axis. The dotted curve shows the result using a pseudo-orbit expansion, to be discussed in the next sub-section. The latter expansion included 26706 pseudo-orbits, with word- length ≤ 19. (From [23].)

Significant improvement in the location of the zeros may be made, however, by making use of the functional relation (7.50), and applying the quantization rule

$$\xi(E) = \Re e \left\{ \exp\left[-i\pi\, \widetilde{N}(E) \right] \zeta_s(E) \right\} = 0\,. \tag{7.51}$$

The calculation of the left-hand side of the above equation for the 49° wedge is shown in Fig. 7.6, with $\zeta_s(E)$ calculated as before from the Euler product formula (7.45). Now the solid curve does not miss any zero in the first hundred eigenvalues, and the resolution is much improved. Moreover, no spurious zeros are introduced. We may take Eq. (7.51) as the semiclassical quantization rule from the zeta function approach, with $\zeta_s(E)$ calculated from the semiclassical formula (7.45).

This apparent success of the quantization rule (7.51) should not blind us to the fact that the Euler product formula Eq. (7.45) is not convergent for real E due to the exponential proliferation of periodic orbits in the case of hard chaos. A damping of the longer periodic orbits may be conveniently incorporated by making E, and therefore the action $S(E)$, complex. The factor $\exp\left(iS(E)\right)$ in the trace formula then damps out the periodic orbits with larger S. Convergence is achieved provided the imaginary part of E is sufficiently large to overcome the exponential growth factor in the number of these orbits. (From the point of view of the Riemann zeta function, this is like taking $(\sigma - 1/2) > 1/2$). This condition is sometimes referred to as "going across the entropy barrier", and given as [15]

$$\Im m\, E > \frac{\hbar}{2}\, h(E)\,, \tag{7.52}$$

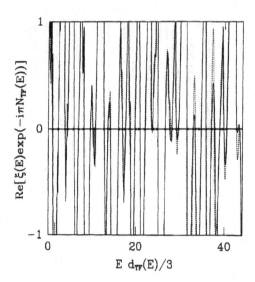

Figure 7.6: The function $\xi(E)$ and its zeros for the 49° wedge, plotted from (7.51), and compared with the exact eigenvalues (shown as solid dots on the horizontal axis). The dotted curve, as in the previous figure, is the pseudo-orbit calculation. (From [23].)

where the topological entropy $h(E)$ was defined in Eq. (7.23). As we have seen earlier, this procedure results in a loss of resolution of energy. One way out may be to use the Dirichlet series rather than the Euler product formula, and then convert it to a convergent Riemann-Siegel form (7.32). Alternately, one may hope to analytically continue ζ_s across the entropy barrier by using the Dirichlet series. We illustrate this by going to the particular case of the Selberg zeta function on the non-Euclidean surface of constant negative curvature, where the algebra is simpler. We follow the treatment of Aurich and Steiner [17].

7.2.2   Pseudo-orbits and the Selberg zeta function

In order to have a Dirichlet series with pseudo-orbits analogous to Eq. (7.24), we have to express the double product $\prod_n \prod_\gamma (1 - \exp[iS_\gamma/\hbar..])$ occurring in Eq. (7.45) in the form $\exp[-i/\hbar \sum_p m_p S_p..]$. The exponents in ζ_s are particularly simple for the motion of a particle on a compact surface of constant negative curvature. It is also of special interest, because the Gutzwiller trace formula in this case becomes exact [18, 19]. An introductory account of this problem is given in detail in chapter 19 of Gutzwiller's book [20], and will not be repeated here. We only mention that the infinitesimal distance squared, ds^2, between two points in this geometry in two dimensions is not $dx^2 + dy^2$, but is given by (taking only $y > 0$),

$$ds^2 = R^2 \frac{dx^2 + dy^2}{y^2}.$$

Here R is a positive constant with the dimension of a length. All classical periodic orbits in this hyperbolic geometry are unstable. In the following, we shall take $R = 1$ and use dimensionless length units. For the quantum problem, the stationary Schrödinger equation may be solved with the appropriate boundary conditions in this geometry. With $\hbar = 2m = 1$, the discrete momenta k_n, $n = 1, \ldots \infty$ are related to the eigenenergies by the relation $E_n = 1/4 + k_n^2$. The full density of states may be expressed *exactly* in the form of a trace formula named after Selberg. It consists of a smooth Thomas-Fermi term promotional to the area of the compact surface, and an oscillatory part that is in the form of Gutzwiller's trace formula. A primitive periodic orbit, labeled by γ, has a geodesic length denoted by l_γ, proportional to its period. The precise form of the Selberg trace formula may be found in [20], and will not be given here. For our purpose, it is important to note that in this model, the Selberg zeta function may be written exactly in the form of a double product, as in Eq. (7.45). It has an infinite number of simple zeros, symmetrically placed on the imaginary axis at $\sigma = 1/2$, which we denote by $s = s_n = 1/2 \pm ik_n$. Note that these may be related to the eigenenergies E_n by the equation $-E_n = s_n(s_n - 1)$. Its Euler product form [compare with Eq. (7.45)] converges only for $\sigma > 1$:

$$\zeta_s(s) = \prod_\gamma \prod_{n=0}^{\infty} \{ 1 - \exp[-(s+n)l_\gamma] \} . \tag{7.53}$$

To convert the infinite product over the index n in Eq. (7.53), we follow [17] and use Euler's identity:

$$\prod_{n=0}^{\infty} (1 - yx^n) = 1 + \sum_{m=1}^{\infty} \frac{(-)^m y^m x^{\frac{1}{2}m(m-1)}}{\prod_{l=1}^{m}(1 - x^l)}, \quad |x| < 1 . \tag{7.54}$$

Substituting $y = \exp(-sl_\gamma)$, $x = \exp(-l_\gamma)$, we obtain

$$\zeta_s(s) = \prod_\gamma \left\{ 1 + \sum_{m=1}^{\infty} \frac{a_{\gamma m}}{[\exp(ml_\gamma)]^s} \right\} . \tag{7.55}$$

In the above,

$$a_{\gamma m} = \frac{(-)^m \exp\left(-\frac{1}{2}m(m-1)l_\gamma\right)}{\prod_{n=1}^{m}[1 - \exp(-nl_\gamma)]} .$$

On expanding the product form in Eq. (7.55), and multiplying out, we obtain an infinite number of terms with a product $\exp[(m_1 + m_2 + m_3 + \ldots)l_\gamma]$ in the denominator, where the dummy indices m_i may take all possible integer values. We may therefore rearrange the terms in the Dirichlet form

$$\zeta_s(s) = 1 + \sum_\alpha \frac{A_\alpha}{(N_\alpha)^s} , \tag{7.56}$$

where the pseudo-orbit α has a length L_α, with

$$N_\alpha = \exp(L_\alpha), \quad L_\alpha = \sum_i m_i l_{\gamma i} . \tag{7.57}$$

The coefficient $A_\alpha = \prod_i a_{m_1 \gamma_i}$. In Eq. (7.56), 1 on the right-hand side comes from taking all $m_i = 0$. The different pseudo-orbits α correspond to different combinations of the integers m_i in Eq. (7.57), and are arranged in the order of increasing pseudo-lengths L_α. Of course, just because $\zeta_s(s)$ has been written in the Dirichlet form (7.56), it does not mean that it converges on the critical line $\sigma = 1/2$. In fact, a careful analysis by Aurich and Steiner [17], using the well-known convergence criteria of a Dirichlet series, shows that it diverges for the present case under consideration. On the other hand, it appears to be conditionally convergent for the case of a planar hyperbola billiard [24], and also for the so-called Artin's billiard [25] on the surface of a constant negative curvature. For such systems, we may immediately apply the quantization condition (7.51) $\xi(k) = 0$ on the critical line, i.e., for $s = 1/2 + ik$. Substituting for ζ_s from Eq. (7.56) in Eq. (7.51), we obtain

$$\xi(k) = \cos\left[\pi \widetilde{N}(E)\right] + \sum_\alpha \frac{A_\alpha}{\sqrt{N_\alpha}} \cos\left[kL_\alpha - \pi \widetilde{N}(E)\right] = 0. \qquad (7.58)$$

This formula looks like the Riemann-Siegel form (7.32), except for the important point that the sum is over an infinite number of pseudo-orbits and not resummed optimally as in Eq. (7.32). A practical application of this formula entails truncating the sum above in some suitable fashion. When such a test is done for the 49° wedge, Eq. (7.58) performs poorly [23]. As a necessary next step, it is desirable to develop [15] a Riemann-Siegel "look-alike" formula, analogous to Eq. (7.32).

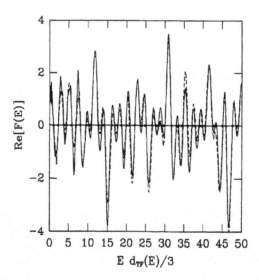

Figure 7.7: Numerical test of the 49° wedge with the Riemann-Siegel look-alike formula (7.59). The dotted curve is the result of taking 26706 pseudo-orbits of word length ≤ 19. The solid curve is the plot due to a smoothed version of the formula [26]. The solid dots on the horizontal axis denote the quantum energy eigenvalues. (From [23].)

The formula (7.58) was obtained for the special case of motion on a surface of constant negative curvature, for which the Gutzwiller trace formula is exact. In the more general dynamical problem, a similar Riemann-Siegel-like formula may be derived [15], with the argument of the cosine term being replaced by $S_n(E) - \pi \widetilde{N}(E)$. $S_n(E)$ is the action of the nth pseudo-orbit.

$$\xi_s(E) = \sum_{n=0}^{n^*} C_n \cos \left[S_n(E)/\hbar - \pi \widetilde{N}(E) \right], \qquad (7.59)$$

where n^* is chosen by the condition given by Eq. (7.33) that includes all pseudo-periods $\mathcal{T} < \mathcal{T}^*$, such that

$$\mathcal{T}_{m^*} = \frac{h}{2} \, \tilde{g}(E), \qquad (7.60)$$

where h is Planck's constant. In a later paper [26], Berry and Keating have modified Eqs. (7.58) and (7.59) to smooth over the truncation over the pseudo-orbits. We do not describe this here. Szeredi and Goodings have tested the quantization condition (7.58) for the wedge billiard. The comparison with the exact eigenvalues for the 49° wedge billiard is shown in Fig. 7.7. The agreement is excellent, and gives some credibility to this approach.

7.3 The scattering matrix method

We have been focusing till now on the bound-state spectrum of a Hamiltonian. The semiclassical quantization of these states was formulated via the bound classical orbits of the particle. In general, however, a potential may have bound as well as scattering states, the latter in the continuum. It is well-known that there is an intimate connection between these two classes of states. For the description of scattering, it is customary to define the so-called Scattering matrix, which relates the incoming wave to the asymptotic outgoing wave after scattering. We shall only be concerned here with elastic scattering, where the energy of the scattered wave remains the same, though in general other quantum numbers in the outgoing channel may be different. The analytical properties of this scattering matrix in the complex momentum plane describe in a unified manner both the bound and the scattering properties of the potential [27]. For example, the poles of the scattering matrix in the upper half of the imaginary axis at $k = i \, k_n$ give the bound state eigenvalues [28] at $E_n = -\hbar^2 k_n^2/2m$.

Here we focus our attention to an approach pioneered by Smilansky [29, 30], where a secular equation involving the scattering matrix is formulated that determines the bound-state energies of the potential. Although of general applicability, the method is most transparent when applied to billiards (with an impenetrable boundary). We first show, in a toy model, that there is a simple connection, at a quantum level, between the eigenvalue spectrum of the billiard and the scattering matrix describing it as an obstacle to incident waves outside the billiard. Consider once again the circle billiard. Quite generally, the same Schrödinger equation describes the stationary quantum states inside and outside the billiard:

$$\left(\nabla^2 + k^2 \right) \Psi = 0, \qquad (7.61)$$

where $k^2 = 2mE/\hbar^2$. In this problem, it is natural to take the origin to be at the center of the circle, and polar coordinates. Writing $\Psi(r, \phi) = \sum_l \psi_l(r, \phi) = R_l(r) \exp(il\phi)$, Eq. (7.61) takes the form of the cylindrical Bessel equation for $R_l(r)$:

$$\frac{d^2R_l}{dr^2} + \frac{1}{r}\frac{dR_l}{dr} + \left(k^2 - \frac{l^2}{r^2}\right)R_l(r) = 0. \tag{7.62}$$

For the inside problem of bound states, the origin is included, and the eigenstate $R_l(r)$ has to be regular. It is therefore taken to be the regular cylindrical Bessel function $J_l(kr)$. Since the wave function has to vanish at the boundary $r = R$ (Dirichlet boundary condition), the eigenvalues are determined by the zeros of the Bessel function:

$$J_l(k_{nl}R) = 0, \quad l = 0, \pm 1, \pm 2, \ldots, \quad n = 1, 2, \ldots. \tag{7.63}$$

Now consider Eq. (7.62) for the scattering problem, with the impenetrable circle as the scatterer of incident waves. Because of circular symmetry, orbital angular momentum l is a good quantum number, and the incoming and outgoing channels carry the same angular momentum l. Since the origin is excluded, both the regular and the irregular solutions, $J_l(kr)$ and $Y_l(kr)$ are now allowed. It is more convenient, however, to take their linear combinations $H_l^\pm = J_l \pm iY_l$, that have the form of outgoing and incoming circular waves for large argument:

$$H_l^\pm(kr) \sim \left(\frac{2}{\pi kr}\right)^{1/2} \exp\left[\pm i\left(kr - l\pi/2 - \pi/4\right)\right]. \tag{7.64}$$

Thus the radial part of the scattering solution may be written as

$$R_l(kr) = aH_l^-(kr) + cH_l^+(kr), \tag{7.65}$$

where a and c depend on k, but are independent of r. The scattering solution (7.65) has to satisfy the boundary condition that it must vanish on the surface of the sphere, the same as the inside solution. The two constants a and c are therefore related, and the solution may be written in the form, apart from an overall constant, as

$$\psi_l(r, \phi) = \left[H_l^-(kr) + S_l(k)H_l^+(kr)\right]\exp(il\phi), \tag{7.66}$$

where the "S-matrix" in the channel l is

$$S_l(k) = \frac{c}{a} = -\frac{H_l^-(kR)}{H_l^+(kR)}. \tag{7.67}$$

From the above, we see that if we set $k = k_{nl}$, and use the bound state condition (7.63) for the inside problem, Eq. (7.67) reduces to

$$S_l(k_{nl}) = 1. \tag{7.68}$$

This may be turned into a secular equation $(S_l(k) - 1) = 0$, whose roots yield the bound states of the interior problem in the partial wave with angular momentum l. In this problem with circular symmetry, l is a good quantum number, and $S_l(k) = \exp(2i\delta_l(k))$, where $\delta_l(k)$ is the scattering phase shift. The full description of the problem, of course, involves all the partial waves with integer l. The secular equation determining the bound states of the circle billiard then takes the form

$$\det\left(1 - \mathcal{S}(E)\right) = 0, \tag{7.69}$$

where we have switched to the energy variable E. Note that the above equation is quantum mechanically exact, and the determinant is of infinite dimension (corresponding to an infinite number of partial waves, with $l = 0, \pm 1, \pm 2, \ldots$). Although we have derived Eq. (7.69) for the circle billiard only, the same equation may be obtained for the more general problem. For a non-concave billiard, the derivation is very similar, except that without circular symmetry, l is no longer a constant of motion. In such a case, there will be nondiagonal matrix elements $\mathcal{S}_{ll'}$ in the S-matrix, and Eq. (7.66) is replaced by

$$\psi_l(r, \phi) \sim H_l^-(kr)\exp(il\phi) + \sum_{l'} \mathcal{S}_{ll'}(kr)H_{l'}^+(kr)\exp(il'\phi). \tag{7.70}$$

The above equation is asymptotic, in the sense that it is valid only outside the smallest circle that encloses the non-concave billiard fully. Also, it is still possible to express the S-matrix in a diagonal form in terms of its eigenphases ϕ_m, $m = 1, 2, \ldots \infty$. Only in the case of circular symmetry, can the index m be identified with the angular momentum quantum number l (it may now take both $\pm$ signs). In the following, we continue to discuss the circular problem for simplicity, keeping in mind that the method is applicable in more general cases.

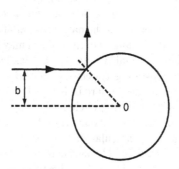

Figure 7.8: The geometrical ray limit of scattering. A particle is incident on an impenetrable circular billiard with impact parameter b.

To make Eq. (7.69) tractable, it is necessary to reduce the dimension of the determinant. To this end, consider Fig. 7.8 where we show an incoming particle with impact parameter b incident on the circle, which has an angular momentum $l = bk$. This is a classical description, valid only in the short wave-length limit $(h/p << R)$ of the quantum mechanical wave picture. In this limit, it is clear that there will be no scattering at all if $b > R$. Even in the wave picture (for longer wave lengths), the scattering phase shifts will be progressively smaller at a fixed energy as the angular

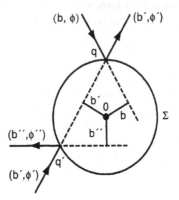

Figure 7.9: The inside-outside duality for specular reflections on the surface Σ of a non-concave billiard. (After [29].)

momentum l is increased. For such l, $S_l \to 1$, with the result that the left-hand side of Eq. (7.69) becomes far too small to be of any practical use. It is therefore necessary to make a semiclassical approximation to Eq. (7.69), and truncate the determinant to a finite dimension Λ by assuming that for l larger than a maximum value, no scattering takes place. This cut-off is actually energy-dependent, though in practice it is often taken to be a constant. In any case, the scattering matrix now is finite dimensional, and with this approximation Eq. (7.69) becomes a semiclassical equation.

Before proceeding further with the mathematical formulation, it may be appropriate to point out why the inside-outside duality of the billiard may be regarded as a Poincaré mapping. It is easy to picture the duality between the classical reflections of a particle inside the billiard with the scattering trajectories outside the billiard. To see this, consider Fig. 7.9 where a non-concave billiard is depicted. We take, as before, the origin to be the center of the smallest covering circle that completely encloses the billiard. An incident particle trajectory on the outside of the billiard may then be specified by the impact parameter b, the perpendicular distance from the origin, and an angle ϕ with respect to a reference direction. The specularly reflected trajectory is similarly specified by (b', ϕ'), as shown in the figure. In order to regard the billiard boundary as a Poincaré surface of section (PSS), consider how the mapping is performed usually. In Fig. 7.10 we choose a line (denoted by Σ), as the PSS in two space dimensions. The trajectory of a particle with constant energy intersects Σ at the point q at time t (we are suppressing the momentum variable here). With a proper choice of Σ, the emerging trajectory re-enters Σ at the point q' from the same direction as before at a later time t'. The unfolding of the continuous motion in a time interval in two dimensions is mapped from point q to point q' on Σ in a discrete time step, with Σ mapping onto itself.

Note that in such a mapping, the emerging trajectory at q reinjects itself on Σ at q' at a later time. Referring now to Fig. 7.9, the same kind of reinjection of the outgoing scattered trajectory (b', ϕ') on to the surface of the billiard may be made at a diametrically opposite point of the billiard, as shown. This turning of the outgoing trajectory to an incoming one is only possible through a dual trajectory inside the billiard. The internal trajectory (b, ϕ) gets reflected to (b', ϕ'), and then at a point across the other side to $(b''\phi'')$. Thus the information about the external scattering paths are contained in the corresponding inside ones. The S-matrix, which describes the outside scattering quantum mechanically (instead of classical rays) may thus be regarded as the quantized version of the Poincaré Scattering Mapping (PSM) on the outside surface. From our discussion here, it is clear that PSM is also equivalent to a quantum mapping of the trajectories in the inside of the billiard, when the PSS is taken to be the boundary of the billiard itself. The latter mapping was proposed by Bogomolny [22], who derived a secular equation similar to Eq. (7.69), but replacing S by a T-matrix. The two approaches are equivalent. It is crucial that the semiclassical approximation be made, so that the dimension of the matrix (S or T) is finite. Since no periodic orbits are involved in the construction of the secular equation in either approach, the convergence problem over their sum (as in the trace formula) is avoided (unless the trace of the matrix is expressed again in terms of the periodic orbits of the inside problem).

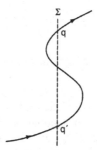

Figure 7.10: Poincaré surface of section (PSS) in two-dimensional motion.

Finally, it is worth noting the close connection of the present approach with the Selberg zeta function. After truncating the S-matrix semiclassically to a finite dimensional matrix of dimension Λ, let us denote $\det(1 - S)$ by ζ_{sc}. We may express it as a diagonal matrix in terms of its eigenphases ϕ_m, (we take $\phi_m \geq 0$) and write

$$\zeta_{sc}(E) = \det(1 - S) = \prod_{m=1}^{\Lambda} [1 - \exp(i\phi_m(E))] \tag{7.71}$$

$$= 2^{\Lambda} \exp\left[\frac{i}{2} \sum_{m=1}^{\Lambda} \phi_m(E) - i\Lambda\frac{\pi}{2}\right] \prod_{m=1}^{\Lambda} \sin\frac{\phi_m(E)}{2}. \tag{7.72}$$

From this expression, we see that the secular equation $\zeta_{sc} = 0$ can only be satisfied when the last factor on the right-hand side of Eq. (7.72) goes to zero, i.e., any of the $\phi_m(E)$ go through 2π as E is varied. Every time this happens, the sign of $\sin(\phi_m/2)$ changes, which means ζ_{sc} gets multiplied by a factor $\exp(i\pi)$. Taking this into account, the imaginary part of the logarithm of ζ_{sc} is given by

$$\Im m \ (\ln\zeta_{sc}(E)) \ = \ \frac{1}{2}\sum_m \phi_m(E) - \pi \sum_{m=1}^{\Lambda} \sum_{n=1}^{\infty} \Theta(E - E_{mn}) - \Lambda\frac{\pi}{2}. \qquad (7.73)$$

In the above, the choice of the negative sign before the step function $\Theta(E)$ is consistent with the requirement that for large E, $\det(1 - \mathcal{S}(E)) \to 0$. Differentiating the above equation with respect to E, we obtain

$$\frac{d}{dE}\Im m \ (\ln\zeta_{sc}(E)) \ = \ \frac{1}{2}\sum_m \frac{d}{dE}\phi_m(E) - \pi \sum_{m=1}^{\Lambda} \sum_{n=1}^{\infty} \delta(E - E_{mn}). \qquad (7.74)$$

This may be recognized, in the limit of $\Lambda \to \infty$, as being of the same form as Eq. (7.38):

$$g(E) - g_R(E) \ = \ -\frac{1}{\pi}\frac{d}{dE}\left[\Im m \ln \zeta_{sc}(E)\right]. \qquad (7.75)$$

Here the resonance density $g_R(E)$ in the continuum is

$$g_R(E) \ = \ \frac{1}{2\pi}\sum_m \frac{d\phi_m}{dE}. \qquad (7.76)$$

This is so since at every resonance the scattering phase shift (for a central potential the phase shift $\delta_l = 2\phi_l$) rises by π, a well-known result in scattering theory. Moreover, there is a theorem in scattering theory, called 'Levinson's theorem' [31], that ensures that the total number of resonances in the continuum (between $0 \le E \le \infty$) equals the total number of bound states in a potential that has vanishing phase shifts as $E \to \infty$. In fact, it is shown by Doran and Smilansky [29] that $g_R(E) \to \tilde{g}(E)$ as $E \to \infty$ (see their figure 5, where the corresponding integrated counting functions for the Sinai billiard are compared in an energy interval). From these considerations, we see that Eq. (7.75) takes exactly the same form as Eq. (7.38) derived for the Selberg zeta function:

$$\delta g(E) \ = \ -\frac{1}{\pi}\frac{d}{dE}\left[\Im m \ln \zeta_{sc}(E)\right]. \qquad (7.77)$$

To evaluate the right-hand side, note that

$$\ln\det \ (1 - \mathcal{S}) = \mathrm{tr}\ln(1 - \mathcal{S}) = -\sum_{n=1}^{\infty}\frac{1}{n}\mathrm{tr}\,\mathcal{S}^n. \qquad (7.78)$$

In the approach pioneered by Smilansky, the right-hand side of the above equation is expressed in a Gutzwiller-like trace formula involving a subset of the periodic orbits within the billiard. This is possible because of duality between the trajectories inside and outside the billiard. We shall not elaborate on this method, since it also suffers from the same convergence problems as the usual trace formula. Instead, we briefly describe a related approach given by Bogomolny [22] that does not involve periodic orbits.

7.4 The transfer-matrix method of Bogomolny

Much effort in semiclassical physics has gone into trying to obtain the quantized energy eigenvalues of a system whose classical analogue is chaotic. This was done in Sect. 7.2 using periodic orbits in the trace formula, and its resummed variants. The main obstacle to this approach was the exponential growth in the number of the primitive periodic orbits with increasing time periods. Periodic orbits with larger and larger time periods need to be included to improve the resolution of the level density in the energy domain. Whereas we could obtain the gross beat structure in the latter for the Hénon-Heiles potential with only the shortest three orbits (see Fig. 5.6), more than 13,000 primitive periodic orbits were needed to obtain the lowest 150 energy levels of the hyperbola billiard [24]. The sophisticated formalism of the dynamical zeta function and the pseudo-orbits had to be used (see Eq. (7.58)) to obtain this result. In most cases, it is more difficult to find the tens of thousands of isolated periodic orbits of a dynamical system than to diagonalize the quantum Hamiltonian. Therefore one may well question the practical utility of this semiclassical approach. Bogomolny's semiclassical transfer-matrix method, on the other hand, does not need periodic orbits at all, and incorporates the Heisenberg uncertainty principle in a natural way. Also, finding the eigenvalues of the system by this method amounts to diagonalization of a 'transfer matrix' on a Poincaré surface of section (PSS) of a lower dimension, so that the basis space is much smaller than that required for the diagonalization of the Hamiltonian in a basis with a higher dimension. In the following, we shall confine our attention to a two-dimensional billiard problem for simplicity, so that the PSS in the configuration space is only one-dimensional.

How is it that the eigenvalues (and even the eigenfunctions) of a two-dimensional system may be recovered by working with a one-dimensional basis? The clue lies in the so-called *boundary integral method*, which is particularly simple for billiards. Using this method, we may obtain the eigenvalues of the billiard from a knowledge of the normal derivative of the wave function defined on the boundary alone, which has one dimension less. This normal derivative is required over all the boundary as a continuous function of the boundary coordinate, and the billiard surface has to be smooth for the method to be applicable. Since the method is often used for computation of eigenvalues, we briefly outline its derivation. From Eq. (3.30), we note that the free Green's function $G_0(\mathbf{r}, \mathbf{r}'; E)$ obeys the differential equation

$$\left(E + \frac{\hbar^2}{2m}\nabla_{\mathbf{r}'}^2\right) G_0(\mathbf{r}, \mathbf{r}'; E) = \delta(\mathbf{r} - \mathbf{r}')\,. \tag{7.79}$$

Note that the boundary condition obeyed by G_0 need not be the same as the eigenfunctions of the billiard. The Schrödinger equation obeyed by the wave function $\Psi_E(\mathbf{r}')$ for the force-free billiard is

$$\left(E + \frac{\hbar^2}{2m}\nabla_{\mathbf{r}'}^2\right) \Psi_E(\mathbf{r}') = 0\,. \tag{7.80}$$

We multiply Eq. (7.79) by $\Psi_E(\mathbf{r}')$, and Eq. (7.80) by $G_0(\mathbf{r}, \mathbf{r}'; E)$, and then subtract one from the other. Now integrating this over the area of the billiard with the boundary

B, we get

$$\frac{\hbar^2}{2m} \int \left[\Psi_E(\mathbf{r}') \nabla^2_{\mathbf{r}'} G_0(\mathbf{r}, \mathbf{r}'; E) - G_0(\mathbf{r}, \mathbf{r}'; E) \nabla^2_{\mathbf{r}'} \Psi_E(\mathbf{r}') \right] d^2 r' = \int \delta(\mathbf{r} - \mathbf{r}') \, \Psi_E(\mathbf{r}') \, d^2 r'. \tag{7.81}$$

We now use Green's theorem to obtain

$$\frac{\hbar^2}{2m} \oint_B \left[\Psi_E(\mathbf{r}') \nabla_{\mathbf{r}'} G_0(\mathbf{r}, \mathbf{r}'; E) \cdot \mathbf{n}' - G_0(\mathbf{r}, \mathbf{r}'; E) \nabla_{\mathbf{r}'} \Psi_E(\mathbf{r}') \cdot \mathbf{n}' \right] dr' = \Psi_E(\mathbf{r}), \tag{7.82}$$

where we have taken the point $\mathbf{r}$ to be inside the billiard, and $\mathbf{n}'$ denotes the outward normal to the boundary B at the point $\mathbf{r}'$. Note that if the point $\mathbf{r}$ is on the boundary ($\mathbf{r} \in B$), the r.h.s. of (7.82) must be multiplied by a factor $1/2$. The equation above is quite general, valid even if a one-body potential were to be added to Eqs. (7.79, 7.80), provided the Green's function is changed appropriately. In general, however, its form is not known. For the force-free billiard problem with a sharp boundary, on the other hand, we know the free Green's function G_0 and, moreover, the first term on the left-hand side of Eq. (7.82) vanishes with the Dirichlet boundary condition $\Psi_E(\mathbf{r}') = 0$, $\mathbf{r}' \in B$. Alternately, with the Neumann boundary condition, it is the second term that is zero. With the Dirichlet boundary condition, Eq. (7.82) reduces to

$$\Psi_E(\mathbf{r}) = -\frac{\hbar^2}{2m} \oint_B G_0(\mathbf{r}, \mathbf{r}'; E) \, \mu(\mathbf{r}') \, dr', \qquad (\mathbf{r} \notin B) \tag{7.83}$$

where

$$\mu(\mathbf{r}') = \nabla_{\mathbf{r}'} \Psi_E(\mathbf{r}') \cdot \mathbf{n}'. \tag{7.84}$$

We see from Eq. (7.83) that we may calculate the wave function inside the billiard from a knowledge of the normal derivative on its boundary, using the free Green's function with an arbitrary boundary condition. In particular, if we let the point $\mathbf{r}$ approach the boundary from the inside, then the left-hand side of Eq. (7.83) goes to zero, and the equation may be used to compute the eigenvalues for the billiard. This shows that it is possible to solve a *two-dimensional* quantum mechanical problem from a *one-dimensional* integration (along the boundary). For later use, we also apply the gradient operator $\nabla_{\mathbf{r}}$ to both sides of Eq. (7.83) to obtain

$$\nabla_{\mathbf{r}} \Psi_E(\mathbf{r}) = -\frac{\hbar^2}{2m} \oint_B \nabla_{\mathbf{r}} G_0(\mathbf{r}, \mathbf{r}'; E) \, \mu(\mathbf{r}') \, dr'. \qquad (\mathbf{r} \notin B) \tag{7.85}$$

If we let $\mathbf{r}$ be on the boundary and take the scalar product of both sides above with $\mathbf{n}$, the outward normal to B at $\mathbf{r}$, we obtain

$$\mu(\mathbf{r}) = -\frac{\hbar^2}{m} \oint_B \left[\nabla_{\mathbf{r}} G_0(\mathbf{r}, \mathbf{r}'; E) \cdot \mathbf{n} \right] \mu(\mathbf{r}') \, dr'. \qquad (\mathbf{r} \in B) \tag{7.86}$$

This equation (with an additional inhomogeneous term) has also been studied by Georgeot and Prange [32] exploiting the solutions of Fredholm integral equation.

The first step in Bogomolny's method consists of a suitable choice of a PSS Σ (not necessarily the boundary) that is well-frequented by classical trajectories. The secular equation for the determination of the eigenvalues is set up on this one-dimensional PSS in the configuration space, and we now denote the coordinate along the PSS Σ by q. Rather than work with a Green's function whose poles correspond to the eigenvalues of the system, Bogomolny sets up a secular equation like (7.69), but with a transfer matrix linking two points on Σ, rather than the scattering matrix of the previous section. One considers those trajectories that take the particle from q' to q on Σ, crossing the PSS only once in between, as shown on the left-hand side of Fig. 7.11. On the one-dimensional PSS we may define, for each point $q \in \Sigma$, a well-behaved continuous function $\psi(q)$ at a fixed energy which is analogous to the function $\mu(\mathbf{r}')$ of Eq. (7.84). Using this continuous function $\psi(q)$, and an equation similar to (7.83), the true wave function $\Psi(\mathbf{r})$ for a point inside the billiard may again be obtained. Note that $\psi(q)$ itself should *not* be regarded as the projection of the billiard wave function on Σ. If, for example, Σ is chosen to be the boundary B itself, $\psi(q)$ is in general nonzero, whereas the actual wave function Ψ on the boundary is zero.

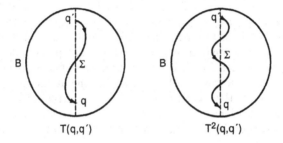

Figure 7.11: The left-hand diagram shows a classical trajectory that arrives at q in the same direction as it left q'. It crosses the PSS Σ once in between in the opposite sense. The contribution to $T(q, q')$ comes from all such trajectories. The right-hand figure depicts a classical path from q' to q that likewise crosses Σ twice. Such trajectories contribute to $T^2(q, q')$.

In the left part of Fig. 7.11, we consider the motion of a particle inside a billiard with boundary B. The quantum mechanical mapping of a particle from $q' \to q$ is defined by the transfer operator $T(q, q'; E)$ in coordinate space through the equation

$$\psi(q) = \int_\Sigma T(q, q'; E)\, \psi(q')\, dq'. \tag{7.87}$$

Note the similarity of this integral equation with (7.86), with $T(q, q'; E)$ corresponding to $[\nabla_{\mathbf{r}} G_0(\mathbf{r}, \mathbf{r}'; E) \cdot \mathbf{n}]$, and $\psi(q')$ to $\mu(\mathbf{r}')$ [33].

This quantum mapping is only possible at some discrete values of the energy E, whereas the semiclassical transfer matrix is defined for continuous E. The semiclassical transfer matrix $T_{sc}(q, q'; E)$ is obtained by summing over all possible classical paths at energy E that go from q' to q, crossing the PSS *in the same direction* at both points, and at no other point during the motion from q' to q. (The same direction means that the component of the momentum normal to the PSS points in the same sense at q' and q.). Although defined for a fixed energy E, it has a form analogous to the Green's function of a one-dimensional problem in the time representation. The semiclassical expression for this Green's function is given by Eq. (5.8) that involves Hamilton's principal function $R(q, q', t)$. Bogomolny [22] derives an expression for the (time-independent) semiclassical transfer matrix that looks exactly similar, except that Hamilton's principal function $R(q, q', t)$ is replaced by the classical action for fixed E, $S(q, q', E)$, evaluated along the actual trajectory (not confined to Σ) from q' to q as shown in Fig. 7.11 (left part). The derivation is lengthy, and will be found in the original paper [22]. The semiclassical expression for $T(q, q'; E)$ for a system having two degrees of freedom is

$$T_{sc}(q, q'; E) = \sum_{cl} \frac{1}{(2\pi i\hbar)^{1/2}} \left| \frac{\partial^2 S(q, q', E)}{\partial q' \partial q} \right|^{1/2} \exp\left[iS(q, q', E)/\hbar - i\nu\pi/2\right]. \quad (7.88)$$

The sum is over all classical trajectories that connect q' to q with one crossing (in the *opposite* sense in between. In the absence of any such trajectory, $T_{sc}(q, q'; E)$ is zero. This expression is valid in a two-dimensional problem, where the PSS is a line. In higher dimensions, the second derivative involving S is replaced by the corresponding determinant, and the pre-factor $(2i\pi\hbar)$ acquires a power $(D-1)/2$, where D is the spatial dimension of the system under consideration. The phase index ν is evaluated separately for each trajectory. It is equal to the number of caustics in the trajectory, plus two at each reflection from the boundary, plus one at the highest point of the trajectory. At a caustic, the second derivative of S occurring in Eq. (7.88) changes sign. For the two-dimensional problem, Eqs.(7.87) and (7.88) imply that $T(q, q'; E)$ (or its semiclassical approximation) has the dimension of an inverse length. Like the S-matrix, the T-matrix is unitary:

$$\int_\Sigma T(q, q''; E)\, T^*(q', q''; E)\, dq'' = \delta(q - q'). \quad (7.89)$$

It also obeys the chain rule (we suppress the fixed energy symbol E):

$$T^l(q, q') = \int_\Sigma T(q, q_{l-1})...T(q_2, q_1)T(q_1, q')\, dq_1\, dq_2 ....dq_{l-1}. \quad (7.90)$$

Thus the left-hand side, $T^l(q, q')$, may be interpreted as the transfer matrix for classical trajectories that cross Σ l times in the opposite sense, as shown in Fig. 7.11 for $l = 2$.

Any practical calculation involving the T-matrix is made by discretizing the continuous variable q. This is done by dividing the line defining the PSS Σ into a number of segments ("cells"), the ith cell having a width Δ_i, and a central point q_i. Let the entire PSS be divided into n such cells; the optimum choice of n will be discussed presently. In order to obtain the secular equation that determines the quantization of energy, we first discretize the quantum mapping defined by Eq. (7.87). For this purpose, it is convenient to construct a set of real basis states ("top-hat" representation):

$$\psi_i(q) = \frac{1}{\sqrt{\Delta_i}} \qquad q \in i^{th} \text{cell},$$

$$= 0 \qquad\qquad \text{otherwise}. \quad (7.91)$$

Note that by construction this is an orthonormal set,

$$\int_\Sigma \psi_j(q)\, \psi_i(q)\, dq = \delta_{ij}. \quad (7.92)$$

We may now express the wave function $\psi(q)$ as a histogram

$$\psi(q) = \sum_i c_i \, \psi_i(q) \, , \tag{7.93}$$

where the coefficients c_i are complex. We may now proceed to obtain the secular equation to obtain the eigenvalues and the unknown coefficients c_i (i.e., the eigenvector). The first step is to substitute (7.93) in Eq. (7.87). Next, we multiply both sides of the equation by $\psi_j(q')$, and integrate over q' on the entire range of Σ. Using the orthogonality of the basis states, we get

$$\sum_i (T_{ji} - \delta_{ji}) \, c_i = 0 \, , \qquad \text{or} \quad \det (T_{ji} - \delta_{ji}) = 0 \, . \tag{7.94}$$

In the above equation, T_{ji} is the *discretized* representation of the transfer matrix, and is given by

$$T_{ji} = \int_\Sigma \, dq' \, dq \, \psi_j(q) \, T(q, q') \, \psi_i(q') \, . \tag{7.95}$$

By substituting Eq. (7.91), we get

$$T_{ji} = \sqrt{\Delta_j \Delta_i} \, T(q_j, q_i) \, . \tag{7.96}$$

Unlike $T(q, q')$, the discretized T_{ji} is dimensionless. The latter is a $n \times n$ matrix, where n is the number of cells in the configuration space spanning the PSS Σ.

For numerical calculation of T_{ji}, the semiclassical approximation given by Eq. (7.88) may be substituted in Eq. (7.95) to obtain (we omit the subscript sc)

$$T_{ji} = \sum_{cl} B_{ji} \exp \left(\frac{i}{\hbar} S_{ji} - i\frac{\pi}{2} \nu_{ji} \right) \, , \tag{7.97}$$

where $S_{ji} = S(q_j, q_i; E)$, and the sum is over the classical trajectories from q_i to q_j considered before. The expression for the transfer amplitude B_{ji} is given by

$$B_{ji} = \sqrt{\Delta_j \Delta_i} \, \frac{1}{(2\pi i\hbar)^{1/2}} \left| \frac{\partial^2 S_{ji}}{\partial q_j \partial q_i} \right|^{1/2} \, . \tag{7.98}$$

Discretization of the continuous coordinate variable q by a finite set of points spoils the unitarity of the T-matrix, and the energies for which the secular equation (7.94) is satisfied are no longer real. The choice of the number n of cells into which the PSS Σ is divided, (as well as the widths of the individual cells) is crucial for the accuracy of the eigenenergies of the system. For a rough criterion of the choice of Δ_i for the ith cell on the PSS in the coordinate space, Bogomolny used the uncertainty principle argument in the *phase space*. The latter also involves the momentum conjugate to q. The width Δ should be such that the corresponding phase space area is equal to Planck's constant h. By the uncertainty principle, quantum mechanics has this inherent fuzziness, so it may not be necessary to set the phase space area much smaller. An estimate for the overall number n of the cells in the coordinate space may also be made by a global counting of states in the phase space. Let the desired energy cut-off in the eigenvalue spectrum be E_{max}. Then the approximate number of states in the range

$(0 \leq E \leq E_{max})$ is given by the Thomas-Fermi estimate $A(E_{max})/h$, where $A(E_{max})$ is the accessible area of the phase space corresponding to one-space dimensional motion. The dimension n of the T-matrix should be at least bigger than A/h. This is far smaller than the size of the Hamiltonian matrix which has to be diagonalized in the direct quantum mechanical method of obtaining the same number of eigenstates. Just as in Eq. (7.71), $\det[\delta_{ij} - T_{ij}]$ may be identified with the Selberg zeta function, and it obeys the same functional equation (7.50). Thus one may find the real solutions of the equation analogous to Eq. (7.51):

$$D(E) = \Re e \left\{ \exp\left[-i\pi \tilde{N}(E) \right] \det\left(\delta_{ij} - T_{ij} \right) \right\} = 0. \qquad (7.99)$$

In the literature, $D(E)$ is often called the functional determinant.

This method of quantization is applicable to integrable as well as nonintegrable problems. The main advantage is that it does not involve a sum over periodic orbits, so one does not have to search for them, or face the problem of the convergence of their sum. Nevertheless, we should remember that Eq. (7.88) is a semiclassical approximation, and is expected to be very accurate only for highly excited states. Szeredi *et al.* [34] have calculated the eigenenergies of the disc and the wedge billiard using this method. First consider the integrable problem of the circle billiard whose exact eigenvalues with the Dirichlet boundary condition are given by Eq. (7.63). The first step is to choose a suitable PSS. This is taken to be the circular boundary itself. Divide the circumference into n equal cells, each of width $\Delta = 2\pi R/n$. There is only one classical trajectory between the central points q_i and q_j of the ith and j^{th} cells, this being the chord of the circle joining the two points. For a fixed energy E, the classical action for this trajectory is the length of the chord times the momentum $p = (2mE)^{1/2}$. In our notation, it is

$$S(q_i, q_j, E) = 2R(2mE)^{1/2}\sin\left(\frac{|q_i - q_j|}{2R} \right). \qquad (7.100)$$

From this, the (absolute value of the) second derivative $\left| \frac{\partial^2 S}{\partial q_i \partial q_j} \right|$ is easily obtained. The Maslov index ν in Eq. (7.88) here is always 2, corresponding to a single reflection on the boundary in going from the ith to the jth cell. There is no caustic along the trajectory. For constructing the T-matrix and its numerical diagonalization, we have to choose the matrix-dimension n. For the fixed energy E, the available area of the phase space associated with the perimetric PSS (between the momenta $-p$ and p) is $(2\pi R) \times 2(2mE)^{1/2}$, so n should at least be this area divided by Planck's constant h. In units of $\hbar = m = 1$, taking $E = 20$ and the radius $R = 2$, we get n to be 25 to the nearest integer. Because of truncation, the roots of the secular equation (7.94) are complex.

Fig. 7.12 shows the absolute value $|\det(\delta_{ij} - T_{ij})|$ for (a) $n = 25$ and (b) $n = 100$. These minima are compared with the zeros corresponding to the exact eigenvalues of the lowest 19 states. Not surprisingly, the naive estimate of $n = 25$ based on the (Thomas-Fermi) phase space consideration is an underestimate for these low-lying levels. Note that (a) fails to produce minima at the 17th and the 18th eigenvalues. The agreement is excellent for (b) even for these low-lying excited states.

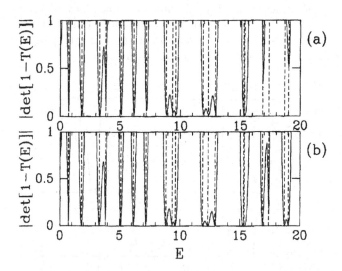

Figure 7.12: Plots of $|\det[1 - T(E)]|$ versus E for the circle billiard with $R = 2$. The dashed lines give the exact energy eigenvalues obtained from (7.63). (a) calculated with a 25×25 matrix, (b) calculated with a 100×100 matrix. (From [34].)

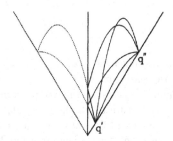

Figure 7.13: The four classical trajectories used by Szeredi *et al.* [34] to calculate the transfer operator for the wedge billiard.

Szeredi *et al.* [34] also applied Bogomolny's T-matrix method to the wedge billiard, defined in Eq. (5.58). The PSS in this case was taken to be along the tilted wall of the wedge, which comes in repeated contact with the bouncing trajectories of the particle. Denoting the distance along the wedge from the vertex by r, the classical turning point for a given energy E is $r_{max} = E/\cos\phi$, since the conjugate momentum is $p_r = [2m(E - r \cos\phi)]^{1/2}$. The distance r_{max}, as before, is divided into n cells, and the width of a cell with the central point at a distance r is chosen so as to keep the phase-space area $p_r \Delta_r$ a constant. Fig. 7.13 shows the four classical trajectories that take the particle from the initial point q' to the final point q'', for each of which the action $S(q'', q', E)$ is calculated. The calculation is straightforward, except that special care has to be taken when the final point coincides with a caustic along the trajectory. This difficulty is avoided in [34] by making use of the scaling property of the action S. The reader may look up the details in in the original paper.

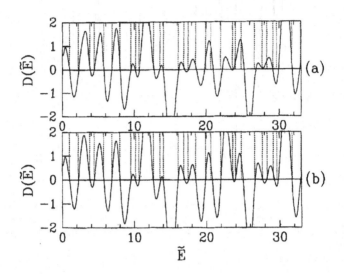

Figure 7.14: Plot of $D(\bar{E})$ as a function of the scaled energy $\bar{E}$ for the wedge billiard with 49°. (a) T-matrix of dimension 25; (b) T-matrix of dimension 150. (From [34].)

A typical result for the lowest eigenvalues of the 49° wedge using Eq. (7.99) is shown in Fig. 7.14. Similar calculations are done for the integrable problem at 45°, and for 41°, where chaotic and regular motion coexists. The T-matrix method is equally applicable in all these cases. A common feature of all these calculations is that the dimension n of the matrix has to be taken about $5 - 10$ times larger than the minimal estimate obtained by dividing the accessible phase space area by Planck's constant h. Using this larger dimensional matrix, the T-matrix method is superior to the "stair-case" quantization rule (5.57), and comparable in accuracy to the zeta function method of the earlier section. As already pointed out before, its main advantage is that one does not have to search for the periodic orbits of motion in the entire energy range of interest. In the T-matrix method, the details of the classical motion do not have to be resolved in an area less than the chosen size of the phase cell on the PSS.

7.5 Diffractive Corrections to the Trace Formula

7.5.1 Introduction

The Gutzwiller trace formula that was discussed at length in Chapters 5 and 6 expressed the semiclassical density of states in terms of classical periodic orbits. This connection between classical motion and quantum level density is gratifying, but of course does not tell the whole story. The trace formula included only *classical* trajectories (from one point in space to another) that are obtained by solving Newton's equations of motion. This is like using geometrical rays instead of waves for obtaining the propagator, and has some inherent limitations. To elaborate on this, recall that the semiclassical method uses the Gutzwiller-Van Vleck propagator from a point $\mathbf{r}'$ to $\mathbf{r}$ in a time interval t, which is given by Eq. (5.8)

$$K_{scl}(\mathbf{r}, \mathbf{r}'; t) = \sum_{class.paths} (2\pi i\hbar)^{-D/2} \sqrt{|\det C|} \exp\left[\frac{i}{\hbar} R(\mathbf{r}, \mathbf{r}', t) - i\kappa\pi/2\right]. \quad (7.101)$$

The notation has been defined in Chapter 5. The evolution of the wave function is then determined by the equation

$$\psi(\mathbf{r}, t) = \int d\mathbf{r}' \, K_{scl}(\mathbf{r}, \mathbf{r}'; t) \, \psi(\mathbf{r}', 0). \quad (7.102)$$

Note that the semiclassical propagator K_{scl} is determined by calculating Hamilton's principal function R over the allowed *classical* paths only. When two points may be connected by more than one classical path (say one by a direct, and the other by a reflected trajectory), interference effects in the propagator are automatically included by the summation over these paths in Eq. (7.101). In this respect, it is more than classical physics. However, diffraction effects are missed, because only classical mechanics is used to find the trajectories. In the exact path integral formulation of quantum mechanics, all trajectories including the nonclassical ones are included, even though the action is calculated along any trajectory using the classical formula. Quantum and classical mechanics parallel wave and geometrical optics. Although geometrical optics is generally satisfactory when the wave length of the wave is much smaller than the size of the object, it is still not adequate in a narrow boundary region between light and shadow of the obstacle. Geometrical optics predicts a discontinuous change in the intensity on the boundary of the shadow, whereas diffraction softens this edge due to the bending of the waves. Waves reach regions not reachable by geometric rays, as shown by the diffraction of a plane wave front by a half-plate in Fig. 7.15. By geometrical optics, there would be complete darkness on the other side of the plate, but not so due to diffraction. This may be understood by Huygens's principle, according to which a wave-front propagates by emanating secondary spherical wavelets from each point. The construction of such wavelets allows some light to bend around the obstacle, as shown in the figure. Similarly, when light is incident on a sphere or a disc, for example, diffraction allows it to bend and creep along the boundary [36]. Even when the wave length is very small compared to the radius, diffraction in the near-forward direction has a huge effect on the scattering cross section [37]. The geometrical ray picture of reflection is also inadequate in scattering from a discontinuity like a knife-edge or a sharp vertex [38]. One cannot use the rule of specular reflection at the discontinuity. Keller [39] developed a geometrical approach embodying diffraction in the form of creeping rays, and this was applied later to extend the semiclassical trace formula [40]. A good presentation of the trace formula in which periodic diffractive orbits are taken into account is given in [41]. Classical mechanics also fails to describe the scattering of a charged particle from a magnetic field that is confined in a solenoid, the so-called Bohm-Aharonov effect [42]. An extreme example of this is the scattering from a singular flux-line [43]. There have been many recent studies on all these topics in two-dimensions. These include wedges or corners [40, 44, 45]; point scatterers [46, 47], discs [48, 49], and scattering from a flux line [50].

Figure 7.15: Diffraction by a half-plane according to the Huygens-Fresnel principle. (After [35].)

In this section, we shall first give an elementary account of quantum scattering theory for a plane wave in two dimensions incident on a single scatterer. We shall apply this to a disc, and show that the scattering cross section for small angles is very different from its classical behavior even in the limit of the de Broglie wave length of the particle being much smaller than the radius of the disc. The narrow region between light and shadow is the so-called penumbra, or the half-shadow. Diffraction effects are particularly significant in this region due to rays that are almost tangent to the curved boundary. This has been investigated for the Sinai billiard which is classically chaotic [51], and for the circular annulus billiard which is integrable [52, 53] and analytically tractable. In this book, we shall discuss the simpler example of scattering from a hard disc to underscore the importance of diffractive effects. In the later parts of this section, we link the scattering amplitude to Keller's diffraction coefficient, and indicate how the semiclassical Green's function for a diffractive periodic orbit is calculated. This paves the way for the extended version of the trace formula taking account of diffractive corrections.

Before deriving the quantum results, let us first consider the scattering of a particle by a fixed (repulsive) potential using classical mechanics. Throughout this section, we restrict our considerations to two spatial dimensions. A particle is incident on a fixed scatterer with a central potential $V(r)$ of finite range. Its impact parameter, as shown in Fig. 7.16, is given by b, and its initial path, outside the field of force, is a straight line. As it approaches the scatterer, its path bends according to Newton's equation of motion, and finally it goes out of the force range in a straight line with the same energy as before at a scattering angle ϕ. The scatterer is infinitely heavy,

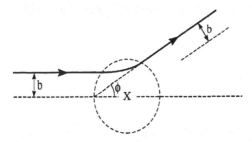

Figure 7.16: Classical potential scattering.

and experiences no recoil. The angular momentum of the particle about the origin remains the same before and after the scattering event. Its direction is perpendicular to the plane, and its magnitude is $m|\mathbf{v}|b$. Consider a line perpendicular to the incident direction. If N particles pass through a unit length of this line per second, then there are $N\,db$ incident particles between b and $b + db$ with angular momentum $m\mathbf{b}\times\mathbf{v}$. The same particles must be scattered between angles $|\phi|$ and $|\phi + d\phi|$, with the same angular momentum about the center. The number of scattered particles per second, by definition, is $N\sigma_k^{cl}(\phi)\,d\phi$, where $\sigma_k^{cl}(\phi)$ is the differential scattering cross section. (In general, of course, it is energy-dependent). From this, we obtain the classical result for planar scattering,

$$\sigma_k^{cl}(\phi) = \left|\frac{db}{d\phi}\right|. \tag{7.103}$$

Given the form of the potential $V(r)$, one has to find the right-hand side of the above equation for a given energy to obtain the classical result.

7.5.2 Quantum theory of scattering

We now derive the general expression for the elastic scattering cross section using quantum mechanics. Next we apply this to the case of hard-disc scattering, and compare it with the classical cross section obtained above. *This will make the point that the quantum mechanical result does not reduce to the classical particle picture even in the high frequency limit.* Using polar co-ordinates (r, ϕ), the incident plane wave of a well-defined momentum $\mathbf{p} = \hbar\mathbf{k}$ may be resolved into a sum of partial waves, each with angular momentum $\hbar l$ perpendicular to the plane:

$$\psi^{inc} = \exp\,(i\mathbf{k}\cdot\mathbf{r}) = \sum_{l=-\infty}^{\infty} i^l\,\exp(il\phi)\,J_l(kr)\,. \tag{7.104}$$

Consider the solution of the Schrödinger equation, as given by Eq. (7.66) for the partial wave with angular momentum l:

$$\psi_l(r, \phi) = A_l\left[H_l^-(kr) + S_l(k)H_l^+(kr)\right]\exp(il\phi)\,, \tag{7.105}$$

where A_l is a constant. To fix it, note that we may rewrite ψ^{inc} given by Eq. (7.104) in the form

$$\psi^{inc} = \frac{1}{2} \sum_l i^l \, e^{il\phi} \, (H_l^+ + H_l^-).$$ (7.106)

So, by choosing $A_l = \frac{1}{2}i^l$, we may express Eq. (7.105) for the partial wave ψ_l as

$$\psi_l = \psi_l^{inc} + \psi_l^{sct} ,$$ (7.107)

where the scattered wave ψ_l^{sct} is given by

$$\psi_l^{sct} = \frac{i^l}{2}(S_l - 1) \, H_l^+ \, e^{il\phi} .$$ (7.108)

Note that ψ_l^{sct} is an outgoing circular wave, as is evident from the the asymptotic expression for H_l^+ given in Eq. (7.64). Using this asymptotic form, and the identity $i^l = \exp(il\pi/2)$, we see that the scattered wave for the lth partial wave is

$$\psi_l^{sct} \simeq \sqrt{\frac{1}{2\pi kr}} \, (S_l - 1) \, e^{i(kr+l\phi-\pi/4)} .$$ (7.109)

Summing the contributions from all the partial waves, the total outgoing scattered wave is often written in the form

$$\psi^{sct} = f_k(\phi) \, \frac{e^{ikr}}{\sqrt{r}} ,$$ (7.110)

where the scattering amplitude $f_k(\phi)$ is

$$f_k(\phi) = \sqrt{\frac{1}{2i\pi k}} \sum_{l=-\infty}^{\infty} (S_l(k) - 1) \, e^{il\phi}$$ (7.111)

$$= \sqrt{\frac{1}{2i\pi k}} \sum_{l=-\infty}^{\infty} 2ie^{i\delta_l} \, sin\delta_l \, e^{il\phi}.$$ (7.112)

In the above, we have used the relation $S_l(k) = \exp(2i\delta_l(k))$, valid for elastic scattering by a central (angle-independent) potential. From Eq. (7.110) we see that the outgoing scattered flux, between ϕ and $\phi + d\phi$, from a large circle of radius R enclosing the scatterer at the origin is $|f_k(\phi)|^2 \, d\phi \, \hbar k/m$. The incident flux within the same angle, from Eq. (7.104), is simply $d\phi \, \hbar k/m$. Therefore the differential scattering cross section, the ratio of the two, is

$$\sigma_k(\phi) = |f_k(\phi)|^2 .$$ (7.113)

The total cross section is obtained by integrating the differential cross section over all angles ϕ.

$$\sigma_k = \int d\phi \, |f_k(\phi)|^2 = \sum_l \frac{4}{k} \, \sin^2 \delta_l(k) .$$ (7.114)

We note that the cross terms with $l \neq l'$ have dropped out on integration. Eqs. (7.104-7.114) are quite general. By taking the imaginary part of $[\exp(-i\pi/4) \, f_k]$ from equation

(7.112) in the forward direction $\phi = 0$, and using Eq. (7.114), we immediately obtain what is called the "Optical theorem":

$$\sigma_k = \sqrt{\frac{8\pi}{k}} \, \Im m \, \left[e^{-i\pi/4} f_k(0) \right].$$
(7.115)

The above relation between the scattering cross section and the forward scattering amplitude is exact.

7.5.3 Scattering by a hard disk

We first calculate the differential scattering cross section from purely classical physics, using the model that the hard disk is infinitely heavy (no recoil), and a particle incident on the disc with impact parameter b (see Fig. 7.17) is specularly reflected with energy conservation. We apply Eq. (7.103) for the classical differential cross section, noting from the diagram that the impact parameter $b = a \cos(\phi/2)$. We then obtain

$$\sigma_k^{cl}(\phi) = \frac{a}{2} \sin\frac{\phi}{2}, \quad -\pi \le \phi \le \pi.$$
(7.116)

For this simple case, the differential cross section is energy-independent, and is zero in the forward direction $\phi = 0$, since the particles cannot go round the disc by diffraction. The total classical cross section is obtained by integrating the differential cross section given by Eq. (7.116):

$$\sigma_k^{cl} = \int_{-\pi}^{\pi} \sigma_k^{cl}(\phi) \, d\phi = 2a.$$
(7.117)

This makes sense classically, since the incident particle is blocked across this length. The corresponding quantity for a sphere is πa^2.

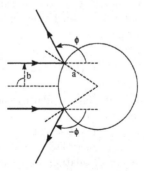

Figure 7.17: Classical hard-disk scattering. Note that for the trajectory in the upper half of the diagram, the classical angular momentum $\mathbf{L} = \mathbf{r} \times \mathbf{p}$ is directed into the plane of the paper by the right-hand rule. By convention, the sign of the angular momentum is then taken to be negative. For this case, the scattering angle, from the initial to the final direction of the ray, is counter clock-wise (positive). For the trajectory shown in the lower half, on the other hand, the angular momentum has a positive sign, and ϕ is negative.

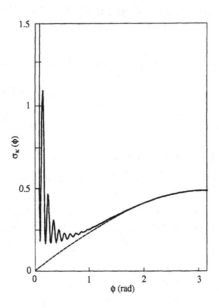

Figure 7.18:   The comparison between classical and quantum differential cross
section on a disc, for $ka = 30$. (Courtesy Nina Snaith.)

Now we apply Eq. (7.114) to the case of hard-disc scattering for the situation where
the incident de Broglie wave length is much smaller than the radius a of the disc, i.e.,
$ka \gg 1$. In this limit, we may have naively expected the classical result (7.117) to hold.
Using the expression (7.67) for $\mathcal{S}_l$, and the asymptotic form of the Hankel functions
given by Eq. (7.64), we get $\delta_l \simeq -(ka - l\pi/2 + \pi/4)$. Furthermore, in the geometrical
limit, the maximum l that can contribute to the sum in Eq. (7.114) is ka, since the
partial waves with $l > ka$ do not get scattered. The sum over l in Eq. (7.114) may be
semiclassically approximated by an integral:

$$\sigma_k = \frac{4}{k} \int_{-ka}^{ka} dl \, \sin^2(ka - l\pi/2 + \pi/4) \simeq 4a \,. \tag{7.118}$$

In the last line, we have dropped the fast oscillating contribution to the integral for
$ka \to \infty$. Eq. (7.118) shows that the cross section is *double* the classical value of $2a$,
so the quantum result does not go over to the classical limit even when the de Broglie
wave length is $\ll a$ [37]. Although Eq. (7.118) has been obtained in a non-rigorous
way, it is confirmed by a numerical calculation using the exact expression (7.114). It
is interesting to compute the quantum mechanical differential scattering cross section
using Eqs. (7.112) and (7.113) for $ka \gg 1$, and compare with the classical result $\sigma_k^{cl}(\phi)$
given by Eq. (7.116). Such a comparison is shown in Fig. 7.18. In this figure, we have
taken $a = 1$ (in arbitrary units), $ka = 30$, and the cut-off for l in the quantum sum
over partial waves is made at $l_{max} = 3\,ka$. We have checked that the convergence of
the l-sum is good, in the sense that there was no visual difference in the diagram when
l_{max} was lowered to $2\,ka$.

We note that the largest discrepancy between the classical and quantum results comes about in the forward direction, and for small-angle scattering. This is due to diffraction. It is worth elaborating by examining the quantum mechanical expression given by Eq. (7.111) in more detail, to find out how diffraction is a "built-in" feature of quantum mechanics. One important feature of Eq. (7.111) is the factor $(S_l(k) - 1)$. For $|l| > (ka)$, there is little scattering, and $S_l(k) \simeq 1$. This makes it reasonable to cut off the summation in Eq. (7.111) at $|l| = ka$, particularly for large ka, when the de Broglie wave length is small compared to the radius (see Problem 7.3). The next task is to examine, for large ka, the relative importance of the two terms in the expression

$$f_k(\phi) \simeq \sum_{l=-ka}^{ka} \sqrt{\frac{1}{2i\pi k}} \, (S_l(k) - 1) \, e^{il\phi}. \tag{7.119}$$

We shall first show that in the illuminated region, for large ϕ, it is the first sum involving $S_l(k)$ that dominates, and its magnitude is the same as the classical value. On the other hand, for grazing angles $\phi \to 0$, both the terms in Eq. (7.119) are important, and should be carefully calculated to obtain interference effects in the cross section. By itself, the truncated l-sum over the second term gives rise to a forward diffraction pattern that yields the right $f_k(0)$ satisfying the optical theorem (7.115); but this pattern is very substantially modified by $S_l(k)$ even for a very small ϕ. We shall now perform a simple semiclassical calculation to demonstrate these points.

We start by evaluating the first sum in Eq. (7.119) by making an approximation which holds when the scattering angle $\phi \gg (ka)^{-1/3}$. We shall show that in the limit of $(ka) \to \infty$, this yields the classical cross section. The second sum with -1 gives a delta function peak at $\phi = 0$ in this limit, and *is omitted* in what follows immediately. We shall come back to it presently. Recall that for hard-disc scattering, the scattering matrix in a partial wave l is given by

$$S_l(k) = - \left(\frac{H_l^-(ka)}{H_l^+(ka)} \right). \tag{7.120}$$

To obtain the sum over l in Eq. (7.111), we first use the Poisson sum formula [cf. Eq. (7.28)]:

$$\sum_{l=-ka}^{ka} f(l) \simeq \sum_{M=-\infty}^{\infty} \int_{-ka}^{ka} f(m) \, \exp(2\pi i M m) \, dm \ . \tag{7.121}$$

Omitting the sum involving -1 in Eq. (7.119), the scattering amplitude $f_k(\phi)$ then takes the form

$$f_k(\phi) = \sqrt{\frac{1}{2i\pi k}} \sum_{M=-\infty}^{\infty} \int_{-ka}^{ka} dm \, S_m(k) \, e^{im(\phi + 2\pi M)}. \tag{7.122}$$

We rewrite this as

$$f_k(\phi) = - \sqrt{\frac{1}{2i\pi k}} \left(\sum_{M=-\infty}^{\infty} I_M \right), \tag{7.123}$$

where I_M is defined as

$$I_M = \int_{-ka}^{ka} dm \, \frac{H_m^-(ka)}{H_m^+(ka)} \exp\left[im(\phi + 2\pi M) \right]. \tag{7.124}$$

The dominant $M = 0$ term I_0 in (7.124) may be evaluated directly along the real axis using the saddle-point approximation. Here we evaluate only this term,

$$I_0(\phi) = -\int_{-ka}^{ka} dm \, S_m(k) \exp \, (im\phi) \,. \tag{7.125}$$

We now examine the illuminated region, which may be reached directly by geometrical rays through specular reflection. We use the Debye approximation [54] in this case, which is an asymptotic expansion for large m, and $ka > m$. The leading term of this series is given by:

$$H_m^\pm(x) \simeq \sqrt{\frac{2}{\pi}} \, \frac{\exp\left[\pm i \left(\sqrt{x^2 - m^2} - m \cos^{-1}\frac{m}{x} - \frac{\pi}{4}\right)\right]}{(x^2 - m^2)^{1/4}} \,. \tag{7.126}$$

In the above, the angle $\cos^{-1}\frac{m}{x}$ is between 0 and $\pi/2$, so m is positive. This expansion for $H_m^\pm(ka)$ is valid only when $|m - ka| > (ka)^{1/3}$. Since the angular momentum m (in units of $\hbar$) is $ka \cos(\phi/2)$, the above approximation is valid when $2(ka)\sin^2(\phi/4) > (ka)^{1/3}$. For small angles ϕ, this means that it is safe to use Eq. (7.126) when

$$\phi \gg (ka)^{-1/3} \,. \tag{7.127}$$

The Debye approximation fails for small angles and grazing incidence, except in the limit of $(ka) \to \infty$, when it fails only in the forward direction. Under the above restriction (with $m > 0$), we may use Eq. (7.126) to calculate the scattering matrix $S_m(k) = -H_m^-(ka)/H_m^+(ka)$. This yields, on absorbing the minus sign as $\exp(-i\pi)$, the WKB result (cf. Eq. (2.153)):

$$S_m(k) \simeq \exp\left[-2i\left(\sqrt{(ka)^2 - m^2} - m \cos^{-1}\frac{m}{ka} + \frac{\pi}{4}\right)\right] \,. \tag{7.128}$$

We now substitute this in Eq. (7.125) to obtain for the illuminated region (7.127)

$$I_0(\phi) = -\int_{-ka}^{ka} \exp\left[iF(k, m)\right] \,, \tag{7.129}$$

where

$$F(k, m) = -2\sqrt{((ka)^2 - m^2)} + 2m \cos^{-1}\left(\frac{m}{ka}\right) - \frac{\pi}{2} + m\phi \,. \tag{7.130}$$

The m-integration is done by the stationary phase method. The stationary point at $m = l$ is determined by

$$\left(\frac{\partial F}{\partial m}\right)_{m=l} = \frac{2l}{\sqrt{(ka)^2 - l^2}} + 2 \cos^{-1}\frac{l}{ka} - \frac{2l}{\sqrt{(ka)^2 - l^2}} + \phi = 0 \,. \tag{7.131}$$

This yields the saddle point solution

$$\frac{\phi}{2} = -\cos^{-1}\frac{l}{ka} \,. \tag{7.132}$$

The principal value of the inverse cosine is in the range 0 and π. Thus, when l is positive, $0 \leq \cos^{-1}(l/ka) \leq \pi/2$, which implies that $-\pi \leq \phi \leq 0$, so ϕ is negative. Similarly, for negative l, ϕ is positive. These results are consistent with the signs in the two cases as illustrated in Fig. 7.17 for classical scattering. Moreover, the phase factor $iF(k, l)$ may be taken out of the integrand, where

$$F(k, l) = -2\sqrt{(ka)^2 - l^2} - \pi/2 = -2ka \sin\frac{\phi}{2} - \pi/2. \qquad (7.133)$$

Retaining up to the quadratic term in the expansion of $F(k, m)$ about $m = l$, we now obtain

$$I_0(\phi) \simeq -\exp\left(-2ika \sin\frac{\phi}{2} - i\pi/2\right) \int_{-ka}^{ka} dm \, \exp\left[-\frac{i(m-l)^2}{ka \sin\frac{\phi}{2}}\right].$$

On taking the limit $ka \to \infty$, the Fresnel integral is easily done. Using Eq. (7.123), we obtain the expression for the scattering amplitude as (still neglecting the $M \neq 0$ terms):

$$f_k(\phi) \simeq \sqrt{\frac{1}{2\pi k}} \, \exp\left(-2ika \sin\frac{\phi}{2}\right) \sqrt{\left(\pi ka \sin\frac{\phi}{2}\right)}. \qquad (7.134)$$

Note that this yields, for the illuminated region, a differential scattering cross section the same as the classical result (7.116). We see from Fig. 7.18 that this agrees with the quantum result for $\phi > 1$ rad, which is about $3 \times (ka)^{-1/3}$ for the case under consideration.

For smaller angles, $S_l(k)$ must be evaluated more carefully. Also, the sum involving -1 in Eq. (7.119) by itself gives a contribution

$$f_k^{(d)}(\phi) = -\sqrt{\frac{1}{2i\pi k}} \, 2ka \left[\frac{\sin(ka\phi)}{ka\phi}\right]. \qquad (7.135)$$

This has the correct value at $\phi = 0$, and satisfies the optical theorem, Eq. (7.115). The numerator in this term is a combination of $\exp(ika\phi)$, and $\exp(-ika\phi)$, a structure that persists, but with modified coefficients when S_l is also included for the smaller angles.

For large (ka) but $\phi \sim (ka)^{-1/3}$ or smaller, a different approximation for the Hankel function is appropriate [55]. For large $|l|$, the leading term in an asymptotic expansion [54] is given by

$$H_l^{\pm}\left(l + x\left(\frac{l}{2}\right)^{1/3}\right) \simeq \left(\frac{2}{l}\right)^{1/3} [Ai(-x) \mp i \, Bi(-x)] + O(l^{-1}), \qquad (7.136)$$

where x is a parameter which is of order 1. For grazing incidence, $l \simeq ka$ (we take positive l for simplicity; appropriate changes for negative l are easily made), and we may replace the argument of the Hankel function in Eq. (7.136) by $(l + x(\frac{ka}{2})^{1/3})$. For a fixed (ka), we may use the variable x instead of l by defining the relation

$$ka = \left[l + x\left(\frac{ka}{2}\right)^{1/3}\right]. \qquad (7.137)$$

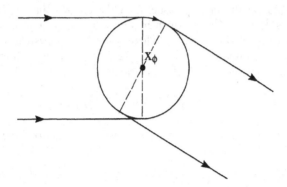

Figure 7.19: Two trajectories with opposite phases, $(ka\phi)$ (the upper creeping ray), and $-(ka\phi)$ (the lower diffracted ray), shown for grazing incidence.

The scattering matrix for a given l is now given by

$$S_l(k) \simeq -\frac{Ai(-x) + i\ Bi(-x)}{Ai(-x) - i\ Bi(-x)} = \exp[\,2i\delta(x)\,]. \tag{7.138}$$

In the above, $\delta(x)$ is the scattering phase shift suffered by the particle for a fixed k as a function of the new variable x that specifies the angular momentum l. In classical specular reflection, the scattering angle ϕ was completely determined by the relation $l = (ka)\cos\phi/2$. This is not the case for diffraction, where one has to span all l (or x) for a given ϕ. In the following, it is more convenient to parameterize ϕ by the relation

$$\phi = \left(\frac{2}{ka}\right)^{1/3}\alpha\,, \tag{7.139}$$

where the real parameter $\alpha > 0$, and we are mostly interested in the vicinity of α around unity. We now consider Eq. (7.111) for the scattering amplitude:

$$f_k(\phi) = \sqrt{\frac{1}{2i\pi k}}\ \sum_{l=-\infty}^{\infty} (S_l(k) - 1)\ e^{il\phi}\,.$$

We may approximate the l-sum by an integral. Further Eqs.(7.137) and (7.139) may be used to note

$$l\phi = ka\phi - x\alpha\,. \tag{7.140}$$

Using these relations, the scattering amplitude may be expressed in an instructive way:

$$f_k(\phi) = \sqrt{\frac{1}{2i\pi k}}\left(\frac{ka}{2}\right)^{1/3}\left[e^{ika\phi}\int_{-\infty}^{\infty}dx\,e^{-i\alpha x}(e^{2i\delta(x)} - 1) + e^{-ika\phi}\int_{-\infty}^{\infty}dx\,e^{i\alpha x}(e^{2i\delta(x)} - 1)\right].$$
$$\tag{7.141}$$

Like Eq. (7.135), this is the superposition of two terms with path lengths $ka\phi$ and $-ka\phi$, but unequal coefficients. The geometrical interpretation of this is shown in Fig. 7.19 for two interfering trajectories in the direction ϕ on either side of the disc.

It is clear from this diagram that whereas the upper ray creeps a path $ka\phi$ along the arc of the disc, the lower one is diffracted at the same angle with a path length shorter by the same amount. For larger angles the creeping ray is rapidly damped (shadow region), and one gets back the result of the illuminated region arising from the classical trajectory. It is clear from Eq. (7.137) that for negative x, $|l| > (ka)$, and therefore the phase shift is nearly zero (see Problem 7.3). In Eq. (7.141), the function $\exp(2i\delta(x)) \to 1$ rapidly for negative x, so the integrand contributes little on this side.

Most of the contribution to the integrals in Eq. (7.137) come from the positive side of x, *close* to $x = 0$. The integrals were performed numerically, and the result is displayed in Fig. 7.20.

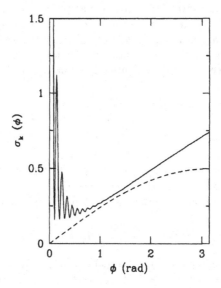

Figure 7.20: Plot of the differential scattering cross section versus ϕ, calculated by taking $|f_k(\phi)|^2$, where $f_k(\phi)$ is computed from Eq. (7.141). The same scale as in Fig. 7.18 is chosen, and $(ka) = 30$. (Courtesy Diptiman Sen.)

In the above calculation, special care has to be taken for very small angle $\phi \sim (ka)^{-1}$, when $\alpha \ll 1$ for large (ka). In this situation, one recovers the result (7.135) by taking the appropriate limit of the phase shift $\delta(x)$ in Eq. (7.141). A comparison of Fig. 7.20 with the quantum mechanical behavior of the cross section in Fig. 7.18 shows excellent agreement for scattering angle $\phi \sim (ka)^{-1/3}$, and even larger angles. It fails for much larger angles because the approximation (7.136) for the Hankel function breaks down in this range. This is the classical illuminated region, where the Debye approximation is appropriate. We shall not attempt to improve the description in the delicate range of overlap of these two.

We next derive the relationship of the Green's function to the scattering amplitude $f_k(\phi)$. This enables us to examine how diffraction effects may be included in the semiclassical Green's function. One advantage of using the Green's function formalism is that Gutzwiller's trace formula may then be extended to include such effects. This will be discussed shortly.

7.5.4   The scattering amplitude and the Green's function

The Green's function $G(\mathbf{r}, \mathbf{r}'; E)$ was defined by Eq. (3.29) as the Fourier transform of the propagator for $\mathbf{r}' \to \mathbf{r}$ in a time interval t, $K(\mathbf{r}', \mathbf{r}; t)$. The semiclassical Gutzwiller-Van Vleck form of the Green's function is the foundation of the trace formula, and is given by Eq. (5.15). For the scattering problem on the disc, or in the complementary problem of the inside, the semiclassical Green's function is particularly simple. We first make the connection between the scattering amplitude defined by Eq. (7.111) and the diffraction coefficient of Keller using the Green's function language [49]. This will enable us to include diffraction effects in the semiclassical trace formula in a concise manner ([41]). Consider first the free quantum mechanical Green's function $G_0(\mathbf{r}, \mathbf{r}'; E)$ in two-dimensional space. Using the definition given in Eq. (3.29), we see that it obeys the differential equation

$$\left(E + \frac{\hbar^2}{2m} \nabla_{\mathbf{r}}^2 \right) G_0(\mathbf{r}, \mathbf{r}'; E) = \delta(\mathbf{r} - \mathbf{r}') . \tag{7.142}$$

Of course, the appropriate boundary condition has to be put in: for outgoing or incoming asymptotic solutions, the exponential factor in $G_0(\mathbf{r}, \mathbf{r}'; E)$ goes like $\exp(\pm ik|\mathbf{r} - \mathbf{r}'|)$, whereas for the inside solution at the eigenvalues $E = E_n$ of a billiard, $G_0(\mathbf{r}, \mathbf{r}'; E)$ should vanish if $\mathbf{r}$ or $\mathbf{r}'$ is on the surface of the billiard. The scattering solutions are known to be

$$G_0^{\pm}(\mathbf{r}, \mathbf{r}'; E) = -\frac{2m}{\hbar^2} \frac{i}{4} H_0^{\pm}(k|\mathbf{r} - \mathbf{r}'|) . \tag{7.143}$$

In the following, we shall consider only $G_0^+(x)$, where $x = (k|\mathbf{r} - \mathbf{r}'|)$, and drop the superscript on G_0^+. Note that as $x \to 0$, $H_0^+(x) \to \frac{2i}{\pi} \ln x$. Using the identity $\nabla^2 \ln x = 2\pi\delta(\mathbf{x})$, we see that Eq. (7.142) is satisfied. Moreover, from the asymptotic form of the Hankel function given by Eq. (7.64), the semiclassical G_0 may be written as [40]

$$G_0^{scl}(\mathbf{r}, \mathbf{r}'; E) = -\frac{2m}{\hbar^2} \frac{i}{4} \left(\frac{2}{\pi k|\mathbf{r} - \mathbf{r}'|} \right)^{1/2} \exp\left[ik|\mathbf{r} - \mathbf{r}'| - i\pi/4 \right] . \tag{7.144}$$

This semiclassical Green's function does not exactly satisfy Eq. (7.142), and terms of the order of $\hbar^0$ remain on the right-hand side of the equation even when $\mathbf{r} \neq \mathbf{r}'$. The semiclassical approximation for the Green's function inside a billiard with the Dirichlet boundary condition is of the same form as (cf. Eq. (5.15)) above [45]:

$$G^{scl}(\mathbf{r}, \mathbf{r}'; E) = -\frac{2m}{\hbar^2} \frac{i}{4} \left(\frac{2}{i\pi kL} \right)^{1/2} \sum_{\mathbf{r}' \to \mathbf{r}} \exp\left[ikL - i\mu\pi/2 \right] . \tag{7.145}$$

The sum above is taken over all classical trajectories going from $\mathbf{r}'$ to $\mathbf{r}$; L is the length of the trajectory, and μ the number of conjugate points. For $\mu = 0$, this reduces exactly to Eq. (7.144).

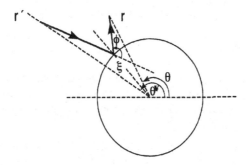

Figure 7.21: Scattering from a hard disk, showing the relation between $(\theta - \theta')$ and ϕ.

Consider a free particle propagating from a point $\mathbf{r}' = (r', \theta')$ to another point $\mathbf{r} = (r, \theta)$, with respect to a fixed origin, as shown in Fig. 7.21. Note that from now on, we denote the polar angle by θ, reserving the symbol ϕ for the scattering angle between the incident and the scattered ray. To make the connection with scattering amplitude, we first make a partial wave decomposition of the Green's function in Eq. (7.142) by writing

$$G_0(\mathbf{r}, \mathbf{r}'; E) = \frac{2m}{\hbar^2} \sum_{l=-\infty}^{\infty} g_l(r, r') \frac{\exp\left[il(\theta - \theta')\right]}{2\pi}. \tag{7.146}$$

A similar decomposition can be done for the Dirac delta function on the right-hand side of Eq. (7.142):

$$\delta(\mathbf{r} - \mathbf{r}') = \frac{1}{r} \delta(r - r') \sum_{l=-\infty}^{\infty} \frac{\exp[il(\theta - \theta')]}{2\pi}. \tag{7.147}$$

Substituting these two expansions in Eq. (7.142), and integrating over the relative angle $(\theta - \theta')$, we obtain the differential equation obeyed by $g_l(r, r')$. For $r \neq r'$, this is just Bessel's equation in the variable r. Therefore $g_l(r, r')$ may be written as a linear combination of the regular and the irregular solutions, and the free Green's function is of the form

$$G_0(\mathbf{r}, \mathbf{r}'; E) = \frac{2m}{\hbar^2} \sum_{l=-\infty}^{\infty} [A(kr')J_l(kr) + B(kr')N_l(kr)] \frac{\exp\left[il(\theta - \theta')\right]}{2\pi}. \tag{7.148}$$

We may determine the dimensionless constants A and B by substituting on the left-hand side the exact solution from Eq. (7.143), and setting $r = 0$, $r' > r$. The coefficient B of the irregular solution must be zero, and $A(kr') = -2\pi \frac{i}{4} H_0^+(kr')$. We thus obtain the partial wave expansion relating to the free Green's function $(r' > r)$:

$$H_0^+(k|\mathbf{r} - \mathbf{r}'|) = \sum_{l=-\infty}^{\infty} H_0^+(kr') \, J_l(kr) \, \exp\left[il(\theta - \theta')\right]. \tag{7.149}$$

More generally, this expression is written as

$$G_0(\mathbf{r},\mathbf{r}',k) = -\frac{2m}{\hbar^2}\frac{i}{8}H_0^+(k|\mathbf{r}-\mathbf{r}'|) = -\frac{2m}{\hbar^2}\frac{i}{4}\sum_{l=-\infty}^{\infty}H_0^+(kr_>)\,J_l(kr_<)\,\exp\left[il(\theta-\theta')\right].$$

$$(7.150)$$

To make a connection with our earlier result on scattering, we place a disc of radius a at the origin in Fig. 7.21. Using the same method as above, and applying the Dirichlet boundary condition, the Green's function $G(\mathbf{r},\mathbf{r}';E)$ for the scattered wave from $\mathbf{r}' \to \mathbf{r}$ is given by (see Problem 7.4)

$$G(\mathbf{r},\mathbf{r}';E) = -\frac{2m}{\hbar^2}\frac{i}{8}\sum_{l=-\infty}^{\infty}\left(H_l^-(kr) + \mathcal{S}_l(ka)H_l^+(kr)\right)H_l^+(kr')\,\exp[il(\theta-\theta')], \quad (7.151)$$

where we have assumed $r' > r$. We now add and subtract the free Green's function given by Eq. (7.143), and use the expansion (7.149), together with the identity that $J_l(kr) = [H_l^+(kr) + H_l^-(kr)]/2$. We then obtain the relation

$$G(\mathbf{r},\mathbf{r}';E) = G_0(\mathbf{r},\mathbf{r}';E) - \frac{2m}{\hbar^2}\frac{i}{8}\sum_{l=-\infty}^{\infty}(\mathcal{S}_l(k)-1)\,H_l^+(kr')\,H_l^+(kr)\,\exp[il(\theta-\theta')].$$

$$(7.152)$$

To relate to the total scattering amplitude $f_k(\phi)$ directly, we now have to take the far-field pattern, i.e., assume that both $r',r \gg a$. This is why Rosenqvist *et al.* [49] call it the small disc problem. Referring to Fig. 7.21, we see that in this limit, $(\theta-\theta') \simeq -(\pi-\phi)$, where ϕ is the scattering angle defined in the previous section. In this limit, we use the asymptotic relation

$$H_l^+(kr) \simeq H_0^+(kr)\,e^{-il\pi/2}. \quad (7.153)$$

We then obtain

$$G(\mathbf{r},\mathbf{r}';E) = G_0(\mathbf{r},\mathbf{r}';E) - \frac{2m}{\hbar^2}\frac{i}{8}H_0^+(kr')\,H_0^+(kr)\sum_{l=-\infty}^{\infty}(\mathcal{S}_l(k)-1)\,\exp[il(\phi+2\pi)].$$

$$(7.154)$$

Using the relation (7.143) between the free Green's function and H_0^+, and Eq. (7.111), we obtain the desired form

$$G(\mathbf{r},\mathbf{r}';E) = G_0(\mathbf{r},\mathbf{r}';E) + G_d(\mathbf{r},\mathbf{r}';E), \quad (7.155)$$

where the diffractive Green's function $G_d(\mathbf{r},\mathbf{r}';E)$ for one scattering center at ξ is given by (assuming $|\mathbf{r}-\xi| \simeq r$ and $r \gg a$)

$$G_d(\mathbf{r},\mathbf{r}';E) = \frac{\hbar^2}{2m}\,G_0(\mathbf{r},\xi;E)\,d(\phi)\,G_0(\xi,\mathbf{r}';E). \quad (7.156)$$

The dimensionless *diffraction coefficient* $d(\phi)$ is given in terms of $f_k(\phi)$:

$$d(\phi) = (2i)^{3/2}\sqrt{\pi k}\,f_k(\phi). \quad (7.157)$$

The physical interpretation of Eq. (7.156) is originally due to Keller [39], and is easily understood by referring to Fig. 7.21. The wave travels freely from $\mathbf{r}'$ to the scattering point ξ, where it gets diffracted, whose sole effect is to multiply the source at the point by the diffraction coefficient $d(\phi)$. The wave travels freely again from the point ξ to $\mathbf{r}$.

The equations (7.156) and (7.157), although derived in the context of the disc, are quite general, and may be applied to any scatterer if the diffractive coefficient is known. A well-studied example is the vertex of a wedge [56], diffraction from which was first studied by Sommerfeld [38]. We shall use this example to illustrate the modification in the density of states due to diffractive orbits in a polygonal billiard in the next subsection. Its diffractive coefficient is given by (see Fig. 7.22)

$$d(\theta - \theta') = -2\frac{\sin(\pi/\nu)}{\nu}\left(\frac{1}{\cos(\pi/\nu) - \cos([\theta - \theta']/\nu)} \pm \frac{1}{\cos(\pi/\nu) + \cos([\theta + \theta']/\nu)}\right),$$
(7.158)

where the angles θ and θ' are shown in the figure, and the wedge angle α is related to ν by the relation $\nu = \alpha/\pi$. In Eq. (7.158), the $\pm$ signs refer to Neumann and Dirichlet boundary conditions. The diffraction coefficient d from a wedge is identically zero if the wedge angle $\alpha = \pi/n$, where n is an integer. We next use Eq. (7.156) to generalize the trace formula, and apply the diffraction coefficient (7.157) to an irregular triangular billiard.

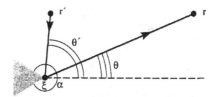

Figure 7.22: Scattering by a wedge. A path connecting $\mathbf{x}'$ to $\mathbf{x}$ via the wedge at point ξ. (After [41].)

7.5.5 Modification to the trace formula

We are now in a position to work out the modification to the trace formula due to "diffractive periodic orbits". The idea is take account of diffraction at isolated scattering centers *locally*, but otherwise describe the trajectories in between the scatterers by the free propagator. This is following the geometric theory of diffraction developed by Keller [39] in wave optics. The derivation of the trace formula may be illustrated [41] by the diagram shown in Fig. 7.23, in which there are three scattering centers placed at points ξ_1, ξ_2, ξ_3 on a plane. A diffractive periodic orbit due to successive diffraction of the wave at the scattering centers is shown. The diffractive Green's function $G_d(\mathbf{r}, \mathbf{r}'; E)$ from a point $\mathbf{r}'$ to a different point $\mathbf{r}$ with diffractions at the wedges ξ_2, ξ_3, and ξ_1, as shown in Fig. 7.23, may be expressed as a straight-forward generalization of Eq. (7.156):

$$G_d(\mathbf{r}, \mathbf{r}'; E) = \left(\frac{\hbar^2}{2m}\right)^3 G_0(\mathbf{r}, \xi_1)\, d_1\, G_0(\xi_1, \xi_3)\, d_3\, G_0(\xi_3, \xi_2)\, d_2\, G_0(\xi_2, \mathbf{r}').$$
(7.159)

In the above equation, the energy E is the same in every Green's function, and has been suppressed for brevity. To obtain its contribution to the level-density, we take the trace of the above Green's function, using the local coordinates (y, z) at each point $\mathbf{r}$ as shown in Fig. 7.23. The coordinate y is transverse to the trajectory of the ray, and the coordinate z is along the path. The trace is formally given by

$$\operatorname{tr} G_d = \int dy\, dz\, G_d(\mathbf{r}, \mathbf{r}; E),$$

$$= \left(\frac{\hbar^2}{2m}\right)^3 d_1 d_2 d_3 \int dy\, dz\, G_0(\mathbf{r}, \xi_1)\, G_0(\xi_1, \xi_3)\, G_0(\xi_3, \xi_2)\, G_0(\xi_2, \mathbf{r}). \quad (7.160)$$

We can place the diffraction coefficients d_i outside the integral sign because these are only dependent on the local coordinates ξ_i at the scattering centers. In the above trace integral, only two of the Green's functions (the first and the last) depend on the integration variable $\mathbf{r}$. Rearranging their order, we therefore have to evaluate the integral

$$I(\xi_2, \xi_1) = \int dy\, dz\, G_0(\xi_2, \mathbf{r})\, G_0(\mathbf{r}, \xi_1). \quad (7.161)$$

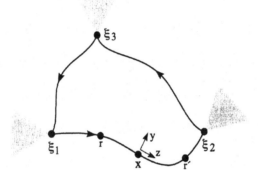

Figure 7.23: A periodic diffractive orbit with three scattering centers. (After [41].)

For a billiard, the semiclassical Green's function (7.145) for G_0 may be taken, but more generally the Gutzwiller form (5.15) derived in Chapter 5 is appropriate. At each point $(\mathbf{r})$ on the trajectory, the transverse y-integral is evaluated by the stationary phase method, while the integration along the trajectory is simply the time-interval T_{21} taken by the particle to go from ξ_1 to ξ_2. Instead of giving the mathematical details that may be found in [41], we mimic the result with the one-dimensional Green's function (3.31) found in Chapter 3. Recall that this is given by

$$G_0(x, x'; E) = \frac{-i\, m}{\hbar^2 k}\, \exp\left(-ik|x - x'|\right). \quad (7.162)$$

Consider the analogue of the trace integral (7.161) for a straight-line propagation $\xi_1 \rightarrow x \rightarrow \xi_2$:

$$I_1(\xi_2, \xi_1) = \int_{\xi_1}^{\xi_2} dx\, G_0(\xi_2, x)\, G_0(x, \xi_1). \quad (7.163)$$

Substituting from Eq. (7.162), we immediately get

$$I_1(\xi_2, \xi_1) = \left(\frac{-i}{\hbar}\right)^2 \left(\frac{m}{\hbar k}\right)^2 (\xi_2 - \xi_1) \exp[-ik(\xi_2 - \xi_1)]. \tag{7.164}$$

Noting that the time of transit from ξ_1 to ξ_2 is $T_{21} = m(\xi_2 - \xi_1)/\hbar k$, and Eq. (7.162), we obtain the result

$$\int_{\xi_2}^{\xi_1} dx\, G_0(\xi_2, x)\, G_0(x, \xi_1) = \left(\frac{T_{21}}{i\hbar}\right) G_0(\xi_2, \xi_1). \tag{7.165}$$

This is exactly the result that one would have obtained by doing the two-dimensional integral (7.161). Note that we started this calculation by placing the point **r** on the path between ξ_1 and ξ_2, but we could also perform a similar calculation from the other two arcs. Adding these contributions, the trace (7.160) of the Green's function for a particular diffractive orbit labeled by γ is given by

$$\operatorname{tr} G_d = \frac{T_\gamma}{i\hbar} \left(\frac{\hbar^2}{2m}\right)^3 d_1 d_2 d_3\, G_0(\xi_1, \xi_3)\, G_0(\xi_3, \xi_2)\, G_0(\xi_2, \xi_1). \tag{7.166}$$

We may use the semiclassical form of the Green's function G_0 given by Eq. (7.145), which could be rewritten as

$$G_0(L) = \frac{2m}{\hbar^2} \left(\frac{1}{8\pi k L}\right)^{1/2} \exp\left[i\left(kL - \mu\frac{\pi}{2} - \frac{3\pi}{4}\right)\right]. \tag{7.167}$$

Collecting the contributions, Eq. (7.166) for the periodic orbit γ of length L_γ and Maslov index μ_γ is given by

$$\operatorname{tr} G_d = \frac{T_\gamma}{i\hbar} \left(\prod_i \frac{d_i}{\sqrt{8\pi k L_i}}\right) \exp\left[i\left(kL_\gamma - \mu_\gamma\frac{\pi}{2} - \frac{3 \times 3\pi}{4}\right)\right]. \tag{7.168}$$

In the above, L_i is the length of the ith arm of the diffractive orbit labeled by γ, and $\sum_i L_i = L_\gamma$. For this particular case, we have assumed that the orbit is not repetitive and has three arms, as shown in the figure. The formula above may, however, be generalized to include repetitions of the primitive orbits as follows. Let us still denote by γ the primitive orbits only, and by T_γ, L_γ the period and the length of the *primitive* orbit. This primitive diffractive orbit suffers diffraction at n_γ successive scattering centers with coefficients $d_{\gamma,i}$, having arm lengths $L_{\gamma,i}$ between them. It may be shown that summing over all the primitive diffractive orbits γ, and their repetitions p, yields the result

$$\operatorname{tr} G_d = \sum_\gamma \frac{T_\gamma}{i\hbar} \sum_{p=1}^{\infty} \left(\prod_{i=1}^{pn_\gamma} \frac{d_{\gamma,i}}{\sqrt{8\pi k L_{\gamma,i}}}\right) \exp\left[ip\left(kL_\gamma - \mu_\gamma\frac{\pi}{2} - n_\gamma\frac{3\pi}{4}\right)\right]. \tag{7.169}$$

In the above equation, μ_γ is the Maslov index of the primitive orbit labeled by γ, and $\sum_{i=1}^{n_\gamma} L_i = L_\gamma$. This is very similar to the Gutzwiller trace formula (5.15) derived for Newtonian orbits earlier. We see that the diffractive contribution from each scattering center gives an additional factor of $(1/\sqrt{k})$ to the trace over the periodic orbit, compared to what is obtained from the classical orbit.

In the following, we shall only take the $p = 1$ primitive contributions into account. The sum over the repetitive orbits may be taken into account by the zeta function technique (see Problem 7.5). The oscillating component of the density of states $g(E)$ is simply the imaginary part of the trace of the Green's function, multiplied by $-\frac{1}{\pi}$ (see Eq. (3.36)). Considering only the the diffractive contribution (7.169). we get

$$\delta g_d(E) = \frac{1}{\pi\hbar} \sum_\gamma T_\gamma \left(\frac{\hbar}{\sqrt{2mE}}\right)^{n_\gamma/2} \left(\prod_{i=1}^{n_\gamma} \frac{d_{\gamma,i}}{\sqrt{8\pi L_i}}\right) \cos\left(kL_\gamma - \mu_\gamma \frac{\pi}{2} - \frac{3n_\gamma\pi}{4}\right). \quad (7.170)$$

The factor of $\sqrt{k}$ in the denominator of Eq. (7.170) for each diffractive center results in an additional $\sqrt{\hbar}$ in the numerator of the level density. Pavloff and Schmit [45] applied this formula, together with the wedge diffraction coefficient (7.158) to study the level density in a triangular billiard with angles $(\pi/4, \pi/6, 7\pi/12)$. Note that diffractive effects are nonzero only from the vertex with the angle $7\pi/12$.

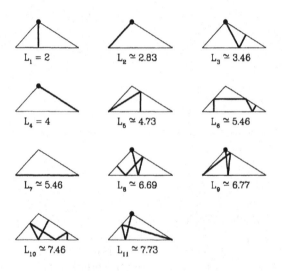

Figure 7.24: The shortest periodic orbits, classical as well as diffractive, are shown in the triangle with angles specified in the text. The diffractive vertex is marked with a black dot. The length of each orbit is given in units of the height of the triangle. Orbits 6 and 10 form families, while 5 and 7 are isolated. (After [45].)

The dominant effect is found by including periodic orbits with only one and two scatterings from this vertex, since each such encounter suppresses the contribution by $\sqrt{\hbar}$. In Fig. 7.24, the shortest 11 periodic orbits are shown that were included in the calculation of the semiclassical $\delta g(E)$. Not all of them are diffractive, but all are self-retracing. It is more convenient to work with the density of states in k-space, since the action is kL_γ, and the Fourier transform with respect to k reveals the relative importance of the different periodic orbits to the level density. In the earlier chapters, we had denoted the level density in k-space by the same symbol $g(k)$ as $g(E)$ in energy.

In this chapter, however, in order to conform with the literature, we shall denote it by $\rho(k)$:

$$\rho(k) = g(E) \frac{dE}{dk} = \hbar v \, g(E) \,, \tag{7.171}$$

where v is the speed of the particle. For an isolated orbit, we may use Eq. (5.36) to obtain $\rho(k)$. In this problem, the isolated orbits have an odd number of bounces. The Maslov index μ is simply twice the number of bounces, since there are no caustics. We then get, for any isolated primitive orbit specified by γ,

$$\delta\rho_\gamma = -\frac{L_\gamma}{2\pi} \cos\left(kL_\gamma\right) . \tag{7.172}$$

In Fig. 7.24, orbits 6 and 10 form families, and cannot be regarded as isolated. This is so because we can slide one leg of the trajectory along a side of the triangle, and still retain the self-tracing shape of the periodic orbit. Such is not the case, for example, for orbit 5, since a slight displacement of the vertical leg of the trajectory along the base would destroy the periodic character of the trajectory. The contribution of a family of such orbits, arising from a continuous symmetry, has been discussed in Sect. 6.1.4. In this problem, the family contribution is given by

$$\delta\rho_\gamma = \sqrt{\frac{kL_\gamma}{2\pi^3}} \, l_\perp \cos\left(kL_\gamma - \pi/4\right) . \tag{7.173}$$

Here L_γ is the length of a primitive orbit of the family, and $l_\perp = l \cos\psi$, where l is the length occupied by the family on a face, and ψ is the angle between the direction of the orbit and the normal to this face. The diffractive orbits shown in Fig. 7.24 have diffraction from one vertex only. Using the general formula (7.170), and Eq. (7.171), the contribution from such a diffractive orbit is given by

$$\delta\rho_d(k) = \frac{L_\gamma}{\pi} \left(\prod_{i=1}^{n_\gamma} \frac{d_{\gamma,i}}{\sqrt{8\pi kL_i}}\right) \cos\left(kL_\gamma - \mu_\gamma\frac{\pi}{2} - \frac{3n_\gamma\pi}{4}\right) . \tag{7.174}$$

Note that since there is only one diffractive vertex, $n_\gamma = 2$ actually corresponds to a repetitive orbit of the primitive one, and it is sufficient to include this, together with the primitive one. The semiclassical and quantum results are compared numerically in [45]. Pavloff and Schmit obtained the first 957 levels numerically, and computed the "regularised" Fourier transform of the quantum $\rho(k)$:

$$F(L) = \int_0^{k_{max}} k \, \rho(k) \, e^{ikL - \alpha k^2} \, dk \,. \tag{7.175}$$

Here k_{max} is the wave number corresponding to the maximum computed energy eigenvalue, and $\alpha = 9/k_{max}^2$. The multiplicative factor of k cancels $(\sqrt{k})^2$ that comes in the denominator of the diffractive repetitive orbit. Fig. 7.25 shows the computed $|F(L)|$ as a function of L, and the quantum and semiclassical results are indistinguishable in the diagram. The contributions of the diffractive orbits (shaded areas) are seen to be important for the excellent agreement.

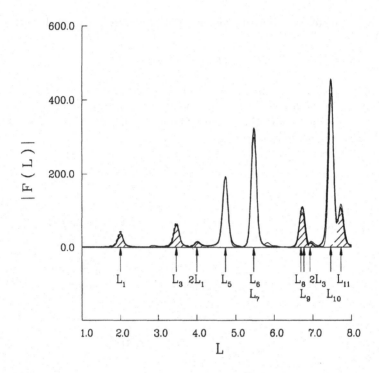

Figure 7.25:   The absolute value $|F(L)|$ of the regularised Fourier transform of $\rho(k)$, defined by Eq. (7.175), as a function of L. The quantum and semiclassical results are hardly distinguishable. The shaded areas show the contribution of the diffractive orbits. (After [45].)

7.5.6   The circular annulus billiard

Till now we have been mostly considering far-field (Fraunhofer-like) scattering. For example, in determining the scattering amplitude $f_k(\phi)$, we assumed a plane incident wave, and the scattered wave being detected at a far distance where the curvature of the wave-front could be neglected. By contrast, we now consider scattering inside the annulus billiard, where near-field (Fresnel) diffraction is important. The annulus consists of an outer and an inner disc having radii of R and a respectively. The particle is confined (both classically and quantum mechanically) in the annular space in between the two discs. Such a system has been considered by Snaith [52] and by Snaith and Goodings [53], whom we follow closely. The advantage in this simple example is that the comparison of the semiclassical with the exact quantum mechanical result may be made to demonstrate the importance of diffraction for grazing incidence. Referring to Fig. 7.26, we note that for waves emanating from a point A at $\theta = 0$ on the boundary of the outer circle, the entire annulus may be divided into three regions. These are the illuminated region $\theta \leq \theta_{ip}$, the half-shadow, or the penumbra $\theta_{ip} \leq \theta \leq \theta_{ps}$, and the

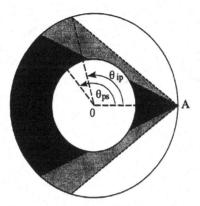

Figure 7.26: The annulus billiard, with an outer circle of radius R, and an inner disc of radius a. For rays starting from a point A at $\theta = 0$ on the rim of the outer circle, the illuminated region, the penumbra and the shadow region are shown by progressively darker shades. The angles θ_{ip} and θ_{ps} delineates the boundaries (somewhat artificially) between the three domains.

shadow region, $\theta > \theta_{ps}$. These angles will be defined later on. If only straight-line rays were allowed from A, then the illuminated region, by simple geometry, will only extend to an angle $\theta = 2\cos^{-1} a/R$, and there would be a shadow for larger angles. Due to the wave nature of the particle in the quantum description, however, there is an extended grey area in between the two regions, since the trajectory can now bend. This is depicted in the left-half of Fig. 7.27, where the bending angle is denoted by γ. From this half of the diagram, simple geometry shows that

$$\frac{\gamma}{2} = \left(\frac{\theta}{2} - \cos^{-1} \frac{a}{R} \right). \tag{7.176}$$

In the penumbra, in addition to the usual bending of the ray (left side of Fig. 7.27), it is also possible to have the sign of the right-hand side of Eq. (7.176) to be reversed, as shown at the right in Fig. 7.27. We had encountered this 'glancing path' [51] also in Fraunhofer scattering (see Fig. 7.19) before.

The semiclassical approximation made in the penumbra is valid when the magnitude of the bending angle is less than $(ka)^{-1/3}$,

$$\left| \cos^{-1}(a/R) - \theta/2 \right| < (ka)^{-1/3}. \tag{7.177}$$

In the shadow, only the left-half of Fig. 7.27 is relevant, and the bending angle γ is larger than $(ka)^{-1/3}$. The creeping rays in the shadow are damped exponentially with the path length along the arc. The precise definitions [53] of the angles θ_{ip} and θ_{ps} are based on certain approximations that are made in the Hankel functions that occur in the expression (7.67) of the scattering matrix, and will be given shortly.

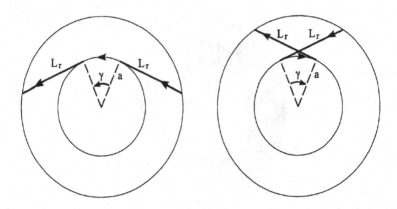

Figure 7.27:   A creeping wave along the arc of the inner disc. On the left, the
bending angle is γ and the total path-length is $(2L_r + a\gamma)$. On the right,
the bending angle is $-\gamma$ and the path-length is $(2L_r - a\gamma)$.

In Fig. 7.28, we show a ray from a point (R, θ') on the outer circle reaching another
point (R, θ) on the outer rim, either by a direct path, or by a reflection from the inner
disc. From this diagram, it may be seen that the angle $(\theta - \theta')$ is different for the
near-field from the scattering angle ϕ defined in the previous discussion. The latter
is just the angle between the incident and the scattered directions, measured in the
counter-clockwise direction. Quite generally, the propagation of a wave from a point
$(\mathbf{r}' = r', \theta')$ to another point $(\mathbf{r} = r, \theta)$ may be described by the Green's function (see
Problem 7.4 given by Eq. (7.151). It will be seen from Eq. (7.151) that the Green's
function vanishes identically when $r = a$, the radius of the inner disc. Note also that
the wave number k, in contrast to the bound-state problem, is a continuous variable.

The quantum Green's function on the boundary of the outer disc may be written
down from Eq. (7.151) by putting $r = r' = R$. This is only a function of the angle $\Delta\theta$
shown in Fig. 7.28, and the wave number k:

$$G(\Delta\theta, k) = -\frac{2m}{\hbar^2} \frac{i}{8} \sum_{l=-\infty}^{\infty} \left[H_l^-(kR) + \mathcal{S}_l(ka)H_l^+(kR) \right] H_l^+(kR) \exp(il\Delta\theta). \quad (7.178)$$

If we now impose the boundary condition that this be zero, then it may only be
satisfied for discrete values k_i which define the bound states. This condition for each
partial wave is

$$H_l^-(k_iR) + \mathcal{S}_l(k_ia)H_l^+(k_iR) = H_l^-(k_iR)H_l^+(k_ia) - H_l^-(k_ia)H_l^+(k_iR) = 0. \quad (7.179)$$

These bound states may be obtained numerically, and compared with those obtained
semiclassically. For the semiclassical approximation, we use a variant of the Poisson
sum formula given by Eq. (3.46):

$$\sum_{l=-\infty}^{\infty} f(l) = \sum_{M=-\infty}^{\infty} \int_{-\infty}^{\infty} f(m) \exp(2\pi i M m) \, dm. \quad (7.180)$$

Applying this formula to Eq. (7.178), and retaining only the dominant $M = 0$ term, we get

$$G_0(\Delta\theta, k) = -\frac{2m}{\hbar^2}\frac{i}{8}\int_{-\infty}^{\infty} dm \left[H_m^-(kR) + S_m(ka)H_m^+(kR)\right] H_m^+(kR) \exp(im\Delta\theta).$$
(7.181)

In evaluating this integral by a contour integration in the complex m-plane, the first term on the right does not contribute because it has no poles. We therefore are left with the expression:

$$G_0(\Delta\theta, k) = -\frac{2m}{\hbar^2}\frac{i}{8}\int_{-\infty}^{\infty} dm\, S_m(ka) \left(H_m^+(kR)\right)^2 \exp(im\Delta\theta).$$
(7.182)

From now on, we shall denote the difference $\Delta\theta$ by θ itself for simplicity of notation.

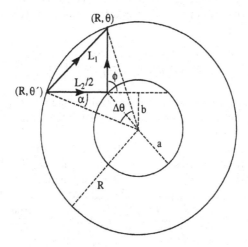

Figure 7.28: Scattering geometry in the annulus billiard.

The semiclassical Green's function for the three different regions shown in Fig. 7.26 may now be obtained from this expression by making the appropriate approximations for the Hankel functions, and performing the m-integration in the complex plane by the saddle-point method. The mathematical details may be found in refs. [36] and [52]. We only summarize the main results.

First we focus on the semiclassical expression for the Green's function in the illuminated region, $\theta \leq \theta_{ip}$. As shown in Fig. 7.28, a ray from a point on the rim of the outer disc may reach another point on the rim of the same disc by a direct straight-line path, of length L_1, or via a path of length L_2 after a reflection from the inner disc. Not surprisingly, an evaluation of Eq. (7.182) for the direct path exactly yields the free Green's function given by Eq. (7.144), with $|\mathbf{r} - \mathbf{r}'|$ replaced by L_1. We rewrite this in a slightly rearranged form:

$$G_0^{(i)}(\theta, k) = -\frac{2m}{\hbar^2}\frac{1}{2}\frac{e^{i\pi/4}}{(2\pi)^{1/2}}\left[\frac{e^{ikL_1}}{(kL_1)^{1/2}}\right].$$
(7.183)

The Green's function for the reflected trajectory of length L_2 is somewhat more complicated. Naively, we may expect it to be of the form (7.156). This is a product of two G_0's of the form (7.183) each with length $L_2/2$, and the diffraction coefficient $d(\phi)$ given by Eq. (7.157). Recalling the first term for the scattering amplitude in Eq. (7.134), we take $|f_k(\phi)| = \sqrt{(a/2 \sin \phi/2)}$ in the expression for $d(\phi)$. This prescription does not give the correct result for the Green's function of the reflected path, since Eq. (7.156) was derived by using plane waves. One has to correct for an "obliquity factor", and an additional negative sign due to the phase change at reflection. From Fig. 7.28, note that

$$R \cos\alpha - \frac{L_2}{2} = a \sin\frac{\phi}{2}. \tag{7.184}$$

In the above, α is the angle between the ray that undergoes reflection and the inward normal to the outer circle, as shown in Fig. 7.28. The correct Green's function in the illuminated region is the sum of the direct and the bounced paths, and is found to be [52]

$$G_0^{(i)}(\theta, k) = -\frac{2m}{\hbar^2} \frac{1}{2} \frac{e^{i\pi/4}}{(2\pi)^{1/2}} \left[\frac{e^{ikL_1}}{(kL_1)^{1/2}} - \frac{e^{ikL_2}}{(kL_2)^{1/2}} \left(\frac{a \sin(\phi/2)}{R \cos\alpha} \right)^{1/2} \right]. \tag{7.185}$$

The expression (7.185) has been found for straight-line ray trajectories of classical mechanics that are the content of the leading-order trace formula. In mathematical terms, it involves, as before, using the Debye approximation, Eq. (7.126), for the Hankel functions in the integral (7.182). In the Debye approximation, the angle $\cos^{-1}\frac{m}{x}$ is between 0 and $\pi/2$, so m is positive. This expansion for $H_m^\pm(ka)$ breaks down near grazing incidence, when $(m - ka) < (ka)^{1/3}$. Since the angular momentum m (in units of $\hbar$) is $kR \cos(\theta/2)$, the maximum permissible angle for the illuminated region is given by the angle $\theta_{i,max} = 2\cos^{-1}[(ka + (ka)^{1/3})/kR]$. Moreover, from Eq. (7.177), in the penumbra region we have $|\cos^{-1}(a/R) - \theta/2| < (ka)^{-1/3}$. This defines the limits of the penumbra region between the angles $(\theta_{p,max}, \theta_{p,min})$ as $2 \cos^{-1}(a/R) \pm 2(ka)^{-1/3}$. The boundary angle θ_{ip} of the illuminated region is defined by the average of the overlapping angles $(\theta_{p,min}, \theta_{i,max})$:

$$\theta_{i,p} = \cos^{-1}\left[(ka + (ka)^{1/3})/kR \right] + \cos^{-1}(a/R) - (ka)^{-1/3}. \tag{7.186}$$

As mentioned already, the bending angle γ is larger than $(ka)^{-1/3}$ in the shadow, so that the minimum angle for the shadow region would be $\theta_{s,min} = 2\cos^{-1}(a/R) + (ka)^{-1/3}$. This is less than $\theta_{p,max}$. Defining the boundary between the penumbra and the shadow by the angle $\theta_{ps} = (\theta_{p,max} + \theta_{s,min})/2$, we get

$$\theta_{ps} = 2 \cos^{-1}(a/R) + \frac{3}{2}(ka)^{-1/3}. \tag{7.187}$$

For the penumbra, the results for grazing incidence on a disc were obtained first by Primack *et al.* [51]. The Debye approximation is no longer valid, and the Hankel function is approximated by the relation

$$H_m^\pm(ka) = 2 \exp(\mp i\pi/3) \left(\frac{2}{m} \right)^{1/3} Ai \left[\exp(\pm 2i\pi/3)(2/m)^{1/3}(m - ka) \right] + O(m^{-1}), \tag{7.188}$$

where $Ai(x)$ is the Airy function regular at $x = 0$. In Eq. (7.126), there was a sharp cut-off for $m = ka$, giving rise to a discontinuous change in the Green's function as a function of the angle. The penumbra contribution corrects for this unphysical behavior. This is similar to Fresnel diffraction at a boundary encountered in optics. To be more specific, this gives rise to exactly the first (direct) term in Eq. (7.185), multiplied by a Fresnel smoothing factor $F_r(k, \theta)$ that varies between 0 and 1. It tends to 1 as θ approaches the illuminated border, and rapidly goes to zero as θ approaches the shadow region. Additionally, there are two other terms in the Green's function for the glancing rays within the penumbra corresponding to the glancing rays shown in Fig. 7.27. For the left-half of Fig. 7.27, note that the total path length for the creeping plus the straight line paths of the ray is $(2L_r + a\gamma)$, while for the right-half, this path length is $(2L_r - a\gamma)$, where $2L_r$ is the length of the straight segments. These two diffractive paths yield two terms in the Green's function proportional to $\exp[ik(2L_r \pm a\gamma_0)]/(kL_r)$. The mathematical derivation is not trivial, and may be found in [52]. The calculated form, in the angular region $\theta_{ip} \leq \theta \leq \theta_{ps}$ is given by [51]

$$G_0^{(p)}(\theta, k) = \frac{2m}{\hbar^2} \left(-\frac{1}{2} \frac{e^{i\pi/4}}{(2\pi)^{1/2}} \left[\frac{e^{ikL_1}}{(kL_1)^{1/2}} F_r(k, \theta) \right] + \frac{(ka)^{\frac{1}{3}} C}{2\pi k L_r} \exp \left[ik(2L_r \pm a\gamma_0) \right] \right).$$

(7.189)

In the above, $F_r(k, \theta)$ is the Fresnel smoothing factor, and the constant C is found to be $[0.99615 \exp(i\pi/3)]$.

Finally, we sketch the steps to obtain the expression for the Green's function in the shadow, for $\theta > \theta_{ps}$. The integral (7.182) may be calculated by considering the complex m-plane, and evaluating the residues at the poles of the integrand. These poles are simple, and are obtained from the zeros of $H_m^+(ka)$. They are shown in Fig. 7.29 on the upper half of the complex plane.

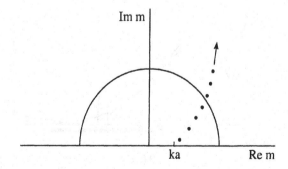

Figure 7.29: Poles of $H_m^+(ka)$ for a fixed ka in the upper half of the complex m-plane. There are similar (simple) poles in the third quadrant in the lower half that are not shown.

For positive $\phi > 0$, the contour may be closed on the upper half of the complex plane by a semicircle of infinite radius, on which the integrand vanishes. There are also zeros of H_m^+ in the lower half of the complex m-plane in the third quadrant, otherwise similarly placed, and not shown in Fig. 7.29. We note in passing that the contribution of the terms with $M > 0$ will be rapidly damped out on this contour. As long as the bending angle γ in Fig. 7.27 is large enough to obey the relation

$$\gamma > (ka)^{-1/3}, \tag{7.190}$$

Nussenzveig [36] shows that only the first few poles nearest to the real axis contribute appreciably to the integral (7.182). The first few complex zeros of $H_m^+(ka)$ that give rise to these poles are given by

$$m_n = ka + (ka/2)^{1/3} x_n \exp(i\pi/3), \tag{7.191}$$

where $-x_n$ is the nth zero of the Airy function ($n = 1, 2, ..$). The expression (7.188) for $H_m^\pm(ka)$ valid near $m = ka$ is needed to evaluate the residues at the poles (7.191). These are given by

$$\frac{H_{m_n}^-(ka)}{\frac{\partial}{\partial m}H_m^+(ka)|_{m=m_n}} \simeq -\frac{\exp(-i\pi/6)}{2\pi}\left(\frac{ka}{2}\right)^{1/3}\frac{1}{[Ai'(-x_n)]^2}. \tag{7.192}$$

On the outer sphere of radius R, we may still use the Debye approximation for the Hankel functions, given by Eq. (7.126). Combining these results in the expression for the Green's function as given by Eq. (7.182), we get, in the shadow ($\theta > \theta_{ps}$)

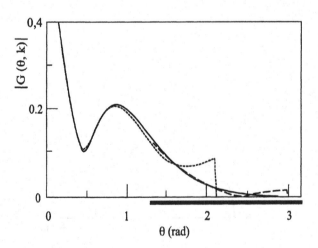

Figure 7.30: The absolute value of the semiclassical Green's function (dotted curve: illuminated region, dashed curve: penumbra and shadow regions) is compared with quantum mechanical result (continuous line), for k=10.189, a=0.5, R=1 in arbitrary units. (From [53].)

$$G_0^{(s)} = -\frac{2m}{\hbar^2}\frac{i}{8}(2\pi i)\sum_n \frac{H_{m_n}^-(ka)}{\frac{\partial}{\partial m}H_m^+(ka)|_{m=m_n}}[H_{m_n}^{(+)}(kR)]^2 \exp(im_n\theta),$$

$$= -\frac{2m}{\hbar^2}\frac{i}{4\pi}e^{-i\pi/6}(ka/2)^{1/3}\sum_n \frac{(k^2R^2-m_n^2)^{-1/2}}{[Ai'(-x_n)]^2}$$

$$\times \exp\left[2i\left(\sqrt{k^2R^2-m_n^2}-m_n\cos^{-1}\frac{m_n}{kR}\right)+im_n\theta\right].$$

The sum in the shadow region is over the first few m_n's given by Eq. (7.191). We see that this sum converges quickly only if ϕ, which is the same as the bending angle γ_0, satisfies the restriction given by Eq. (7.190). Note also that the phase factor $ka\phi$ is the wave number times the path length along the arc of the bent ray. Moreover, since the poles at m_n given by Eq. (7.191) are complex, there is a rapid attenuation of these creeping rays with increasing bending angle. This is also the reason why only the $M = 0$ term in the Poisson sum dominates. We display the absolute value of the semiclassical Green's function in the range of $0 \leq \theta \leq \pi$ in Fig. 7.30, and compare it with the quantum mechanical result (drawn in a continuous curve).

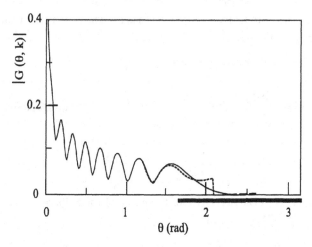

Figure 7.31: The same as Fig. 7.30, but for $k = 50$. In both figures, the dark segment drawn along the θ-axis shows the range $\theta_{ip} \leq \theta \leq \theta_{ps}$. (From [53].)

The semiclassical G_0^i given by Eq. (7.185) for the illuminated region is drawn by the dotted curve, and gets abruptly cut off at θ_{ip}. The expressions G_0^p and G_0^s for the penumbra and the shadow are drawn together in the dashed curve. Because different approximations have been used for the Hankel functions in the illuminated and the penumbra regions, there is a discontinuity in the curve at θ_{ip}. The angular range $\theta_{ip} \leq \theta \leq \theta_{ps}$ is shown on the horizontal axis by a heavy line. In Fig. 7.30, $k = 10.189$, $a = 0.5$, and $R = 1$ in arbitrary units, and $\hbar = m = 1$. The comparison is repeated in Fig. 7.31, with $k = 50$, but otherwise the same a and R. In the latter curve, the penumbra region (shown by the dark line on the θ-axis) shrinks considerably, showing that the shadow contribution, G_0^s, is negligible. It is this underlined range of θ that embraces the penumbra. Recall that for the far-field scattering cross section near forward angles, it was again the penumbra that came in, and required the largest diffractive corrections. The dotted curve in the illuminated region is essentially the classical approximation. Fig. 7.31 shows that even when $ka \gg 1$, quantum corrections persist in the penumbra.

7.6 Problems

(7.1) Using the relationship between the functions $\zeta(s)$ and $\xi(s)$ defined in Eq. (7.4), and the functional relation in Eq. (7.5), show that on the critical line

$$\zeta(1/2 - it) = (\pi)^{-it} \frac{\Gamma(1/4 + it/2)}{\Gamma(1/4 - it/2)} \, \zeta(1/2 + it) \, .$$

Let us define

$$\zeta(1/2 + it) = Z(t) \exp[-i\bar{\theta}(t)] \, ,$$

where both $Z(t)$, $\bar{\theta}(t)$ are real, and $Z(t)$ may change sign. Note further that $\zeta(1/2 - it)$ is the complex conjugate of $\zeta(1/2 + it)$. Using Stirling's formula for $\ln \Gamma(z)$, show that

$$\bar{\theta}(t)/\pi = (t/2\pi) \ln(t/2\pi) - (t/2\pi) + 7/8 + \mathcal{O}[t^{(-1)}] + \dots \, . \tag{7.193}$$

(7.2) ζ_s **for the inverted oscillator**
For a two-dimensional potential $V(x, y) = \frac{1}{2}m\left(\omega_1^2 x^2 - \omega_2^2 y^2\right)$, we had derived Eq. (5.68) in chapter 5 for the oscillating part of the level density

$$\delta g_{sc} = \frac{1}{\hbar\omega_1} \sum_{k=1}^{\infty} (-)^k \frac{\cos\left(\frac{2\pi k E}{\hbar\omega_1}\right)}{\sinh\left(\frac{\pi k \omega_2}{\omega_1}\right)} \, . \tag{7.194}$$

This may be related to the derivative of the phase of the Selberg zeta function $\zeta_s(E)$ using the relation (7.38)

$$\delta g_{sc}(E) = -\frac{1}{\pi} \frac{d}{dE} \Im m \ln \zeta_s(E) \, .$$

Using the two equations above, show that

$$\zeta_s(E) = \prod_{l=0}^{\infty} \left\{ 1 + \exp\left[\frac{2\pi i}{\hbar\omega_1} \left(E + i(l + 1/2) \, \hbar\omega_2\right) \right] \right\} \, .$$

From the zeros of this function, find the quantized energies, and interpret the result physically.

(7.3) Scattering amplitude and barrier penetration
(a) The (quantum mechanical) scattering amplitude of a particle is given in Eq. (7.111), where the scattering matrix $\mathcal{S}_l(k)$ for hard-disc scattering is a ratio of the Hankel

functions, Eq. (7.120). By using asymptotic expansions [54] of the Hankel functions (fixed k, $|l| \gg ka$), show that

$$(S_l(k) - 1) \simeq -i \left(\frac{eka}{2l} \right)^{2l}.$$ (7.195)

This shows that the scattering amplitude falls off rapidly if the impact parameter b is much larger than the disc radius a.

(b) In the one-dimensional barrier penetration problem, the magnitude of the transmission *amplitude* in the WKB approximation is given by (see Eq. (2.86) for the transmission probability $T = |t|^2$):

$$|t| = \exp \left[-\frac{1}{\hbar} \int_{x_1}^{x_2} dx \, \sqrt{(2m(V(x) - E)} \right],$$ (7.196)

where x_1, x_2 are the classical turning points where $V = E$, and $x_2 > x_1$. In the scattering problem discussed in (a), for the radial coordinate $r > a$, the particle experiences a centrifugal barrier $\hbar^2 l^2 / (2mr^2)$ before it encounters the hard wall at $r = a$. Use the formula for $|t|$ to obtain the transmission amplitude for the radial motion at fixed $|l|$, taking into account that there is a reflection at the hard edge, and the particle retraces its path. Verify that one then gets for $|t|$ the same result as $|S_l(k) - 1|$ given by Eq. (7.195) in part (a).

(7.4) Prove Eq. (7.151) for the Green's function of the annular billiard. Hints (from [52]): Start with Eq. (7.142) in the text, defining $G_0(\mathbf{r}', \mathbf{r}; E)$, where $E = \hbar^2 k^2 / 2m$. Taking advantage of circular symmetry, make the expansion

$$G_0(\mathbf{r}', \mathbf{r}; E) = \sum_{l=-\infty}^{\infty} g_l(r', r) \frac{e^{il(\theta - \theta')}}{2\pi}.$$ (7.197)

Further, expand the Dirac delta function on the right-hand side of Eq. (7.142) :

$$\delta(\mathbf{r} - \mathbf{r}') = \frac{1}{r} \delta(r - r') \frac{1}{2\pi} \sum_{l=-\infty}^{\infty} e^{il(\theta - \theta')}.$$ (7.198)

Hence find the radial equation obeyed by $g_l(r', r)$. Its homogeneous form is just the differential equation for Bessel's cylindrical functions $J_l(kr)$ and $N_l(kr)$. Apply the boundary condition that g_l has to vanish on the inner disc of radius a, and is an outgoing circular wave for $r_> \to \infty$. For further help, you may consult the book by Arfken [57].

(7.5) The Ruelle zeta function
Consider Eq. (7.169) for the trace of the diffractive Green's function, tr $G_d(k)$, where k is the wave number. By defining (for a primitive periodic orbit γ) the quantity

$$t_\gamma(k) = \left(\prod_{i=1}^{n_\gamma} \frac{d_{\gamma,i}}{\sqrt{8\pi k L_{\gamma,i}}} \right) \exp \left[i \left(k L_\gamma - \mu_\gamma \frac{\pi}{2} - n_\gamma \frac{3\pi}{4} \right) \right],$$ (7.199)

show that

$$\text{tr } G_d(k) = -\frac{im}{\hbar^2 k} \sum_\gamma L_\gamma \frac{t_\gamma}{(1 - t_\gamma)}.$$ (7.200)

Note that $dt_\gamma/dk = iL_\gamma t_\gamma - t_\gamma n_\gamma/(2k)$, and the second term is small for large k. Neglecting this small term, show that

$$\text{tr}\, G_d(k) \simeq \frac{m}{\hbar^2 k} \frac{d}{dk} \ln \left[\prod_\gamma (1 - t_\gamma) \right]. \tag{7.201}$$

The product $\prod_\gamma (1 - t_\gamma(k))$ is also defined as $\zeta_r^{-1}(k)$, where $\zeta_r(k)$ is the Ruelle Zeta function [49, 58, 59]. A similar procedure may also be adopted for summing geometrical repetitive orbits. The combined Ruelle zeta function is a product of the geometrical and the diffractive one.

(7.6) Scattering of anyons

In problem (3.4), we had discussed the bound state spectrum of two anyons (an anyon is a charged particle with a flux line (of arbitrary strength) attached to it. Here we consider the scattering of two "free" anyons. Using the same notation as of problem (3.4), the Hamiltonian (in relative coordinates) is

$$H = \frac{p_r^2}{2} + \frac{(p_\phi - \hbar\alpha)^2}{2r^2}. \tag{7.202}$$

Show, following the procedure of Sect. 7.5.2, that the phase shift δ_l suffered in the partial wave l is

$$\begin{aligned}
\delta_l(k) &= \frac{\alpha}{2}\pi, \quad l > 0, \\
&= -\frac{\alpha}{2}\pi, \quad l \leq 0.
\end{aligned} \tag{7.203}$$

Thus the scattering amplitude is given by

$$f_k(\phi) = \frac{1}{\sqrt{2\pi ik}} \left[\sum_{l \geq 1} e^{il\phi}(e^{i\alpha\pi} - 1) + \sum_{l \geq 0} e^{-il\phi}(e^{-i\alpha\pi} - 1) \right]. \tag{7.204}$$

This may be summed analytically to yield the famous Bohm-Aharonov scattering amplitude [60, 61]:

$$f_k(\phi) = -\frac{e^{i\phi/2} \sin(\alpha\pi)}{\sqrt{2\pi ik} \sin(\phi/2)}. \tag{7.205}$$

In calculating the differential scattering cross section for identical particles, one should use $|f_k(\phi) \pm f_k(\phi - \pi)|^2$ for bosons/ fermions. Consequently, note that the differential cross-section is symmetric about $\phi = \pi/2$. An asymmetry is introduced, however, when the anyons scatter via a potential [62].

(7.7) Scattering of a particle by a flux line in a disc

Consider a particle confined in a circular billiard with a singular flux-line at the center. The Hamiltonian within the disc is the same as Eq. (7.202), but now the coordinate r refers to the particle coordinate.

(a) Noting that the wave function must vanish on the boundary of the disc, state how Eq. (7.63) determining the eigenvalues is modified.

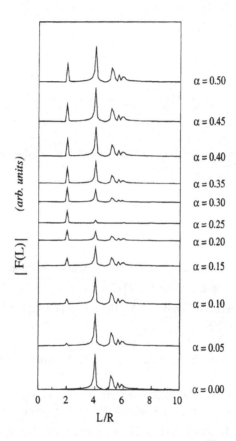

Figure 7.32: Fourier transform of the level density $\delta g\left(\sqrt{E}\right)$ (Gaussian averaged with $\gamma = 0.6/R$) of the Aharonov-Bohm disk billiard for various flux strengths α, plotted versus orbit length L/R. (After [50].)

(b) The quantum density of states $g(E)$ for this problem was calculated by Reimann *et al.* [50] and compared with the semiclassical result. In Fig. 7.32, the Fourier spectra of $\delta g(\sqrt{E})$ for various flux strengths α are plotted as functions of L/R, where R is the radius of the disc. Correlate the location of the peaks in the the Fourier amplitudes $|F(L)|$ with the shortest classical periodic orbits for the case $\alpha = 0$, i.e., in the absence of the flux line. When α is nonzero, comment on the appearance of the peak at $L = 2R$.

The variation in the height of this peak may be fitted by a function $\sin^2(\pi\alpha)$, as shown in Fig. 7.33. Explain this behavior, commenting on the nature of the orbit.

(c) Explain the disappearance of most of the Fourier peaks for the flux strength $\alpha = 0.25$, including the near disappearance of the one at $L = 4R$.

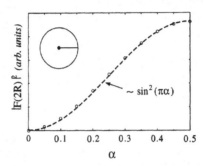

Figure 7.33:   The squared Fourier amplitude of the half-diameter orbit, plotted versus flux strength α. (After [50].)

Bibliography

[1] H. M. Edwards: *Riemann's Zeta Function* (Academic Press, New York, 1974).

[2] G. H. Hardy, Comptes Rendus **CLVIII**, 280 (1914).

[3] R. K. Bhaduri, Avinash Khare, S. M. Reimann, and E. L. Tomusiak, Ann. Phys. (N. Y.) **254**, 25 (1997).

[4] M. V. Berry, in: *Quantum Chaos and Statistical Nuclear Physics*, Lecture Notes in Physics 263, ed. by T. H. Seligman and H. Nishioka (Springer Verlag, Berlin, 1986), pp. 1-17.

[5] A. M. Odlyzko, AT&T Bell Lab., preprint (1989). For a general discussion, see M. L. Mehta: *Random Matrices* (Academic Press, New York, second edition, 1991).

[6] A. Bohr and B. R. Mottelson, it Nuclear Structure, Vol. I (W. Benjamin, New York, 1969).

[7] E. P. Wigner, in Proc. Fourth Can. Math. Congress (1957).

[8] F. J. Dyson and M. L, Mehta, J. Math. Phys. **4**, 701 (1963).

[9] R. U. Haq, A. Pandey and O. Bohigas, Phys. Rev. Lett. **48**, 1086 (1982).

[10] S. W. McDonald and A. N. Kaufman, Phys. Rev. Lett. **42**, 1189 (1979).

[11] O. Bohigas, M. J. Giannoni, and C. Schmit, Phys. Rev. Lett. **52**, 1 (1984).

[12] M. Gutzwiller: *Quantum Chaos*, The Scientific American, January 1992, p. 78.

[13] *Nobel Symposium on Quantum Chaos*, eds. K.-F. Berggren and S. Åberg; Physica Scripta **T90** (2001).

[14] H.-J. Stöckmann: *Quantum Chaos: An Introduction* (Cambridge University Press, 1999)

[15] M. V. Berry and J. P. Keating, J. Phys. **A 23**, 4839 (1990).

[16] A. Selberg, J. Indian Math. Soc. **20**, 47 (1956).

[17] R. Aurich and F. Steiner, Proc. Roy. Soc. (London), **A 437**, 693 (1992).

[18] M. Gutzwiller, Phys. Rev. Lett. **45**, 150 (1980).

[19] M. Gutzwiller, Physics Scripta **T 9**, 184 (1985).

[20] M. Gutzwiller: *Chaos in Classical and Quantum Mechanics* (Springer-Verlag, New York, 1990) p. 362.

[21] A. Voros, J. Phys. **A 21**, 685 (1988).

[22] E. B. Bogomolny, Nonlinearity **5**, 805 (1992).

[23] T. Szeredi and D. A. Goodings, Phys. Rev. **E 48**, 3529 (1993).

[24] M. Sieber and F. Steiner, Phys. Rev. Lett. **67**, 1941 (1991).

[25] C. Matthies and F. Steiner, Phys. Rev. **A 44**, R7877 (1991).

[26] M. V. Berry and J. P. Keating, Proc. Roy. Soc. (London) **A 437**, 151 (1992).

[27] D. W. L. Sprung, Hua Wu, and J. Martorell, Am. J. Phys. **64**, 136 (1996).

[28] R. G. Newton: *Scattering Theory of Waves and Particles* (McGraw-Hill, New York, 1966).

[29] E. Doran and U. Smilansky, Nonlinearity **5**, 1055 (1992).

[30] C. Rouvinez and U. Smilansky, J. Phys. **A 28**, 77 (1995).

[31] N. Levinson, Kgl. Dan. Vid. Selsk., Mat.-Fys. Medd. **25**, No. 9 (1949); M. Wellner, Am. J. Phys. **32**, 787 (1964).

[32] B. Georgeot and R. E. Prange, Phys. Rev. Lett. **74**, 2851 (1995).

[33] P. A. Boasman, Nonlinearity **7**, 485 (1994).

[34] T. Szeredi, J. H. Lefébvre, and D. A. Goodings, Nonlinearity **7**, 1463 (1994).

[35] H. M. Nussenzveig: *Diffraction Effects in Semiclassical Scattering* (Cambridge University Press, Cambridge, 1992).

[36] H. M. Nussenzveig, Ann. Phys. (N. Y.) **34**, 23 (1965).

[37] H. S. W. Massey and Mohr, Proc. Roy. Soc. (Lond.) **A 141**, 434 (1934).

[38] A. Sommerfeld, Math. Ann. **47**, 317 (1896); *Optics* (Academic Press, New York, 1954).

[39] B. R. Levy and J. B. Keller, Comm. Pure and Appl. Math. **12**, 159 (1959).

[40] G. Vattay, A. Wirzba, and P. E. Rosenqvist, Phys. Rev. Lett. **73**, 2304 (1994).

[41] H. Bruus and N. D. Whelan, Nonlinearity **9**, 1023 (1996).

[42] Y. Aharonov and D. Bohm, Phys. Rev. **115**, 485 (1959).

[43] M. Brack, R. K. Bhaduri, J. Law, M. V. N. Murthy, and Ch. Maier, Chaos **5**, 317 (1995); Erratum **5**, 707 (1995).

[44] N. D. Whelan, Phys. Rev. **E 51**, 3778 (1995).

[45] N. Pavloff and C. Schmit, Phys. Rev. Lett. **75**, 61 (1995), Erratum **75**, 3779 (1995).

[46] P. Seba, Phys. Rev. Lett. **64**, 1855 (1990).

[47] G. Date, S. R. Jain, and M. V. N. Murthy, Phys. Rev. **E 51**, 198 (1995).

[48] G. Vattay, A. Wirzba, and P. E. Rosenqvist, J. Stat. Phys. **83**, 243 (1996).

[49] P. Rosenqvist, N. D. Whelan, and A. Wirzba, J. Phys. **A 29**, 5441 (1996).

[50] S. M. Reimann, M. Brack, A. G. Magner, J. Blaschke, and M. V. N. Murthy, Phys. Rev. **A 53**, 39 (1996).

[51] H. Primack, H. Schanz, U. Smilansky, and I. Ussishkin, Phys. Rev. Lett. **76**, 1615 (1996).

[52] N. C. Snaith, B.Sc. thesis, McMaster University (1996, unpublished).

[53] N. C. Snaith and D. A. Goodings, Phys. Rev. **E 55**, 5212 (1997).

[54] M. Abramowitz and I. A. Stegun: *Handbook of Mathematical Functions* (Dover, New York, 1970) p. 366.

[55] Diptiman Sen, private communication (1996).

[56] N. D. Whelan, Phys. Rev. Lett. **76**, 2605 (1996).

[57] G. Arfken: *Mathematical Methods for Physicists* (Academic Press, New York, 1985).

[58] D. Ruelle: *Statistical Mechanics, Thermodynamic formalism* (Addison Wesley, Reading, Mass., 1978).

[59] P. Cvitanovic, P. E. Rosenqvist, and G. Vattay, Chaos **3**, 623 (1993).

[60] Y. Aharonov and D. Bohm, Phys. Rev. **115**, 485 (1959).

[61] C. R. Hagen, Phys. Rev. **D 41**, 2015 (1990).

[62] A. Suzuki, M. K. Srivastava, R. K. Bhaduri, and J. Law, Phys. Rev. **B 44**, 10731 (1991).

Chapter 8

Shells and periodic orbits in finite fermion systems

In this final chapter, we want to present some applications of the methods developed earlier in this book to many-body systems in three different domains of physics, namely atomic nuclei (Sect. 8.1), metal clusters (Sect. 8.2), and semiconductor quantum dots (Sect. 8.3). We shall mainly focus on the manifestation of periodic orbits through shell effects, but we also give an example of Strutinsky's shell-correction method. No completeness in the presentation of the selected topics has been attempted. What we want to convey with these examples is the beauty of some physical phenomena and of their simple description by semiclassical methods.

8.1 Shells and shapes in atomic nuclei

Atomic nuclei exhibit all the complexities of a finite many-fermion system. A nucleus is composed of two kinds of nucleons, the protons and neutrons, each of them consisting of three quarks which are the constituents of all strongly interacting particles. In our present understanding, the strong interactions are dominated by the characteristics of quantum chromodynamics (QCD), the modern quantum field theory of quarks and their interactions through the exchange of gluons. (For an introduction to modern particle physics, see Ref. [1], and to models of the nucleon, see Ref. [2].) If one is interested in low-energy excitations, which on the nuclear scale means energies of a few MeV, the quark-gluon degrees of freedom may be ignored for most considerations, and the meson-exchange theory still applies. In this picture, the strong nuclear forces between the nucleons are described by the exchange of pions, as first proposed by Yukawa [3] (see also Sect. 4.1), and by heavier mesons, as later elaborated in great detail. As a result, one can describe the nucleon-nucleon interaction in the non-relativistic domain by a two-body potential which, however, is highly nonlocal and non-central, and depends on the spin and isospin degrees of freedom of the nucleon (see Ref. [4] for an introduction to the nucleon-nucleon interaction). Its exact form at short distances is unknown, unlike the Coulomb potential which dominates atomic and molecular systems.

For these reasons the nucleus confronts the physicist with a double difficulty: first, the basic interaction is not exactly known and even in its simplest forms very complicated, and second, one has to tackle a finite interacting many-body system. The latter problem is, of course, common to all disciplines in physics, chemistry and astrophysics, where one deals with interacting particles. Powerful techniques of reducing the many-body problem to that of a system of non-interacting particles in an average potential have been developed, and constitute the essence of the so-called *mean-field theories*. In these, all interactions between the particles have been incorporated into an average potential (or mean field), in which the particles move independently. The quantum-mechanical description is then governed by the one-body Schrödinger equation, as assumed in the previous chapters of this book. In the Appendix A, we give a short summary of the two most commonly used versions of the mean-field approach, the Hartree-Fock theory and the density functional theory. One consequence of the many-body origin of the mean field is that it depends on the density of the particles itself, so that it has to be determined in a self-consistent way. In nuclear physics, a further difficulty comes from the complexity of the nucleon-nucleon interaction that makes it necessary to introduce the so-called "effective interactions" which incorporate various many-body correlations through an explicit dependence on the local density $\rho(\mathbf{r})$ [5, 6, 7]. Therefore, even in the mean-field approach, the calculations for large systems containing many particles become very time consuming, in particular when the systems have no spatial symmetries. (For a review on Hartree-Fock calculations with effective interactions in nuclei, see Quentin and Flocard [8].)

A considerable progress in the understanding of the single-particle aspects in nuclear physics had been made already in 1949 by the successful formulation of the nuclear shell model by Goeppert-Mayer [9] and by Haxel, Jensen, and Suess [10]. The mean-field potential was parameterized (independently for neutrons and protons) by a Woods-Saxon form. It included a strong spin-orbit coupling term, which was the essential step to understand the "magic numbers" N_0 and Z_0 that characterize the neutron and proton numbers, respectively, of the most stable nuclei. These magic numbers appear in the form of minima in the total nuclear binding energies (for an example, see Fig. 4.3), and as sharp steps in the separation energies of neutrons and protons (i.e., the energies required to remove one neutron or proton from the nucleus). The potentials in Refs. [9, 10] were assumed to have spherical symmetry. Then, the single-particle states are characterized by the angular-momentum quantum numbers j, $j_z = m$, and a radial quantum number n_r. Filled j-shells with their $(2j + 1)$ degeneracy always give a spherical probability distribution. Furthermore, these states group into bunches or "main shells", consisting of several nearly-degenerate levels of different j and n_r, which is typical for three-dimensional potentials. When these main shells are filled by one kind of nucleons, one arrives at the corresponding magic numbers N_0 or Z_0. With the development of the effective interactions and the Hartree-Fock calculations already mentioned, it later became possible to justify the Woods-Saxon form of the mean field from a self-consistent microscopic basis.

Not all nuclei are spherical, however. In the independent-particle picture of a self-bound system, the occurrence of deformations is easily understood by what in molecules corresponds to the Jahn-Teller effect [11]. When a spherical j-shell is only partially filled, (i.e., the number of particles in that shell is less than its degeneracy $2j+1$), its wavefunction cannot be uniquely defined. Nature tends to avoid such degenerate situations, and in the case of nuclei the easiest way is to break the spherical symmetry of the mean field. In order to describe this effect in the independent-particle model, Nilsson [12] developed a deformed shell model that has become famous as the "Nilsson model", and consists of an axially deformed harmonic-oscillator potential, a spin-orbit term, and a term proportional to ℓ^2 which simulates a steeper surface. Only one spheroidal deformation parameter ϵ was used in this model (cf. the axially deformed harmonic oscillator in Sect. 1.2.4). The equilibrium deformation ϵ_0 of each nucleus was then determined by minimizing the sum of occupied single-particle levels, E_{sp} (3.39), with respect to ϵ. This procedure could account for the correct magnitudes and systematics of the nuclear ground-state deformations [13].

An immediate consequence of a static deformed shape is the existence of low-lying rotational excitations. When it became possible to excite rotational states in nuclei systematically by scattering of α particles (i.e., He nuclei) and other ions (see Alder *et al.* [14] for a review of the Coulomb excitation process), ample evidence accumulated for the majority of the nuclei to have deformed equilibrium shapes. The simultaneous description of collective rotations and vibrations in terms of a few shape parameters, and of the single-particle degrees of freedom coupled to them, led to the formulation of the nuclear "collective model" by Rainwater, Bohr, and Mottelson [15]. For detailed accounts of the various nuclear models and their balanced roles in accounting for a wealth of experimental data, we refer the interested reader to standard text books on nuclear structure [16, 17, 18].

The particular aspect of quantum shell effects and their interplay with nuclear deformations was emphasized by Strutinsky with the introduction of his shell-correction method [19, 20] that proved particularly successful in connection with the nuclear fission process. The shell-correction method has been discussed in Sect. 4.7 in connection with the Strutinsky energy averaging which is a numerical alternative to the extended Thomas-Fermi model. In Sect. 8.1.2, we shall present an example of its application to the calculation of nuclear fission barriers. Before that, we report in Sect. 8.1.1 some early attempts to explain the nuclear ground-state deformations in the framework of the periodic orbit theory. A recent application of the periodic orbit theory to the mass asymmetry in nuclear fission will be discussed in Sect. 8.1.3.

8.1.1 Nuclear ground-state deformations

Most nuclei have deformed ground states that in a rough approximation may be described by an axially symmetric ellipsoid with a typical axis ratio of ~ 1.2 to 1.3. More sophisticated deformation parameters involving higher multipoles are required, however, to account for detailed spectroscopic information. The most successful studies of nuclear shapes, both for ground-state deformations and for the shapes occurring in nuclear fission, have been made using Strutinsky's shell-correction method which has been described in Sect. 4.7. For a nice overview on nuclear shapes and shells, we refer the interested reader to a review of Ragnarsson *et al.* [21].

Since in nuclei there are two kinds of nucleons, separate energy shell-corrections must be evaluated for both. One therefore writes the total nuclear binding energy in the shell-correction approach in the the following form

$$E_{tot}(N, Z; def) = E_{LDM}(N, Z; def) + \delta E_n(N; def) + \delta E_p(Z; def). \quad (8.1)$$

Hereby N and Z are the neutron and proton numbers, respectively. (We give in Appendix A.3 the justification for this approximation from the point of view of self-consistent mean-field theory and also the reason why the neutron and proton contributions may be added separately, although the total binding energy is not linear in these contributions.) Both the liquid drop model (LDM) energy E_{LDM} and the individual shell-corrections δE_n, δE_p depend also on the shape of the nucleus which is usually defined in terms of a set of suitable deformation parameters, denoted here schematically by "def". When using Eq. (8.1) for the evaluation of the ground-state energy of a given nucleus (N, Z), E_{tot} has to be minimized with respect to all deformation parameters. By varying these parameters away from their equilibrium values, one obtains multi-dimensional deformation energy surfaces which play an important role for nuclear dynamics, such as collective vibrations, or the fission process discussed in the following two sections.

Strutinsky and coworkers [22, 23] were the first to apply the periodic orbit theory to spherical and deformed shell-model potentials relevant to nuclear physics, after extending the Gutzwiller theory to take correctly into account the degeneracies of the orbits due to the continuous symmetries (see Sect. 6.1). In Ref. [23], they investigated the leading contributions from the shortest periodic orbits to the oscillating parts of the level densities $\delta g(E)$ and to the energy shell-corrections δE for deformed harmonic-oscillator potentials and ellipsoidal cavities. In particular, they investigated the locations of the minima of these quantities, corresponding to the ground-state deformations, as functions of both the deformation and the Fermi energy or particle number, respectively.

To illustrate their procedure we show in Fig. 8.1 a contour plot of the neutron energy shell-correction δE_n as a function of the neutron number N and the deformation parameter η which here is the ratio of semi-axes of a spheroidally deformed Woods-Saxon potential. (More specifically, the axis ratio is that of the spheroid given by the equipotential surface with half the central value of the potential. The potential used in Ref. [23] also included a realistic spin-orbit term.) The contours show the shell-corrections $\delta E_n(N, \eta)$ obtained quantum-mechanically from the single-particle spectrum of the Woods-Saxon potential according to Eq. (4.189). The shaded areas are those where δE_n is negative and contain the local minima. The black dots with values $1 < \eta \lesssim 1.3$ correspond to the experimentally known deformations of the ground states; the one with an error bar around $\eta \simeq 2$ corresponds to the experimental deformation of the fission isomer in ^{240}Pu (see the following section).

What is striking from Fig. 8.1 is not only the periodicity of the shell effects in the particle number N for a fixed value of the deformation η, but also the appearance of rather well-defined and smoothly varying slopes of the ground-state valleys in the (N, η) plane. It is these slopes that were investigated semiclassically in Ref. [23] and that are discussed further here.

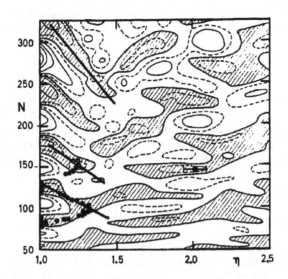

Figure 8.1: Neutron energy shell-correction $\delta E_n(N, \eta)$ for a realistic nuclear Woods-Saxon potential (including spin-orbit term), plotted versus neutron number N and axis ratio η of the spheroidally deformed potential (equidistance: 2.5 MeV, areas with negative values are shaded). Dots indicate experimental deformations, the heavy bars are the semiclassical predictions for the loci of the ground-state minima using the leading classical periodic orbit families in a spheroidal cavity. (From Ref. [23].)

In Sect. 5.5, we had derived the following semiclassical expression for the shell-correction energy:

$$\delta E \simeq \Im m \sum_{\Gamma} \mathcal{A}_{\Gamma} \left(\frac{\hbar}{T_{\Gamma}} \right)^2 e^{i[S_{\Gamma}/\hbar + \phi]}. \tag{8.2}$$

Here Γ refers to the individual periodic orbits or orbit families (including their repetitions); $\mathcal{A}_{\Gamma}$, S_{Γ}, T_{Γ} are the amplitude, the action and the period of the orbit, respectively, (all of them evaluated at the Fermi energy μ), and the extra phase ϕ in the exponent contains the Maslov index. All these quantities have been specified in Chapters 5 and 6. We now assume for the following that just one single orbit (or orbit family) Γ is mainly responsible for the value of δE in (8.2). Then, the locus in the (N, η) plane where δE is stationary is essentially given by the condition that the action S_{Γ} of that orbit be stationary, in the spirit of the stationary phase approximation. This yields the condition

$$\delta S_{\Gamma}(N, \eta) = \frac{\partial S_{\Gamma}}{\partial N} \delta N + \frac{\partial S_{\Gamma}}{\partial \eta} \delta \eta = 0. \tag{8.3}$$

Now we exploit the relation $\partial N / \partial \mu = g(\mu)$ between the particle number and the level density to write

$$\frac{\partial S_{\Gamma}}{\partial N} \delta N = \frac{\partial S_{\Gamma}}{\partial \mu} \frac{1}{g(\mu)} \delta N. \tag{8.4}$$

Next, using the fact that the derivative of the action with respect to the energy (here the Fermi energy) gives the period: $\partial S_\Gamma(\mu)/\partial\mu = T_\Gamma$, we finally get

$$\frac{\partial N}{\partial\eta} = -\frac{1}{T_\Gamma}\,g(\mu)\,\frac{\partial S_\Gamma}{\partial\eta}. \tag{8.5}$$

Eq. (8.5) gives directly the slopes of the conditional extrema, and thus also the minima, in the (N,η) plane in terms of the total level density, and the period T and the quantity $\partial S/\partial\eta$ of the leading periodic orbit (family) Γ. The latter quantity measures the rate of change of its action with increasing deformation and is easily evaluated for a given orbit. If more than one orbit (family) is important for the shell-correction, the above formula will have to include the corresponding summations:

$$\frac{\partial N}{\partial\eta} = -\frac{g(\mu)}{(\sum_\Gamma c_\Gamma T_\Gamma)}\,\sum_\Gamma c_\Gamma\frac{\partial S_\Gamma}{\partial\eta}, \tag{8.6}$$

where c_Γ equals $i/\hbar$ times the terms summed over in (8.2).

It is already most interesting to watch the sign of $\partial S_\Gamma/\partial\eta$. In fact, from the negative slopes $\partial N/\partial\eta$ seen in Fig. 8.1 we can conclude that here the action of the leading orbit(s) must *increase* with deformation. As Strutinsky et al. have shown [23], the opposite is true in an axially deformed harmonic-oscillator potential: there the corresponding slopes are positive and thus the leading orbits have a decreasing action with growing η. The explanation is that in a potential with steep walls, the leading orbits at moderate prolate deformations are those lying in a plane containing the symmetry axis, whereas in harmonic-oscillator potentials, which have no well-defined surface region, the most important orbits lie in the equatorial plane perpendicular to the surface.

Strutinsky and coworkers [22, 23] did not, in fact, evaluate the periodic orbits in the smooth Woods-Saxon potential, which can only be found numerically. Instead, they considered a spheroidal cavity with infinitely steep, reflecting walls and a constant volume (independent of deformation). Its classical orbits are very similar to those in a smooth potential with a sufficiently steep surface. (This has later been verified in the context of metal cluster physics, see Sect. 8.2 below.) The periodic orbits in a spheroid with not too large prolate deformation fall into two classes: planar orbits lying in a plane containing the symmetry axis (let us call them "type A" orbits), and orbits that lie in the central equatorial plane perpendicular to the symmetry axis ("type B"). At larger deformations ($\eta \gtrsim 1.6$), non-planar orbits also exist, but these are longer than the planar ones. There is also an isolated pendular orbit along the symmetry axis which has very little weight because of its non-degeneracy. (For a complete classification of all periodic orbits in a spheroidal cavity, see Ref. [24].) The type B orbits are simply those of the circular billiard which forms the cross section in the equatorial plane, i.e., the diameter and polygonal orbits discussed in Sect. 6.1.3 (see Fig. 6.6). They have a one-dimensional degeneracy corresponding to rotations around the symmetry axis. The type A orbits, however, are two-fold degenerate. Within their plane of motion, they form degenerate families of polygons with 3, 4, 5, etc. reflections, as illustrated in Fig. 6.16 for the two-dimensional elliptic billiard. In addition, each family may be rotated continuously around the symmetry axis. This two-dimensional degeneracy comes from the fact that the spheroidal cavity is integrable. For this reason, the orbits of type A, although longer than the equatorial polygons for prolate spheroids, have larger amplitudes in the trace formula by a factor $(pR/\hbar)^{1/2}$ than the type B orbits (see the general discussion in Sect. 6.1.4). Only for much larger deformations (with η of the order 2 or more), the latter are so much shorter that they become more important [25, 26].

It is clear, now, that the lengths and thus the actions of the leading type A orbits increase with prolate deformation ($\eta > 1$). Furthermore, it turns out that the quantity $\partial S/\partial \eta$ is the same for all of them [25, 26]. Thus, even if more than one of them contribute to δE, the slopes $\partial N/\partial \eta$ are still proportional to the quantity $\partial S/\partial \eta$ of just one of them. (In Ref. [23], only the 4-cornered rhombus family was considered.) The slopes (8.5) thus obtained are shown in Fig. 8.1 by the heavy bars. They fit the actual slopes of the quantum-mechanically calculated valleys surprisingly well – considering the simplified potential and the fact that no spin-orbit interaction was included. The latter, of course, affects the positions of the minima as functions of N, i.e., the magic numbers N_0, so that these had to be adjusted. But the slopes of the valleys and their rough dependence on the size of the system agree almost quantitatively. This pioneering result demonstrates convincingly that the gross-shell structure leading to the experimental nuclear ground-state deformations may be explained semiclassically from a few leading classical orbits.

For spheroidal harmonic-oscillator potentials the situation is different, as already stated above. There, the Lissajous-type periodic orbits lying in the equatorial plane have the largest degeneracies. Their actions decrease with increasing $\eta > 1$, and the corresponding slopes of the minimum valleys are positive. The same type of quantitative agreement as that shown in Fig. 8.1 was achieved in Ref. [22, 23] for the corresponding calculations.

Frisk [25] has taken up the calculations for spheroidal cavities, extending them also to oblate deformations, using the method of Berry and Tabor [27] discussed in Sect. 2.7 to evaluate the explicit trace formula for the level density. For prolate deformations he confirmed the results of Strutinsky et al. and pointed out the importance of the equatorial orbits at large deformations. For the latter, he was using the trace formula given by Balian and Bloch [28] (see also Sect. 8.1.3 below for the expressions), since they cannot be described by the method of Berry and Tabor in its standard form. For oblate deformations, Frisk found much less pronounced shell effects and weaker slopes of the minimum valleys. This is explained by the fact that for oblate deformations ($\eta < 1$), several orbits of comparable lengths contribute and these have much weaker deformation dependences $\partial S/\partial \eta$ than on the prolate side. This observation was taken in Ref. [25] as an explanation for the fact that there exist many more nuclei with prolate than with oblate deformations. Extensions of these semiclassical investigations of the shell structure of spheroids, and of non-integrable cavities with various multipole deformations, are presently being carried on [26, 29].

8.1.2 The double-humped fission barrier

When a heavy nucleus fissions, its shape undergoes dramatic changes. As one excites
a nucleus by bombarding it with gamma rays, neutrons, or other particles, it begins
to oscillate anharmonically first mainly in the quadrupole mode, but then it slowly
develops a waist. Eventually, it constricts itself at the waist line to break up into (usually) two fragments. This has been shown in the famous paper of Bohr and Wheeler
[30], who explained the fission process by the liquid drop model (LDM) in which the
nucleus is considered as a uniformly charged liquid. (See, e.g., Wilets [31] and chapter
12 of Ref. [17] for easily readable introductions on nuclear fission.) In detailed numerical LDM studies by Cohen and Swiatecki [32], the nuclear shape was expanded in a
multipole series, writing the radius vector R that points towards the surface as

$$R(\theta) = \frac{R_0}{\xi} \left[1 + \sum_{\lambda=2,4,\dots} a_\lambda P_\lambda(\cos\theta) \right]. \tag{8.7}$$

Hereby R_0 is the radius of the undeformed sphere, $\xi(a_2, a_4, \dots)$ is a normalization factor
used to keep the volume constant (simulating the incompressibility of the liquid drop),
and $P_\lambda(x)$ is the Legendre polynomial of order λ. The set of deformation parameters
$(a_2, a_4, \dots)$ span a multi-dimensional space. In this space one has to search the conditional shapes that lead in the energetically most favourable way from the ground state
minimum – which is always spherical in the LDM – over a saddle point to the valley
that leads down to fission. Due to the charged protons, the fragments will repel each
other, releasing a kinetic energy of about 200 MeV per event. Within the LDM, axially
symmetric and left-right symmetric shapes with even λ are always most favourable.
To find the optimal sequence of shapes leading to the lowest saddle-point energy, i.e.
the lowest height of the fission barrier, Cohen and Swiatecki [32] used multipole orders
up to $\lambda = 18$.

In Ref. [33], the optimized shapes determined by Cohen and Swiatecki were re-
parameterized by a simple functional form of the surface, given in cylindrical coordinates (ρ, z, ϕ) by $\rho^2(z)$ as a fourth-order polynomial in z:

$$\rho^2(z) = (c^2 R_0^2 - z^2) \left[A + \alpha(z/cR_0) + B(z/cR_0)^2 \right]. \tag{8.8}$$

Here c is a dimensionless parameter that measures the half-length of the liquid drop
along the symmetry axis (i.e., the z axis) in units of the spherical radius R_0; the
parameters A and B allow for elongation and neck formation, and $\alpha \neq 0$ yields left-
right asymmetric shapes corresponding to odd multipoles in the expansion (8.7). The
volume of the liquid drop is fixed by the condition

$$\pi \int_{-cR_0}^{+cR_0} \rho^2(z)\, dz = \frac{4\pi}{3} R_0^3, \tag{8.9}$$

which imposes the constraint $A = 1/c^3 - B/5$. For $\alpha = B = 0$, Eq. (8.8) describes
spheroids with an axis ratio $\eta = c^{3/2}$, $B > 0$ allows for the formation of a neck. For
convenience, a linear transformation from the parameter B to a new neck parameter h
is done

$$B = 2h + \frac{1}{2}(c - 1), \tag{8.10}$$

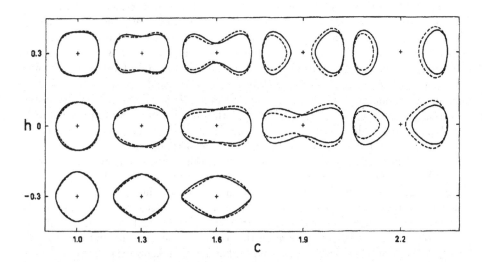

Figure 8.2: Typical axially symmetric nuclear shapes occurring in the nuclear fission process. $2c$ is the length of the nucleus along the symmetry axis in units of the spherical radius R_0, and h regulates the neck formation. The symmetric shapes (solid lines) along $h = 0$ correspond to the adiabatic path to fission for actinide nuclei, resulting from the liquid drop model. The dashed shapes have left-right asymmetry. (From [33].)

leading to the family of shapes shown in Fig. 8.2. By rotating these shapes around the symmetry axis (horizontal in the figure), one obtains the surface of the liquid drop. The (c, h) parameters are chosen in such a way that for typical actinide nuclei, the line $h = 0$ roughly follows the series of shapes leading over the LDM barrier along the way of steepest descent, as determined in [32]. The dashed shapes are obtained for $\alpha \neq 0$ which gives left-right asymmetric shapes which become important when the shell effects are included (see the next section). The separation of the shapes into two touching fragments, i.e. scission, occurs at $h = (5/2c^3) - (c - 1)/4$.

In the calculations of Ref. [33], the droplet model of Myers and Swiatecki [34] was used for the LDM energy, and Woods-Saxon type shell-model potentials (including spin-orbit interaction) were used to obtain the shell-correction energies of the neutrons and protons; the Woods-Saxon potentials were deformed in such a way that the loci of the steepest points in the surface coincide with the shapes of the liquid drop shown in Fig. 8.2. As a representative example, we show in Fig. 8.3 on the next page the deformation energy surface obtained for the nucleus ^{240}Pu, plotted as a function of c and h. The upper right panel contains the LDM deformation energy $E_{LDM}(N, Z; c, h)$. At the left edge ($c = 1$, $h = 0$), we see the spherical ground-state minimum whose energy is normalized to zero, and with increasing c along $h = 0$ one goes over the saddle which has a height of about 3.8 MeV. The two left panels contain the individual shell-correction energies of the protons (Z=94) and neutrons (N=146). Both show pronounced shell effects, with valleys that lead away from the spherical shape. When adding up all three contributions according to Eq. (8.1), one obtains the landscape shown in the lower right panel of Fig. 8.3. Due to the shell effects, the ground-state energy is seen to be lowered by about 2.2 MeV; the minimum is now located at a deformed shape with $c \simeq 1.2$ and $h \simeq -0.12$. Moreover, the topology of the deformation energy surface is changed rather dramatically by the shell effects, leading to the existence of two saddle points separated by a second minimum which is located at a deformation with $c \simeq 1.42$ and $h \simeq 0$ and an energy relative to the ground state of about 3 MeV. Going along the way of steepest descent from the ground state over the two saddle points via the second minimum, the nucleus thus exhibits a *double-humped fission barrier*, as a result of the microscopic shell effects, rather than the smooth single-humped barrier resulting from the liquid drop model. Along this path the deformation energy has the systematic pattern shown in Fig. 8.4, which is typical for many actinide nuclei.

A lot of interesting physics is connected with this double-humped fission barrier. It would lead too far to discuss all the nice features connected with the existence of a shape isomer, such as the spectroscopy within the second minimum and the resonant tunneling through the double barrier. Much of this is discussed extensively in the two reviews on the double-humped fission barrier given in Ref. [35]. But first of all, the very existence of a second minimum on the way to fission gave the first quantitative explanation of the short-lived fission isomers that had been known experimentally since 1962 [36], but not theoretically understood. The discovery of this second minimum in

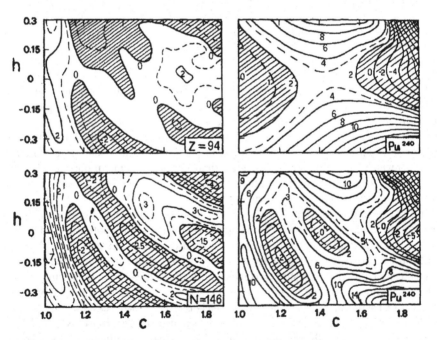

Figure 8.3: The deformation energy surface of the ^{240}Pu nucleus. Solid lines have an equidistance of 2 MeV, dashed lines one of 1 MeV. *Upper right:* LDM energy E_{LDM}; *upper left:* proton shell-correction δE_p; *lower left:* neutron shell-correction δE_n; *lower right:* total energy (8.1). The LDM ground-state energy at $(c = 1, h = 0)$ has been fixed at zero energy. In the right panels, regions below 2 MeV and in the left panels, those below 0 MeV are shaded. (From [33].)

the deformation energy of heavy nuclei by Strutinsky in 1966 [19], who used a simple Nilsson-model potential [12], therefore led to a triumph of the shell-correction method which was later consolidated in calculations with more sophisticated nuclear shell-model potentials [33, 37, 38, 39]. Note that the isomers occur at shapes with an axis ratio η close to 2.

One more dramatic shell structure effect is the asymmetry in the mass distribution of the fission fragments, which for many actinide nuclei is peaked around a mass ratio of ~ 1.35 - 1.45. This cannot be explained within the classical liquid drop model which always yields equal fragments for heavy nuclei. In fact, at any point in the (c, h) surface, the LDM energy would be increased by switching on the left-right asymmetry parameter $\alpha \neq 0$, leading to the dashed shapes shown in Fig. 8.2. However, it turns out that the shell effects favour such asymmetric shapes in the region of the second barrier and beyond it, leading to an overall lower total energy until the point of scission where two unequal fragments emerge. We shall discuss this topic and its semiclassical description in the following section. Nonaxial deformations also lead to a slight reduction of the total energy near the barrier tops. These effects are illustrated schematically in Fig. 8.4. (See Ref. [40] for a short review on barrier systematics and for further references on fission).

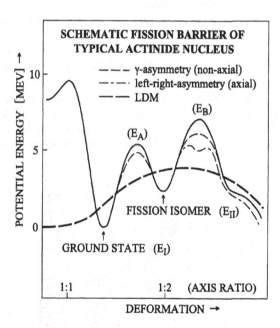

Figure 8.4: Systematic picture of the double-humped fission barrier of a typical actinide nucleus. The solid line is found with axially and left-right symmetric shapes. Nonaxial deformations tend to lower the barrier tops, and left-right asymmetric deformations lower the second barrier and the region beyond it, leading to unequal fission fragments. The heavy dashed line is the LDM barrier. (From [40].)

8.1.3   The mass asymmetry in nuclear fission

The mass asymmetry of the fission fragments, found experimentally for many actinide nuclei, had been a long-standing puzzle since it could not be explained by the liquid drop model. Early attempts to associate them to an instability of the nucleus against octupole deformations [41, 42] failed, partially because simple octupole deformations were not sufficient as it turned out later, and partially because no realistic enough models for the nuclear binding energy including shell effects were available at their time. It is only through the advent of Strutinsky's shell-correction method – and roughly a decade later, through the progress made with Hartree-Fock calculations using realistic effective interactions – that a quantitative account could be given of the microscopic shell effects in the total nuclear binding energies, especially at such large deformations as they occur in the fission process.

Möller and Nilsson [43] pointed out, using the shell-correction method and the Nilsson model, that a suitable linear combination of P_3 and P_5 deformations in the general multipole expansion (8.7) leads, indeed, to an instability of the total nuclear energy at the deformation of the fission isomer and the region beyond it. In the parametrization (8.8), the same was found to happen when the parameter $\alpha \neq 0$ was turned on in the shell-correction calculations with deformed Woods-Saxon potentials [44]. In Fig. 8.5 we show an excerpt of the deformation energy surface of ^{240}Pu, shown as a function of elongation c and asymmetry α, taken at the fixed value $h=0$ of the neck parameter which corresponds to the fission isomer. We see that the isomer minimum is unstable against α, and that the second fission barrier is lowered by about 2.5 MeV when going along the path of steepest descent (shown by the dashed curve in the figure).

In Fig. 8.5, we see only the onset of the mass asymmetry for shapes that are not yet fissioned, since along $h=0$ the scission point occurs only for $c \gtrsim 2$ (see Fig. 8.2). But it turns out that the gain in energy due to left-right asymmetric shapes continues all the way down the fission barrier, until the moment of scission where the nucleus is

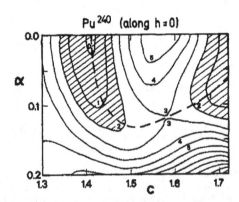

Figure 8.5: The onset of mass asymmetry in the fission of ^{240}Pu. Shown is the total nuclear energy $E_{LDM} + \delta E_n + \delta E_p$ (8.1), evaluated as in the lower right panel of Fig. 8.3, but here versus elongation c and mass asymmetry α for $h=0$ (equidistance 1 MeV, regions below 2 MeV shaded). (From [44].)

breaking into two fragments of different size. Dynamically, there exists a whole variety of fission paths, leading to a continuous mass distribution of the fragments. The maxima of this distribution correspond typically to a mass ratio of $\sim$ 1.3 to 1.5 for the lighter actinide nuclei [33, 40]. This asymmetry is thus a clear consequence of the quantum shell structure.

Shell effects exist, of course, also in the fragments themselves, and in many cases the most probable mass ratios can also be explained in terms of the magic nucleon numbers of the (spherical or deformed) fragments, i.e., in their enhanced stabilities. It is highly interesting and non-trivial, however, that roughly the same ratio of the nascent fragments is found in the static deformation energies, already quite early along the way of steepest descent and quite a way before the fragments are actually separated.

We shall in the following give a short account of a recent semiclassical description [45] of this shell effect, using the shortest periodic orbits in a simplified potential. Replacing the Woods-Saxon potential by a cavity with reflecting walls, we can use directly the shapes defined by Eq. (8.8) for the boundary of the cavity. In the region of deformations where the mass asymmetry sets in (see Fig. 8.5), the most important periodic orbits can be expected to be those lying in equatorial planes perpendicular to the symmetry axis. Obviously, only those equatorial planes contain periodic orbits which lie at a stationary point of the shape function $\rho(z)$, i.e., at the solutions z_i of $\rho'(z_i) = 0$, which in the parametrization (8.8) leads to a cubic equation. As long as the shapes are not necked in, there is only one solution, after the neck has evolved there are three solutions. This is shown in Fig. 8.6, where we have plotted the shape corresponding to the isomer minimum on the left, and a shape occurring after passing over the saddle (cf. Fig. 8.5) on the right. The vertical lines give the shortest periodic orbits in the corresponding two-dimensional billiards which both are non-integrable, the asymmetric one actually being highly chaotic; all these orbits are thus isolated. When rotating the shape function around the z axis, the periodic orbits are not only the linear diameters and their repetitions, but all polygons that appear in a circular billiard of the corresponding radius and are characterized by the pairs of integers (v, w) discussed in Sect. 6.1.3. These orbits all have a one-dimensional degeneracy due to rotation around the symmetry axis.

Balian and Bloch [28] have evaluated the contributions to the trace formula from these equatorial periodic orbits in any axially deformed cavity. The decisive quantity

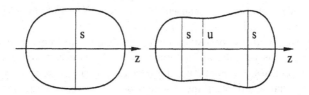

Figure 8.6: Two typical shapes occurring during the fission of ^{240}Pu. *Left side:* at the isomer minimum with c=1.42, h=0, and α=0. *Right:* beyond the outer barrier, at c=1.7, h=0, and $\alpha = 0.08$. Vertical lines indicate the shortest orbits in the corresponding two-dimensional billiards (s = stable, solid lines; u = unstable, dashed line). This system is quite chaotic and all orbits are isolated.

which regulates their stability is a parameter A_{vw} given by the ratio of the two principal curvature radii at the reflection points, R_1 and R_2, and the numbers v, w :

$$A_{vw} = 1 - \frac{2R_1}{R_2} \sin^2 \varphi_{vw}, \qquad \varphi_{vw} = \pi \frac{w}{v}. \qquad (v \geq 2w) \qquad (8.11)$$

Hereby R_2 is the curvature radius in the plane containing the symmetry axis, and $R_1 = \rho(z_i)$ is the radius of the equatorial circle. Depending on the sign and the absolute value of A_{vw}, the stability factor $D_{vw}^{1/2}$ in the denominator of the trace formula is given by [28]

$$
\begin{aligned}
A_{vw} > 1: & \quad A_{vw} = \mathrm{Cosh}\chi, & D_{vw}^{1/2} &= 2\,\mathrm{Sinh}(v\chi/2), \\
|A_{vw}| < 1: & \quad A_{vw} = \cos\chi, & D_{vw}^{1/2} &= 2i\,\sin(v\chi/2), & (\chi < \pi) \\
A_{vw} < -1: & \quad A_{vw} = -\mathrm{Cosh}\chi, & D_{vw}^{1/2} &= \left\{ \begin{array}{ll} 2i^v\,\mathrm{Sinh}(v\chi/2) & v \text{ even} \\ 2i^v\,\mathrm{Cosh}(v\chi/2) & v \text{ odd} \end{array} \right\} \text{ for}
\end{aligned} \qquad (8.12)
$$

In terms of the quantity $D_{vw}^{1/2}$, the contributions $\delta g_{diam}(E)$ from the diameter orbits and $\delta g_{poly}(E)$ from the polygons to the trace formula are then [28]

$$
\begin{aligned}
\delta g_{diam}(E) &= \Re e \frac{1}{\sqrt{\pi k}} \sum_{w \geq 1} (-1)^w \frac{R_1^{3/2}}{(2w D_{vw})^{1/2}} e^{i(4wkR_1 + \pi/4)}, \\
\delta g_{poly}(E) &= \Re e \frac{1}{\sqrt{\pi k}} \sum_{w \geq 1} \sum_{v > 2w} i^v \frac{2(R_1 \sin \varphi_{vw})^{3/2}}{(v D_{vw})^{1/2}} e^{i(2vkR_1 \sin \varphi_{vw} + \pi/4)}, \quad (8.13)
\end{aligned}
$$

where Dirichlet boundary conditions have been assumed and $\hbar^2/2m = 1$.

At the deformations where the three orbit planes coincide (i.e., cases where all three solutions z_i of $\rho'(z_i) = 0$ are real and identical), bifurcations of the single equatorial orbits take place. At these points the amplitudes in the trace formula (8.13) become singular, as always in connection with orbit bifurcations. To regularize these singularities, a suitable uniform approximation was used in [45].

In Fig. 8.7 we show a contour plot of the averaged level density $\delta g(E_F)$, obtained with the formulae (8.13) above, as a function of c and α along $h=0$ in the same region of deformations as seen in Fig. 8.5. Only the primitive orbits ($w=1$) of the families with v up to 3, lying in the equatorial planes illustrated in Fig. 8.6, have been included. The Fermi energy E_F is chosen to have some $N \sim 150$ particles in the cavity, which corresponds roughly to the neutron number of ^{240}Pu. The contributions of the square orbits already affect the results by less than a few percent, so that the neglect of the longer orbits is justified and no extra averaging is necessary. It is, in fact, the triangular orbit families that are required to give the correct instability against the mass asymmetry and that are dominating the final results. Even leaving out the diameter orbit contributions, the topology in the (c, α) plane would be roughly the same. (The rapid convergence of the sums in (8.13) was checked in [46] against quantum-mechanical results of δg using the same cavity model, including a coarse-graining corresponding to the smoothing caused in nuclei by the pairing interactions.) A three-dimensional perspective view of the shell-correction energy δE, obtained according to Eq. (8.2) with $\mathcal{A}_\Gamma$ taken from (8.13), is shown in Fig. 8.8 for the same deformation region. On its left

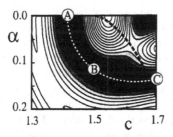

Figure 8.7: Contour plot of the semiclassical level density $\delta g\,(E_F)$ (8.13) (in arbitrary units), plotted like the total energy in Fig. 8.5. The black dashed line is the locus of the bifurcation points of the equatorial orbits. Along the white dashed line, the actions of the shortest periodic orbits are constant. (From [45].)

side we see the cavity shapes at the three points A, B, C of deformation space which are also marked in Fig. 8.7.

The qualitative agreement of the contour plot in Fig. 8.7 with that of the total energy shown in Fig. 8.5, which was evaluated with quantum-mechanical shell-corrections from realistic Woods-Saxon potentials including spin-orbit terms, is quite remarkable. In particular, the onset of the asymmetry instability at the isomer minimum, located exactly at c=1.42 as in the quantum calculations, is quantitatively correct. To understand some of the remaining differences, recall that the semiclassical expression for the shell-correction energy δE is directly proportional to the level density $\delta g\,(E_F)$, as long as only one periodic orbit family is dominating (see Sect. 5.5). Noting furthermore that the leading equatorial orbits here all have the same deformation dependence, it is therefore not surprising that the shell-correction energy δE has the same deformation behavior as $\delta g\,(E_F)$. Furthermore, a look at Fig. 8.5 tells us that the neutron shell-correction is far more important than that of the protons. The LDM energy sur-

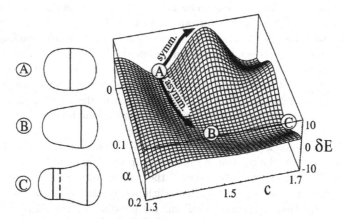

Figure 8.8: *Right:* Three-dimensional perspective view of the semiclassical shell-correction energy δE around the second fission barrier. *Left:* Shapes along the asymmetric fission path, taken at the three points A, B, C in deformation space which are marked also in Fig. 8.7 (vertical lines like in Fig. 8.6). (From [45].)

face is also rather smooth in the region of interest, but as a function of α it increases. Adding the LDM energy will thus reduce the magnitude of the α deformation. The latter is, indeed, slightly too large in the semiclassical result, as seen in Fig. 8.7. Finally, the neglected spin-orbit interaction affects mostly the position of the magic numbers, i.e., it rescales the N dependence of the shell effects, but it does not appear to change appreciably the deformation behavior of the shell effects. This had already been observed by Strutinsky et al. [23], who also replaced the realistic nuclear potentials by cavities with reflecting walls (see the discussion of Fig. 8.1 above).

Analogously to the discussion in Sec. 8.1.1, we can define the fission valley in the (c, α) plane (at constant $h = 0$) by the condition that $\delta E(c, \alpha)$ be stationary. From our semiclassical results, we can relate this directly to the stationary condition $\delta S = 0$ for the actions of the shortest equatorial orbits which leads, indeed, to the white dashed line in Fig. 8.7 that follows exactly the bottom of the asymmetric fission valley. We have thus found a surprisingly simple semiclassical explanation of a quantum effect in a complex physical system, leading to the mass asymmetry of the fission fragments, in terms of the least action principle applied to the shortest periodic orbits of the system.

It is worth looking back at the original quantum-mechanical investigations of this shell effect. Already in 1961, Johansson [42] had pointed at the importance of single-particle orbits with $n_z = 0$, in the notation of the asymptotic Nilsson quantum numbers[1] $[Nn_z\Lambda\Omega]$ (see Sect. 1.2.4), for a possible instability of the nucleus towards asymmetric distortions. Indeed, after the onset of the mass asymmetry in fission had been obtained in shell-correction calculations with the Nilsson model [43], Gustafsson et al. [47] traced the effect down to two sets of neutron single-particle orbits which are particularly sensitiv to the combined P_3 and P_5 deformations, denoted here by ϵ_{35}. Whereas most other single-particle energies depend little or not at all on ϵ_{35}, a group of levels $[N0\Lambda\Omega]$ decreases and a corresponding group of levels $[(N+1)1\Lambda\Omega]$ increases rather clearly with increasing asymmetry ϵ_{35}. This occurs through a strong coupling by the asymmetric part of the Nilsson Hamiltonian between pairs of such states whose matrix elements obey the selection rules $\Delta N = \Delta n_z = \pm 1$ and $\Delta\Lambda = \Delta\Omega = 0$. In particular, near the Fermi energy corresponding to $\sim 140 - 150$ neutrons, such pairs of levels can be found which repel each other, whereby the upwards bending levels lie above and the downwards bending levels lie below the Fermi energy. The net result is that the sum of occupied single-particle energies, which contains the main shell effect of the neutrons, becomes lower for $\epsilon_{35} \neq 0$ than in the symmetric case: hence the decrease of the total binding energy. (Part of the energy gain in the neutron shell-correction energies is, of course, counterbalanced by the increase of the average LDM energy; in the present case the shell effect is, however, stronger.)

Now, the down-going $[N0\Lambda\Omega]$ states were found to have the maxima of their probability densities along the equator of the nucleus in the $z=0$ plane, i.e., exactly in the plane where the shortest periodic orbits of the symmetric nucleus are located. This suggests a connection between the single-particle states responsible for the asymmetry effect on the quantum-mechanical level and the periodic orbits causing the onset of asymmetry in the semiclassical picture. In Ref. [48], this connection has been confirmed within the deformed cavity model discussed above. In Fig. 8.9 we see the lowest eigenenergies of the quantum states i with angular momentum $\Lambda = 4$, plotted versus

---

[1] Here Λ is the absolute value of the l_z quantum number.

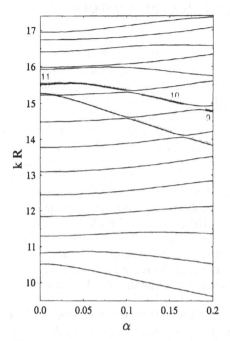

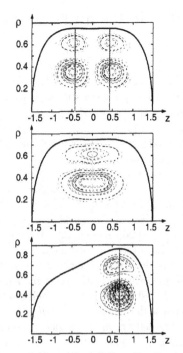

Figure 8.9: Quantum levels $k_i R$ for the lowest states i (numbered the from the bottom) with $\Lambda = 4$ at $c = 1.53$, $h = 0$, plotted versus asymmetry α. The emphasized portions connected by avoided crossings denote some of the diabatic states whose wavefunctions are concentrated on the planes containing the shortest periodic orbits. (From [48].)

Figure 8.10: Probability distributions $|\psi_i(z, \rho)|^2$ for selected quantum states with $\Lambda = 4$ at $c = 1.53$, $h = 0$. The heavy black lines give the boundary of $\rho(z)$. Grey vertical lines indicate the planes z_i of the shortest periodic orbits. *Top:* state $i{=}11$ at $\alpha = 0.0$. *Center:* state $i{=}10$ at $\alpha = 0.0$. *Bottom:* state $i{=}7$ at $\alpha = 0.2$. (From [48].)

the asymmetry parameter α at the elongation $c = 1.53$ (with $h = 0$) which corresponds to the symmetric outer fission barrier seen in Fig. 8.8. We clearly recognize some diabatic levels, connected by avoided crossings, which depend more strongly on α than the remaining levels. As demonstrated in Fig. 8.10, their probability amplitudes $|\psi_i(z, \rho)|^2$ have their maxima on the planes containing the shortest periodic orbits discussed above. The state $i = 10$, which sits on an unstable orbit at $\alpha = 0$, is an example of the "scars" that have been much discussed in connection with the quantum-classical correspondence for chaotic systems (cf. Sect. 1.5.2).

The classical phase space of the fissioning nucleus is, indeed, highly chaotic, as can be seen on the relevant Poincaré surfaces of section (cf. also [49]). The shortest periodic orbits discussed above live in some small regular islands in the middle of a chaotic see, and performing an approximate EBK quantization of the regular motion around these orbits [48] reproduces rather accurately diabatic quantum levels as those seen in Fig. 8.9. It is precisely this small regular part in phase space that is responsible for the strong shell effect leading to asymmetric fission. – We thus observe in this system the peaceful coexistence of chaotic dynamics with ordered motion, leading to shell effects with far-reaching consequences.

8.2 Shells and supershells in metal clusters

Metal clusters have become an object of intensive studies, both for technological applications and in basic research (see the recent review article by de Heer [50] on the experimental aspects of simple metal clusters). In their pioneering work in 1984, Knight *et al.* [51] discovered a striking regularity in the abundance spectrum of a beam of alkali clusters: the enhanced stability of species with "magic numbers" of valence electrons, $N_n = 2, 8, 40, 58, \ldots$, that could be associated with filled spherical shells in an average potential felt by the electrons. Independently, Ekardt [52] predicted in 1984 the same shell structure, based on self-consistent calculations with the spherical jellium model in which the valence electrons move in the field of a positively charged sphere representing the averaged density distribution of the ions. Soon thereafter, many experimental groups confirmed these findings, and the important role of the shell effects due to the valence electrons in metal clusters was well established. Clemenger [53] adapted the nuclear Nilsson model [12], omitting the spin-orbit interaction for which no evidence could been found in metal clusters. With this, he was able to interpret the finer structures in the mass spectra, occurring between the magic numbers, as due to spheroidally deformed shapes of the clusters. An early account of the experimental developments and the establishment of a shell model for simple metal clusters may be found in the review article by de Heer *et al.* [54].

The jellium model neglects the explicit structure of the ionic charge distribution within the clusters. It does therefore not apply to very small, crystal-like clusters at low temperatures, for which quantum-chemical *ab initio* methods have been successfully used for reproducing detailed experimental properties (for a review, see Ref. [55]). For sufficiently hot clusters, where many ionic configurations are mixed by their temperature motion, the jellium model can be used also for rather small systems, as demonstrated recently by probing their optical response explicitly as a function of temperature [56]. Also, in very large metal clusters containing many thousands of atoms, the ionic structure effects appear to dominate the energetics, and thus their stability. This is partially due to the temperature effect on the valence electrons: with increasing size, the distance between the electronic single-particle levels decreases and thus even moderate temperatures have an increasing smoothing effect (see Fig. 8.12 and the related discussion below). For sodium clusters with $N \gtrsim 1500$ atoms, an icosahedral packing of the ions has been found by Martin *et al.* [57], through the appearance of new sequences of magic numbers which are different from those of the electronic shells discussed below. Similar new magic numbers related to so-called "atomic shells" were also found for other types of metal clusters (see the recent review article by Martin [58]).

For sizes in between, however, covering the range of metal clusters with tens to several thousand atoms, the electronic shell effects do play an important role. Here the jellium model has been remarkably successful in reproducing many average cluster properties, in particular those related to the electronic shell effects. The neglect of the ionic structure in this model is helped by the fact that the de Broglie wave length of the valence electrons at the Fermi energy is larger than the average distance of the ions. This can easily be estimated by using the Thomas-Fermi expression of the electronic density ρ_e in terms of the Fermi wave number: $\rho_e = k_F^3/(3\pi^2)$. Equating the electronic density ρ_e with the negative ionic bulk density ρ_I (in the interior of the cluster, the electric charges must cancel), which is usually expressed in terms of the Wigner-Seitz radius r_s by $\rho_I = 3/(4\pi r_s^3)$, we can relate k_F to r_s. In this way we find that the de Broglie wave length is

$$\lambda = 2\pi/k_F = (32\pi^2/9)^{1/3} r_s = 3.274 \, r_s, \qquad (8.14)$$

and thus ~ 1.6 times the average distance $2\,r_s$ between the ions. Therefore the electrons do not "see" the details of the ionic distributions. The self-consistent determination of their mean potential, which includes the effects of the mutual Coulomb repulsion, is done in the framework of the density functional theory using the local density approximation (see Appendix A.2). For an overview of the jellium model (including extensions for deformed systems) and its application to simple metal clusters, we refer the reader to a recent review [59].

The pronounced shell structure with its consequences, including the deformations and the occurrence of collective vibrations, bears many resemblances with the corresponding properties of atomic nuclei. Not surprisingly, metal clusters have therefore attracted the attention of many nuclear physicists (see, e.g., Ref. [60]). Nuclei are limited in size, due to their decay by fission as a consequence of the Coulomb repulsion between the protons. Metal clusters, however, can be made neutral and are stable for any size. They present thus an ideal instrument to study the long-range periodicity of shell effects. In Sect. 6.1.6, we have discussed this aspect of the shell structure for particles confined to a spherical box potential: the beating amplitude of the oscillating level density which had been found by Balian and Bloch [28] in their early development of the periodic orbit theory. Nishioka, Mottelson, and Hansen [61] extended this work to realistic Woods-Saxon potentials which were fitted to the self-consistent mean fields obtained in jellium model calculations [52, 62]. Employing the method of Berry and Tabor [27], they also found the beating amplitude of the shell oscillations in both the level density and the energy shell-corrections, and proposed that these "supershells" should be observable in metal clusters.

This prediction triggered an enormous effort to observe the electronic shell structure in increasingly larger metal clusters. In Fig. 8.11 on the following page, we show the mass abundance spectra from an adiabatic expansion source using sodium clusters, observed by Bjørnholm *et al.* [63]. The upper part contains the total mass yield I_N as a function of the number of atoms N (which is the same as the number of valence electrons since Na has the valence 1). Besides a gross distribution which is determined by the particular experimental conditions (temperature, pressure and form of the nozzle through which the sodium vapor expands into a vacuum), we clearly see some superimposed fine structure with maxima at the magic numbers $N_n = 92, 138, 198$, etc. These numbers are exactly the ones found in the jellium model [52] or shell-model [61] calculations for filled major spherical shells. We see, however, that the amplitude of these oscillations vanishes with increasing size. In order to focus on the shell oscillations, the lower part of Fig. 8.11 contains the first differences of the logarithms of the *relative* yields, obtained by dividing the curve I_N by its suitably fitted average, and further averaged over a range $N \pm K_0$ with $K_0 = 0.03N$ in order to eliminate statistical fluctuations (for details, see Bjørnholm *et al.* [63]). In this way the oscillations are dramatically magnified and can be seen to persist up to sizes of $N \sim 1600$; the statistics become, however, rather bad. The reason for looking at the logarithmic differences $\Delta_1 \ln I_N$ is the following. The expanding cluster beam is originally rather hot and cools down, mainly by evaporating single neutral atoms from the individual clusters. If this

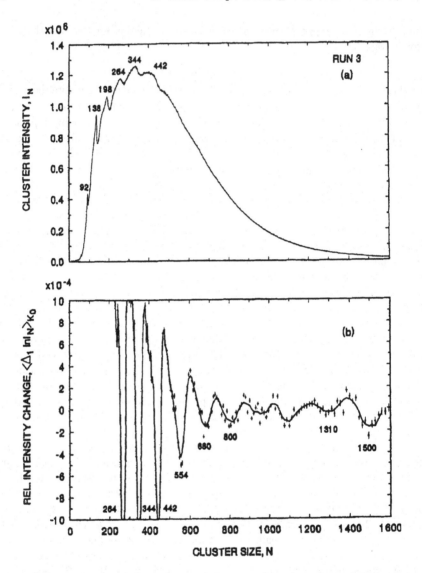

Figure 8.11: Total yields I_N (upper part) and their logarithmic differences (lower part; see text and Ref. [63] for definition) of Na clusters from an adiabatic expansion source. (Courtesy S. Bjørnholm.)

could be described by an evaporative ensemble at thermodynamical equilibrium (which is not really correct; see the discussion further below), the abundances would be dominated by a Boltzmann factor:

$$I_N \sim \exp[-\Delta_1 F(N-1)/k_B T], \qquad (8.15)$$

where $F(N)$ is the total free Helmholtz energy of a cluster with N atoms, $\Delta_1 F(N)$

is its first difference, $\Delta_1 F(N) = F(N+1) - F(N)$, and T is the temperature. Since $\Delta_1 F(N)$ is (up to a constant) equal to the free dissociation energy of one neutral atom, the Boltzmann factor (8.15) is a measure of the relative stability of the cluster N against evaporation of a monomer, i.e., against the process $Na_N \longrightarrow Na_{N-1} + Na_1$. This is why we observe maxima of the abundance curve I_N in the upper part of Fig. 8.11 for the spherically magic clusters which are particularly stable. The quantity $\Delta_1 \ln I_N$ now becomes proportional to the negative second difference of the free energy,

$$\Delta_1 \ln I_N \sim -\Delta_2 F(N) = 2F(N) - F(N+1) - F(N-1). \tag{8.16}$$

The quantity $\Delta_2 F(N)$, on the other hand, can easily be understood to have maxima for the magic clusters. If we think of the total binding energy at $T=0$, where $F(N) = E(N)$, plotted versus the particle number N, its curve has pronounced minima at the magic numbers N_n (see, e.g., the nuclear binding energies shown in Fig. 4.3 at the end of Chapter 4). Hence its second differences, which are equal to the curvatures for large N, have maxima. This is why the magic numbers appear as sharp minima in the lower part of Fig. 8.11, at least for smaller N where the amplitude of the oscillations is large.

The fast decrease of the amplitude of the shell oscillations with increasing size N of the clusters is due to the temperature. In order to make the oscillations experimentally detectable, one *must* start from hot clusters, because it is the evaporation mechanism that leads to the enhanced abundances of the most stable species, as we have just seen. On the other hand, a finite temperature has a smoothing effect on the level density and the quantities derived thereof, as we have discussed in Sects. 3.2.5 and 5.5. This is most directly shown in Fig. 3.3 for the three-dimensional harmonic oscillator, where the trace formula for the level density is known analytically. We have seen there that a temperature T (measured in energy units, putting $k_B = 1$) of about half the level spacing $\hbar\omega$ is sufficient to completely wipe out the shell effects, so that the exact temperature-averaged level density becomes identical with its smooth extended Thomas-Fermi part. Consequently, also the total free energy $F(N)$ becomes smooth, i.e., the shell-corrections $\delta F(N)$ are smoothed out. As shown in Eq. (3.77) for the harmonic oscillator, the decrease of the leading oscillating term is governed by a temperature-dependent damping factor

$$f(T) = \frac{\tau_n}{\text{Sinh}(\tau_n)} \sim \exp(-\tau_n), \qquad \tau_n = (2\pi^2 nT/\hbar\omega) \tag{8.17}$$

which falls off exponentially with T at large temperatures $T \gg \hbar\omega$. Here n is the repetition number of the periodic orbit. The same factor will also suppress the leading terms in the shell-correction energy $\delta E(N)$ evaluated at $T=0$. As discussed in Sect. 5.5, this factor (8.17) is the same also for other potentials in the semiclassical approximation, but τ_n has to be replaced by $\tau_\Gamma = \pi T T_\Gamma / \hbar$, T_Γ being the action of the (n times repeated) periodic orbit, see Eq.(5.54). Still, the harmonic oscillator result gives us a guide to the rate of suppression of the shell-correction energy by the temperature. Since for a saturated system, the Fermi energy is constant as a function of N, we see from Eq. (4.200) in Problem 4.2 that $\hbar\omega$ must be proportional to $N^{-1/3}$. Hence the exponential factor in (8.17) goes like

$$f(T, N) \sim \exp(-cTN^{-1/3}), \tag{8.18}$$

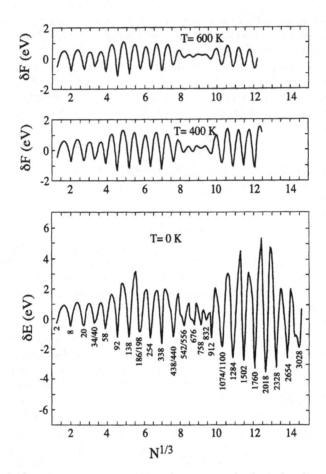

Figure 8.12:   Oscillating part of total free energy of spherical sodium clusters, obtained in self-consistent jellium calculations at finite temperatures, plotted versus cube root of the number N of valence electrons. (After [65].)

with c being a constant characteristic of the potential (or, equivalently, of the leading periodic orbits).

In Fig. 8.12, we show self-consistent microscopic results obtained for sodium clusters in the spherical jellium model at finite temperatures [64, 65]. Hereby the clusters were assumed to form a *canonical ensemble* and all relevant quantities were derived from the exactly calculated canonical partition function. This means that the number N of electrons in each cluster was determined exactly, not only on the average like in a grand-canonical system; for finite systems this makes a difference (see Ref. [64] for a detailed discussion and a fast numerical recipe for calculating the canonical partition function). The figure shows the oscillating shell-correction part of the total free energy $F(N)$ at the temperatures $T = 0$, 400, and 600 K, plotted versus the cube-root $N^{1/3}$ of the particle number which is proportional to the cluster radius: $R = r_s N^{1/3}$. (Here R is the radius of the spherical jellium background density distribution). In the lower part at $T=0$, we see the regular shell structure with its sharp minima at the magic numbers N_n. This part of the figure is very similar to the result obtained by Nishioka *et al.* [61] using a Woods-Saxon potential, confirming the latter from the self-consistent point of view. It exhibits not only the strict regularity of the main shells, lying equidistant on the scale $N^{1/3}$ as discussed further below, but also the supershell beating pattern known from the work of Balian and Bloch. At higher temperatures, these essential features are preserved, but the amplitude of the oscillations is reduced.

As we have just seen, however, the cluster abundance spectra are not related directly to the total energies $F(N)$ but rather to their first or second differences. In Ref. [64] it was shown that the differences $\Delta_1 F(N)$ and $\Delta_2 F(N)$ decrease even faster with temperature, so that at $T \simeq 600$ K, hardly any oscillations can be seen. This appeared at first discouraging for the quest for the supershells. Nevertheless, a joined effort of several experimental groups made the experimental observation of the supershells possible, first for sodium clusters [66] (for the largest sizes after an extra heating of the cluster beam by a laser [67]), and later also for lithium [68] and gallium clusters [69]. The only experiment in which the beating amplitude of the shell oscillations could be put into direct evidence was that of Pedersen *et al.* [66], as shown in Fig. 8.13. Here the logarithmically derived relative abundances are plotted like in the lower part of Fig. 8.11, but they have been multiplied here by an extra factor $N^{1/2} \exp(cN^{1/3})$. The

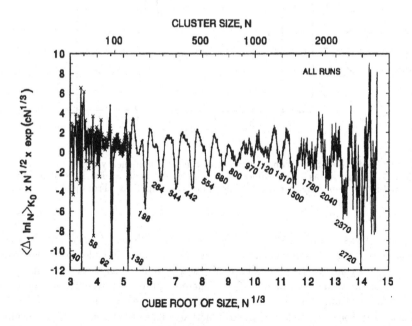

Figure 8.13: Experimental observation of the supershells. Shown are the logarithmically derived relative yields of sodium clusters from an adiabatic expansion source as in Fig. 8.11 (below), but scaled by an extra factor $N^{1/2} \exp(cN^{1/3})$ (see text for its explanation). (From [66].)

exponential factor compensates exactly for the estimated temperature smoothing factor (8.18), whereby the constant c, which here includes the temperature factor T, was adjusted to optimize the result. The extra factor $N^{1/2}$ was used to compensate for the estimated decrease of the shell-corrections at $T=0$ with increasing N [66]. (The power in this factor depends actually on the form of the potential. This is, however, immaterial here since the dominating scaling effect is brought about by the exponential factor.)

We have seen in Sect. 6.1.6 that in the case of the spherical cavity, the beating amplitude of the shell oscillations is a result of the interference of the two leading families of periodic orbits: the squares and the triangles. As shown by Nishioka *et al.* [61], the periodic orbits in a smooth Woods-Saxon potential have the same behavior, although they are not exactly polygons but rather have rounded corners due to the smoothness of the potential: their amplitudes still have similar ratios as in the spherical cavity. Fig. 8.13 therefore is a direct experimental manifestation of the role of two leading orbit families in a finite many-body system.

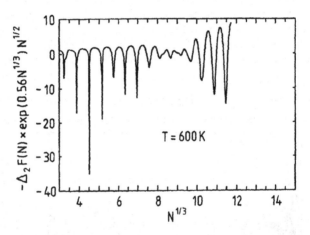

Figure 8.14: Negative second difference of the total free energy $F(N)$ of sodium clusters, calculated as in the spherical jellium model at $T = 600$ K, scaled and plotted in the same way as the experimental data in Fig. 8.13. (From [65].)

Fig. 8.14 shows a similar plot of the negative second differences $-\Delta_2 F(N)$ obtained in the self-consistent jellium calculations of Ref. [65], scaled and plotted in the same way as the experimental yields in Fig. 8.13 (the constant c was chosen within about 10 percent of the value used in Ref. [66]). There is quite some resemblance between the two figures, although many crude assumptions have been made in the model calculations. The least justifiable of them is actually that of an evaporative ensemble in thermodynamical equilibrium, leading to Eq. (8.16). The clusters are certainly no longer in a thermodynamic equilibrium after the first evaporation events. And more than one atom is evaporated by many clusters before they are cooled down. Since each cluster size is behaving differently, following its own chain of evaporations, one cannot speak of any well-defined temperature common to all clusters in the beam. As a matter of fact, the detailed forms of the experimental abundance spectra and their relation to the total electronic binding energies are still rather poorly understood, in spite of serious efforts to explain them by evaporation calculations, using both analytical and numerical (Monte Carlo) methods [70]. Therefore one cannot expect more than a qualitative agreement between theory and experiment at the present stage. It should also be mentioned that the values of the adjusted constants c in the exponential scaling factor used in the above two figures would lead, using the harmonic-oscillator estimate (8.17), to an effective temperature about 2 – 3 times larger than the estimated values of $\sim$ 400 - 500 K for the cluster beam in the experiments. A detailed discussion of the interpretation of the mass abundances obtained in evaporation experiments may be found in a recent review by Bjørnholm *et al.* [71].

The features that do not depend on the details (experimental or theoretical), however, are the periodicity of the main shells, which turns out to be universal and can be obtained under quite varying experimental conditions, and the fact that the peaks in the abundances coincide with the spherically magic clusters. The strict equidistance of the magic numbers on the scale $N^{1/3}$ is also well established, and it turns out to be independent of the species of metal clusters investigated. Both facts can be understood easily on the grounds of the periodic orbit theory, assuming only that the magic numbers are given by minimizing the total energy (at $T=0$), as we now want to show. We start from the trace formula for the spherical cavity, Eq. (6.39) as derived by Balian and Bloch [28]. Taking only the triangle and the square orbit families (without repetitions, i.e., with $w=1$ only), the level density reads (up to an overall phase)

$$\delta g\left(E\right) \simeq \frac{1}{E_0}\sqrt{\frac{kR}{\pi}}\left\{A_3\sin(kL_3 + \frac{\pi}{4}) + A_4\sin(kL_4 + \frac{3\pi}{4})\right\}, \qquad (8.19)$$

where the dimensionless amplitudes are $A_3 = (\sqrt{3}/2)^{1/2}$ and $A_4 = (\sqrt{2}/2)^{1/2}$, the lengths of the orbits are $L_3 = 3\sqrt{3}R$ and $L_4 = 4\sqrt{2}R$, $k = \sqrt{2mE}/\hbar$ is the wave number of the particle, and $E_0 = \hbar^2/(2mR^2)$ is the energy unit. Using the addition theorem of the sin function, the above can be rewritten as

$$\delta g\left(E\right) \simeq \frac{2}{E_0}\sqrt{\frac{kR}{\pi}}A_3\cos(k\bar{L})\cos(k\Delta L + \pi/4), \qquad (8.20)$$

neglecting an oscillating term proportional to $(A_3 - A_4)/2A_3$ which will contribute less than 5 percent. Hereby the two lengths appearing in the cos functions are

$$\bar{L} = \frac{1}{2}(L_3 + L_4), \quad \Delta L = \frac{1}{2}(L_4 - L_3). \qquad (8.21)$$

Eq. (8.20) now exhibits precisely the beating level density: apart from a smooth prefactor which goes like $E^{1/4}$, the term $\cos(k\bar{L})$ contains rapid oscillations, giving the main shells, and the term $\cos(\Delta L + \pi/4)$ gives the beating amplitude. We next use the estimate (8.2) for the energy shell-correction and proceed exactly in the same way, to find

$$\delta E \simeq \frac{4}{27}A_3\,E_F\,\cos(k_F\bar{L})\cos(k_F\Delta L + \pi/4), \qquad (8.22)$$

where E_F is the Fermi energy and k_F the Fermi wave number, and a neglected term proportional to $L_3^2(A_3/L_3^2 - A_4/L_4^2)/2A_3$ contributes less than ~ 11 percent. Looking for the minima of the rapid oscillations, we get the condition

$$k_F \bar{L} = (2n + 1)\pi, \qquad n = 0, 1, 2, \ldots \qquad (8.23)$$

This is, actually, as if we had quantized the average action of the triangle and squared orbits in the old Bohr-Sommerfeld way, writing $\hbar k_f \bar{L} = \bar{S}_{34} = h(n + 1/2)$. Inserting now the Thomas-Fermi value of k_F given in Eq. (8.14) and the actual value of the average length of the triangle and square orbits, $\bar{L} = 5.427 R = 5.427 r_s N^{1/3}$, we get a quantization condition for the "magic radii" $R_n = r_s N_n^{1/3}$ corresponding to the clusters with filled spherical shells:

$$R_n = r_s N_n^{1/3} = \frac{1}{2}(2n + 1) 0.6033 r_s. \qquad (8.24)$$

For the lowest values of n, this semiclassical prediction cannot be expected to hold. If one takes the experimental values N_n from the minima in Fig. 8.13 and plots the R_n versus n, these are found to form a straight line for $n \geq 6$, indeed. The slope of this line is experimentally 0.607 ± 0.006, in excellent agreement with (8.24). Note that the value 0.603 of the predicted slope does not depend on the nature of the cluster, specified here by the Wigner-Seitz radius r_s, when one plots just the dimensionless values of $N_n^{1/3}$ versus the shell number n. This was, indeed, found for all experiments with Na [66, 67], Li [68], and Ga clusters [69] within the same error bars [72]. The slope found from the microscopic jellium results in Fig. 8.12 is exactly 0.607. This shows that both the jellium model and the cavity model in this respect give excellent approximations. For a three-dimensional harmonic-oscillator potential, the slope would be 0.693 – which clearly rules out this kind of shell-model potential for large clusters (see also [59]). (The harmonic oscillator, however, is a good model for the lowest shells, as is evident from the lowest magic numbers up to 40.) In the large clusters with sizes up to $N \sim 3000$, it is thus the average length $\bar{L}$ of the dominating classical triangle and square orbits that determines quantitatively the main-shell spacing in this interacting fermion system.

Around $n \sim 13 - 15$, the straight line $N_n^{1/3}(n)$ is slightly broken and then continues parallel to the original line with the same slope up to the largest measured shells $n \sim 23$. This break of the line occurs at the first minimum of the beating amplitude, where the sequence of minima undergoes a phase change of π. According to Eq. (8.22) with $\Delta L = 0.2304 R$, this would lie at a radius $R_0 = 10.66 r_s$. From Fig. 8.13 we see that for Na clusters the beat minimum is found near $R \sim 9.5 r_s$, which is not much different in view of the uncertainty in the exact determination of the scaling factor used in that figure. (Using a slightly different power of N than $1/3$ in the exponential factor might easily shift the position of the beat minimum by 10 percent.) It should also be borne in mind that at the energies where the interference of the triangle and square orbits is destructive, the contributions from other orbits are no longer negligible. Some of them will affect the level density near those energies appreciably and make a more exact determination of the beat minimum difficult. In the Ga results, the first beat minimum occurs at even larger radii. This has been brought into connection with a possible softer surface of the mean field valid for these clusters [69]. An extensive discussion of the supershells and the scaling of the beat minima with the Wigner-Seitz radius, as predicted by the jellium model, may be found in a recent paper by Koch and Gunnarsson [73].

Another interesting topic in metal cluster physics, which again has a close analog in nuclei, concerns the optical response. The photoabsorption cross sections of many metal clusters exhibit a pronounced resonance, often called the "Mie resonance", which can be associated to a collective vibration of the electrons against the ionic background. The analog to this in nuclear physics are the so-called "giant dipole resonances", which are collective vibrations of the neutrons against the protons. We will not discuss this phenomenon in more detail here, but instead refer the reader to the text books and review articles quoted earlier. But we like to mention that the description of collective dynamical phenomena by semiclassical methods including classical orbits has so far not received too much attention. To quote just some first steps in this direction (without being complete), we note that the spherical cavity model of Balian and Bloch has been applied to include shell effects in semiclassical response functions for nuclei [74] and for metal clusters [75]. A more systematic ansatz towards a semiclassical formulation of response functions involving periodic orbits, but without assuming the integrability of the underlying independent-particle system, has recently been proposed by Gaspard and Jain [76]. The intrinsic complexity of quantum-mechanical dynamical calculations seems to make it worth while developing the periodic orbit theory also in this direction. We note that in Chapter 1, we have reported some interesting experiments involving the collective transport of electrons in nanostructures. The semiclassical description of shell effects in these phenomena via classical orbits, which need not be periodic in this case, is a field of active research.

8.3 Conductance oscillations in a circular quantum dot

In Chapter 1 we have already encountered the fascinating physics of two-dimensional nanostructures. As a further example, we shall discuss in this final section a direct application of some of the simple models that we have used in Chapter 6 to illustrate the periodic orbit theory. We refer to recent experiments by Persson *et al.* [77], in which the conductance of a large quantum dot in an external magnetic field was found to exhibit pronounced oscillations. What we will discuss here is that these oscillations can be understood as shell effects in the density of states of the electrons in the quantum dot. (For an introductory overview of electronic shells in large quantum dots, we refer to an article by Lindelof *et al.* [78].)

The experimental situation is displayed schematically in Fig. 8.15 on the next page. The arrangement is a typical partially doped GaAs/GaAlAs heterostructure, already discussed in Sect. 1.5, by which a 2-dimensional electron gas (2DEG) is formed. On the top layer, one applies an electric voltage V_{g1} to some thin metallic films, the so-called "Schottky gates" (shown in black), which are shaped such that the static electric field that penetrates down to the plane of the 2DEG confines the electrons to a circular domain. These "trapped" electrons are separated from the surrounding electron gas by a barrier. In a so-called "open quantum dot", the gaps between the Schottky gates are made large enough so that the barrier is lowered near the gaps, and the electrons can leak out over the corresponding saddles of the confining potential. The experiment is done for a "clean" dot at very low temperature, so that the electrons inside the dot can be assumed to move freely in the confining potential, with a mean-free path and a coherence length much larger than the dimensions of the dot. Near two opposite gaps, one applies two quantum point contacts with a voltage difference and sends an electric current through the device, and thus its conductance can be measured. A homogeneous magnetic field of strength B may furthermore be applied in the direction perpendicular to the plane of the 2DEG. By varying the applied gate voltage V_{g1}, the effective radius of the confining domain can be regulated. The experiments to be discussed here concern a "closed" dot. Therefore the electrons providing the current have to tunnel through the confining barriers at both ends. For resonant tunneling, the conductance is then proportional to the density of electronic states inside the dot, multiplied by an exponential barrier penetration factor. The latter is smoothly varying with energy; the oscillating part of the conductance is thus proportional to the oscillating part $\delta g\left(E_F\right)$ of the level density, taken at the Fermi energy.

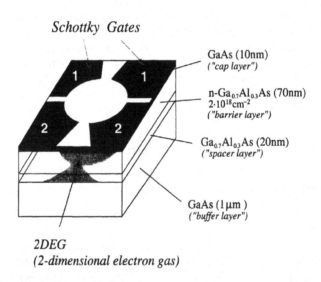

Figure 8.15: Schematic picture of a semiconductor heterostructure with a 2-dimensional electron gas confined by Schottky gates to a circular domain, a so-called "quantum dot". (After [77].)

In the experiments of Persson *et al.* [77], the parameters were chosen such that about 1000 - 1500 electrons were trapped in the dot at a density of $\sim 4 \times 10^{15}$ m^{-2}; the radius was estimated to be $R \sim 400$ nm. The de Broglie wave length of an electron at the Fermi energy was of the order of 60 nm, large enough for the electronic motion to average over the finer details of the surrounding ionic structures and eventual impurities, but short enough for many wavelengths to fit into the confinement region. Therefore a semiclassical description of the electronic shell effects should be reasonable.

The conductance was then measured as a function of the applied gate voltage V_{g1}, and hence the effective radius, and of the applied magnetic field strength. Fig. 8.16 shows a gray-scale plot of the measured conductance, from which a smooth background has been subtracted, versus V_{g1} and B in a region of relatively small field strengths. White areas show high conductance and dark regions show low conductance, respectively. We recognize a regular pattern of fluctuations, with a roughly constant amplitude and more or less well-defined periods in both directions.

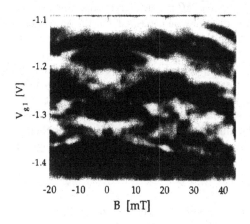

Figure 8.16: Gray-scale plot of the measured conductance oscillations as a function of magnetic field B (in milli-Tesla, horizontal direction) and gate voltage V_{g1} (in Volt, vertical direction). White regions correspond to high conductance and dark regions to low conductance. The Fermi energy is $E_F \approx 13.2\text{meV}$ and the Temperature is $T \approx 25\text{mK}$. (After [77].)

We now ask the question: Which form of the potential $V(r)$ felt by the electrons (we take it to be spherical) leads to an oscillating level density $\delta g\,(E_F)$ that follows the pattern shown in Fig. 8.16 as a function of gate voltage and magnetic field? The answer to this question was studied in Ref. [79] using the periodic orbit theory for two model potentials. The dependence on the gate voltage V_{g1} can, as stated above, be translated into a variation of the effective radius R of the dot; in the region shown, the relation between δV_{g1} and δR was estimated to be linear [77]. Note that, in the language of the mean field theory presented in Appendix A, $V(r)$ is the *total* potential which includes the external confining potential $U(r)$, the Hartree-potential $V_H(r)$ coming from the Coulomb repulsion between the electrons, and the exchange (plus correlation) potential. The external confining potential $U(r)$ is that of the confining electric field of the Schottky gates; it has its minimum at the center of the dot and may be taken to be harmonic. (This seems to be confirmed from many other experiments.) With increasing number N of electrons, the Coulomb repulsion and exchange-correlation have an increasing effect. For large N, we expect the repulsion to be important and its effect to result in a potential $V(r)$ that is flat in the interior and has steep walls, pretty much like it was found in the three-dimensional metal clusters discussed in the previous section. As we have seen there, for large numbers N the cavity model with infinitely steep walls gives, in fact, an excellent model for the semiclassical study of the electronic shell effects. In the limit of small N, where the electron-electron interaction is weak compared to the confining potential, the harmonic approximation of $U(r)$ itself may provide a guide for $V(r)$.

Let us first study the steep-wall limit, which in the present geometry corresponds to a disk billiard. The trace formula for $\delta(E)$ of this system without magnetic field has been given in Sect. 6.1.3. The effect of the magnetic field may be studied to lowest order of B in perturbation theory. When the field is weak enough, so that the cyclotron radius $R_c = \sqrt{2m^* E_F}/eB$ is much larger than the dot radius, the curvature of the periodic orbits in the disk may be neglected and the only linear term modifying the trace formula is a change in the actions S_Γ. For an effective mass $m^* = 0.067\, m_e$, valid for this semiconductor (m_e is the free electron mass), for a Fermi energy $E_F \sim 13\,\mathrm{meV}$ and the range of B fields shown in Fig. 8.16 we have, indeed, $R_c \gtrsim 7R$. This leads us to the perturbative treatment of the periodic orbit theory in the presence of weak magnetic fields, which we have discussed extensively in Sect. 6.2.3. As shown there, for systems with simple axial $U(1)$ symmetry in a homogeneous B field parallel to the symmetry axis, the only change in the trace formula is then a simple modulation factor $\cos\Phi_\Gamma$ under sum over periodic orbits Γ (which are singly-degenerate families). Thereby Φ_Γ is the magnetic flux through the area $\mathcal{A}_\Gamma$ surrounded by the (unperturbed) periodic orbit, taken in units of $\hbar c/e$:

$$\Phi_\Gamma = \frac{eB}{\hbar c}\mathcal{A}_\Gamma\,, \qquad \mathcal{A}_\Gamma = \frac{1}{2}vR^2 \sin(2\pi w/v)\,. \tag{8.25}$$

Here (v, w) are the numbers which characterize the periodic orbits in the disk (see Sect. 6.1.3). As discussed in Sect. 6.2.3, the origin of this extra modulation factor is physically the Aharonov-Bohm phase that adds itself to the classical action, with opposite signs for identical periodic orbits running in opposite directions (cf. also Problem 3.4 and Ref. [80]).

For each periodic orbit (family) summed over in the trace formula, we have thus

$$\delta g_\Gamma(E_F) \sim \cos\left(\frac{eB}{\hbar c}\mathcal{A}_\Gamma\right) \sin\left(k_F L_\Gamma - \sigma_\Gamma \frac{\pi}{2}\right)\,, \tag{8.26}$$

where k_F is the Fermi wave number and L_Γ are the lengths (6.12) of the periodic orbits. Since the latter are proportional to R, varying R at constant Fermi energy is the same as varying the wave number k_F at fixed R. The oscillations in the vertical direction in Fig. 8.16, say at $T = 0$, reflect thus nothing but the main-shell oscillations of $\delta g(E_F)$, plotted versus $k_F R = \sqrt{E_F/E_0}$ as shown in Fig. 6.8. The main-shell spacing, i.e., the distance between the minima of δg which is constant on this scale, is mainly determined by the average length of the shortest orbits with highest degeneracies. We have discussed this extensively for the supershells in metal clusters in the previous section. Note that in the disk, however, all orbits have the same degeneracy 1, so that here the diameter and the triangle orbits have the leading amplitudes – in contrast to the spherical cavity, where the diameter orbit has a very small amplitude and the triangle and square amplitudes are leading ones (compare the two Fourier spectra shown in Figs. 6.9 and 6.14). Using the average lengths of the diameter orbits $L_{21} = 4R$ and the triangle orbits $L_{31} = 3\sqrt{3}R$, and the above estimate of the Fermi energy, this leads to a constant radius increment $\Delta R \simeq 0.009\,\mu$m, in good agreement with the period of the oscillations in the V_{g1} direction which translate [77] to $\Delta R \simeq 0.007\,\mu$m.

The new degree of freedom of the magnetic field strength B, now, leads to oscillations which have, according to Eq. (8.26), a constant amplitude and a period

$$B_0 = 2\pi\hbar c/e\mathcal{A}_\Gamma. \qquad (8.27)$$

The constant amplitude is, indeed, what we seem to observe in the experiment. The period B_0 of the magnetic field oscillations, now, is given not by the lengths, but by the areas $\mathcal{A}_\Gamma$ of the leading (unperturbed) orbits. The (unperturbed) diameter thus does not contribute, since it includes a zero area, so that the leading term determining B_0 should be the triangle orbit (3,1). [Note, from Fig. 6.9, that the amplitude of the square orbit (4,1) is already appreciably smaller. The exact amplitude ratio depends, of course, on the range γ chosen for the coarse-graining.] From Fig. 8.16, we estimate B_0 to be $\sim 40\,$mT which, using Eqs. (8.25,8.27), yields $R \simeq 300\,$nm, in reasonable agreement with the experimental estimate of the radius given above. This shows that we can estimate the radius of the quantum dot directly from the period of the B-field oscillations, through the area of the leading periodic orbit.

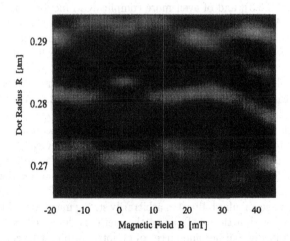

Figure 8.17: Gray-scale plot of the semiclassical level density δg, obtained by summing the terms in (8.26) with their correct amplitudes given in Eq. (6.14), of the disk billiard at fixed Fermi energy $E_F = 13.2\,$meV, shown versus dot radius R (vertical direction) and magnetic field strength B (horizontal direction). (From Ref. [79].)

Let us now see what the complete trace formula for the disk billiard predicts, after summing over all contributions in (8.26) with the amplitudes given in Eq. (6.14). The result, coarse-grained in k space with a Gaussian width $\gamma = 0.15/R$ (see Sect. 5.5), is shown in Fig. 8.17, in the form of a gray-scale plot similar to the experimental one above, versus radius R and B. It exhibits, indeed, a pattern very similar to that of the experimental conductance.

When taking a harmonic-oscillator potential $V(r)$ as a total mean field for the electrons and fitting the frequency ω to the experimental radius R, one finds also oscillations with the correct period in the vertical direction. But the amplitude of the B-field oscillations is found to rapidly decrease with increasing B. This is so, because the modulation factor for weak magnetic fields is different for an isotropic harmonic oscillator. The periodic orbits in this potential (in two dimensions) have a $SU(2)$ symmetry and thus a two-fold degeneracy. As discussed in the Problems 3.3 and 6.2, the modulation factor then becomes a spherical Bessel function $j_0(eB\mathcal{A}_\Gamma/\hbar c)$ (see also Refs. [72, 79]). Clearly, this gives the wrong B dependence to the amplitudes of the B-oscillations, as compared to experiment. This rules out the harmonic-oscillator model for the total field $V(r)$ of the electrons in the present case.

We have discussed here another nice example of the manifestation of classical periodic orbits in an interesting physical many-body system. The agreement is less quantitative than for the supershells in metal clusters, but still, we have seen how the methods of semiclassical physics allow us to perform quick and easy estimates of observable quantum effects, using only the two or three leading shortest orbits in the system. It may also be said that the exact quantum-mechanical calculation for the level density of the disk billiard in a magnetic field is straight-forward in principle, but requires considerable numerical effort, since it involves the zeros of Bessel functions for $B = 0$ (see Sect. 2.6.3), and of even more complicated functions for $B > 0$ (see, e.g., Refs. [81, 82]). And when only a coarse-grained level density has been observed as in the present experiment, so that most of the quantal fine structures in the spectrum cannot be resolved, a full-fledged quantum calculation is really like shooting mosquitoes with canons.

Let us speculate what will happen at large field strengths. We have already presented similar calculations [83] for metal clusters in Sect. 6.2.3 (see Fig. 6.20). Unfortunately, even the largest available metal clusters are still too small compared to the magnetic length l_B [see Eq. (6.78)] for the strongest experimentally available B fields, for their effects to be observable there. Nanostructures are several orders of magnitude larger, and here the effects of strong magnetic fields are a field of active current research (see also Sect. 1.5). The experiments on the above circular quantum dot have also been carried to stronger fields [77], and their analysis in terms of POT is still under way. A trace formula for the circular billiard in arbitrarily strong magnetic fields B has recently been derived [81] within the POT, using the methods described in Sect. 6.1.4. The polygons shown in Fig. 6.6 become pieces of cyclotron orbits with constant curvature. They can still be classified by the same pairs of vertex numbers v and winding numbers w. But since the time-reversal symmetry is broken by the external field, each of them breaks up into pairs with opposite curvatures for opposite orientations of the particle current. At specific field strengths and energies for $R_c < R$, some orbits cease to exist. For instance, a curved triangular orbit (3,1) cannot close any more if R_c is too small. Thus, there occur bifurcations at specific points in the E, B space for each family of orbits (v, w). Furthermore, for $R_c < R$ there exist complete cyclotron orbits within the boundary; these have a two-fold degeneracy corresponding to translations in the two spatial degrees of freedom. When these cyclotron orbits graze the boundary, some special problems arise. We refer to the paper of Blaschke et al. [81] for the details.

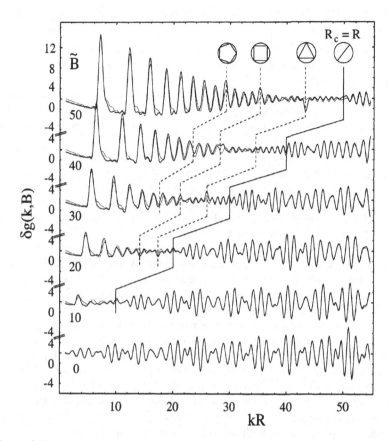

Figure 8.18: Level density $\delta g(k, B)$ of the circular disk in a homogeneous magnetic field B, plotted for various reduced field strengths $\tilde{B}$ versus kR. The coarse-graining corresponds to a Gaussian averaging over k with a width $\gamma \sim 0.4/R$. The thin lines give the quantum-mechanical, the heavy lines the semiclassical POT result. No difference can be seen for $R_c < R$, except at $\tilde{B} = 50$ (the slight disagreement for $kR > 52$ is due to the truncation of the quantum spectrum). In the POT results, orbits with up to 10 reflections are included (lowest harmonics only.) The staggered solid and dashed lines indicate the locations where $R_c = R$, and where bifurcations of the orbits with 3, 4, and 5 reflections occur, respectively. Note the appearance of the Landau levels in the strong-field limit, seen as the high peaks for low kR. (From [81].)

In Fig. 8.18 we show the coarse-grained oscillating level density $\delta g(k, B)$ versus kR for a set of increasing reduce field strengths $\tilde{B} = BeR^2/\hbar c$. The heavier lines give the POT results, the thinner lines the exact quantum-mechanical ones. For practically all cases with $R_c > R$, no difference can be seen between the two. But also for very strong fields with $R_c < R$, very good agreement is reached. The staggered solid lines connects the points (k, B) where $R_c = R$. The dashed staggered lines connect the bifurcation points for the triangle, square and pentagon orbits. (In the corresponding symbols at the top, only the topology or these orbits is shown, not their real curved shape.) The bifurcations have not been handled correctly in Ref. [81]; the results near these points have just been slightly smoothed and interpolated. This is a point where the new uniform approximations discussed in Sect. 6.3.3 should be applied in future research. In the strong-field limit, i.e., for small kR at fixed large $\tilde{B}$, one sees the development of the Landau quantization (cf. Problem 1.3). This is the situation where the field is so strong that the effect of the boundary becomes relatively small. In the POT language, the two-fold degenerate cyclotron orbits then are the leading terms in the trace formula, which in the limit $B \to \infty$ yields the exact Landau levels $\hbar\Omega_c(n + 1/2)$, see Eq. (1.131).

We end this section on quantum dots with an interesting speculation [84]. The Schottky gates (see Fig. 8.15) may also be shaped to form an equilateral triangular confinement domain. If it is a closed dot, the steep-wall model is the triangular billiard discussed in Sects. 3.2.7 and 6.1.2, for which we have an exact trace formula in Eq. (3.102), and whose coarse-grained level density exhibits nice beats, see Fig. 6.4. More interestingly, however, one may separate the three gates a little, so that it becomes an open dot. The confinement potential then will have three saddle points, is harmonic near the centre, and in a good approximation given by the Hénon-Heiles potential (5.73) discussed in Sect. 5.6.4, whose level density also exhibits beats (see Fig. 5.6) that can be explained by the three shortest isolated periodic orbits using Gutzwiller's trace formula. The lithography for the "Hénon-Heiles dot" has already been constructed and experiments for measuring its conductance are presently under way in Copenhagen [84]. The attractive aspect of this potential is that it exhibits mixed classical dynamics. Near the bottom of the well, the motion is quasi-regular, and near the saddle-point energy it becomes chaotic (see Sect. 5.6.4 and, for a similar potential with four saddles, Sect. 6.2.4). The "Hénon-Heiles dot" therefore is a device that should allow one to study order and chaos in one and the same system, tunable from one limit to the other by varying the Fermi energy of the trapped electrons. (We recall that in Sect. 1.5.1 we have discussed experiments on both ordered and chaotic quantum dots, but these were two separate systems.) If a magnetic field is furthermore applied to the Hénon-Heiles dot, the beat structure in its level density will be modified appreciably (see Ref. [85]). The period of the B-oscillations for weak fields should allow one to measure the area surrounded by the orbit C (see Fig. 5.5) which is the only short orbit to enclose a finite area $\mathcal{A}$. – Thus, the Hénon-Heiles potential, which was introduced by astrophysicists [86] over 30 years ago to model a galaxy in outer space, has found its way into the world of nanostructure physics.

Another nice example of Aharonov-Bohm oscillations in the magneto-conductance of a two-dimensional electron gas was observed in an experiment with a mesoscopic channel containing two antidots, exposed to an external magnetic field [87]. The quantum oscillations could be well reproduced [88, 89] (cf. also section 3.6.3 in [90]) using the periodic orbits in this classically quite chaotic system. An interesting feature thereby was that the systematic changes in the phases of the AB oscillations under variations of the magnetic field strength and the applied gate voltage exhibited some characteristic dislocations. These could be related to bifurcations of the leading periodic orbits – showing that orbit bifurcations can, indeed, have experimentally observable consequences!

Bibliography

[1] F. Halzen and A. D. Martin: *Quarks and Leptons: An Introductory Course in Modern Particle Physics* (John Wiley & Sons, New York, 1984).

[2] R. K. Bhaduri: *Models of the Nucleon* (Addison-Wesley, Reading, USA, 1988).

[3] H. Yukawa, Proc. Phys. Math. Soc. of Japan **17**, 48 (1935).

[4] G. E. Brown and A. D. Jackson: *The Nucleon-Nucleon Interaction* (North-Holland, Amsterdam, Oxford, 1976).

[5] T. H. R. Skyrme, Phil. Mag. **1**, 1043 (1956).

[6] D. Vautherin and D. M. Brink, Phys. Rev. **C 5**, 626 (1972).

[7] D. Gogny, Nucl. Phys. **A 237**, 399 (1975).

[8] P. Quentin and H. Flocard, Ann. Rev. Nucl. Part. Sci. **28**, 523 (1978).

[9] M. Goeppert-Mayer, Phys. Rev. **75**, 1969 (1949); bf 78, 16 (1950).

[10] O. Haxel, J. H. D. Jensen, and H. E. Suess, Phys. Rev. **75**, 1766 (1949); Z. Phys. **128**, 295 (1950).

[11] H. A. Jahn and E. Teller, Proc. Roy. Soc. (London) **A 161**, 220 (1937).

[12] S. G. Nilsson, Mat.-Fys. Medd. Dan. Vid. Selsk. **29**, no. 16 (1955).

[13] B. R. Mottelson and S. G. Nilsson, Phys. Rev. **99**, 1615 (1955); Mat.-Fys. Skr. Dan. Vid. Selsk. **1**, no. 8 (1959).

[14] K. Alder, A. Bohr, T. Huus, B. Mottelson, and A. Winther, Rev. Mod. Phys. **28**, 432 (1956).

[15] J. Rainwater, Phys. Rev. **79**, 432 (1959); A. Bohr, Mat.-Fys. Medd. Dan. Vid. Selsk. **26**, no. 14 (1952); A. Bohr and B. R. Mottelson, Mat.-Fys. Medd. Dan. Vid. Selsk. **27**, no. 16 (1953).

[16] A. Bohr and B. R. Mottelson: *Nuclear Structure, Vols. I, II* (W. A. Benjamin, New York, 1969; Reading, 1975).

[17] M. A. Preston and R. K. Bhaduri: *Structure of the Nucleus* (Addison-Wesley, Reading, 1975).

[18] P. Ring and P. Schuck: *The Nuclear Many Body Problem* (Springer Verlag, New York, 1980).

[19] V. M. Strutinsky, Sov. J. Nucl. Phys. **3**, 449 (1966) [Yad. Fiz. **3**, 614 (1966)]; Ark. Fys. **36**, 629 (1966); Nucl. Phys. **A 95**, 420 (1967).

[20] V. M. Strutinsky, Nucl. Phys. **A 122**, 1 (1968).

[21] I. Ragnarsson, S. G. Nilsson, and R. K. Sheline, Phys. Rep. **45**, 1 (1978).

[22] V. M. Strutinsky, Nukleonika (Poland) **20**, 679 (1975); V. M. Strutinsky and A. G. Magner, Sov. J. Part. Nucl. **7**, 138 (1976) [Elem. Part. & Nucl. (Atomizdat, Moscow) **7**, 356 (1976)].

[23] V. M. Strutinsky, A. G. Magner, S. R. Ofengenden, and T. Døssing, Z. Phys. **A 283**, 269 (1977).

[24] H. Nishioka, N. Nitanda, M. Ohta, and S. Okai, Mem. Konan Univ., Sci. Ser. **38**, 1 (1991).

[25] H. Frisk, Nucl. Phys. A 511, 309 (1990).

[26] A. G. Magner, S. N. Fedotkin, F. A. Ivanyk, P. Meier, M. Brack, S. M. Reimann, and H. Koizumi, Ann. Phys. (Leipzig) 6, 555 (1997).

[27] M. V. Berry and M. Tabor, Proc. R. Soc. Lond. A 349, 101 (1976).

[28] R. Balian and C. Bloch, Ann. Phys. (N. Y.) 69, 76 (1972).

[29] P. Meier, M. Brack, and S. C. Creagh, Z. Phys. D 41, 281 (1997).

[30] N. Bohr and J. A. Wheeler, Phys. Rev. 56, 426 (1939).

[31] L. Wilets: Theories of nuclear fission (Clarendon Press, Oxford, 1964).

[32] S. Cohen and W. J. Swiatecki, Ann. Phys. (N. Y.) 22, 406 (1963).

[33] M. Brack, J. Damgård, A. S. Jensen, H. C. Pauli, V. M. Strutinsky, and C. Y. Wong, Rev. Mod. Phys. 44, 320 (1972).

[34] W. D. Myers and W. J. Swiatecki, Ann. Phys. (N. Y.) 55, 395 (1969).

[35] W. J. Swiatecki and S. Bjørnholm, Phys. Rep. 4C, No.6 (1972); S. Bjørnholm and J. E. Lynn, Rev. Mod. Phys. 52, 725 (1980).

[36] S. M. Polikanov, V. A. Druin, V. A. Karnaukhov, V. L. Mikheev, A. A. Pleve, N. K. Skobelev, V. G. Subbotin, G. M. Ter-Akop'yan, and V. A. Fomichev, Sov. Phys. JETP 15, 1016 (1962).

[37] S. G. Nilsson, C. F. Tsang, A. Sobiczewski, Z. Szymański, S. Wycech, C. Gustafsson, I.-L. Lamm, P. Möller, and B. Nilsson, Nucl. Phys. A 131, 1 (1969).

[38] M. Bolsterli, E. O. Fiset, J. R. Nix, and J. L. Norton, Phys. Rev. C 5, 1050 (1972); J. R. Nix, Ann. Rev. Nucl. Sci. 22, 65 (1972).

[39] V. V. Pashkevich, Nucl. Phys. A 169, 275 (1971).

[40] M. Brack, in: Physics and Chemistry of Fission 1979, International Symposium held in Jülich in 1979 (IAEA Vienna, 1980), Vol. I, p. 227. For further references on fission barrier calculations, see many other contributions to this conference.

[41] K. Lee and D. R. Inglish, Phys. Rev. 108, 774 (1957).

[42] S. A. E. Johansson, Nucl. Phys. 22, 529 (1962).

[43] P. Möller and S. G. Nilsson, Phys. Lett. 31 B, 283 (1970).

[44] H. C. Pauli, T. Ledergerber, and M. Brack, Phys. Lett. 34 B, 264 (1971).

[45] M. Brack, S. M. Reimann, and M. Sieber, Phys. Rev. Lett. 79, 1817 (1997).

[46] M. Brack, P. Meier, S. M. Reimann, and M. Sieber, in: Similarities and differences between atomic nuclei and clusters, eds. Y. Abe et al. (American Institute of Physics, 1998) p. 17.

[47] C. Gustafsson, P. Möller, and S. G. Nilsson, Phys. Lett. 34 B, 349 (1971).

[48] M. Brack, M. Sieber, and S. M. Reimann, in: Nobel Symposium on Quantum Chaos, eds. K.-F. Berggren and S. Åberg; Physica Scripta T90, 146 (2001).

[49] K. Arita and K. Matsuyanagi, Prog. Theor. Phys. (Japan) 91, 723 (1994).

[50] W. A. de Heer, Rev. Mod. Phys. 65, 611 (1993).

[51] W. D. Knight, K. Clemenger, W. A. de Heer, W. A. Saunders, M. Y. Chou, and M. L. Cohen, Phys. Rev. Lett. 52, 2141 (1984).

[52] W. Ekardt, Phys. Rev. Lett. **52**, 1925 (1984); Phys. Rev. **B 29**, 1558 (1984).

[53] K. Clemenger, Ph.D. thesis, University of California, Berkeley (1985, unpublished); Phys. Rev. **B 32**, 1359 (1985).

[54] W. A. de Heer, W. D. Knight, M. Y. Chou, and M. L. Cohen, in *Solid State Physics, Vol. 40*, eds. H. Ehrenreich and D. Turnbull (Academic Press, New York, 1987), p. 93.

[55] V. Bonačić-Koutecký, P. Fantucci, and J. Koutecký, Chem. Rev. **91**, 1035 (1991).

[56] Ch. Ellert and H. Haberland, Phys. Rev. Lett. **75**, 1731 (1995).

[57] T. P. Martin, T. Bergmann, H. Göhlich, and T. Lange, Chem. Phys. Lett. **172**, 209 (1990); Z. Phys. **D 19**, 25 (1991).

[58] T. P. Martin, Phys. Rep. **273**, 199 (1996).

[59] M. Brack, Rev. Mod. Phys. **65**, 677 (1993); for a more popular introduction to the physics of metal clusters, see M. Brack, The Scientific American, December 1997, p. 50.

[60] Topical issue on *Nuclear Aspects of Simple Metal Clusters*, eds. C. Bréchignac and Ph. Cahuzac, Comm. At. Mol. Phys. **31** (1995).

[61] H. Nishioka, K. Hansen, and B. R. Mottelson, Phys. Rev. **B 42**, 9377 (1990).

[62] M. Brack, Phys. Rev. **B 39**, 3533 (1989).

[63] S. Bjørnholm, J. Borggreen, O. Echt, K. Hansen, J. Pedersen, and H. D. Rasmussen, Phys. Rev. Lett. **65**, 1627 (1990); Z. Phys. **D 19**, 47 (1991).

[64] M. Brack, O. Genzken, and K. Hansen, Z. Phys. **D 21**, 65 (1991); [see also Z. Phys. **D 19**, 51 (1991) for an earlier short report].

[65] O. Genzken and M. Brack, Phys. Rev. Lett. **67**, 3286 (1991).

[66] J. Pedersen, S. Bjørnholm, J. Borggreen, K. Hansen, T. P. Martin, and H. D. Rasmussen, Nature **353**, 733 (1991).

[67] T. P. Martin, S. Bjørnholm, J. Borggreen, C. Bréchignac, P. Cahuzac, K. Hansen, and J. Pedersen, Chem. Phys. Lett. **186**, 53 (1991).

[68] C. Bréchignac, Ph. Cahuzac, F. Carlier, M. de Frutos, and J.-Ph. Roux, Phys. Rev. **B 47**, 2271 (1993); and earlier references quoted therein.

[69] M. Pellarin, B. Baguenard, C. Bordas, M. Broyer, J. Lermé, and J. L. Vialle, Phys. Rev. **B 48**, 17645 (1993).

[70] U. Näher and K. Hansen, J. Chem. Phys. **101**, 5367 (1994); K. Hansen, Surf. Rev. Lett. (Singapore) **3**, 597 (1996).

[71] S. Bjørnholm, J. Borggreen, H. Busch, and F. Chandezon, in: *Large Clusters of Atoms and Molecules*, ed. T. P. Martin (Kluwer, Dordrecht, 1996), p. 111.

[72] M. Brack, S. C. Creagh, P. Meier, S. M. Reimann, and M. Seidl, in: *Large Clusters of Atoms and Molecules*, ed. T. P. Martin (Kluwer, Dordrecht, 1996), p. 1.

[73] E. Koch and O. Gunnarsson, Phys. Rev. **B 54**, 5168 (1996).

[74] A. G. Magner, S. M. Vydrug-Vlasenko, and H. Hoffmann, Nucl. Phys. **A 524**, 31 (1991).

[75] A. Dellafiore, F. Matera, and D. M. Brink, Phys. Rev. **A 51**, 914 (1995); see also some earlier papers quoted therein.

[76] P. Gaspard and S. R. Jain, Pramana - J. Phys. **48**, 503 (1997).

[77] M. Persson, J. Pettersson, B. von Sydow, P. E. Lindelof, A. Kristensen, and K. Berggreen, Phys. Rev. **B 52**, 8921 (1995); and earlier references quoted therein.

[78] P. E. Lindelof, P. Hullmann, P. Bøggild, M. Persson, and S. M. Reimann, in: *Large Clusters of Atoms and Molecules*, ed. T. P. Martin (Kluwer, Dordrecht, 1996), p. 89.

[79] S. M. Reimann, M. Persson, P. E. Lindelof, and M. Brack, Z. Phys. **B 101**, 377 (1996).

[80] S. M. Reimann, M. Brack, A. G. Magner, J. Blaschke, and M. V. N. Murthy, Phys. Rev. **A 53**, 39 (1996).

[81] J. Blaschke, Diploma thesis, Regensburg University (1995, unpublished); J. Blaschke and M. Brack, Phys. Rev. **A 56**, 182 (1997); Physica **E1**, 288 (1997).

[82] K. Tanaka, Ph.D. thesis, McMaster University (1997, unpublished); K. Tanaka, Ann. Phys. (N. Y.) **268**, 31 (1998).

[83] K. Tanaka, S. C. Creagh, and M. Brack, Phys. Rev. **B 53**, 16050 (1996).

[84] P. Bøggild, A. Kristensen, P. E. Lindelof, S. M. Reimann, and C. B. Sørensen, in: *23rd International Conference on the Physics of Semiconductors (ICPS), Berlin 1996* (World Scientific, Singapore, 1997), p. 1533.

[85] M. Brack, J. Blaschke, S. C. Creagh, A. G. Magner, P. Meier, and S. M. Reimann, Z. Phys. **D 40**, 276 (1997).

[86] M. Hénon and C. Heiles, Astr. J. **69**, 73 (1964).

[87] C. Gould *et al.*, Phys. Rev. **B 51**, 11213 (1995); G. Kirczenov *et al.*, Phys. Rev. **B 56**, 7503 (1997).

[88] J. Blaschke, Ph.D. thesis (Regensburg University 1999), available at ⟨http://www.joachim-blaschke.de⟩.

[89] J. Blaschke and M. Brack, Europhys. Lett. **50**, 294 (2000).

[90] M. Brack, in: *Atomic Clusters and Nanoparticles*, Les Houches Summer School (NATO Advanced Study Institute), Session LXXIII, eds. C. Guet *et al.* (Springer-Verlag, Berlin, 2001), pp. 161-219.

Chapter 9

Concluding remarks

The Gutzwiller trace formula has played a prominent role in the study of quantum systems whose classical analogues are chaotic. Although of measure zero in a sea of chaos in the classical phase space, the periodic orbits still open a classical window to view the quantum behavior of such systems. One challenge has been to try to obtain the quantized levels of simple systems, by using sophisticated mathematical techniques for summing the divergent series in the trace formula. These studies have included the hyperbolic billiard and some other variations, and the wedge billiard which is not force-free. Brave attempts, with impressive success for a class of states in the Helium atom, also fall in this category. Such studies fall under the heading of 'quantization of nonintegrable systems', and some of them have been described at length in Chapter 7.

The other area of study, more rewarding in our view, is the semiclassical description of diverse physical phenomena by making use of only a few orbits with short periods. Often, on the quantum level, a few single-particle states tilt the balance of the energetics. In the trace formulae, the shortest orbits with the highest degeneracies often govern the gross-shell structure (with a poor resolution in energy, but good enough to reflect the shell effects).

One of the earliest applications of the trace formula in this connection was made by Strutinsky and collaborators in their studies of deformed atomic nuclei. Recently, some progress has been made in the understanding of the asymmetric fission of nuclei in a similar way. As has been emphasized in Chapter 4, it is the shell structure in the single-particle quantal dynamics that determines the nuclear shapes and the detailed structure of the nuclear fission barriers; examples for this have been given in Chapter 8. In the fission process, there is turmoil in the motion of the particles as the nucleus tears apart. However, the fate of the fragments seems to be decided by only a few periodic orbits much earlier, when the neck is about to form. This fascinating topic is also described in Chapter 8. A semiclassical description of tunneling in a chaotic system has been made in a different context by Creagh and Whelan, as mentioned in passing in Chapter 1. We see here a possibility of extending and applying their approach to the half-lives of fissioning nuclei, which are a result of tunneling through a complicated multi-dimensional barrier of a classically chaotic system.

A recent revival of the periodic-orbit connection to magic numbers has taken place in the field of metal clusters. In Chapter 8 we have seen the beautiful experimental realization of Balian and Bloch's supershells for particles in a spherical box. But there is also a close connection between periodic orbits and the mathematical theory of numbers and the Riemann zeta function, as discussed in Chapter 7.

The same physics of the dominance of short periodic orbits in finite-sized systems makes the trace formula a powerful tool in the study of mesoscopic devices. We have seen in Chapter 1 that chaotic orbits left a mark in the measured magnetoresistance of high-mobility heterostructures. One could also study in this connection the tunneling through a double well, and the role of chaotic periodic orbits. In Chapter 8, further results have been given on the magnetic response in quantum dots. Since the shapes of these devices can be controlled to a large degree, we are able to study the response of quantum dots to external stimuli (light, electric and magnetic fields) for both integrable and chaotic dynamics. This is a field full of promise of more and better experimental results, and the trace formula with a few orbits included may play a central role. Already it is being applied to a systematic formulation of transport properties, as mentioned at the end of Chapter 1. Another related area will be the study of the role of periodic orbits to the time-dependent response of complex systems, including collective effects. Some steps in this direction have been mentioned at the end of our tale on the electronic supershells in Chapter 8.

We should mention that till now, less attention has been paid to the wave function and the propagation of wave packets using the classical-orbit approach. Exceptions have been the pioneering studies of McDonald, Heller, Tomsovic, Bogomolny, and Berry and collaborators. The effect of a set of periodic orbits on the quantal probability may be systematically studied through the Gutzwiller approach, as was outlined briefly in Chapter 1 while discussing Bogomolny's classic paper. The propagation of wave packets in atoms was also described in the introduction. Stroud and his collaborators continue to make progress in studying atoms excited by short-pulsed lasers. Periodic orbits have, in fact, become a natural tool in the analysis of the Rydberg states. Fascinating quantal behavior in the propagation of wave packets, like the phenomenon of recurrence, is explained by a semiclassical analysis. This area has the prospect of giving a better understanding of the correspondence principle and its ramifications.

On the theoretical side, apart from the problem of convergence of the trace formula, more work has to be done in the description of mixed systems. For quite some time, the trace formula of Gutzwiller for hyperbolically unstable systems on one hand, and the Berry-Tabor approach to integrable systems on the other hand seemed to be the only two working realizations of the periodic orbit theory, with Balian and Bloch's formalism, specialized to cavities, lying somewhere in between. The early achievements of the Strutinsky group in extending the Gutzwiller approach to systems with continuous symmetries, and the modern versions of this given by Creagh and Littlejohn in a more mathematical form, have brought invaluable progress. Trace formulae exist now for almost all forms of (partial) symmetries. Still, major problems persist. Stable orbits are prone to bifurcate with increasing time periods, leading to the break-down of the usual trace formulae, as is also the case with symmetry breaking. These two mechanisms lie at the root of the difficulties with the semiclassical description of mixed systems. New uniform approximations have been developed which allow one to handle the breaking of the simplest symmetries and all generic types of bifurcations, as long as they are well separated. Some progress has been made with bifurcations of codimension two, but more complicated situations such as several close-lying bifurcations, or bifurcation cascades cumulating at one particular value of a given parameter, pose serious challenges to the specialist. The full semiclassical quantization of systems with mixed classical dynamics is therefore still far far from being realizable – and it will perhaps remain a dream.

Another interesting development on the theoretical side has been the inclusion of diffractive orbits in the trace formula. Diffractive orbits are specially important for grazing angles of encounter and, of course, play a leading role in scattering in the forward direction. This is discussed in some detail in Chapter 7. Considerable progress has been made recently in this area of research, specially exploiting the close connection (and duality) between scattering states and the bound states. Smilansky and collaborators have applied these ideas to billiards with success.

With all the classical orbits in our minds, let us not forget the merits of the extended Thomas-Fermi model in determining average properties of finite interacting fermion systems in a variational way. In Chapter 4 we have shown an impressive result of the efforts by the Pearson group to fit the binding energies of 1492 nuclei by only eight parameters of a realistic effective nucleon-nucleon interaction.

Our efforts to point out the pivotal role of periodic orbits in a one-body potential that describes, in the mean-field approximation, an interacting many-body system could not have been anticipated better than by G. Lamé in 1852, whom we have quoted in Sect. 3.2.7. After having derived, in his *Leçons sur la théorie mathématique de l'élasticité des corps solides*, the eigenfrequency spectra of various simple systems such as the vibrating string, and the rectangular and triangular elastic membranes, he writes (in our own translation from the French original):

The object of this lesson will no doubt appear of little interest to the engineers who are specially interested in elastic equilibrium. But, apart from the fact that it is often necessary to study the effect of vibrations upon certain constructions, has the time not come to ask oneself if the molecular state of the solids whose repose seems to us most established, is really a static one; if it is not, on the contrary, the result of very rapid vibrations, which never stand still? All leads us to think, in fact, that the relative rest of the molecules in a solid is only a very exceptional case, a pure abstraction, a vain illusion ("une chimère") perhaps. This idea might appear singular; but, patience, before long the new way of teaching Mechanics will bear its fruits; one will want to explain everything by the motion, by the work, and this very idea will become commonplace. From this point of view, everything that concerns the vibrating states merits to be studied with care, in order to prepare the way to these future explanations. Now, as we shall see, the vibrations of the solids of which no dimension is very small, lead to the same problems of analysis, to the same discussions, as the vibrating string and the elastic membrane; there was thus a real interest to treat, as completely as possible, these two first examples. These considerations seem to us to put beyond doubt the usefulness of the study of vibrations; and, let us repeat it, this study, once recognized to be necessary, would be superficial and incomplete, had we not the access to the properties of quadratic forms of integer numbers, to this theory of numbers, so often disapproved of by the distractors from pure science, by the exclusive practitioners.

Finally, after hard and long labor in presenting a disciplined account of the developments of semiclassical physics in autonomous systems, we may indulge in the luxury of airing a few wild questions. What roles, we may ask, do classical orbits play in macroscopically exhibited quantal phenomena like superconductivity and superfluidity? When there is large-length quantal coherence in the wave function, does it imply very ordered classical motion? Quantum statistics can cause blocking of states, how do they affect classical motion of particles? We know experimentally that dramatic changes in the momentum distribution of atoms in a trap take place in Bose condensation. How does this affect the periodic orbits?

To conclude on a wilder note, consider the following. A creature living on a two-dimensional flat space which happens to be the Poincaré surface of section of a three-dimensional classical system seems to witness quantum-like behavior. Individual events hit at random, but a regular pattern or island may appear after many hits. Reminiscent are the quantal hits of a self-interfering photon or electron from the famous double-slit experiment. Is there a hint here that there may be a connection between dimensional reduction and quantum effects?

Appendix A

The self-consistent mean field approach

In this appendix, we give a brief sketch of the mean field approach which is used to reduce the many body problem of N interacting particles to that of N independent particles moving in a self-consistent average potential or 'mean field'. The idea is to incorporate the sum of all two body interactions which act on a given particle into this mean field. When the particles are indistinguishable, the field is the same for all particles. Their motion is then determined by solving the Schrödinger equation for the mean field. In order to determine it in an optimal way, one uses the variational principle. The price one has to pay for reducing the many body problem is that the mean field itself depends on the wavefunctions or the density of the independent particles. This is what is meant by the word 'self-consistent', and it is usually achieved by solving the Schrödinger equation iteratively.

We start from the Hamiltonian of a system of N identical fermions moving in an external potential $U(\mathbf{r})$ and interacting through a two-body potential $V^{(2)}(\mathbf{r}_1, \mathbf{r}_2)$:

$$\hat{H} = \sum_{i=1}^{N} \left\{ \frac{\mathbf{p}_i^2}{2m} + U(\mathbf{r}_i) \right\} + \frac{1}{2} \sum_{i,j(\neq i)=1}^{N} V^{(2)}(\mathbf{r}_i, \mathbf{r}_j) \,. \tag{A.1}$$

The exact wavefunction $\Psi(\mathbf{r}_1, \mathbf{r}_2, \ldots, \mathbf{r}_N)$ belonging to this Hamiltonian can in general not be calculated. From it we define the *one-body* or *single-particle density matrix* by

$$\rho_1(\mathbf{r}, \mathbf{r}') = N \int d^3 r_2 \int d^3 r_3 \ldots \int d^3 r_N \, \Psi^*(\mathbf{r}', \mathbf{r}_2, \ldots, \mathbf{r}_N) \, \Psi(\mathbf{r}, \mathbf{r}_2, \ldots, \mathbf{r}_N) \,. \tag{A.2}$$

Its diagonal part is the density $\rho(\mathbf{r})$, which shall be normalized to the number N of particles:

$$\rho(\mathbf{r}) = \rho_1(\mathbf{r}, \mathbf{r}) \,, \qquad \int \rho(\mathbf{r}) d^3 r = N \,, \tag{A.3}$$

For the sake of simplicity, we do not exhibit the spin degrees of freedom, since they are not used explicitly in this book. They would render the expressions for the exchange (Fock) terms given in Appendix A.1 more complicated. Apart from that, they simply result in a spin degeneracy factor 2, to be included in all summations over the single-particle labels i used below, as long as the interaction does not depend on the spin. (Note, however, that in the Chapters 4 and 8, we have made this factor explicit in the summations over i and included it in the definitions of the local densities.)

In the following two sections, we give the basic equations that determine the mean field in the two most common approaches, the Hartree-Fock theory and the density functional theory. In both of them, the density $\rho(\mathbf{r})$ is written in terms of a set of orthonormal single-particle wavefunctions $\psi_i(\mathbf{r})$

$$\rho(\mathbf{r}) = \sum_{i=1}^{N} |\psi_i(\mathbf{r})|^2 \,, \tag{A.4}$$

where the sum goes over the lowest occupied states in the mean field. The determination of this field, now, depends on the method one is using.

A.1   Hartree-Fock theory

In the Hartree-Fock (HF) theory, the ground-state wavefunction of the N-fermion system is approximated by a Slater determinant Φ, built from a complete orthogonal set $\{\psi_\alpha(\mathbf{r})\}$ of single-particle wavefunctions:

$$\Phi(\mathbf{r}_1, \mathbf{r}_2, ..., \mathbf{r}_N) = \frac{1}{\sqrt{N!}} \det |\psi_i(\mathbf{r}_j)|_{i,j=1,2,...,N} \,. \tag{A.5}$$

This wavefunction has the correct antisymmetry under exchange of any pair of fermions and thus respects the Fermi-Dirac statistics. The density matrix (A.2) then takes the form

$$\rho_1^{HF}(\mathbf{r}, \mathbf{r}') = \sum_{i=1}^{N} \psi_i^*(\mathbf{r}')\psi_i(\mathbf{r}) \,, \tag{A.6}$$

from which Eq. (A.4) follows. The choice of the single-particle wavefunctions ψ_i is determined by the variational principle: one makes the expectation value of the total Hamiltonian (A.1) between the Slater determinants (A.5) stationary with respect to small variations of the wavefunctions ψ_i. The condition of their orthogonalization is enforced by means of Lagrange multipliers E_i. The variational principle reads

$$\frac{\delta}{\delta\psi_i^*(\mathbf{r})} \left[\langle\Phi|\hat{H}|\Phi\rangle - E_i \int |\psi_i(\mathbf{r})|^2 d^3r \right] = 0 \,. \tag{A.7}$$

It leads to a set of coupled integro-differential equations of Schrödinger type, but with a nonlocal part of the potential:

$$\left\{ \hat{T} + U(\mathbf{r}) + V_H(\mathbf{r}) \right\} \psi_i(\mathbf{r}) + \hat{V}_F \psi_i(\mathbf{r}) = E_i \psi_i(\mathbf{r}) \,. \tag{A.8}$$

These equations are called the *Hartree-Fock equations*. Here $V_H(\mathbf{r})$ is the local 'Hartree' or 'direct' potential

$$V_H(\mathbf{r}) = \int \rho(\mathbf{r}')V^{(2)}(\mathbf{r}, \mathbf{r}')d^3r' \,. \tag{A.9}$$

When $V^{(2)}$ is the Coulomb interaction, $V_H(\mathbf{r})$ is just the classical Coulomb potential associated to the density $\rho(\mathbf{r})$. In (A.8), $\hat{V}_F$ is the nonlocal 'Fock' or 'exchange' potential. It is an integral operator and originates from the antisymmetrization of the wavefunction Φ. It is defined (apart from spin complications) by the way in which it acts on the wavefunction to its right:

$$\hat{V}_F \psi_i(\mathbf{r}) = -\frac{1}{2} \int \rho_1^{HF}(\mathbf{r}, \mathbf{r}') V^{(2)}(\mathbf{r}, \mathbf{r}')\, \psi_i(\mathbf{r}') d^3 r' . \qquad (A.10)$$

$\hat{V}_F$ is a nonlocal potential because its effect on the wavefunction ψ_i at the point $\mathbf{r}$ depends on its values at all other points $\mathbf{r}'$ in space, over which one has to integrate in Eq. (A.10). Since both V_H and $\hat{V}_F$ depend on the wavefunctions $\psi_i(\mathbf{r})$ of *all* particles, the HF equations (A.8) are nonlinear and must be solved self-consistently; this is usually done iteratively. The biggest numerical complications arise from the integral operator $\hat{V}_F$. The lowest energy obtained after convergence is usually called the 'HF energy' E_{HF}, and the corresponding Slater determinant is denoted by $|HF\rangle$:

$$E_{HF} = \min_{\{\Phi\}} \langle \Phi | \hat{H} | \Phi \rangle = \langle HF | \hat{H} | HF \rangle . \qquad (A.11)$$

The sum of Hartree and Fock potentials in (A.8) is usually referred to as the 'HF potential':

$$\hat{V}_{HF} = V_H + \hat{V}_F . \qquad (A.12)$$

The HF energy may be broken up into its different contributions by writing

$$E_{HF} = \int \left\{ \tau(\mathbf{r}) + U(\mathbf{r})\rho(\mathbf{r}) + \frac{1}{2}V_H(\mathbf{r})\rho(\mathbf{r}) \right\} d^3 r + E_x , \qquad (A.13)$$

where the kinetic energy density $\tau(\mathbf{r})$ is given by

$$\tau(\mathbf{r}) = \frac{\hbar^2}{2m} \sum_{i=1}^{N} |\nabla \psi_i(\mathbf{r})|^2 , \qquad (A.14)$$

and E_x is the exchange energy

$$E_x = -\frac{1}{4} \int \int \rho_1^{HF}(\mathbf{r}, \mathbf{r}') V^{(2)}(\mathbf{r}, \mathbf{r}') \rho_1^{HF}(\mathbf{r}', \mathbf{r}) d^3 r' d^3 r . \qquad (A.15)$$

The kinetic energy density (A.14) may be obtained from the HF one-body density matrix (A.6) by

$$\tau(\mathbf{r}) = \frac{\hbar^2}{2m} \nabla_\mathbf{r} \cdot \nabla_{\mathbf{r}'} \rho_1^{HF}(\mathbf{r}, \mathbf{r}')\Big|_{\mathbf{r}'=\mathbf{r}} , \qquad (A.16)$$

and the local density $\rho(\mathbf{r})$ is, according to Eq. (A.3), given by

$$\rho(\mathbf{r}) = \rho_1^{HF}(\mathbf{r}, \mathbf{r}')\Big|_{\mathbf{r}'=\mathbf{r}} . \qquad (A.17)$$

With all these ingredients, the total HF energy can be expressed entirely as a functional of the one-body density matrix:

$$E_{HF} = E_{HF}[\rho_1^{HF}] . \qquad (A.18)$$

For later reference, we rewrite ρ_1^{HF} in matrix notation, omitting the subscript '1' and the superscript 'HF', and represent it in an arbitrary single-particle basis $\{\alpha, \beta, \ldots\}$

$$\rho_{\alpha\beta} = \sum_{i=1}^{N} \langle\alpha|i\rangle\langle i|\beta\rangle. \tag{A.19}$$

The HF energy is then given by

$$E_{HF}[\rho] = \mathrm{tr}\,(\mathcal{T} + \mathcal{U})\,\rho + \frac{1}{2}\mathrm{tr}\,\rho\,(\mathrm{tr}\,\mathcal{V}\rho), \tag{A.20}$$

where $\mathcal{T}$, $\mathcal{U}$ are the one-body and $\mathcal{V}$ the antisymmetrized two-body matrix elements of the kinetic energy, the external potential and the two-body potential, respectively:

$$\mathcal{T}_{\alpha\beta} = \langle\alpha|\hat{T}|\beta\rangle, \quad \mathcal{U}_{\alpha\beta} = \langle\alpha|U|\beta\rangle, \quad \mathcal{V}_{\alpha\beta,\gamma\delta} = \langle\alpha\beta|V^{(2)}|\widetilde{\gamma\delta}\rangle. \tag{A.21}$$

It is a well-known feature of the self-consistent mean-field theory that the total energy is not equal to the sum of the occupied single-particle energies E_i. (This holds also for the density functional theory described in the next section.) Indeed, from Eqs. (A.8,A.13) one easily verifies that

$$E_{HF} = \sum_{i}^{N} E_i - \frac{1}{2}\langle HF|\hat{V}_{HF}|HF\rangle. \tag{A.22}$$

Strictly speaking, the above expressions for the V_H, $\hat{V}_F$, and E_x contain unphysical contributions due to the interaction of the particle in the ith state with itself which should have been omitted, as indicated by the condition $j\neq i$ in Eq. (A.1). However, in taking the sum of the direct and exchange terms in E_{HF}, these contributions cancel exactly. Leaving them out would make both potentials state dependent – as is the case in simple Hartree theory – and thus render the HF equations still more complicated to solve. It is therefore standard praxis to include them into both potentials. Classically, the inclusion of the self interaction in the Coulomb potential poses no problem for a continuous density distribution $\rho(\mathbf{r})$. This point becomes of crucial importance when different approximations are made to the direct and the exchange terms of the energy, as it happens in the density functional theory with the local density approximation.

The HF formalism has been used successfully in all branches of physics. In nuclei, the two-body interactions employed in the HF framework depend on the density of the nucleons (see the discussion and the references in Sect. 8.1). We make use of HF results for nuclei in Chapters 4 and 8.

A.2   Density functional theory

The density functional theory (DFT) goes beyond the HF approach in the sense that correlations are taken into account which are not contained in the HF energy (A.13). With 'correlation energies' we mean here all those contributions to the total energy which come from wavefunctions that are not simple Slater determinants (A.5). (Sometimes, the exchange contributions within HF are also called 'Pauli correlations'.)

The basic idea of the DFT is almost as old as quantum mechanics and has already been used by Thomas [1] and Fermi [2] in their famous work, namely to calculate the total energy of a system by an integral over an expression depending only on the local ground-state density $\rho(\mathbf{r})$ (cf. Sect. 4):

$$E_{tot} = E[\rho] = \int \mathcal{E}[\rho(\mathbf{r})] \, d^3r \; . \tag{A.23}$$

Mathematically speaking, the energy is assumed to be a functional of $\rho(\mathbf{r})$, denoted by $E[\rho]$. The formal basis of the ensuing theory was laid by Hohenberg and Kohn [3] in a theorem which they proved for a non-degenerate electronic system. A more general proof, independent of ground-state degeneracy and of the so-called V-representability assumed by Hohenberg and Kohn, was given by Levy [4]. The Hohenberg-Kohn theorem states that the exact ground-state energy of a correlated electron system is a functional of the density $\rho(\mathbf{r})$, and that this functional has its variational minimum when evaluated for the exact ground-state density. This is expressed by the variational equation

$$\frac{\delta}{\delta\rho(\mathbf{r})} \left[E[\rho(\mathbf{r})] - \mu \int \rho(\mathbf{r}) \, d^3r \right] = 0 \,, \tag{A.24}$$

using the Lagrange multiplier μ to fix the number of particles N according to (A.3). If the exact functional $E[\rho]$ were known – which it is not – the solution of Eq. (A.24) would lead to the knowledge of the exact ground-state energy and density.

It is useful to break up the energy functional (A.23) into several parts, by writing

$$E[\rho] = T_s[\rho] + \int \rho(\mathbf{r}) \left\{ U(\mathbf{r}) + \frac{1}{2} V_H[\rho(\mathbf{r})] \right\} d^3r \; + E_{xc}[\rho] \,. \tag{A.25}$$

Hereby $T_s[\rho]$ contains that part of the kinetic energy which corresponds to a system of independent particles with density ρ. The external potential energy and the Hartree energy are obvious and the same as in Eq. (A.11) for the HF energy. The last term in (A.25) is the so-called *exchange-correlation energy*. It contains the exchange part of the energy, i.e. E_x in Eq. (A.15) above, plus all the contributions due to the correlations, coming from the fact that the exact wavefunction is *not* a Slater determinant. (Note that in DFT, no assumption is made at all about the *wavefunction* of the total system.) E_x contains, in particular, also a correlation part coming from the total kinetic energy.

$E_{xc}[\rho]$ is in general not exactly known for finite interacting fermion systems (with the exception of simple model cases with suitably chosen forms of the interaction). The same holds for the kinetic energy functional $T_s[\rho]$ which is not known explicitly for many-fermion systems. Note that in Chapter 4 we derive an approximate semiclassical functional $T_{ETF}[\rho]$ that allows one to perform the variational principle (A.24) directly on the local density $\rho(\mathbf{r})$ (see Sect. 4.4.2). But the functional $T_{ETF}[\rho]$ cannot describe quantum shell effects, and the ETF variational method therefore only yields average results.

In order to avoid the difficulty of finding an explicit density functional for the kinetic energy, Kohn and Sham [5] proposed to write the density $\rho(\mathbf{r})$ in the form of Eq. (A.4) in terms of some trial single-particle wavefunctions $\psi_i(\mathbf{r})$. This is, in fact, possible for any non-negative normalizable density [6]. The non-interacting part of the kinetic energy density can then be given in the form $\tau(\mathbf{r})$ (A.14) in terms of the same $\psi_i(\mathbf{r})$. The variation (A.24) of the energy functional can now be done through variations of the trial functions $\psi_i(\mathbf{r})$ with a constraint on their norms, as in the HF variation principle (A.7), except that $\langle\Phi|\hat{H}|\Phi\rangle$ here is replaced by $E[\rho]$ (A.23). This leads to the widely used *Kohn-Sham (KS) equations*:

$$\left\{\hat{T} + V_{KS}(\mathbf{r})\right\}\psi_i(\mathbf{r}) = E_i\psi_i(\mathbf{r}),\qquad\text{(A.26)}$$

in which the local potential $V_{KS}(\mathbf{r})$ is a sum of three terms:

$$V_{KS}(\mathbf{r}) = V_{KS}[\rho(\mathbf{r})] = U(\mathbf{r}) + V_H[\rho(\mathbf{r})] + V_{xc}[\rho(\mathbf{r})].\qquad\text{(A.27)}$$

The first two of them are the same as above, and the third term is the variational derivative of the exchange-correlation energy:

$$V_{xc}[\rho(\mathbf{r})] = \frac{\delta}{\delta\rho(\mathbf{r})}E_{xc}[\rho].\qquad\text{(A.28)}$$

Like the HF equations, the KS equations (A.26) are nonlinear due to the density dependence of V_{KS} in (A.27). The important difference, however, is that the potential $V_{KS}(\mathbf{r})$ is local and the KS equations therefore are much easier to solve.

In principle, the Hohenberg-Kohn theorem thus allows one to transform the full many-body problem for the ground state of a correlated fermion system to the much simpler self-consistent one-body variational equations (A.24) or (A.28), since no approximation has been used up to this point. Practically, however, the exchange and correlation contributions can only be evaluated approximately. It is thus a matter of state-of-the-art of the DFT to use more or less fancy approximations to the exchange-correlation functional. The simplest and most frequently applied functionals for $E_{xc}[\rho]$ make use of the *local density approximation* (LDA). Here one performs more or less sophisticated many-body calculations for a hypothetical infinite system with constant density ρ. The resulting energy per particle is used to extract the corresponding xc-part $e_{xc}(\rho)$ which is a function of the variable ρ. The LDA for a finite system with variable density $\rho(\mathbf{r})$ then consists in assuming the local xc-energy density to be that of the corresponding system with density $\rho = \rho(\mathbf{r})$:

$$E_{xc}^{LDA}[\rho] = \int \rho(\mathbf{r})e_{xc}(\rho(\mathbf{r}))\,d^3r\;.\qquad\text{(A.29)}$$

This formalism can be extended to take spin degrees of the particles into account, by introducing a spin-up density and a spin-down density. This leads to the 'local spin density' (LSD or LSDA) formalism.

Another extension of the DFT concerns the inclusion of a finite temperature $T > 0$. Mermin [7] derived the Hohenberg-Kohn theorem and the Kohn-Sham formalism at

$T > 0$ for a grand canonical system. Evans [8] extended it to canonical systems. In essence, one goes over from the (internal) energy $E[\rho]$ (A.25) to the free energy $F[\rho]$

$$F[\rho] = E[\rho] - TS_s[\rho],\tag{A.30}$$

where S_s is the noninteracting part of the entropy [see Eq. (4.95)]. The exchange-correlation energy $E_{xc}[\rho]$ will, in general, depend on T explicitly (i.e., not only through the density). The Kohn-Sham formalism then is obtained by including into the definition of the densities (A.4,A.14) the finite-temperature occupation numbers n_i

$$\rho(\mathbf{r}) = \sum_i |\psi_i(\mathbf{r})|^2 n_i, \qquad \tau(\mathbf{r}) = \frac{\hbar^2}{2m} \sum_i |\nabla \psi_i(\mathbf{r})|^2 n_i, \qquad \sum_i n_i = N,\tag{A.31}$$

and by minimizing $F[\rho]$ with respect to both the ψ_i and the n_i. Since S_s does not depend explicitly on the wavefunctions ψ_i, the variation of the latter gives exactly the same form (A.26) of the KS equations, the only difference being that the potential V_{KS} becomes temperature dependent. Variation of the n_i gives their explicit form in terms of the E_i; the result depends on whether one treats the system as a canonical or a grand canonical ensemble. In the latter case, where the chemical potential μ is used to constrain the average particle number N, one obtains the familiar Fermi occupation numbers given in Eq. (4.94). For an extensive discussion of the finite-temperature DFT, we refer to a review article by Gupta and Rajagopal [9].

The DFT has had a considerable success in many branches of physics. For a comprehensive discussion of its foundations, the formalism, and applications in atomic, molecular and solid state physics, we refer to the text book of Dreizler and Gross [10]. The DFT has been used in the self-consistent jellium model calculations discussed in Chapters 4 and 8.

A.3 The Strutinsky energy theorem

The shell-correction method discussed in Sect. 4.7.2 was based by Strutinsky [11] on the Hartree-Fock approach. We shall here give a brief derivation and discussion of the basic energy theorem used in this context.

Let us assume that we can split the one-body density matrix ρ into a smooth and an oscillating part:

$$\rho = \tilde{\rho} + \delta\rho.\tag{A.32}$$

The precise definition of the average part $\tilde{\rho}$ does not matter in principle. The important assumption is that the total energy as a functional of $\tilde{\rho}$ is smooth in the same sense as an extended Thomas-Fermi energy or a Strutinsky-averaged energy (see Chapter 4). To fix the idea [12, 13], we define it here by a Strutinsky averaging (see Sect. 4.7.1) of ρ in the matrix representation (A.19)

$$\tilde{\rho}_{\alpha\beta} = \sum_{i=1} \langle \alpha | i \rangle \langle i | \beta \rangle \tilde{n}_i\tag{A.33}$$

in terms of the averaging numbers $\tilde{n}_i$ defined in Eq. (4.177). We now make a Taylor expansion of the total HF energy (A.20) around $\tilde{\rho}$:

$$
\begin{aligned}
E_{HF}[\rho] &= E_{HF}[\tilde{\rho}] + \mathrm{tr}\, \frac{\partial E_{HF}}{\partial \rho}\bigg|_{\tilde{\rho}} \delta\rho + \mathcal{O}[(\delta\rho)^2] \\
&= \tilde{E}_{HF} + \delta_1 E + \mathcal{O}[(\delta\rho)^2].
\end{aligned}
\tag{A.34}
$$

The first term represents, by assumption, the smooth part of the HF energy. In order to bring the second term into a simple form, we note that

$$
\frac{\partial E_{HF}}{\partial \rho_{\beta\alpha}} = \langle \alpha | \hat{T} + U + \hat{V}_{HF} | \beta \rangle
\tag{A.35}
$$

is the matrix element of the total HF one-body operator in Eq. (A.8) which we just call the (one-body) HF Hamiltonian $\hat{H}_{HF}$. We can therefore write the second term in the expansion (A.34) as [cf. Eq. (4.190)]

$$
\delta_1 E = \mathrm{tr}\, \hat{H}_{HF}[\tilde{\rho}]\, \delta\rho = \sum_{i=1}^{N} \tilde{E}_i - \sum_i \tilde{E}_i \tilde{n}_i .
\tag{A.36}
$$

Hereby, $\tilde{E}_i$ are the eigenvalues of the *averaged* HF Hamiltonian, given by the equation

$$
\hat{H}_{HF}[\tilde{\rho}]\, \tilde{\psi}_i = \tilde{E}_i \tilde{\psi}_i .
\tag{A.37}
$$

We arrive thus at the following expression for the total HF energy

$$
E_{HF} = \tilde{E}_{HF} + \delta_1 E + \mathcal{O}[(\delta\rho)^2] ,
\tag{A.38}
$$

usually referred to as the *Strutinsky energy theorem*. The name was introduced by Bethe [14] who showed that it also holds for effective two-body potentials (the so-called "effective interactions" in nuclear physics) that are density dependent. What this theorem tells us is that up to terms of second order in $\delta\rho$, the oscillating part of the total HF energy is contained in the sum of occupied energy levels of the *averaged* HF mean field. The oscillating term $\delta_1 E$, which contains all contribution of first order in $\delta\rho$, is thus identical with the shell-correction energy δE defined in Eq. (4.190) of Sect. 4.7.1. When derived from the HF theory as here, it is evaluated in terms of the average part of the HF potential. In practical applications, the spectrum $\tilde{E}_i$ is replaced by that of a phenomenological shell-model potential, and the average total HF energy $\tilde{E}_{HF}$ is taken from the liquid drop model, as discussed in Sects. 4.7.1 and 8.1.1.

We note that the Strutinsky energy theorem holds also within the density functional theory. The basic point is that one uses a variational procedure, in which the mean field is given by the variational derivative of the total energy with respect to the density or density matrix. In the DFT, the key relation is Eq. (A.28) which shows that the mean field contribution from the exchange-correlation energy (which formally is the same as the HF potential energy obtained with a density dependent two-body interaction) is given by the variational derivative with respect to ρ. Similarly, Strutinsky and collaborators [15] showed that the energy theorem also applies within the Landau-Migdal theory of interacting quasiparticles. Discussions of the theorem in the framework of the Hartree-Fock-Bogolyubov theory for pairing interactions were given in Refs. [13, 16].

The Strutinsky method can therefore be applied straightforwardly to any fermion system, provided that the expansion (A.34) converges, i.e., that the neglected terms of second order in $\delta\rho$ are sufficiently small. This convergence has extensively been tested for nuclei by Brack and Quentin [13, 17] with realistic density-dependent effective interactions. As an example we show in Fig. A.1 the deformation energy of a typical rare-earth nucleus. The thin solid line shows the exact HF energy as a function of the total mass quadrupole moment of the nucleus. The heavy line is the energy $\tilde{E}_{HF}$, as obtained from the self-consistently Strutinsky-averaged density matrix (A.33). The thin dashed line is the 'ideal' shell-correction approximation, in which $\delta_1 E$ was evaluated from the average HF fields (for both neutrons and protons), and the higher-order terms of the expansion (A.34) were neglected. The figure shows that they amount to less than ~ 1 MeV at all deformations, which is small enough with respect to the main shell oscillations. The same kind of convergence was found also for many other nuclei with different effective interactions. A comprehensive review on the Strutinsky energy theorem and its numerical verifications has been given in Ref. [13].

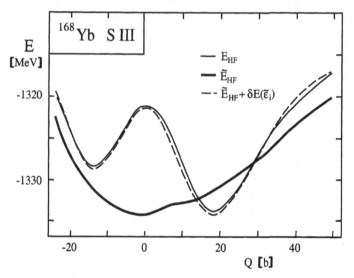

Figure A.1: Deformation energy (in MeV) of the nucleus ^{168}Yb, shown versus the total mass quadrupole moment Q (in barns). *Thin solid line:* exact total Hartree-Fock (HF) energy, obtained with the Skyrme SIII interaction. *Heavy solid line:* average total HF energy obtained through self-consistent Strutinsky averaging. *Thin dashed line:* approximation given by the Strutinsky energy theorem (A.38). (After [17].)

Appendix B

Inverse Laplace transforms

We give here a few simple rules and examples of inverse Laplace transforms. More examples may be found, e.g., in Ref. [18]. A standard reference for the theory of Laplace transforms is the book by Van der Pohl and Bremmer [19]. The folding theorem reads

$$\int_0^E g_1(E - E') g_2(E') \, dE' = \mathcal{L}_E^{-1} [Z_1(\beta)Z_2(\beta)]. \tag{B.1}$$

Other general relations are

$$
\begin{aligned}
g(E) &= \mathcal{L}_E^{-1} [Z(\beta)], \\
g(E - \alpha) &= \mathcal{L}_E^{-1} \left[e^{-\alpha\beta} Z(\beta) \right], \\
\frac{d}{dE} g(E) &= \mathcal{L}_E^{-1} [\beta Z(\beta)], \\
-E g(E) &= \mathcal{L}_E^{-1} \left[\frac{d}{d\beta} Z(\beta) \right], \\
\int_0^E g(E') dE' &= \mathcal{L}_E^{-1} \left[\frac{1}{\beta} Z(\beta) \right],
\end{aligned}
\tag{B.2}
$$

Some special cases often used in this book are

$$
\begin{aligned}
\delta(E) &= \mathcal{L}_E^{-1}[1], \\
\frac{1}{\Gamma(\mu)} E^{\mu-1} &= \mathcal{L}_E^{-1}[\beta^{-\mu}], \qquad (\mu > 0) \\
\frac{1}{\sqrt{\pi k}} \sin(2\sqrt{kE}) &= \mathcal{L}_E^{-1} \left[\frac{1}{\beta^{3/2}} e^{-k/\beta} \right]. \\
\left(\frac{E}{k} \right)^{\mu/2} J_\mu(2\sqrt{kE}) &= \mathcal{L}_E^{-1} \left[\frac{1}{\beta^{\mu+1}} e^{-k/\beta} \right]. \qquad (\mu > -1)
\end{aligned}
\tag{B.3}
$$

Strictly speaking, the left sides in the last three relations above have to be multiplied by a step function $\Theta(E)$ which cuts the resulting functions below $E = 0$. This need not be done in connection with two-sided Laplace transforms [19], for which the lower integration limits in Eqs. (B.1), (B.2) are $-\infty$, and the upper limit in Eq. (B.1) is $+\infty$.

Appendix C

More about the monodromy matrix

The stability of periodic orbits is rarely discussed in the standard textbooks on classical mechanics. An important quantity which contains the stability information is the monodromy matrix. The book by Yakubovich and Starzhinskii on *Linear differential equations with periodic coefficients* [20] gives an extensive mathematical discussion of the subject. We have in this appendix excerpted those definitions and relations which are the most important for the present context, and give an example that is easy to calculate.

C.1 Linear differential equations with periodic coefficients

We investigate the solutions of the following system of D linear differential equations for the coordinate vector $\mathbf{q}(t) = \{q_i(t)\}$, $i = 1, 2, ..., D$:

$$\dot{\mathbf{q}} = A(t)\,\mathbf{q}, \tag{C.1}$$

where $A(t)$ is a $D \times D$ matrix with T-periodic elements:

$$A_{ij}(t + T) = A_{ij}(t). \tag{C.2}$$

We take D linearly independent solutions $\mathbf{q}^{(j)}$, $(j = 1, 2, ..., D)$ of (C.1) and build a matrix $X(t)$, the so-called *matrizant*:

$$X_{ij}(t) = \{q_i^{(j)}(t)\} \tag{C.3}$$

which fulfills

$$\frac{d}{dt}X(t) = A(t)\,X(t) \tag{C.4}$$

with the initial condition

$$X(0) = I, \tag{C.5}$$

431

where I is the $D \times D$ unit matrix. One finds that $X(t)$ has the following important property:

$$X(t + T) = X(t) X(T).$$ (C.6)

The matrizant at the time of the period T is called the *monodromy matrix*[1]:

$$M = X(T).$$ (C.7)

The so-called *characteristic equation* of (C.1) is the equation that diagonalizes M:

$$\det |M - \lambda_i I| = 0.$$ (C.8)

The eigenvalues λ_i are called the *multipliers* of (C.1), and the set $\{\lambda_i\}$ is the *spectrum* of (C.1). One also shows that

$$\mathbf{q}(t) = X(t) \mathbf{q}(0).$$ (C.9)

Thus, $X(t)$ can be considered as a kind of "propagator" for the time development of $\mathbf{q}(t)$. For any multiplier λ_i there is a solution $\mathbf{q}(t)$ with

$$\mathbf{q}(t + T) = \lambda_i \mathbf{q}(t),$$ (C.10)

from which it follows (with $t=0$) that

$$(M - \lambda_i I) \mathbf{q}(0) = 0,$$ (C.11)

so that $\mathbf{q}(0)$ are the eigenvectors of M.

The *Floquet-Lyapounov theorem* states that the matrizant $X(t)$ can be written as

$$X(t) = F(t) e^{Kt}, \quad \text{with} \quad M = e^{KT},$$ (C.12)

where $F(t)$ is T-periodic:

$$F(t + T) = F(t), \quad F(0) = I.$$ (C.13)

If $A(t)$ is real, then M is also real but K is in general complex. The eigenvalues κ_i of the constant matrix K, given by

$$\det |K - \kappa_i I| = 0, \quad \lambda_i = e^{\kappa_i T},$$ (C.14)

are called the *characteristic exponents* of the system (C.1). Assuming the κ_i to be non-degenerate, one finds that there exists a fundamental set $q_i^{(j)}$ of linearly independent solutions of (C.1) which can be written in the form

$$\mathbf{q}^{(j)}(t) = e^{\kappa_j t} \mathbf{f}_j(t),$$ (C.15)

where

$$\mathbf{f}_j(t) = F(t) \mathbf{q}^{(j)}(0), \quad \mathbf{f}_j(t + T) = \mathbf{f}_j(t).$$ (C.16)

We see from (C.15) that the characteristic exponents κ_j are crucial for the stability of the solution $\mathbf{q}^{(j)}(t)$. As soon as one of the κ_i has a positive real part, the corresponding solution is unstable.

---

[1] *monodromy*, from Greek, means "one run (around the track)".

C.2   Hamiltonian equations

Let us start from a harmonic Hamiltonian

$$H = \frac{1}{2m} \sum_i p_i^2 + \frac{1}{2} \sum_{ij} V_{ij}(t) q_i q_j , \qquad (i,j = 1, 2, ..., D) \qquad (C.17)$$

with T-periodic coefficients $V_{ij}(t+T) = V_{ij}(t)$. Hamilton's equations (2.12) can then be written as a system of $2D$ linear differential equations

$$\frac{d}{dt} \begin{pmatrix} \mathbf{q} \\ \mathbf{p} \end{pmatrix} = \begin{pmatrix} 0 & \frac{1}{m} I \\ -\mathcal{V} & 0 \end{pmatrix} \begin{pmatrix} \mathbf{q} \\ \mathbf{p} \end{pmatrix} , \qquad (C.18)$$

where $\mathcal{V}$ is the coefficient matrix V_{ij} of the potential energy in (C.17), and 0 here is the $D \times D$ matrix of zeros. We then have the *Lyapounov-Poincaré theorem:*

The monodromy matrix of Hamilton's equations in the form (C.18) is symplectic and thus has eigenvalues λ_i which are symmetric about the unit circle in the complex plane (i.e., they are pair wise inverse). If the matrix in (C.18) is real, then the λ_i are also symmetric with respect to the real axis. The solutions of (C.18) are stable if and only if all the λ_i lie on the unit circle itself.[2]

It can be shown that the same holds also for (not necessarily canonical Hamiltonian) systems with *time-reversal symmetry* – this is the so-called *Lyapounov theorem*. A consequence of these theorems is that the determinant of M is always unity:

$$\det M = 1 . \qquad (C.19)$$

Note that these theorems also hold for the most general Hamiltonian which is bilinear in the q_i and p_i.

C.2.1   Example: Two-dimensional harmonic oscillator

As a simple example, we study the two-dimensional anisotropic harmonic-oscillator potential (see also Ref. [21]):

$$H = \frac{1}{2}(p_x^2 + p_y^2) + \frac{1}{2}(\omega_x^2 x^2 + \omega_y^2 y^2) . \qquad (C.20)$$

Since this Hamiltonian separates in the two coordinates, it is trivial to find the solutions of the equations of motion (C.18), which are just the harmonic functions. The matrizant fulfilling (C.5) is then immediately found to be

$$X(t) = \begin{pmatrix} \cos(\omega_x t) & 0 & \frac{1}{\omega_x}\sin(\omega_x t) & 0 \\ 0 & \cos(\omega_y t) & 0 & \frac{1}{\omega_y}\sin(\omega_y t) \\ -\omega_x \sin(\omega_x t) & 0 & \cos(\omega_x t) & 0 \\ 0 & -\omega_y \sin(\omega_y t) & 0 & \cos(\omega_y t) \end{pmatrix} . \qquad (C.21)$$

---

[2]We omit here some subtleties concerning the multiplicity of the λ_i and the simplicity of the elementary divisors of the characteristic equation (C.8).

This system has two basic periods $T_x = 2\pi/\omega_x$ and $T_y = 2\pi/\omega_y$; they correspond to one-dimensional oscillations along the principal axes. The monodromy matrices $M_x = X(T_x)$ and $M_y = X(T_y)$ for these linear periodic orbits have the usual form given in Eq.(5.45). The stability matrix for the n times repeated x orbit is immediately read off to be

$$\widetilde{M}_x^n = \begin{pmatrix} \cos(n2\pi\omega_y/\omega_x) & \frac{1}{\omega_y}\sin(n2\pi\omega_y/\omega_x) \\ -\omega_y\sin(n2\pi\omega_y/\omega_x) & \cos(n2\pi\omega_y/\omega_x) \end{pmatrix}, \tag{C.22}$$

and that for the y orbit is obtained by exchanging the indices x and y. The eigenvalues of the stability matrices for $n = 1$ are

$$\lambda_x = \exp\left(\pm 2\pi i\omega_y/\omega_x\right), \qquad \lambda_y = \exp\left(\pm 2\pi i\omega_x/\omega_y\right). \tag{C.23}$$

Thus we see that these one-dimensional orbits are stable since the eigenvalues (C.23) lie on the unit circle. For the case of an *irrational* frequency ratio ω_x/ω_y, they are the only periodic orbits with a finite length. The stability angle then never becomes zero or π, so that these orbits are isolated and Gutzwiller's trace formula can be applied. This is demonstrated in Sect. 5.6.2

C.3 Non-linear systems and the Poincaré variational equations

We now start from a holonomic Hamiltonian

$$H(\mathbf{q}, \mathbf{p}) = \frac{1}{2}\sum_i p_i^2 + V(\mathbf{q}) \tag{C.24}$$

which is not necessarily quadratic in the q_i. (We assume the kinetic energy to be quadratic in the p_i just for simplicity here; any general dependence of the p_i and q_i is allowed in principle.) Hamilton's equations (2.12) are then in general non-linear. Assume that we have found a T-periodic solution

$$q_i = q_i^{(0)}(t) = q_i^{(0)}(t + T), \qquad p_i = p_i^{(0)}(t) = p_i^{(0)}(t + T), \tag{C.25}$$

and that $V(\mathbf{q})$ in Eq. (C.24) is holomorphic around $V_0 = V(\mathbf{q}^{(0)})$. Let us investigate the stability of this solution with respect to some perturbations $\delta\mathbf{q}, \delta\mathbf{p}$:

$$\mathbf{q} = \mathbf{q}^{(0)} + \delta\mathbf{q}, \qquad \mathbf{p} = \mathbf{p}^{(0)} + \delta\mathbf{p}. \tag{C.26}$$

Substituting (C.26) into the Hamilton equations (2.12) and expanding up to second order with respect to the perturbations, we find a set of $2D$ *linear* differential equations for the $\delta\mathbf{q}, \delta\mathbf{p}$ which are of exactly the same form as (C.18), here called the *Poincaré variational equations*:

$$\frac{d}{dt}\begin{pmatrix} \delta\mathbf{q} \\ \delta\mathbf{p} \end{pmatrix} = \begin{pmatrix} 0 & I \\ -\mathcal{V}^0 & 0 \end{pmatrix}\begin{pmatrix} \delta\mathbf{q} \\ \delta\mathbf{p} \end{pmatrix}, \tag{C.27}$$

with the T-periodic Hessian matrix

$$\mathcal{V}^0(t) = V_{ij}^{(0)}(t) = \left.\frac{\partial^2 V(\mathbf{q})}{\partial q_i \partial q_j}\right|_{\mathbf{q}_0(t)}. \tag{C.28}$$

We can now apply the Lyapounov-Poincaré theorem to derive the stability exponents from the monodromy matrix of the system (C.27), in order to study the stability of the periodic solution (C.25).

C.4 Calculation of the monodromy matrix M

The stability matrix $\widetilde{M}$, which is the nontrivial part of the monodromy matrix M as decomposed in Eq. (5.45), is an essential ingredient of the Gutzwiller trace formula (5.36). In order to calculate M for a given periodic orbit, we now make use of the linearized equation of motion (C.27), which describes the time propagation of a small perturbation $(\delta\mathbf{q}(t), \delta\mathbf{p}(t))$ around a periodic orbit in phase space. Note that (C.27) is of the same form as Eq. (C.1), but in the $2D$-dimensional phase space. Applying the mathematics of Sect. C.1, we define the matrizant $X(t)$ for the perturbation by

$$\begin{pmatrix} \delta\mathbf{q}(t) \\ \delta\mathbf{p}(t) \end{pmatrix} = X(t) \begin{pmatrix} \delta\mathbf{q}(0) \\ \delta\mathbf{p}(0) \end{pmatrix}. \tag{C.29}$$

Using Eq. (C.27) it is easily seen that the matrizant fulfills the differential equation

$$\frac{d}{dt}X(t) = \begin{pmatrix} 0 & I \\ -\mathcal{V}^0(t) & 0 \end{pmatrix} X(t) \tag{C.30}$$

with the initial conditions

$$X_{mn}(0) = \delta_{mn}. \qquad (m, n = 1, 2, \ldots, 2D) \tag{C.31}$$

By definition, the monodromy matrix of a given periodic orbit is now the matrizant evaluated at the period T: $M = X(T)$, see Eq. (C.7). Note that the Hessian matrix $\mathcal{V}^0$ appearing in (C.30) is calculated along the periodic orbit.

The most straightforward way to obtain M numerically for systems with smooth potentials is to solve Eq. (C.30) for all components of $X(t)$ along with the Hamiltonian equations of motion. Instead of $2D$, this gives $2D \times (2D + 1)$ first-order differential equations; on modern computers with lots of memory, this is not a problem. Alternatively, one may also use the so-called *monodromy method* developed by Baranger *et al.* [22], in which a fast numerical calculation of M is done on a discretized time mesh *after* having obtained a periodic solution of the equations of motion of a given system. The method of these authors also allows for a fast numerical determination of the periodic orbits themselves.

A method for calculating directly the stability matrix $\widetilde{M}$ for two-dimensional billiards is discussed in the following section.

C.5   Calculation of $\widetilde{M}$ for two-dimensional billiards

A wide class of model potentials in which classical motion is studied are the so-called billiards consisting of infinitely deep potentials with steep walls. Classically, the particles are reflected at the boundary according to the law *angle of reflection = angle of incidence*; quantum mechanically, one imposes Dirichlet boundary conditions on the wave functions. Such potentials do not allow one to calculate the monodromy matrix as outlined in the above section, and different methods must be used. We shall in the following outline the method described by Berry [23].

We discuss here only two-dimensional billiards, limited by a closed curve $y = f(x)$ or, in polar coordinates, $r = r(\varphi)$. At each point of reflection, we define the following variables illustrated in Fig. C.1: the angle Ψ as the direction of the local tangent given by $\tan \Psi = dy/dx = f'(x)$, the arc length s of the limiting curve, integrated from some given reference point (e.g., the point with $\Psi = \pi/2$, lying on the positive x-axis), and the local curvature radius R. We assume that the curve $r(\varphi)$ is *not convex* as seen from the inside of the billiard, so that $R(\Psi)$ can be used to characterize the curve uniquely.

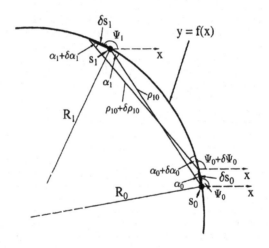

Figure C.1: Dynamics in a two-dimensional billiard. Part of a classical trajectory from point s_0 to point s_1 with path length ρ_{10}, and the corresponding path after a small variation δs_0.

We then have

$$R(\Psi) = \frac{ds}{d\Psi}; \qquad s(\Psi) = \int_{\pi/2}^{\Psi} R(\Psi')\, d\Psi'. \tag{C.32}$$

The total length L of the circumference is

$$L = \oint ds = \int_0^{2\pi} R(\Psi)\, d\Psi. \tag{C.33}$$

The direction of the particle after the impact is measured by the angle α of the outgoing direction with respect to the local tangent. It follows from the law of reflection, as also seen in Fig. C.1, that

$$\Psi_1 - \alpha_1 = \Psi_0 + \alpha_0 . \tag{C.34}$$

Using $dx = ds \cos \Psi$ and $dy = ds \sin \Psi$, we get the changes of the x- and y-coordinates for the trajectory from point s_0 to s_1:

$$\Delta x = x_1 - x_0 = \int_{\Psi_0}^{\Psi_1} R(\Psi) \cos \Psi \, d\Psi ,$$

$$\Delta y = y_1 - y_0 = \int_{\Psi_0}^{\Psi_1} R(\Psi) \sin \Psi \, d\Psi , \tag{C.35}$$

and the slope of this part of the trajectory is thus

$$\tan(\Psi_0 + \alpha_0) = \frac{\Delta y}{\Delta x} . \tag{C.36}$$

The equations (C.34)–(C.36) allow one to uniquely determine the coordinates of the new point (Ψ_1, α_1) from those of the old point (Ψ_0, α_0).

The momentum p of the particle in the tangential direction, which does not change at the reflection, is (with the mass set to $m=1$)

$$p = \cos \alpha . \tag{C.37}$$

For calculating trajectories, the pair of variables (Ψ, α) is more convenient, as shown above, but for the calculation of the monodromy matrix it is better to use the variables (s, p) which constitute a convenient phase space.

Each orbit is uniquely described by a succession of mappings in phase space: $(s_0, p_0) \to (s_1, p_1) \to \dots$. In the Poincaré surface of section (s, p) one obtains a point (s_i, p_i) at each reflection. The mapping

$$(s_i, p_i) \to (s_{i+1}, p_{i+1}) \tag{C.38}$$

is usually non-linear and therefore it cannot be described by a 2×2 matrix. However, this mapping is *area preserving* which we shall presently show.

Let us consider one portion of a trajectory, i.e. one bounce from s_0 to s_1 (see Fig. C.1). The length of the segment is ρ_{10}. If we make an infinitesimal shift δs_0 along the boundary, the next point will be at $s_1 + \delta s_1$. The lowest order change of ρ_{10} will then be perpendicular to ρ_{10} itself; its value is

$$\delta \rho_{10} = \rho_{10}(\delta \Psi_0 + \delta \alpha_0) = \delta s_0 \sin \alpha_0 + \delta s_1 \sin \alpha_1 . \tag{C.39}$$

Using this relation together with $\delta \Psi_i = \delta s_i / R(\Psi_i)$ and $\delta \alpha_i = -\delta p_i / \sin \alpha_i$, it is straightforward to obtain the *linear* transformation of the small changes:

$$\begin{pmatrix} \delta s_1 \\ \delta p_1 \end{pmatrix} = m_{10} \begin{pmatrix} \delta s_0 \\ \delta p_0 \end{pmatrix} , \tag{C.40}$$

where

$$m_{10} = \begin{pmatrix} \dfrac{\rho_{10}}{R_0 \sin \alpha_1} - \dfrac{\sin \alpha_0}{\sin \alpha_1} & -\dfrac{\rho_{10}}{\sin \alpha_0 \sin \alpha_1} \\[2ex] -\dfrac{\rho_{10}}{R_0 R_1} + \dfrac{\sin \alpha_1}{R_0} + \dfrac{\sin \alpha_0}{R_1} & \dfrac{\rho_{10}}{R_1 \sin \alpha_0} - \dfrac{\sin \alpha_1}{\sin \alpha_0} \end{pmatrix} , \tag{C.41}$$

and R_i are the curvature radii at the two points: $R_i = R(\Psi_i)$. One readily verifies that the determinant of m_{10} is unity:

$$\det |m_{10}| = \frac{\partial(s_1, p_1)}{\partial(s_0, p_0)} = 1, \qquad (C.42)$$

showing that the mapping (C.40) is area preserving, indeed.

Periodicity of an orbit means that after N reflections it returns to the same point in phase space:

$$(s_N, p_N) = (s_0, p_0). \qquad (C.43)$$

It is clear from the above that the monodromy matrix for such an orbit is just the product of the N corresponding mappings (C.40) of the deviations:

$$\widetilde{M} = m_{0,N-1} \cdot m_{N-1,N-2} \cdot \ldots \cdot m_{32} \cdot m_{21} \cdot m_{10}. \qquad (C.44)$$

In fact, this is already the *stability matrix*, defined as in Eq. (5.45) by splitting off the monodromy matrix the submatrix containing the trivial eigenvalues 1. This comes out because of the special simplicity of the billiard dynamics described by the pair of one coordinate (s) and one momentum (p).

According to the criteria discussed in Sec. 5.4 (see also Table 5.1), the conditions for the stability of an orbit are

$$\begin{aligned}
|\mathrm{tr}\widetilde{M}| > 2, & \qquad \text{(instability)} \\
|\mathrm{tr}\widetilde{M}| < 2, & \qquad \text{(stability)} \\
|\mathrm{tr}\widetilde{M}| = 2. & \qquad \text{(marginal stability)}
\end{aligned} \qquad (C.45)$$

Let us finally consider the special case of a straight-line trajectory between reflections at normal incidence, $\alpha_0 = \alpha_1 = \pi/2$. If the length of the trajectory is ρ and the curvature radii at both end points are equal, $R_0 = R_1 = R$, then the corresponding infinitesimal mapping matrix is

$$m_{R\rho} = \begin{pmatrix} \frac{\rho}{R} - 1 & -\rho \\ \frac{2}{R} - \frac{\rho}{R^2} & \frac{\rho}{R} - 1 \end{pmatrix}, \qquad (C.46)$$

and the stability matrix $\widetilde{M}$ of the corresponding periodic two-bounce orbit is simply the square of (C.46) which can easily be diagonalized. The conditions (C.45) for the stability of this orbit become

$$\begin{aligned}
\rho > 2R, & \qquad \text{(instability)} \\
\rho < 2R, & \qquad \text{(stability)} \\
\rho = 2R. & \qquad \text{(marginal stability)}
\end{aligned} \qquad (C.47)$$

C.5.1 Example: Elliptic billiard

We consider an elliptic billiard with semiaxes A and B in x- and y-direction, respectively, and the foci at $\pm C$ on the x-axis:

$$x = A\cos\varphi, \qquad y = B\sin\varphi, \qquad (B \le A) \qquad (C.48)$$

with

$$A = C/\epsilon, \quad B = C\sqrt{1-\epsilon^2}/\epsilon; \qquad A^2 - B^2 = C^2, \qquad (C.49)$$

where ϵ is the numerical eccentricity. Note that φ is *not* the polar angle of the point (x,y). The usual representation of the ellipse,

$$r(\phi) = \frac{A(1-\epsilon^2)}{1+\epsilon\cos\phi}, \qquad (C.50)$$

involves polar coordinates (r,ϕ) whose origin is one of the foci and not the origin of the x,y-plane (see Problem 2.2).

The relation between φ and the angle Ψ is given by

$$\tan\Psi = -\frac{B}{A}\cot\varphi, \qquad (C.51)$$

and the curvature radius R is given by

$$R(\varphi) = \frac{1}{AB}\left[B^2\cos^2\varphi + A^2\sin^2\varphi\right]^{3/2}. \qquad (C.52)$$

The motion in this billiard is integrable: besides the energy there is a second constant of motion which is given by

$$F(s,p) = \frac{p^2 - \epsilon^4\cos^2\Psi(s)}{1 - \epsilon^4\cos^2\Psi(s)}. \qquad (C.53)$$

The geometric meaning of $F(s,p)$ is the product of the angular momenta of the particle about the two foci: $F = L_1 \cdot L_2$. In the spherical limit $\epsilon = 0$ it becomes proportional to the energy.

There are two types of orbits: a) "librations" with zero average angular momentum (cf. Fig. C.3 in Problem C.2), and b) "rotations" with nonzero average angular momentum. Their trajectories cover only a part of the available area in the ellipse; the caustics which give the boundaries of that area are hyperbolae for the librations and ellipses for the rotations, both confocal with the boundary. The periodic orbits of both types form families of non-isolated orbits which can be classified in a way similar to the families of polygons in the circular billiard (cf. Fig. 6.16). The only isolated periodic orbits are the straight-line librations along the symmetry axes of the ellipse. The longer one is always unstable. The shorter one is stable except at isolated bifurcation points ϵ_{nm}, where it becomes marginally stable. The bifurcations of the short diameter orbit are discussed in Problem 5.3 and in Sect. 6.2.2. For an exhaustive classification of the periodic orbits in the elliptic billiard, we refer to Ref. [24].

C.6   Problems

(C.1) Rhombus orbit in the elliptic billiard

Calculate the stability matrix for the rhombus orbit, shown in Fig. C.2, of an elliptic billiard with eccentricity ϵ. The result is

$$\widetilde{M} = \begin{pmatrix} 1 & -2A\left(1-\epsilon^2\right)\left(2-\epsilon^2\right)^{3/2} \\ 0 & 1 \end{pmatrix}, \tag{C.54}$$

and obviously has the eigenvalues $\lambda = 1$, independently of the eccentricity. This orbit therefore must belong to a family of degenerate orbits. (Some of its other members are illustrated on the right-hand side of Fig. 6.16.)

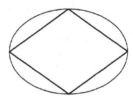

Figure C.2: Rhombic periodic orbit in an elliptic billiard.

(C.2) "Butterfly" and "bird" orbits in the elliptic billiard

In the elliptic billiard, the short diameter orbit has a bifurcation point at $\epsilon = 1/\sqrt{2}$ (see Problem 5.3 and Sect. 6.2.2). At this point, a new family of periodic orbits is "born" in the form of "butterflies" or a V-shaped "bird", shown in Fig. C.3 for a larger deformation with $\epsilon = 0.8$. The dotted line shows the hyperbolic caustic of this family which is confocal with the elliptic boundary. At $\epsilon = 1/\sqrt{2}$, these orbits all collapse into the twice repeated short diameter orbit.

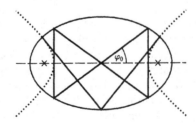

Figure C.3: "Bird" and "butterfly" orbits in an elliptic billiard with eccentricity $\epsilon = 0.8$. The dotted hyperbola is the caustic of the degenerate family to which these two orbits belong.

(a) Using the relations given above in Sect. C.5.1 and elementary planar geometry, show that the angle φ_0 which by Eq. (C.48) defines the upper right reflection point of the bird orbit is given by

$$\sin \varphi_0 = -1 + 1/\epsilon^2 .\qquad(C.55)$$

This equation has a real solution only for $\epsilon \geq 1/\sqrt{2}$. Show that the length of this orbit is $L_V = 4A/\epsilon$, where A is the longer semiaxis of the boundary, and becomes equal to 4B for $\epsilon = 1/\sqrt{2}$.

(b) Calculate the monodromy matrix of the bird orbit and show that it is equal to the unit matrix. Hence, the bird is marginally stable, as it must be for a member of a degenerate family of orbits.

(c) Determine the geometry of the symmetric butterfly orbit shown in Fig. C.3. Show that its two diagonals lie on the asymptotes of the hyperbolic caustic, that their opening angle is twice the angle φ_0 given in (C.55), and that its length is equal to the length $L_V = 4A/\epsilon$ of the bird.

(C.3) Straight-line orbits in a billiard with asymmetric deformation
Let the boundary of a two-dimensional billiard be given by

$$y(x) = (1 - x^2)(1 + \alpha\, x) .\qquad(C.56)$$

For $\alpha = 0$, this gives the integrable circular billiard which is treated in Sects. 2.6.3 and 6.1.3. For $\alpha \neq 0$, the system becomes non-integrable; it is asymmetric with respect to the y axis but retains reflection symmetry at the x axis and time-reversal symmetry. For large α the dynamics becomes quite chaotic, similar to the oval billiard treated in Ref. [25]. In Fig. C.4, we show the shortest periodic orbits with two reflection points in this billiard, obtained for $\alpha = 0.61$. There are four of them; the vertical orbit and the two diagonal orbits, degenerate by reflection at the x axis, are isolated (dashed lines), and the horizontal orbit (solid line) is marginally stable.

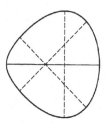

Figure C.4: Shortest periodic orbits with two reflection points in the billiard (C.56), obtained for $\alpha = 0.61$. The vertical and diagonal orbits (dahed lines) are isolated, whereas the horizontal orbit (solid line) is marginally stable.

Calculate the stability matrix for the horizontal orbit and show that it is marginally stable ($\mathrm{tr}\widetilde{M} = 2$) for any value of α.

Note: The breaking up of the degenerate family of diameter orbits of the circular billiard into three isolated orbits and one marginally stable orbit is quite unusual.

The uniform approximation for U(1) symmetry breaking, treated in Sect. 6.3.1, does therefore not apply here. In Ref. [26], trace formulae for billiards with small multipole deformations have been derived in first order of the perturbation theory. However, it is shown there that the diameter orbit is not affected to first order in an octupole deformation, which corresponds to (C.56) for small α, so that one would have to go to second-order perturbation theory in order to describe the properties of the orbits shown in Fig. C.4.

Appendix D

Calculation of Maslov indices for isolated orbits

The Maslov index is an essential ingredient of the periodic orbit theory, since it governs the interference of the different periodic orbit contributions to the trace formulae and hence the quantum ocillations in the level density and the other quantities derived from it. The name "Maslov index" in connection with the Gutzwiller trace formula for non-integrable systems is perhaps somewhat unfortunate. Originally, the Maslov index was introduced in the framework of the WKB and later the EBK quantization of integrable systems, and is defined in Eqs. (2.104), (2.105) and (2.119). For the mathematical definition of this classical Maslov index in the context of symplectic geometry, we refer to the textbooks of Arnold [27] and de Gosson [28]. For the original theory of Maslov in the context of semiclassical physics, the courageous reader may consult the textbook by Maslov and Fedoriuk [29]. The quantity σ appearing in the Gutzwiller trace formula and its extensions for systems with continuous symmetries is a different index, as becomes evident from its definitions given in Chapters 5 and 6. In the literature on the periodic orbit theory, its name has, however, established itself as "Maslov index", and we have in this book adhered to this convention.

Before presenting prescriptions for the calculation of Maslov indices, we venture to present here a conjecture, which is the result of numerical experience with many systems over many years: *For two-dimensional Hamiltonian systems with smooth potentials, and for two-dimensional billiards with smooth boundaries, the Maslov index of an isolated orbit is always* even *if the orbit is* hyperbolically unstable $(\mathrm{tr}\widetilde{M} > 2)$, *and it is* odd *if the orbit is* stable or inverse-hyperbolically unstable $(\mathrm{tr}\widetilde{M} < 2)$. *This holds both for primitive orbits and for all their repetitions.* Consequently, the only change of the Maslov index of a given orbit, which can take place when the energy or any smooth parameter of the system is varied, happens when the value of $\mathrm{tr}\widetilde{M}$ goes through $+2$, i.e., when the orbit bifurcates. (These changes are given explicitly, for all generic bifurcations, in the papers by Sieber and Schomerus [30].) The above conjecture seems to have been proven recently by Sugita [31], who has given a general formula for the Maslov index of a periodic orbit. He makes, however, use of a winding number k for whose calculation no concrete prescriptions are given.

The Maslov index for isolated periodic orbits in two-dimensional systems has been thoroughly discussed by Creagh *et al.* [32] and, for unstable orbits, in D-dimensional systems by Robbins [33]. We have in this book adapted their notation, which is consistent with that of Gutzwiller's original papers (cf. Chapter 5), and shall in the following sketch their methods of calculating the Maslov index in different situations. As we have seen in Sect. 5.3, the total Maslov index appearing in Gutzwiller's trace formula (5.36) is a sum of two contributions: $\sigma = \mu + \nu$ [cf. Eq. (5.37)], whereof the first part μ is the index occurring in the semiclassical expression of the Green's function $G_{scl}(\mathbf{r}, \mathbf{r}'; E)$ (5.15) and counts the number of conjugate points of a given orbit at fixed energy. The second contribution ν arises when taking the trace of $G_{scl}(\mathbf{r}, \mathbf{r}'; E)$ in order to arrive at the level density $g(E)$ (or its oscillating part). Whereas both μ and ν may depend on the starting point along a periodic orbit, their sum has been shown for unstable orbits to be a topological invariant [32, 33]. In fact, σ is a winding number which can be calculated directly from the eigenvectors of the stability matrix.

D.1   Isolated orbits in smooth potentials

We first discuss the case of smooth potentials, assuming the classical Hamiltonian to be of the form $H = T + V(\mathbf{r})$. We assume that the periodic orbits have been found, by integration of the equations of motion analytically or numerically, and that the matrizant $X(t)$ for a given orbit is available (see Appendix C). Two different procedures are used to calculate the Maslov index, depending on whether the orbit is stable or unstable.

D.1.1   Unstable orbits

For an unstable orbit it is possible to obtain the total Maslov index σ at once. For that purpose one uses at time $t = 0$, at a suitable starting point of the orbit (which can be any point, but not a turning point where the orbit in coordinate space has a cusp!), the eigenvector $\mathbf{e}_0 = (e_x, e_y, e_{p_x}, e_{p_y})$ of the monodromy matrix M that corresponds to the unstable eigenvalue, and the flow vector of the Hamiltonian $\mathbf{f}_0 = (\dot{x}, \dot{y}, \dot{p}_x, \dot{p}_y)$. One then propagates these two vectors once around the orbit using the matrizant $X(t)$:

$$\mathbf{e}(t) = X(t)\,\mathbf{e}_0, \qquad \mathbf{f}(t) = X(t)\,\mathbf{f}_0, \qquad\qquad (D.1)$$

which yields two time dependent vectors $\mathbf{e}(t) = e_i(t)$ and $\mathbf{f}(t) = f_i(t)$ $(i = 1, 2, 3, 4)$. Next one defines two real (2x2) matrices $U(t)$ and $V(t)$ by

$$U(t) = \begin{pmatrix} e_1(t) & f_1(t) \\ e_2(t) & f_2(t) \end{pmatrix} \qquad V(t) = \begin{pmatrix} e_3(t) & f_3(t) \\ e_4(t) & f_4(t) \end{pmatrix}. \qquad (D.2)$$

The Maslov index σ_n for the n times repeated periodic orbit is then found as the winding number $\sigma_n = \mathcal{W}[C(nT)]$ of the complex quantity

$$C(t) = \{\det[U(t) - iV(t)]\}^2, \qquad\qquad (D.3)$$

obtained by following its path in the complex plane during n periods, i.e., over a time interval nT. Fig. D.1 shows this path for the orbit A in the Hénon-Heiles potential (see Sect. 5.6.4) at the scaled energy $e=0.9$ where it is unstable. Although the path is not closed, its winding number is clearly $\mathcal{W}[C(T)] = \sigma = 5$.

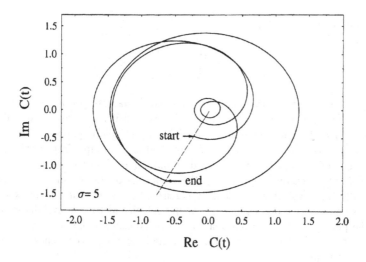

Figure D.1: Illustration of the determination of the Maslov index for an unstable isolated orbit (after [34]). The start is at some time t_0 and the end after one period T: $t = t_0 + T$.

D.1.2 Stable orbits

For stable orbits, σ is in general not an integer winding number, but it may be found as the integer part of a quantity that can be interpreted as a (non-integer) winding number (see the example in Sect. D.1.3 below). Using the methods discussed in Ref. [32], the following are well-working recipes for the separate computation of its two contributions μ and ν (see also Ref. [34]).

The calculation of μ is obtained by a similar procedure as that outlined above for σ of an unstable orbit. One takes the flow vector $\mathbf{f}_0$ and combines it with the vector $\mathbf{e}_0 = (0, 0, 1, -\dot{x}/\dot{y})$, if $\dot{y} \neq 0$ at the starting point or, equivalently, $\mathbf{e}_0 = (0, 0, -\dot{y}/\dot{x}, 1)$, if $\dot{x} \neq 0$ at the starting point. One propagates these two vectors along the orbit using Eq. (D.1) and defines U according to Eq. (D.2). The Maslov index μ is then found as the number of zeros which det$U$ assumes during one full period, hereby not counting the zero that appears trivially at the starting point.

The index ν is given by the sign of the quantity w defined by

$$w = \frac{\mathrm{tr}\widetilde{M} - 2}{b}, \tag{D.4}$$

where b is the upper right element of the stability matrix:

$$\widetilde{M} = \begin{pmatrix} a & b \\ c & d \end{pmatrix}, \tag{D.5}$$

which is also given by [cf. Eq. (5.40) in Sect. 5.4]

$$b = \frac{\partial r_\perp(t = T)}{\partial p_\perp(t = 0)}. \tag{D.6}$$

Hereby, $p_\perp(t)$ and $r_\perp(t)$ are the momentum and coordinate, respectively, perpendicular to the orbit [see Eq. (5.21) in Sect. 5.2]. If w is positive, we have $\nu=0$ and if w is negative, then $\nu=1$. The calculation of b (D.6) in principle requires a transformation to the intrinsic coordinate system of the orbit and is not simple in general. However, only the sign of b must be known for determining ν. Usually, this can be found numerically rather easily when solving the equations of motion on a computer and following the time evolution of a slightly perturbed periodic orbit on the screen: one gives it a small positive perpendicular starting velocity at time t_0 and determines the sign of the perpendicular coordinate after one revolution around the perturbed orbit, i.e. at time $t_0 + T$. Numerical inaccuracies normally do not disturb since only the sign of b has to be found.

Figure D.2 illustrates these methods of determining μ and ν for the example of the orbit A of the Hénon-Heiles potential (see Sect. 5.6.4) at the scaled energy $e=0.8$ where it is stable. In the upper part, one starts a trajectory from the center ($u = v = 0$) at an angle slightly larger than 30 degrees (at which the periodic orbit A is located as a straight line). After one approximate period, the trajectory ends on the upper side of the unperturbed periodic orbit, and hence both $\delta r_\perp(t = T)$ and $\delta p_\perp(t = 0)$ are positive, so that $b > 0$. Since $\mathrm{tr}\widetilde{M} - 2$ is always negative for a stable orbit, one finds $w < 0$ and thus $\nu = 1$. On the lower part of Fig. D.2, the quantity $\det U(t)$ is shown for the same stable orbit over one time period T and seen to have $\mu = 4$ zeros (not counting the zero at the starting time t_0).

An important point is to realize that the Maslov index σ is not necessarily the same for repeated periods, i.e., σ_n need not be proportional to the repetition number n. This holds, in fact, only for unstable orbits; for stable orbits it is in general not the case, as the following example of the harmonic oscillator will demonstrate.

Note that for neutrally stable orbits which have $|\mathrm{tr}\widetilde{M}|=2$, the above methods are not guaranteed to work. But since Gutzwiller's trace formula fails anyhow for these orbits, one has to resort altogether to different methods for deriving a semiclassical trace formula and its ingredients. It should also be noted that in all the above computations of σ, μ and ν, one may *not* start at any turning point of a given periodic orbit, since singularities or discontinuities can occur at turning points in some of the calculated quantities.

Finally, we point out that both above recipes for obtaining μ and ν can also be used for unstable orbits, their sum giving the same result as the winding number of $C(t)$ given in Eq. (D.3).

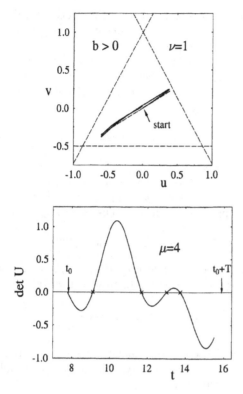

Figure D.2: Illustration of the determination of the Maslov index for a stable isolated orbit: $\sigma = \mu + \nu$. *Top:* determination of ν. *Bottom:* determination of μ. (After [34].)

D.1.3   Example: Two-dimensional harmonic oscillator

We shall illustrate the above method for the stable isolated orbits of the two-dimensional harmonic oscillator (C.20) with irrational frequency ratio, for which the Maslov indices can be computed analytically [21]. We denote them by σ_n where n is the cycle number, i.e. the number of repeated fundamental periods, and write $\sigma_n = \mu_n + \nu_n$. The two primitive periodic orbits are the one-dimensional oscillations along the principal axes with periods $T_x = 2\pi/\omega_x$ and $T_y = 2\pi/\omega_y$; their stability matrices are given in Appendix C.2.1. We shall calculate σ_n for the mode in the x direction.

The index μ_n is determined from $\det U(t)$, Eq. (D.2), whereby the starting vector $\mathbf{e}_0$, as discussed in Appendix D.1.2 for stable orbits, is chosen at $x = y = 0$ with $\dot{x} = 1$, $\dot{y} = 0$, so that $\mathbf{e}_0 = (0,0,0,1)$. The flow vector then is $\mathbf{f}_0 = (1,0,0,0)$. Using Eq. (D.1), we get $e_1(t) = f_2(t) = 0$, $e_2(t) = \sin(\omega_y t)/\omega_y$ and $f_1(t) = \cos(\omega_x t)$, so that we find $\det U(t)$ for the x-mode to be (up to an irrelevant constant factor)

$$\det U_x(t) = \sin(\omega_y t)\cos(\omega_x t). \tag{D.7}$$

Now μ_n is given as the number of zeros that $\det U_x(t)$ goes through during n periods T_x of the x mode, not counting the zero appearing at $t=0$, i.e. the number of zeros in the time interval $t \in (0, nT_x]$. Corresponding to the two factors in Eq. (D.7), we get μ_n as a sum of two contributions. The first counts the number of zeros of the cosine function and is simply equal to twice the number of cycles, $2n$. The second is the number of zeros of the sine function in the interval $(0, 2\pi n/\omega_x]$ which is the integer part of the number $2n\omega_y/\omega_x$. We thus have

$$\mu_n = 2n + [2n\omega_y/\omega_x], \tag{D.8}$$

where $[x]$ denotes the integer part of x.

The index ν_n is determined as $\nu_n = 0$ for $w > 0$ and $\nu_n = 1$ for $w < 0$, where w is given by Eq. (D.4) and the quantity $b = \partial r_\perp(nT_x)/\partial p_\perp(0)$ appearing therein can be read off (C.22) as the upper right element. Thus we get, up to an irrelevant constant factor,

$$b = \sin(n2\pi\omega_y/\omega_x) = \det U_x(nT_x). \tag{D.9}$$

Since the trace of $\widetilde{M}_x^n$ for isolated stable orbits is always less than 2, we find that $\nu_n = 1$ for $b > 0$ and $\nu_n = 0$ for $b < 0$. Using Eq. (D.9) we can write

$$\nu_n = \frac{1}{2}\left\{1 + (-1)^{[2n\omega_y/\omega_x]}\right\}. \tag{D.10}$$

Note that whenever the second contribution to μ_n in Eq. (D.8) is odd, ν_n equals zero, and when it is even, ν_n equals unity. The total Maslov index σ_n^x is therefore always odd and can be rewritten as

$$\sigma_n^x = 2n + o_n = 2n + 2[n\omega_y/\omega_x] + 1, \tag{D.11}$$

with o_n being an odd integer.

Correspondingly, the Maslov index σ_n^y of the y orbit is obtained simply by interchanging x and y on the right-hand side of Eq. (D.11).

n	1	2	3	4	5	6	7	8	9	10
σ_n^x	3	5	7	11	13	15	19	21	23	27
σ_n^y	9	17	25	33	41	49	57	67	75	83

Table D.1: Maslov indices σ_n^x and σ_n^y for the first ten cycles of the two isolated linear orbits in a 2-dimensional harmonic oscillator potential with frequency ratio $\omega_x : \omega_y = \pi$.

As an illustration, we give in Table D.1 the Maslov indices σ_n^x and σ_n^y, found for the first ten cycles in the case of a frequency ratio $\omega_x : \omega_y = \pi$. Note that they are neither multiples of n, nor are they growing linearly in n.

We see from this example that μ_n and ν_n, as well as their sum σ_n, do *not* repeat themselves for consecutive cycles, but thus must be calculated separately for each cycle and accumulated as it was done above. This has not been made very clear in the paper by Creagh *et al.* [32] and does not seem to be widely known. But we should also warn the reader that several different definitions, and methods of computation, for Maslov indices of stable orbits can be found in the literature. As mentioned in the introduction to this appendix, the definitions and rules given here are consistent with the work of Gutzwiller discussed in detail in Chapter 5.

D.2   Isolated orbits in billiards

In principle, the formal background used in the previous section applies also to billiard systems with reflecting boundaries. The numerical recipes for calculating $C(t)$ and $U(t)$ may, however, become critical near the reflection points, due to the discontinuous change of the perpendicular component of the momentum. It is therefore safer always to calculate σ as the sum of $\mu + \nu$. The first part μ can be evaluated geometrically using the law of spectral reflection at the boundaries. Hereby we just have to remember that, as has become evident in Sect. 2.4.2, each hard-wall reflection with Dirichlet boundary conditions gives a phase shift of $-\pi$, and thus a contribution $\Delta\mu = 2$. For von Neumann boundary conditions, the phase shift is zero, i.e., $\Delta\mu = 0$ at the reflection points. (This is exactly like in quantum mechanics.) Focal points in the interior of the billiard give a contribution $\Delta\mu = 1$ each. They are found by starting from a periodic orbit, then slightly changing its initial momentum, and looking for those points in which all neighboring perturbed orbits cross the original orbit, as shown in the example of Fig. 5.1 in Sect. 5.

The numerical method given in Sect. D.1.2 to find the second part ν can be used also in billiards. After a slight change of the initial momentum one looks for the change of b (D.6) after one unperturbed period and uses Eq. (D.4) to determine $\nu = 0$ for $w > 0$ and $\nu = 1$ for $w < 0$.

Bibliography

[1] L. H. Thomas, Proc. Camb. Phil. Soc. **23**, 195 (1926).

[2] E. Fermi, Z. Phys. **48**, 73 (1928).

[3] P. Hohenberg and W. Kohn, Phys. Rev. **136**, B864 (1964).

[4] M. Levy, Proc. Natl. Acad. Sci. (USA) **76**, 6062 (1979).

[5] W. Kohn and L. J. Sham, Phys. Rev. **140**, A1133 (1965).

[6] T. L. Gilbert, Phys. Rev. **B 12**, 2111 (1975).

[7] N. D. Mermin, Phys. Rev. **137**, 1441A (1965).

[8] R. Evans, Adv. in Phys. **28**, 143 (1979).

[9] U. Gupta and A. K. Rajagopal, Physics Reports **87**, 259 (1982).

[10] R. M. Dreizler and E. K. U. Gross, *Density Functional Theory* (Springer-Verlag, Berlin, 1990).

[11] V. M. Strutinsky, Nucl. Phys. **A 122**, 1 (1968).

[12] M. Brack, J. Damgård, A. S. Jensen, H. C. Pauli, V. M. Strutinsky, and C. Y. Wong, Rev. Mod. Phys. **44**, 320 (1972).

[13] M. Brack and P. Quentin, Nucl. Phys. **A 361**, 35 (1981).

[14] H. A. Bethe, Ann. Rev. Nucl. Sci. **21**, 93 (1971).

[15] G. G. Bunatian, V. M. Kolomietz, and V. M. Strutinsky, Nucl. Phys. **A 188**, 225 (1972).

[16] V. M. Kolomietz, Sov. J. Nucl. Phys. **18**, 147 (1974).

[17] M. Brack and P. Quentin, in: *Nuclear Self-consistent Fields*, eds. G. Ripka and M. Porneuf (North-Holland, 1975) p. 353.

[18] M. Abramowitz and I. A. Stegun: *Handbook of Mathematical Functions* (Dover Publications, 91h printing, New York, 1970).

[19] B. Van der Pohl and H. Bremmer: *Operational Calculus* (Cambridge University Press, Cambridge, 1955).

[20] V. A. Yakubovich and V. M. Starzhinskii: *Linear differential equations with periodic coefficients*, Vols. 1 and 2 (J. Wiley & Sons, New York, 1975).

[21] M. Brack and S. R. Jain, Phys. Rev. **A 51**, 3462 (1995).

[22] M. Baranger, K. T. R. Davies, and J. H. Mahoney, Ann. Phys. (N. Y.) **186**, 95 (1988).

[23] M. V. Berry, Eur. J. Phys. **2**, 91 (1981).

[24] S. Okai, H. Nishioka, and M. Ohta, Mem. Konan Univ., Sci. Ser. **37**, 29 (1990).

[25] M. Sieber, J. Phys. **A 30**, 4563 (1997).

[26] P. Meier, M. Brack, and S. C. Creagh, Z. Phys. **D 41**, 281 (1997).

[27] V. I. Arnold: *Mathematical Methods of Classical Mechanics*, 2nd edition, Graduate Texts in Mathematics (Springer-Verlag, 1989).

[28] M. A. de Gosson: *The Principles of Newtonian and Quantum Mechanics: The need for Planck's Constant h*, Imperial College Press (World Scientific, Singapore, 2002).

[29] V. P. Maslov and M. V. Fedoriuk: *Semi-Classical Approximations in Quantum Mechanics* (Reidel, Boston, 1981).

[30] M. Sieber, J. Phys. **A 29**, 4715 (1996); H. Schomerus and M. Sieber, J. Phys. **A 30**, 4537 (1997); M. Sieber and H. Schomerus, J. Phys. **A 31**, 165 (1998).

[31] A. Sugita, Phys. Lett. **A 266**, 321 (2000); Ann. Phys. (N. Y.) **288**, 277 (2001).

[32] S. C. Creagh, J. M. Robbins, and R. G. Littlejohn, Phys. Rev. **A 42**, 1907 (1990).

[33] J. M. Robbins, Nonlinearity 4, 343 (1991).

[34] M. Brack, R. K. Bhaduri, J. Law, M. V. N. Murthy, and Ch. Maier, Chaos 5, 317 (1995); Erratum: Chaos 5, 707 (1995).

Index

abbreviated action, 59
action integral, 7, 59
   and Aharonov-Bohm phase, 140, 289
   and Bohr-Sommerfeld quantization, 121
   and EBK wavefunction, 82
   for box potential, 122, 127
   for harmonic oscillator, 119
   Hamilton's principal function, 65
   in Green's function, 218
   in trace formula, 224
   of free particle, 59
   of periodic orbits, 221
   principle of least action, 217
   stationary points, 220
action-angle variables, 80, 270
adiabatic invariance, 9
   bouncing ball, 10
   pendulum, 11
Aharonov-Bohm phase, 36
   and scattering amplitude, 372
   in disk billiard, 38, 373
   in harmonic oscillator, 140
   in quantum dot, 406
   in stadium billiard, 38
Airy function
   connection formula, 70
   plot, 70
antidot superlattice, 44
anyons, 138, 372
atomic photoabsorption
   quasi-Landau resonances, 32
autonomous system, 57

beat
   in Hénon-Heiles system, 241, 296
   in metal cluster mass yields, 400
   in spherical cavity, 276
   in triangular billiard, 259

Bernoulli numbers, 180, 205
Berry-Tabor method, 89
   and disk billiard, 264
   and elliptic billiard, 281, 284
   and rectangular billiard, 275
   and trace formula, 271
   and triangle billiard, 257
bifurcations
   and uniform approximations, 227, 306
   in elliptic billiard, 245, 281
   in Hénon-Heiles potentials, 240, 294
Bloch density
   classical, 114
   definition, 113, 150
   for harmonic oscillators, 141
   Wigner transform, 150
   Wigner-Kirkwood expansion, 154
Bloch equation, 152
Bogomolny's functional determinant, 340
Bogomolny's transfer operator, 336
Bohm potential, 68
Bohr-Sommerfeld quantization
   and action integral, 121
   and Thomas-Fermi level density, 121
Bose-Einstein condensation, 176, 207
boundary conditions
   Dirichlet, 77, 126, 184, 330
   Neumann, 77, 128, 184
boundary integral method, 335
box, one-dimensional
   exact trace formula, 122

canonical
   momentum, 58
   transformation, 60, 79
canonical ensemble, 398, 425
chaos
   classical, 30
   nearest-neighbor spacings, 318

signatures in quantum systems, 3, 38, 319, 393
transition to, 98
charged particle in magnetic field, 51
circular annulus billiard, 362
coarse-graining of level density, 229
conductance
    Aharonov-Bohm oscillations, 49, 407
    universal fluctuations, 38
conductivity
    Drude model, 45
conjugate points
    and variational principle, 217
    at fixed energy, 219
    at fixed time interval, 216
    stadium billiard, 3
connection formula, 70
continuous symmetries, 266
corner correction, 131, 185
correspondence principle
    and generating function, 63
    and Rydberg atom, 20
Coulomb potential
    exact trace formula, 140
    one-dimensional, 108
    three-dimensional, 86
    two-dimensional, 83, 106
cyclotron frequency, 46
cyclotron orbits, 408

Debye approximation, 93, 350
deformations
    of atomic nuclei
        fission shapes, 385
        mass asymmetry, 388
        octupole, 388
    of metal clusters, 394
degeneracy of periodic orbit, 220, 249
    and continuous symmetry, 266
    and Gutzwiller amplitude, 223, 268
    in disk billiard, 262
    in elliptic billiard, 281
    one-dimensional, 252
    removal, 251
density function
    definition, 113, 419
    ETF expansion, 156, 173

relation to Bloch density, 154
    Thomas-Fermi expression, 146
density functional theory, 423
    and ETF model, 148
density matrix, single-particle, 113, 149
    from many-body wavefunction, 419
    from Bloch density, 116
    in HF theory, 420
    semiclassical (ETF), 161
    Strutinsky averaged, 192
density variational method, 162, 423
diffraction coefficient, 356
    wedge billiard, 357
diffractive creeping wave, 364
disk billiard
    and quantum dots, 406
    Maslov index, 306
    semiclassical trace formula, 262
    with magnetic flux line, 373
    WKB quantization, 88
double-humped fission barrier, 384

EBK method, 80
    N-torus, 81
    disk billiard, 88
    hydrogen atom, 83, 86
    torus quantization, 81
    wave function, 81
electron affinity, 168
electron gas, two-dimensional, 36, 45, 403, 410
electronic shells
    in metal clusters, 168
        magic radii, 402
        supershells, 394, 400
    in quantum dots, 406
elliptic billiard
    definition, 439
    degenerate orbits, 281, 440
    isolated orbits, 245
    level density, 284
end-point corrections, 258, 275, 298
energy averaging
    of semiclassical level density, 230
    Strutinsky method, 189
entropy, 171
    and Strutinsky averaging, 192

barrier, 325
density, 172
density functional, 175
ETF expansion, 174
topological, 318
ergodic motion
 and destruction of tori, 100
 in magnetic field, 30
Ericson fluctuations, 37
Euler-Lagrange equation, 163
Euler-MacLaurin expansion, 180
 and Poisson formula, 182
evaporative ensemble, 396
exchange-correlation energy, 423
exchange-correlation potential, 424
extended Thomas-Fermi (ETF) model
 and Strutinsky averaging, 197
 at finite temperature, 169, 202
 at zero temperature, 155
 energy, 157
 entropy density, 173, 175
 free energy density, 173, 175
 kinetic energy density
  $\hbar$ expansion, 156
  functional of ρ, 158, 175
 relativistic, 161
 turning point problem, 158

family of degenerate orbits, 249
 integration, 266
Farey fan, 134, 136
Fermi function, 203, 204
 generalized, 164
Fermi integrals, 173, 204
 asymptotic expansion, 175
Fermi-Dirac occupation numbers, 171
finite-temperature level density, 171
fission of nuclei, 384
 mass asymmetry, 388
Fock (exchange) potential, 421
Fourier transform
 of propagator, 115, 217
 photoionization of hydrogen, 36
Fredholm integral equation, 336
free (Helmholtz) energy, 171
 and Strutinsky averaging, 192
 density, 172, 173

density functional, 175
ETF expansion, 174
Fresnel integral, 92, 217, 298

Garton-Tomkin resonances, 33, 52
generating function, 60
 and quantum wave function, 63
giant dipole resonance, 403
grand canonical ensemble, 171, 192, 425
grazing orbits, 258, 275
Green's function, 7
 and level density, 116
 definition, 115
 diffractive, 359
 for annulus billiard, 364
 for free particles, 115
 partial wave expansion, 356
 phase-space representation, 267
 semiclassical, 218
gross-shell effects
 and short periodic orbits, 214
 in coarse-grained level density, 229
group
 Abelian, 268
 measure, 266
 non-Abelian, 271
Gutzwiller trace formula, 224

Haar measure, 266
Haldane hierarchy, 136
Hall resistance, 46
Hamilton equations of motion, 59
Hamilton's principal function, 7
 in Van Vleck propagator, 215
 relation to action, 59, 65
 second variation, 216
 variational principle, 58
Hamilton-Jacobi
 equation, 65
 theory, 63
Hamiltonian
 (super)integrable, 60
 cyclic, 60
 equations of motion, 59
 relation to Lagrangian, 59
harmonic oscillator
 and Nilsson model, 18

anisotropic, 16, 132, 136, 235, 243
at finite temperature, 124, 204
axially deformed, 17
coherent and squeezed states, 50
cranked, 132
in magnetic field, 132, 137, 307
inverted, 236
linear, 118
magic numbers, 17
Maslov index, 447
monodromy matrix, 433
motion in phase space, 13
spherical, 122
    ETF expansion at $T=0$, 201
    ETF expansion at $T > 0$, 204
with magnetic flux line, 139
Hartree-Fock equations, 420
Hartree-Fock theory, 420
nuclear deformation energy, 166
Hénon-Heiles potential, 238, 308
and quantum dots, 410
quartic, 293
uniform approximations, 303
heterostructure, high-mobility, 37, 403
Hill-Wheeler box, 184
Husimi transform, 200
Huygens' principle, 343
hydrogen atom, 86
Bohr radius, 28
Bohr-Sommerfeld quantization, 86
circular wave functions, 20
exact trace formula, 140
in magnetic field, 27
    periodic orbits, 36
Kepler orbits, 22
Poincaré surface of section, 30
Rydberg states, 20
scale-invariant Hamiltonian, 29
two-dimensional, 83, 106
wave packets, 20
hyperbolic instability, 225, 228

incompressibility, 202
inhomogeneity correction, 159
integrable system, 60, 82, 90, 270
inverse Laplace transform, 116
of Bloch density, 116, 154

ionization potential, 168

Jacobian determinant, 218, 252, 256, 268
Jahn-Teller effect, 197, 378
jellium model, 394

KAM theorem, 105
Kepler orbits, 107
kinetic energy density, 154, 421
    density functional, 158
        numerical tests, 161
        validity at $T=0$, 176
    ETF expansion, 156
Kirkwood gaps, 105
Kohn-Sham equations, 424

Lagrange equations of motion, 58
Lagrangian
    equation of motion, 58
    particle in magnetic field, 51
    relation to Hamiltonian, 59
Landau levels, 27, 51, 134, 137, 409
Landau orbits, 137, 292
Langer modification, 77
Laplace transform
    examples, 429
    finite-temperature level density, 172
    folding theorem, 429
    inverse, 116
    level density, 113
least action principle, 217
leptodermous expansion, 157, 168
level density
    and phase space, 15
    at finite temperature, 171, 204, 231
    coarse-graining, 230
    definition, 112
    diffractive contribution, 361
    ETF expansion, 156
    Gaussian folding, 230
    in k space, 113
    relation to Green's function, 116
    relation to partition function, 116
    smooth and oscillating parts, 117
    Strutinsky averaging, 190
    Thomas-Fermi expression, 121, 145
    Weyl expansion, 184

level spacing, average, 230
liquid drop model (LDM), 157, 196, 380, 384
local density approximation, 424
Lyapounov
    exponent, 228
    theorem, 433

magic numbers
    in deformed harmonic oscillator, 17
    in metal clusters
        atomic shells, 394
        electronic shells, 168, 394, 398
        magic radii, 402
    in nuclei, 168, 378, 383
magic radii, 402
magnetic flux line
    in disk billiard, 373
    in harmonic oscillator, 139
magnetic length, 28, 287, 292
magnetic susceptibility, 291
magnetoresistance
    ballistic, in cavity, 37
marginal stability, 228
Maslov index
    for disk billiard, 306
    for integrable systems, 271
    for isolated orbits, 224, 443
    in EBK quantization, 82
    in WKB quantization, 77
    numerical calculation
        in billiards, 449
        in smooth potentials, 444
mass spectra
    atomic nuclei, 167
    metal clusters, 396, 399
Maupertuis principle, 217
mesoscopic systems, 36, 403
metal clusters, 168, 394
Mie resonance, 403
mixed system, 227, 235, 238, 295
modulation factor, 279
molecular vibrations, 417
monodromy matrix
    and stability matrix, 227
    definition, 432
    numerical calculation

        in billiards, 436
        in smooth potentials, 435

nanostructure
    antidot superlattice, 44
    quantum dot, 403
    tunneling diode, 42
nearest-neighbor spacings, 318
    and random matrix, 319
    GOE, 318
    GUE, 318
    Wigner distribution, 318
neutral stability, 228
Nilsson model
    asymptotic quantum numbers, 18
    for metal clusters, 394
    for nuclei, 18, 379, 387, 392
non-periodic closed orbits, 287
nuclear deformations, 378
nuclear fission
    fission barriers, 384, 387
    fission isomer, 387, 391
    mass asymmetry of fragments, 388
nuclear forces, 147, 377
nuclear shapes, 377
    for fission, 384
    in ground state, 379
nuclear surface energy, 163, 202
number theory, 417

occupation numbers
    Fermi-Dirac, 171, 192
    finite temperature, 425
    Strutinsky averaging, 192
old quantum theory, 9, 14, 19, 107
overlapping resonances, 100

partial resummation
    of Wigner-Kirkwood series, 162
partition function
    definition, 113
    for harmonic oscillator
        1-dimensional, 118
        D-dimensional, 122
    Wigner-Kirkwood expansion, 152
path integral method, 4, 215
    propagator, 6

periodic orbit, 58
　　action, 221
　　bifurcation, 227
　　diffractive, 358
　　period, 223
　　primitive, 223
　　resonant torus, 92
　　resurgence, 321
　　stability, 225, 227
　　stationary phase approximation, 220
　　torus structure, 85
perturbation theory
　　classical, 99
　　for weak magnetic field, 287, 307
　　quantum mechanical, 278, 279
　　semiclassical, 279
phase velocity, 66
phase-coherence length, 37
Planck's constant
　　and scaled variables, 29
　　numerical value, 14
Poincaré surface of section
　　Hénon-Heiles system, 240, 295
　　hydrogen atom in magnetic field, 30
Poincaré variational equations, 434
Poincaré-Hopf theorem, 81
Poisson bracket, 60
Poisson formula, 90, 119
　　2-dimensional, 90
　　and Euler-MacLaurin series, 182
propagator, single-particle
　　folding property, 115, 215, 251
　　quantum mechanical, 5, 8, 114
　　semiclassical, 216
　　Van Vleck approximation, 7, 215
pseudo-orbits, 318, 326

quantum dot
　　conductance oscillations, 403
　　periodic orbits, 406
quantum Hall effect, 136
quantum stair-case function, 15, 233
quasi-periodic orbit, 86, 92

random-matrix theory, 319
rectangular billiard
　　exact trace formula, 94

semiclassical trace formula, 273
　　in magnetic field, 289
rectangular three-dimensional box
　　exact trace formula, 126
　　semiclassical trace formula, 308
　　Weyl formula, 181
recurrence of a wave packet, 23
recurrence period
　　experimental determination, 24
reflection rule for billiard orbits, 254
resonant perturbation, 98
　　and the asteroid belt, 105
　　destruction of tori, 100
　　Walker-Ford model, 100
resonant tori
　　destruction, 98, 278
　　stationary phase approximation, 92
response function, 403
resurgence, 321
Riemann zeta function, 182, 312
　　asymptotic density of zeros, 315
　　Euler product, 312
　　functional relation, 322
　　functional relationship, 312
　　GUE distribution of zeros, 319
　　phase of, 314
　　Riemann conjecture, 313
　　trace formula for zeros, 315
　　zeros of, 313
Riemann-Siegel formula, 321
　　look-alike, 328
Ruelle zeta function, 371
Rydberg atoms, 19
　　spectrum in magnetic field, 33

scaling
　　in bouncing ball, 49
　　in Hénon-Heiles potential, 238, 294
　　in magnetic field, 27
scars in wave functions, 39
　　and resonant tunneling, 42
　　in fissioning nucleus, 393
　　in helium atom, 39
　　in oval billiard, 41
scattering
　　amplitude and barrier penetration,
　　　371

amplitude, 346
amplitude of disk, 351
at grazing incidence, 352
classical, 345
cross section, 346
diffraction pattern by disk, 348
hard disk, 347
impact parameter, 331
inside-outside duality, 332
optical theorem, 347
penumbra and shadow, 363
quantum mechanical, 345
resonance and phase shift, 334
semiclassical Green's function, 354
scattering matrix, 330
and bound states, 331
and Selberg zeta function, 333
of disk, 330
Schottky gates, 404, 410
Selberg zeta function, 323
inverted oscillator, 370
and scattering matrix, 333
functional relation, 324
pseudo-orbits, 327
quantization rule, 325
zeros and quantization, 324
shape of a drum, 186
shell effects
in level density, 214
in metal cluster mass yields, 396
in nuclear binding energies, 167
in nuclear deformation energies, 379, 384
perturbative inclusion, 166, 427
temperature dependence, 126, 231, 397
shell gaps
and coarse-graining, 230
in axial harmonic oscillator, 17
in cranked harmonic oscillator, 134
shell-corrections
in nuclear binding energy, 167, 380
in nuclear fission barriers, 165, 384
Strutinsky method, 195
shell-model potential, 196, 197
for atomic nuclei, 378, 380

for metal clusters, 394
Shubnikov-de Haas oscillations, 49
Slater approximation, 161
Slater determinant, 420
spherical cavity
and metal clusters, 395
in magnetic field, 291
plus multipole deformations, 286
semiclassical trace formula, 275
spheroidal cavity
and atomic nuclei, 382
and periodic orbits, 251, 268
spin-orbit interaction, 141, 227
stability matrix, 223
and monodromy matrix, 227
for 2-dim. isolated orbits, 225
for harmonic oscillator, 433
in billiards, 438
in trace formula, 224
stair-case
function, 15, 232
quantization, 232, 342
stationary phase approximation, 92, 217, 298
and periodic orbits, 92, 220, 252
and resonant tori, 92
break-down, 223, 226, 278
stationary phase assumption, 252
Stirling's formula, 314
Strutinsky method, 188
energy averaging
and large-N expansion, 194
and temperature averaging, 192
relation to ETF model, 197
self-consistent, 427
energy theorem, 425
shell-correction, 195, 233
superdeformed nuclei, 18
supershells, 275
and periodic orbits, 401
in metal clusters, 394, 400
surface tension of Fermi liquid, 202
symmetry
breaking, 242, 278
degeneracy of periodic orbits, 266
group, 266

symmetry breaking
    and uniform approximation, 300, 303
    perturbative trace formula, 279
symmetry-reduced trace formulae, 232,
    272

temperature averaging, 124
    and Strutinsky averaging, 192
    of shell effects, 126, 231, 397
topological sum, 90
torus quantization, 78
trace formula
    $\hbar$ corrections, 90, 227, 258, 275
        importance for gross-shell structure,
        260, 276
        importance for quantization, 261
    and spin-orbit interaction, 227
    at finite temperature, 124, 231
    convergence, 97, 123, 229
    convoluted by Gaussian, 230
    diffractive orbits, 359
    for broken symmetry, 279
    for energy shell-correction, 234
    for integrable systems, 271
    for isolated orbits, 34, 224
        stable orbits, 226
        unstable orbits, 225
    for number stair case function, 232
    for Riemann zeros, 315
    for singly degenerate orbits, 252
    for spectrum with one quantum num-
        ber, 120
    general structure, 118
    relativistic, 227
    symmetry reduced, 232, 272
transmission probability, 73
triangular billiard
    and quantum dots, 410
    degenerate orbits, 254, 307
    diffractive orbits, 360
    exact trace formula, 128
    isolated orbit, 244, 259
    semiclassical trace formula, 253, 307
tunneling
    in double well, 73
    in quantum diode, 42
turning point, classical, 68, 157

uniform approximation, 227, 242, 278,
    300
universal conductance fluctuations, 38

Van Vleck propagator, 2, 215
variational density
    asymptotic fall-off, 163, 203
    parametrized, 164
variational principle
    classical, by Hamilton, 58
    Euler-Lagrange equation, 163
    in density functional theory, 424
    in ETF model, 162
    in Hartree-Fock theory, 420
vibrating membrane, 129, 417

wave packet
    radially localized, 26
    recurrence period, 24
weak localization, 37
    line-shape of magnetoresistance, 38
wedge billiard
    average number of states, 202
    diffraction coefficient, 357
    quantum stair-case function, 233
Weizsäcker correction, 159
Weyl expansion, 183
    and black-body radiation, 186
Wigner function, 148
Wigner-Kirkwood expansion, 151
    of Bloch density, 154
    of partition function, 152
    for bosons, 206
Wigner-Seitz radius, 395, 402
WKB method, 67
    transmission amplitude, 371
    and Hamilton-Jacobi equation, 68
    double well, 74
    in one dimension, 68
    Langer modification, 78
    Maslov index, 77
    quantization, 71
    radial motion, 75
    transmission probability, 73
    wave function, 70, 78
Woods-Saxon potential
    for atomic nuclei, 160, 200, 378, 385
    for metal clusters, 395

Printed in the United States
by Baker & Taylor Publisher Services